Communications
in Computer and Information Science   1879

Rationale

The CCIS series is devoted to the publication of proceedings of computer science conferences. Its aim is to efficiently disseminate original research results in informatics in printed and electronic form. While the focus is on publication of peer-reviewed full papers presenting mature work, inclusion of reviewed short papers reporting on work in progress is welcome, too. Besides globally relevant meetings with internationally representative program committees guaranteeing a strict peer-reviewing and paper selection process, conferences run by societies or of high regional or national relevance are also considered for publication.

Topics

The topical scope of CCIS spans the entire spectrum of informatics ranging from foundational topics in the theory of computing to information and communications science and technology and a broad variety of interdisciplinary application fields.

Information for Volume Editors and Authors

Publication in CCIS is free of charge. No royalties are paid, however, we offer registered conference participants temporary free access to the online version of the conference proceedings on SpringerLink (http://link.springer.com) by means of an http referrer from the conference website and/or a number of complimentary printed copies, as specified in the official acceptance email of the event.

CCIS proceedings can be published in time for distribution at conferences or as post-proceedings, and delivered in the form of printed books and/or electronically as USBs and/or e-content licenses for accessing proceedings at SpringerLink. Furthermore, CCIS proceedings are included in the CCIS electronic book series hosted in the SpringerLink digital library at http://link.springer.com/bookseries/7899. Conferences publishing in CCIS are allowed to use Online Conference Service (OCS) for managing the whole proceedings lifecycle (from submission and reviewing to preparing for publication) free of charge.

Publication process

The language of publication is exclusively English. Authors publishing in CCIS have to sign the Springer CCIS copyright transfer form, however, they are free to use their material published in CCIS for substantially changed, more elaborate subsequent publications elsewhere. For the preparation of the camera-ready papers/files, authors have to strictly adhere to the Springer CCIS Authors' Instructions and are strongly encouraged to use the CCIS LaTeX style files or templates.

Abstracting/Indexing

CCIS is abstracted/indexed in DBLP, Google Scholar, EI-Compendex, Mathematical Reviews, SCImago, Scopus. CCIS volumes are also submitted for the inclusion in ISI Proceedings.

How to start

To start the evaluation of your proposal for inclusion in the CCIS series, please send an e-mail to ccis@springer.com.

Zhiwen Yu · Qilong Han · Hongzhi Wang ·
Bin Guo · Xiaokang Zhou · Xianhua Song ·
Zeguang Lu
Editors

Data Science

9th International Conference
of Pioneering Computer Scientists, Engineers
and Educators, ICPCSEE 2023
Harbin, China, September 22–24, 2023
Proceedings, Part I

 Springer

Editors
Zhiwen Yu
Harbin Engineering University
Harbin, China

Qilong Han
Harbin Engineering University
Harbin, China

Hongzhi Wang
Harbin Institute of Technology
Harbin, Heilongjiang, China

Bin Guo
Northwestern Polytechnical University
Xi'an, China

Xiaokang Zhou
Shiga University
Shiga, Japan

Xianhua Song
Harbin University of Science and Technology
Harbin, China

Zeguang Lu
National Academy of Guo Ding Institute
of Data Science
Beijing, China

ISSN 1865-0929   ISSN 1865-0937 (electronic)
Communications in Computer and Information Science
ISBN 978-981-99-5967-9   ISBN 978-981-99-5968-6 (eBook)
https://doi.org/10.1007/978-981-99-5968-6

This Springer imprint is published by the registered company Springer Nature Singapore Pte Ltd.
The registered company address is: 152 Beach Road, #21-01/04 Gateway East, Singapore 189721, Singapore

Paper in this product is recyclable.

Preface

As the chairs of the 8th International Conference of Pioneering Computer Scientists, Engineers and Educators 2023 (ICPCSEE 2023, originally ICYCSEE), it is our great pleasure to welcome you to the conference proceedings. ICPCSEE 2023 was held in Harbin, China, during September 22–24, 2023, and hosted by the Harbin Engineering University, the Harbin Institute of Technology, the Northeast Forestry University, the Harbin University of Science and Technology, the Heilongjiang Computer Federation, and the National Academy of Guo Ding Institute of Data Sciences. The goal of this conference series is to provide a forum for computer scientists, engineers, and educators.

This year's conference attracted 244 paper submissions. After the hard work of the Program Committee, 66 papers were accepted to appear in the conference proceedings, with an acceptance rate of 27%. The major topic of this conference is data science. The accepted papers cover a wide range of areas related to basic theory and techniques of data science including mathematical issues in data science, computational theory for data science, big data management and applications, data quality and data preparation, evaluation and measurement in data science, data visualization, big data mining and knowledge management, infrastructure for data science, machine learning for data science, data security and privacy, applications of data science, case studies of data science, multimedia data management and analysis, data-driven scientific research, data-driven bioinformatics, data-driven healthcare, data-driven management, data-driven e-government, data-driven smart city/planet, data marketing and economics, social media and recommendation systems, data-driven security, data-driven business model innovation, and social and/or organizational impacts of data science.

We would like to thank all the Program Committee members, a total of 261 people from 142 different institutes or companies, for their hard work in completing the review tasks. Their collective efforts made it possible to attain quality reviews for all the submissions within a few weeks. Their diverse expertise in each research area helped us to create an exciting program for the conference. Their comments and advice helped the authors to improve the quality of their papers and gain deeper insights.

We thank the team at Springer, whose professional assistance was invaluable in the production of the proceedings. A big thanks also to the authors and participants for their tremendous support in making the conference a success.

Besides the technical program, this year ICPCSEE offered different experiences to the participants. We hope you enjoyed the conference.

July 2023

Rajkumar Buyya
Hai Jin
Zhiwen Yu
Qilong Han
Hongzhi Wang
Bin Guo
Xiaokang Zhou

Organization

General Chairs

Rajkumar Buyya   University of Melbourne, Australia
Hai Jin   Huazhong University of Science and Technology,
     China
Zhiwen Yu   Harbin Engineering University, China

Program Chairs

Qilong Han   Harbin Engineering University, China
Hongzhi Wang   Harbin Institute of Technology, China
Bin Guo   Northwestern Polytechnical University, China
Xiaokang Zhou   Shiga University, Japan

Program Co-chairs

Rui Mao   Shenzhen University, China
Min Li   Central South University, China
Wenguang Chen   Tsinghua University, China
Wei Wang   Harbin Engineering University, China

Organization Chair

Haiwei Pan   Harbin Engineering University, China

Publications Chairs

Xianhua Song   Harbin University of Science and Technology,
     China
Zeguang Lu   National Academy of Guo Ding Institute of Data
     Science, China

Secretary General

Zeguang Lu   National Academy of Guo Ding Institute of Data
                                Science, China

Under Secretary General

Xiaoou Ding   Harbin Institute of Technology, China

Secretary

Zhongchan Sun   National Academy of Guo Ding Institute of Data
                                Science, China
Dan Lu   Harbin Engineering University, China

Executive Members

Cham Tat Huei   UCSI University, Malaysia
Xiaoju Dong   Shanghai Jiao Tong University, China
Lan Huang   Jilin University, China
Ying Jiang   Kunming University of Science and Technology,
                                China
Weipeng Jiang   Northeast Forestry University, China
Min Li   Central South University, China
Junyu Lin   Institute of Information Engineering, CAS, China
Xia Liu   Hainan Province Computer Federation, China
Rui Mao   Shenzhen University, China
Qiguang Miao   Xidian University, China
Haiwei Pan   Harbin Engineering University, China
Pinle Qin   North University of China, China
Xianhua Song   Harbin University of Science and Technology,
                                China
Guanglu Sun   Harbin University of Science and Technology,
                                China
Jin Tang   Anhui University, China
Ning Wang   Xiamen Huaxia University, China
Xin Wang   Tianjin University, China
Yan Wang   Zhengzhou University of Technology, China
Yang Wang   Southwest Petroleum University, China
Shengke Wang   Ocean University of China, China

Yun Wu	Guizhou University, China
Liang Xiao	Nanjing University of Science and Technology, China
Junchang Xin	Northeastern University, China
Zichen Xu	Nanchang University, China
Xiaohui Yang	Hebei University, China
Chen Ye	Hangzhou Dianzi University, China
Canlong Zhang	Guangxi Normal University, China
Zhichang Zhang	Northwest Normal University, China
Yuanyuan Zhu	Wuhan University, China

Steering Committee

Jiajun Bu	Zhejiang University, China
Wanxiang Che	Harbin Institute of Technology, China
Jian Chen	ParaTera, China
Wenguang Chen	Tsinghua University, China
Xuebin Chen	North China University of Science and Technology, China
Xiaoju Dong	Shanghai Jiao Tong University, China
Qilong Han	Harbin Engineering University, China
Yiliang Han	Engineering University of CAPF, China
Yinhe Han	Chinese Academy of Sciences, China
Hai Jin	Huazhong University of Science and Technology, China
Weipeng Jing	Northeast Forestry University, China
Wei Li	Central Queensland University, China
Min Li	Central South University, China
Junyu Lin	Chinese Academy of Sciences, China
Yunhao Liu	Michigan State University, USA
Zeguang Lu	National Academy of Guo Ding Institute of Data Sciences, China
Rui Mao	Shenzhen University, China
Qiguang Miao	Xidian University, China
Haiwei Pan	Harbin Engineering University, China
Pinle Qin	North University of China, China
Zheng Shan	PLA Information Engineering University, China
Guanglu Sun	Harbin University of Science and Technology, China
Jie Tang	Tsinghua University, China
Tian Feng	Chinese Academy of Sciences, China
Tao Wang	Peking University, China

Program Committee

Xiping Duan	Harbin Normal University, China
Xiaolin Fang	Southeast University, China
Ming Fang	Changchun University of Science and Technology, China
Jianlin Feng	Sun Yat-sen University, China
Jing Gao	Dalian University of Technology, China
Yu Gu	Northeastern University, China
Qi Han	Harbin Institute of Technology, China
Meng Han	Georgia State University, USA
Qinglai He	Arizona State University, USA
Wei Hu	Nanjing University, China
Lan Huang	Jilin University, China
Hao Huang	Wuhan University, China
Feng Jiang	Harbin Institute of Technology, China
Bin Jiang	Hunan University, China
Cheqing Jin	East China Normal University, China
Hanjiang Lai	Sun Yat-Sen University, China
Shiyong Lan	Sichuan University, China
Hui Li	Xidian University, China
Zhixu Li	Soochow University, China
Mingzhao Li	RMIT University, China
Peng Li	Shaanxi Normal University, China
Jianjun Li	Huazhong University of Science and Technology, China
Xiaofeng Li	Sichuan University, China
Zheng Li	Sichuan University, China
Min Li	Central South University, China
Zhixun Li	Nanchang University, China
Hua Li	Changchun University of Science and Technology, China
Rong-Hua Li	Shenzhen University, China
Cuiping Li	Renmin University of China, China
Qiong Li	Harbin Institute of Technology, China
Yanli Liu	Sichuan University, China
Hailong Liu	Northwestern Polytechnical University, China
Guanfeng Liu	Macquarie University, Australia
Yan Liu	Harbin Institute of Technology, China
Zeguang Lu	National Academy of Guo Ding Institute of Data Sciences, China
Binbin Lu	Sichuan University, China
Junling Lu	Shaanxi Normal University, China
Jizhou Luo	Harbin Institute of Technology, China

Li Mohan	Jinan University, China
Tiezheng Nie	Northeastern University, China
Haiwei Pan	Harbin Engineering University, China
Jialiang Peng	Norwegian University of Science and Technology, Norway
Fei Peng	Hunan University, China
Jianzhong Qi	University of Melbourne, Australia
Shaojie Qiao	Southwest Jiaotong University, China
Qingliang Li	Changchun University of Science and Technology, China
Zhe Quan	Hunan University, China
Yingxia Shao	Peking University, China
Wei Song	North China University of Technology, China
Yanan Sun	Oklahoma State University, USA
Minghui Sun	Jilin University, China
Guanghua Tan	Hunan University, China
Yongxin Tong	Beihang University, China
Xifeng Tong	Northeast Petroleum University, China
Vicenç Torra	Umeå University, Sweden
Leong Hou	University of Macau, China
Hongzhi Wang	Harbin Institute of Technology, China
Yingjie Wang	Yantai University, China
Dong Wang	Hunan University, China
Yongheng Wang	Hunan University, China
Chunnan Wang	Harbin Institute of Technology, China
Jinbao Wang	Harbin Institute of Technology, China
Xin Wang	Tianjin University, China
Peng Wang	Fudan University, China
Chaokun Wang	Tsinghua University, China
Xiaoling Wang	East China Normal University, China
Jiapeng Wang	Harbin Huade University, China
Huayu Wu	Institute for Infocomm Research, China
Yan Wu	Changchun University of Science and Technology, China
Sheng Xiao	Hunan University, China
Ying Xu	Hunan University, China
Jing Xu	Changchun University of Science and Technology, China
Jianqiu Xu	Nanjing University of Aeronautics and Astronautics, China
Yaohong Xue	Changchun University of Science and Technology, China

Li Xuwei	Sichuan University, China
Mingyuan Yan	University of North Georgia, USA
Yajun Yang	Tianjin University, China
Gaobo Yang	Hunan University, China
Lei Yang	Heilongjiang University, China
Ning Yang	Sichuan University, China
Xiaochun Yang	Northeastern University, China
Bin Yao	Shanghai Jiao Tong University, China
Yuxin Ye	Jilin University, China
Xiufen Ye	Harbin Engineering University, China
Minghao Yin	Northeast Normal University, China
Dan Yin	Harbin Engineering University, China
Zhou Yong	China University of Mining and Technology, China
Lei Yu	Georgia Institute of Technology, USA
Ye Yuan	Northeastern University, China
Kun Yue	Yunnan University, China
Peng Yuwei	Wuhan University, China
Xiaowang Zhang	Tianjin University, China
Lichen Zhang	Shaanxi Normal University, China
Yingtao Zhang	Harbin Institute of Technology, China
Yu Zhang	Harbin Institute of Technology, China
Wenjie Zhang	University of New South Wales, Australia
Dongxiang Zhang	University of Electronic Science and Technology of China, China
Xiao Zhang	Renmin University of China, China
Kejia Zhang	Harbin Engineering University, China
Yonggang Zhang	Jilin University, China
Huijie Zhang	Northeast Normal University, China
Boyu Zhang	Utah State University, USA
Jian Zhao	Changchun University, China
Qijun Zhao	Sichuan University, China
Bihai Zhao	Changsha University, China
Xiaohui Zhao	University of Canberra, Australia
Jiancheng Zhong	Hunan Normal University, China
Fucai Zhou	Northeastern University, China
Changjian Zhou	Northeast Agricultural University, China
Min Zhu	Sichuan University, China
Yuanyuan Zhu	Wuhan University, China
Wangmeng Zuo	Harbin Institute of Technology, China

Contents – Part I

Applications of Data Science

Construction of Software Design and Programming Practice Course
in Information and Communication Engineering 3
    Zhigang Yang, Yahui Shen, Lin Hou, Tao Chen, and Yulong Qiao

A Self-Attention-Based Stock Prediction Method Using Long Short-Term
Memory Network Architecture 12
    Xiaojun Ye, Beixi Ning, Pengyuan Bian, and Xiaoning Feng

CAD-Based Research on the Design of a Standard Unit Cabinet
for Custom Furniture of the Cabinet Type 25
    Guo Lan and Fan Weiquan

An Improved War Strategy Optimization Algorithm for Big Data Analytics 37
    Longjie Han, Hui Xu, and Yalin Hu

Research on Path Planning of Mobile Robots Based on Dyna-RQ 49
    Ziying Zhang, Xian Li, and Yuhua Wang

Small Target Helmet Wearing Detection Algorithm Based on Improved
YOLO V5 60
    Jiajing Hu, Junqiu Li, and Qinghui Zhang

Research on Dance Evaluation Technology Based on Human Posture
Recognition 78
    Yanzi Li, Yiwen Zhu, Yanqing Wang, and Yiming Gao

Multiple-Channel Weight-Based CNN Fault Diagnosis Method 89
    *Peng Xu, Xinyu Liu, Junyu Lin, Zhongyu Lu, Fengming Li,
    and Husheng Gou*

Big Data Management and Applications

Design and Implementation of Key-Value Database for Ship Virtual Test
Platform Based on Distributed System 109
    Qingyu Meng, Kejia Zhang, Haiwei Pan, Maocai Yuan, and Baoying Ma

Big Data Mining and Knowledge Management

Research on Multi-Modal Time Series Data Prediction Method Based
on Dual-Stage Attention Mechanism 127
   Xinyu Liu, Yulong Meng, Fangwei Liu, Lingyu Chen, Xinfeng Zhang,
   Junyu Lin, and Husheng Gou

Prediction of Time Series Data with Low Latitude Features 145
   Haoran Zhang, Haifeng Guo, Donghua Yang, Mengmeng Li, Bo Zheng,
   and Hongzhi Wang

Lightweight and Efficient Attention-Based Superresolution Generative
Adversarial Networks 165
   Shushu Yin, Hefan Li, Yu Sang, Tianjiao Ma, Tie Li, and Mei Jia

The Multisource Time Series Data Granularity Conversion Method 182
   Chongyang Leng, Qilong Han, and Dan Lu

Outlier Detection Model Based on Autoencoder and Data Augmentation
for High-Dimensional Sparse Data 192
   Haitao Zhang, Wenhai Ma, Qilong Han, and Zhiqiang Ma

Dimension Reduction Based on Sampling 207
   Zhuping Li, Donghua Yang, Mengmeng Li, Haifeng Guo, Tiansheng Ye,
   and Hongzhi Wang

Complex Time Series Analysis Based on Conditional Random Fields 221
   Yanjie Wei, Haifeng Guo, Donghua Yang, Mengmeng Li, Bo Zheng,
   and Hongzhi Wang

Feature Extraction of Time Series Data Based on CNN-CBAM 233
   Jiaji Qin, Dapeng Lang, and Chao Gao

Optimization of a Network Topology Generation Algorithm Based
on Spatial Information Network 246
   Peng Yang, Shijie Zhou, and Xiangyang Zhou

Data Visualization

MBTIviz: A Visualization System for Research on Psycho-Demographics
and Personality 259
   Yutong Yang, Xiaoju Dong, Xuefei Tian, Yanling Zhang, and Meng Zhou

Data-Driven Security

Distributed Implementation of SM4 Block Cipher Algorithm Based
on SPDZ Secure Multi-party Computation Protocol 279
   Xiaowen Ma, Maoning Wang, and Zhong Kang

DP-ASSGD: Differential Privacy Protection Based on Stochastic Gradient
Descent Optimization 298
   Qiang Gao, Han Sun, and Zhifang Wang

Study on Tourism Workers' Intercultural Communication Competence 309
   Dan Xian

A Novel Federated Learning with Bidirectional Adaptive Differential
Privacy 329
   Yang Li, Jin Xu, Jianming Zhu, and Youwei Wang

Chaos-Based Construction of LWEs in Lattice-Based Cryptosystems 338
   *Nina Cai, Wuqiang Shen, Fan Yang, Hao Cheng, Huiyi Tang,
   Yihua Feng, Jun Song, and Shanxiang Lyu*

Security Compressed Sensing Image Encryption Algorithm Based
on Elliptic Curve 350
   Anan Jin, Xiang Li, and Qingzhi Xiong

Infrastructure for Data Science

Two-Dimensional Code Transmission System Based on Side Channel
Feedback 363
   Han Sun, Qiang Gao, and Zhifang Wang

An Updatable and Revocable Decentralized Identity Management Scheme
Based on Blockchain 372
   Zhiping Wang, Meijiao Duan, and Maoning Wang

Cloud-Edge Intelligent Collaborative Computing Model Based on Transfer
Learning in IoT 389
   Yang Long and Zhixin Li

Design and Validation of a Hardware-In-The-Loop Based Automated
Driving Simulation Test Platform 404
   Kaichao Zheng, Xianbin Xue, He Li, and Guoliang Cheng

Machine Learning for Data Science

Improving Transferability Reversible Adversarial Examples Based
on Flipping Transformation 417
   Youqing Fang, Jingwen Jia, Yuhai Yang, and Wanli Lyu

Rolling Iterative Prediction for Correlated Multivariate Time Series 433
   Peng Liu, Qilong Han, and Xiao Yang

Multimedia Data Management and Analysis

Video Popularity Prediction Based on Knowledge Graph and LSTM
Network 455
   Pingshan Liu, Zhongshu Yu, Yemin Sun, and Mingjun Xi

Design and Implementation of Speech Generation and Demonstration
Research Based on Deep Learning 475
   Wanyu Luo, Yanqing Wang, Yujia Liu, and Yiqin Xu

Testing and Improvement of OCR Recognition Technology
in Export-Oriented Chinese Dictionary APP 487
   *Qingpei Yang, Xinrui Wan, Hongjun Chen, Mengjiao Wu, Yao Wu,
   and Jiyun Qiao*

Author Index 495

Contents – Part II

Data-Driven Healthcare

Better Fibre Orientation Estimation with Single-Shell Diffusion MRI
Using Spherical U-Net 3
   Hang Zhao, Chengdong Deng, Yu Wang, and Jiquan Ma

A Method for Extracting Electronic Medical Record Entities by Fusing
Multichannel Self-Attention Mechanism with Location Relationship
Features 13
   Hongyan Xu, Hong Wang, Yong Feng, Rongbing Wang,
   and Yonggang Zhang

Application of Neural Networks in Early Warning Systems for Coronary
Heart Disease 31
   Yanhui Fang, Wei Fang, and Weizhen Yang

Research on Delivery Order Scheduling and Delivery Algorithms 40
   Qian Hong and Yue Wang

Research on Gesture Recognition Based on Multialgorithm Fusion 62
   Yuanyuan Zhu, Yanqing Wang, Xiaofeng Gao, and Lihan Liu

Design of Fitness Movement Detection and Counting System Based
on MediaPipe 77
   Yinan Chen and Xia Liu

Data-Driven Smart City/Planet

MetaCity: An Edge Emulator with the Feature of Realistic Geospatial
Support for Urban Computing 95
   Lin Wu, Guogui Yang, Ying Qin, Baokang Zhao, Xue Ouyang,
   and Huan Zhou

Multifunctional Sitting Posture Detector Based on Face Tracking 116
   Zhaoning Jin, Jiahan Wei, Zhiyan Yu, and Yang Zhou

Real-Time Analysis and Prediction System for Rail Transit Passenger
Flow Based on Deep Learning 130
   Xujun Che, Gang Cen, Shuhui Wu, Jiaming Gu, and Keying Zhu

Research on Driver Monitoring Systems Based on Vital Signs and Behavior
Detection 139
   Man Niu, Yanqing Wang, Xinya Shu, and Xiaofeng Gao

People Flow Monitoring Based on Deep Learning 153
   Xinran Wang, Yanqing Wang, Yiqing Xu, and Tianxin Wang

Research on the Influencing Factors of Passenger Traffic at Sanya Airport
Based on Gray Correlation Theory 168
   Yuanhui Li, Haiyun Han, Zhipeng Ou, and Wen Zhao

Qinghai Embroidery Classification System and Intelligent Classification
Research 174
   Xiaofei Lin, Zhonglin Ye, and Haixing Zhao

A Machine Learning-Based Botnet Malicious Domain Detection
Technique for New Business 191
   Aohan Mei, Zekun Chen, Jing Zhao, and Dequan Yang

GPDCCL: Cross-Domain Named Entity Recognition with Span-Based
Domain Confusion Contrastive Learning 202
   Ye Wang, Chenxiao Shi, Lijie Li, and Manyuan Guo

Data Analyses and Parallel Optimization of the Regional Marine
Ecological Model 213
   Yanqiang Wang, Jingjing Zheng, Tianyu Zhang, Peng Liang, and Bo Lin

Comparison of Two Grey Models' Applicability to the Prediction
of Passenger Flow in Sanya Airport 225
   Yinan Chen and Yuanhui Li

Research on High Precision Autonomous Navigation of Shared Balancing
Vehicles Based on EKF-SLAM 237
   Xinyu Cheng, Yanqing Wang, Dingdong Guo, Xuewei Li, and Yiming Gao

Social Media and Recommendation Systems

Link Prediction Based on the Relational Path Inference of Triangular
Structures 255
   Xin Li, Qilong Han, Lijie Li, and Ye Wang

Research on Link Prediction Algorithms Based on Multichannel Structure
Modelling 269
   Gege Li, Lin Zhou, Zhonglin Ye, and Haixing Zhao

Modal Interactive Feature Encoder for Multimodal Sentiment Analysis   285
   Xiaowei Zhao, Jie Zhou, and Xiujuan Xu

Type-Augmented Link Prediction Based on Bayesian Formula   304
   Ye Wang, Enze Luo, Lijie Li, and Wenjian Tao

Multitask Graph Neural Network for Knowledge Graph Link Prediction   318
   Ye Wang, Jianhua Yang, Lijie Li, and Jian Yao

A Short Text Classification Model Based on Chinese Part-of-Speech
Information and Mutual Learning   330
   Yihe Deng and Zuxu Dai

The Analysis of Phase Synchronisation in the Uniform Scale-Free
Hypernetwork   344
   Juan Du, Xiujuan Ma, Fuxiang Ma, Bin Zhou, and Wenqian Yu

LGHAE: Local and Global Hyper-relation Aggregation Embedding
for Link Prediction   364
   Peikai Yuan, Zhenheng Qi, Hui Sun, and Chao Liu

Efficient s-Core Community Search on Attributed Graphs   379
   Yuesheng Fu and Ruilu Sun

Education Using Big Data, Intelligent Computing or Data Mining, etc.

A Study on English Classroom Learning Anxiety of Private Vocational
College Students   393
   Han Wu

Research on the Construction of a Data Warehouse Model for College
Student Performance   408
   Juntao Chen, Jinmei Zhan, and Fei Tian

Research and Application of AI-Enabled Education   420
   Zhanquan Wang, Yuxin Tian, Rui Chen, and Linghe Kong

A Study on the Online Teaching Input of Higher Education Teachers
Based on K-Means Analysis   433
   Zhi-peng Ou, Han Zhang, and Xia Liu

Research and Exploration on Innovation and Entrepreneurship Practice
Education System for Railway Transportation Specialists 447
  Jiu Yong, Jianwu Dang, Yangping Wang, Jianguo Wei, and Yan Lu

Author Index 457

Applications of Data Science

Construction of Software Design and Programming Practice Course in Information and Communication Engineering

Zhigang Yang[✉], Yahui Shen, Lin Hou, Tao Chen, and Yulong Qiao

College of Information and Communication Engineering, Harbin Engineering University, Harbin 150001, China
zgyang@hrbeu.edu.cn

Abstract. Innovation and entrepreneurship education is becoming one of the important goals of higher education in China. According to the requirements of the Ministry of Education to deepen the reform of innovation and entrepreneurship education in colleges and universities, we propose a core principle of "from basis to comprehensiveness, and then to innovation" to construct the software design and programming practice course in the information and communication engineering discipline. We have integrated specialized knowledge teaching with innovation and entrepreneurship training, implemented a series of experimental projects that integrate theory with practice, and explored mixed teaching methods and diversified examination methods. In this paper, details about software design and programming practice course construction are shown from aspects of teaching content, teaching methods, and examination methods. These course construction experiences will benefit teachers who engage in innovative practical courses in the information and communication engineering discipline.

Keywords: Software design and programming practice · Couse construction · Information and communication engineering · Comprehensive practice · Innovation and entrepreneurship

1 Introduction

In recent years, China has made remarkable progress in science and technology in all fields, and higher education is providing strong support for national innovation-driven development. Innovation and entrepreneurship education is becoming one of the important goals of higher education. In March 2019, the Ministry of Education issued a notice on deepening the reform of innovation and entrepreneurship education in model universities. It is emphasized that colleges and universities should actively optimize the specialized curriculum, explore and enrich innovation and entrepreneurship education resources for various specialized courses, integrate specialized knowledge teaching and innovation and entrepreneurship training, enhance students' interests and abilities for specialized research and development, and lay a solid foundation for students to engage in innovation and entrepreneurship activities based on their major [1]. In September

Z. Yu et al. (Eds.): ICPCSEE 2023, CCIS 1879, pp. 3–11, 2023.
https://doi.org/10.1007/978-981-99-5968-6_1

2021, the General Office of the State Council issued guidance on further supporting innovation and entrepreneurship among college students. The document clarifies the requirements for university teachers to improve the teaching ability of innovation and entrepreneurship education and integrates advanced academic and practical experience into classroom teaching by reforming teaching methods and examination methods [2].

To strengthen the cultivation of innovative students [3–6], relying on the College of Information and Communication Engineering at Harbin Engineering University, we reform the software design and programming practice course. The information and communication engineering discipline includes two specialties: electronic information engineering and communication engineering. The software design and programming practice is designed for undergraduate students in these two specialties in the sixth semester with 48 class hours, belonging to a comprehensive practice course on innovation and entrepreneurship.

The main teaching objective of the course is to enable students to have the basic ability to design software according to the project requirements, the ability to write well-structured codes using C language, and the basic ability to debug and test programs. On the basis of consolidating the learning content of forward specialized courses, students can master the basic ideas of software system development from software requirements, software design, and coding to software testing by combining their specialized knowledge with practical project design. Furthermore, the curriculum is set with the purpose of training students' ability to discover, analyse and solve problems independently and stimulating students' creative thinking. Students' interest in programming methods combined with engineering applications will increase greatly, and their ability to work independently and soft development skills will improve greatly. In addition to these benefits, it lays a solid foundation for students to quickly adapt to research work in the future.

From the whole curriculum system of the information and communication engineering discipline, software design and programming practice is the only software practice course that plays an important role in connecting the forward and backward courses. The course not only needs to further deepen the knowledge in forward courses, such as "computational thinking" but also needs to strengthen integration with the core specialized courses, such as digital signal processing, pattern recognition, and machine learning. It will also lay a foundation for subsequent specialized elective courses, such as image processing and radar principles, and graduation projects. Therefore, we carry out a comprehensive reform to the software design and programming practice course from three aspects: teaching content, teaching method, and examination method.

This paper is organized as follows. Section 2 introduces the challenges for course construction. Section 3 shows the education objects of the software design and programming practice course. Section 4 depicts the details of the course construction. Section 5 offers conclusions and future works.

2 Challenges

World-famous universities have offered software practice courses in information fields. For example, the Massachusetts Institute of Technology (MIT) has set up a course named introduction to computer science and programming for nonprogrammers [7], which mainly includes basic knowledge of programming, computational concepts, software engineering, algorithm technology, data types and recursion, and corresponding experiments. Stanford University offers a course named software development for scientists and engineers [8], which mainly covers the basic usage of program language, data structure, and related knowledge of software engineering. The experiment matching the theoretical curriculums is to construct campus shared resources. Some well-known universities in China also offer software courses in information fields [9–12]. These courses mainly focus on the theory and application of computer software design and are not closely related to specialized knowledge of electronic information engineering and communication engineering.

Considering the actual situation of our College of Information and Communication Engineering, the software design and programming practice course involves not only specialized knowledge of information and communication engineering but also specific software programming practice, which has the characteristics of combining theory with practice. Through the teaching practice and research analysis of the course team over 5 years, we find the following problems in the current course.

(1) Students' weak software programming foundation restricts the development of the comprehensive practice of innovation and entrepreneurship.

Students majoring in electronic information engineering and communication engineering only learn an advanced programming language in the first academic year, but the software design and programming practice course starts in the sixth semester. In the two-year interval, there are no other courses and practices in software programming, and students' software programming ability declines without continuous practice. Moreover, as noncomputer major students, they have not learned data structure and software engineering courses and lack relevant programming ideas and concepts. All these factors result in the weak foundation of students' software programming, which restricts the development of the comprehensive practice of innovation and entrepreneurship. This course construction will adopt a step-by-step and simple-to-complex experimental design to solve this problem.

(2) The traditional classroom teaching and final examination mode is not conducive to students' independent learning and independent thinking.

Before this practice course, students mainly studied theoretical knowledge through basic courses and core specialized courses. Teachers usually use classroom teaching and final examinations, and students have adapted to these methods. However, the teaching and examination methods for comprehensive practice courses are quite different from those for theoretical courses [13, 14]. Especially in this software design and programming practice course, many knowledge points require students to acquire through self-directed learning, most designs require students to acquire through independent thinking, and the way to ultimate realization may also be different for everyone. This course construction will adopt mixed teaching methods and diversified examination modes to solve this problem.

(3) The combination of experimental teaching and practical engineering is not close enough, which restricts the cultivation of students' innovative ability and practical ability.

As a comprehensive practice course of innovation and entrepreneurship, the experimental teaching of the software design and programming practice course is the basis for combining specialized knowledge and practice and is also an important link for cultivating students' ability to practice and innovate. Therefore, experimental teaching and practical engineering should be closely related. However, in the cultivation of students majoring in electronic information engineering and communication engineering, teaching tends to be academic, which restricts the cultivation of students' innovative ability and practical ability [15]. In the course construction, we will extract some content from actual engineering projects as the experimental teaching content and strengthen the background introduction of these projects to stimulate students' interest in innovation and entrepreneurship.

3 Education Objectives

Our College of Information and Communication Engineering is accredited by the Chinese Engineering Education Accreditation Association (CEEAA) [16] to ensure the education quality of undergraduate students. CEEAA defines twelve Educational Objectives that are fulfilled by the sum total of all the courses. In our software design and programming practice course, six objectives are fulfilled. The following list describes the specific education objectives.

(1) With the theoretical support of specialized knowledge, through the study and practice of this course, students can fully understand and master the principles and methods involved in relevant courses and have the ability to analyse and solve problems in practical engineering projects with specialized knowledge.
(2) Students can design systems, functional modules, and processes for specific needs and realize the construction of communication systems or signal processing systems.
(3) Students can use the corresponding specialized knowledge and software development methods to design experimental programs, analyse and interpret data with software, and study system problems to obtain reasonable and effective conclusions.
(4) Students can select reasonable software development tools for specific signal processing engineering problems in the field of electronics and communication, use tools to predict, simulate and measure problems, and understand their limitations.
(5) Students are able to effectively communicate with peers on the design scheme, research methods, and technical routes of electronic and communication systems and clearly express or accurately respond to instructions.
(6) Students' self-learning awareness and ability are developed through the design of experimental projects in the extracurricular time of this course, and students can have the ability to continuously learn and adapt to development.

4 Course Construction

4.1 Overview of Course Construction

The core principles of our course construction can be concluded as "from basis to comprehensiveness, and then to innovation". The specific meaning is first consolidating students' basic programming skills, second combining programming with specialized knowledge, and finally encouraging students to participate in innovation competitions. Based on this principle, we reformed the software design and programming practice course from aspects of teaching content, teaching methods, and examination methods.

4.2 Teaching Content

In view of the weak software programming foundation of students majoring in electronic information engineering and communication engineering, the experimental content is divided into three projects of gradual difficulty: the basic project of programming language, the basic project of human-computer interaction, and the innovative comprehensive practice project, as shown in Fig. 1.

The first basic project of programming language includes basic data operation, module design, program structure design, and so on, with the purpose of strengthening students' previous programming foundation. In addition, this project is a connection between the forward courses and our course.

The second basic project of human-computer interaction covers an object-oriented development environment, common controls of the human-computer interaction interface, and so on, with the purpose of enabling students to master basic knowledge of software design. This project is a transition from basic programming languages to software design and development.

The third innovative comprehensive practice project is based on the former two projects and is a relatively difficult one. It is combined with specialized knowledge and practical engineering to achieve relatively complete software for radar signal processing, image signal processing, audio signal processing, or communication systems, which is a design research experiment of high difficulty. The third project can be decomposed into data acquisition, data display, data transformation, data analysis, and data evaluation, following the concept of step-by-step and simple-to-complex design. Not only at the software design level, these functions are closely linked but also at the specialized knowledge level, these data processing steps are closely linked.

After learning the software design and programming practice course, students' innovative and entrepreneurial thinking is initially developed. However, innovation and entrepreneurship education for students should not be limited to this course. After class, we continuously guide students to participate in undergraduate innovation and entrepreneurship training programs and various science and technology competitions, such as China international college students' "Internet+" innovation and entrepreneurship competition [17], "Challenge Cup" national undergraduate extracurricular academic science and technology works competition [18], and national undergraduate innovation and entrepreneurship training program [19], further strengthening the integration of innovation and entrepreneurship education and specialized education.

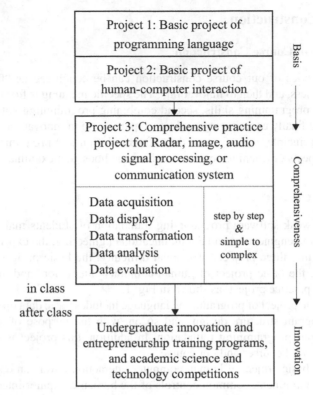

Fig. 1. Teaching contents

4.3 Teaching Methods

The knowledge points in the teaching contents are complex and require repeated practice. Due to the limitation of in-class hours, it is difficult to consider both the depth and breadth of the teaching content in the classroom. Therefore, we adopt mixed teaching methods, as shown in Fig. 2. We only use a small amount of in-class hours to teach basic knowledge points. For the rest of the in-class hours, we adopt teaching methods that include teacher-student interaction, student-student interaction, and discussion. To obtain a better teaching effect, "microlectures" are recorded for students to learn independently. At the same time, this course adopts 1:1 in-class and after-class hours to ensure that students have enough practice time.

4.4 Examination Methods

For different experimental projects, we adopt diversified examination methods, as shown in Fig. 3. For basic projects 1 and 2, students randomly draw questions from the examination database and then conduct on-site programming and oral defense, while teachers assess students according to their operations, codes, and answers to questions. This assessment mode largely avoids the drawback of being easy to copy in software programming experiments and urges students to carry on comprehensive learning. For the

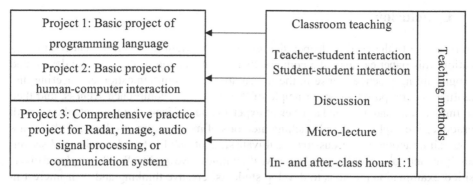

Fig. 2. Teaching Methods

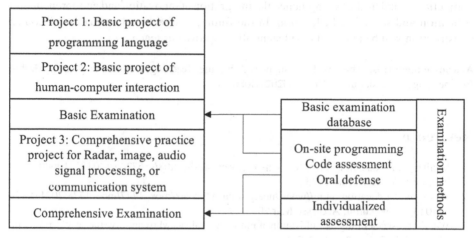

Fig. 3. Examination Methods

third comprehensive practice project, students choose one of the given projects from radar, image, audio single processing, or communication systems according to their own situation and then independently analyse the problem, design the framework, and implement it by programming. Finally, students demonstrate the designed software through on-site demonstration, explanation, and oral defense, while teachers evaluate students based on their operations, codes, and answers to questions. It is noted that software with the same functionality may not be implemented in a unique way, and the implementation methods of different personnel may also be different. Therefore, students who have outstanding performance in functions, implementation approaches, modular design, and other aspects should also be fully affirmed; that is, we need to carry out an individualized assessment.

5   Conclusion

In response to the need to deepen the reform of innovation and entrepreneurship education, this paper carries out a comprehensive construction of the software design and programming practice course in the information and communication engineering discipline. We proposed a core principle of "from basis to comprehensiveness, and then to innovation" and designed a series of experimental projects that integrate theory with practice. We explored mixed teaching methods of microlecture, in-class and after-class hours, interaction, and discussion to cultivate students' ability to independently discover, analyse, and solve problems and improve their innovative quality. We also explored diversified examination methods to develop students' creative thinking and their interest in programming. After class, we continued to guide students to participate in innovation and entrepreneurship training programs and various academic science and technology competitions and further strengthened the integration of innovation and entrepreneurship education and specialized education. In the future, feedback and evaluation of course construction will be obtained to substantially improve our teaching.

Acknowledgements. This work is supported by the Teaching Reform Project of Harbin Engineering University under Grant JG2021B0806.

References

1. Ministry of Education. http://www.moe.gov.cn/srcsite/A08/s5672/201904/t20190408_377040.html. Accessed Mar 2023
2. Ministry of Education. http://www.moe.gov.cn/jyb_xxgk/moe_1777/moe_1778/202110/t20211013_571909.html. Accessed Mar 2023
3. Wu, Z.H.: Accelerating the cultivation of innovative talents who can solve the "neck" problem. China High. Educ. **12**, 7–8 (2021)
4. Li, S.H., Li, X.H.: Study on the trinity of "student-academic-discipline" to cultivate students' innovation ability. China High. Educ. **8**, 53–54 (2020)
5. Xi, Y., Chen, X., Li, Y.: Exploration and practice for the cultivation mode of college students' innovation ability. In: Zeng, J., Qin, P., Jing, W., Song, X., Lu, Z. (eds.) ICPCSEE 2021. CCIS, vol. 1452, pp. 456–464. Springer, Singapore (2021). https://doi.org/10.1007/978-981-16-5943-0_37
6. Zheng, Q.: Creating a space for cultivating top-notch innovative talents in basic disciplines with "no ceiling". China High. Educ. **12**, 27–29 (2022)
7. MIT, Introduction to Computer Science and Programming for Non-Programmers. https://ocw.mit.edu/courses/6-00sc-introduction-to-computer-science-and-programming-spring-2011. Accessed Mar 2023
8. Stanford University, Software Development for Scientists and Engineers. https://explorecourses.stanford.edu. Accessed Mar 2023
9. Peking University, Educational Program and Curriculum of Undergraduate in School of Electronics Engineering and Computer Science. https://eecs.pku.edu.cn/info/1083/5225.htm. Accessed Mar 2023
10. Southeast University, Educational Program and Curriculum of Undergraduate in School of Information Science and Engineering. https://radio.seu.edu.cn/19230/list.htm. Accessed Mar 2023

11. Shanghai Jiao Tong University, Educational Program and Curriculum of Undergraduate in School of Electronic Information and Electrical Engineering. https://bjwb.seiee.sjtu.edu.cn/bkjwb/list/1338-1-20.htm. Accessed Mar 2023

12. Nanjing University. Educational Program and Curriculum of Undergraduate in School of Electronic Science and Engineering. https://ese.nju.edu.cn/22551/list.htm. Accessed Mar 2023

13. Zhang, W., Wang, R., Tang, Y., Yuan, E., Wu, Y., Wang, Z.: Research on teaching evaluation of courses based on computer system ability training. In: Zeng, J., Qin, P., Jing, W., Song, X., Lu, Z. (eds.) ICPCSEE 2021. CCIS, vol. 1452, pp. 434–442 (2021). Springer, Singapore. https://doi.org/10.1007/978-981-16-5943-0_35

14. Li, Y.Q., Li, F., Wang, T.: Research on comprehensive evaluation system of teaching quality of college teachers from the perspective of new engineering. Res. High. Educ. Eng. **S1**, 289–291 (2019)

15. Shu, Y., et al.: Bottom-up teaching reformation for the undergraduate course of computer organization and architecture. In: Mao, R., Wang, H., Xie, X., Lu, Z. (eds.) ICPCSEE 2019. CCIS, vol. 1059, pp. 303–312. Springer, Singapore (2019). https://doi.org/10.1007/978-981-15-0121-0_23

16. CEEAA. China engineering education accreditation association. http://www.ceeaa.org.cn. Accessed Mar 2023

17. China International college students' "Internet+" innovation and entrepreneurship competition. https://cy.ncss.cn/home. Accessed Mar 2023

18. "Challenge Cup" National undergraduate extracurricular academic science and technology works competition.https://www.tiaozhanbei.net/. Accessed Mar 2023

19. National undergraduate innovation and entrepreneurship training program. http://gjcxcy.bjtu.edu.cn/index.aspx. Accessed Mar 2023

A Self-Attention-Based Stock Prediction Method Using Long Short-Term Memory Network Architecture

Xiaojun Ye[1], Beixi Ning[2], Pengyuan Bian[2(✉)], and Xiaoning Feng[1]

[1] College of Computer Science and Technology, Harbin Engineering University, Harbin, China
[2] College of Information and Communication Engineering, Harbin Engineering University, Harbin, China
lekaendari@gmail.com

Abstract. The ability to analyze the trend of the stock market has always been paid high attention to. A large number of machine learning technologies have been used for stock analysis and prediction. The traditional time series prediction models, including RNN, LSTM and their deformed bodies, show the problems of gradient disappearance and low efficiency in long-span prediction. This paper proposes a long-term and short-term memory network architecture, which based on Encoder and Decoder Stacks and self-attention mechanism, replacing the feature extraction part of traditional LSTM through self-attention mechanism and provides interpretable insights into the dynamics of time. Through the results of simulation experiments, this paper shows the comparison of stock prediction effects through using RNN, Bi-LSTM and Encoder and Decoder-Attention-LSTM models. The experimental task shows that the prediction accuracy of this model is improved by an order of magnitude compared with the traditional LSTM-like model, and can achieve high accuracy when the epoch is small.

Keywords: Data mining · Stock Market Prediction · Attention · LSTM · Transformer

1 Introduction

Stock prediction is one of the important topics in the market economy research. In the stock market, the correct investment decision depends on the correct prediction [1]. Effective extraction of feature information from historical data is one of the key parts of stock prediction [2]. The extracted features can significantly affect the accuracy of pattern recognition tasks in data mining [3]. In the past financial market transactions, the market traders used the market activity data to draw the stock trend chart, deduced the characteristics of the stock data through technical analysis [4], and used it to predict the future stock trend in order to optimize the trading strategy. However, such a decision-making method has some subjectivity and with greater decision-making risks [4]. Recently years, with the rapid development of artificial intelligence and other technologies. The deep learning model is considered as the best information extractor and classifier for financial market trend prediction using a large amount of dynamic information [5].

© The Author(s), under exclusive license to Springer Nature Singapore Pte Ltd. 2023
Z. Yu et al. (Eds.): ICPCSEE 2023, CCIS 1879, pp. 12–24, 2023.
https://doi.org/10.1007/978-981-99-5968-6_2

Among the many deep-learning neural network tools used, the classic tools used for stock prediction include recurrent neural network (RNN) and bidirectional short-term memory network (Bidirectional LSTM). In the process of long sequence training, RNN and LSTM will cause gradient disappearance and gradient explosion after multi-stage reverse propagation [6]. In recent years, the self-attention mechanism of Transformer model [7, 8] has shown good prediction effect for time series prediction problems, and is expected to achieve accurate long series prediction of stock data.

We conducted in depth research on the problems of using deep learning models for stock prediction. The tissue structure of the paper is as follows. Section 2 studied these models in depth and analyzed the network structure they used for stock forecasting. Section 2.5 detailed introduce the network structure of the Encoder and Decoder-Attention-LSTM (EDA-LSTM) model. Section 3 shows the steps and results of the simulation experiment. Section 4 describes the conclusions of this paper and proposes the direction of future research and improvement.

The EDA-LSTM model proposed in this paper, is better than the traditional time series prediction model in terms of efficiency and accuracy, and can be used to improve the accuracy and time efficiency of the existing time series data prediction model. The accurate stock data prediction realized by the model provides the general users and market traders with the estimation of the entire path, allowing them to optimize their actions in multiple steps in the future. Meanwhile, the potential value of the self-attention mechanism in capturing the individual dependence between the output and input of a long series of time series is verified [9].

2 Related Work

In recent years, researchers have been exploring neural networks with more accurate expression ability and richer types as experimental models for stock prediction. Recurrent Neural Network (RNN) can complete the processing of relevant time series at each time step and save the whole state of stock data series [10]. It is the main network for processing time series data in deep learning. Nagarjun Y adav Vanguri et al. proposed an effective strategy for accurate stock market prediction using the newly prepared Adaptive Competitive Feedback Particle Swarm Optimization (CFPSO) method, achieving a training result with a minimum MSE of 0.095 and a minimum RMSE of 0.213 [11]. Weiling Chen et al. extracted the emotional features and Latent Dirichlet Allocation (LDA) features in the news information released by the official media account, and input them into the new hybrid model of RNN-boost to predict the trend of Shanghai-Shenzhen 300 Stock Index (HS300) [12]. Ruixun Zhang et al. proposed a new Deep and Wide Neural Networks (DWNN) architecture, which adds the convolution layer of CNN to the hidden state transmission process of RNN, extracts the relevant characteristics of different RNN models running along time steps, and forecasts the trading data of 12 different stocks on the Shanghai Stock Exchange (SSE), The experimental results show that the prediction error of DWNN model is 30% lower than that of general single-chain RNN model [13].

However, for the long-term dependence of data, RNN gradient is dominated by the near-range gradient, which makes it difficult for the model to learn the long-distance dependence. LSTM outperforms RNN on vanishing gradient problems and retains the benefits of RNN on time-series variables [14]. Shuyan Liu proposed to combine deep learning models (such as RNN, LSTM and GRU) with dimensionality reduction methods (such as PCA, 2d-PCA, 2d2d-PCA and SAE) to avoid over-fitting problems. The experimental results show that the best combination of IBM stock prediction is 2-Directional 2-Dimensional PCA (2d2d-PCA) with LSTM model, which is based on R-Square (R2) with an accuracy of 91.77% [15]. Naijie Gu et al. proposed a stock prediction model consisting of a Deep Belief Network (DBN) represented by learning potential features and a Long Short-Term Memory (LSTM) network used to utilize long-term relationships in the trading history, which uses intra-day data instead of daily data for training, enriching sample information and eliminating the variance of training data caused by changes in the financial environment [16]. Neema Singh et al. proposed a hybrid model followed by Recurrent Neural Network - Long Short-Term Memory (RNN-LSTM) to predict a next-day closing price of SBI, and showed good prediction effect [17]. Xiaojiang Wen et al. proposed a new stock forecasting model based on BERT and LSTM, calculated investor sentiment before the opening of the market by fine-tuning BERT model, then aggregated the calculated investor sentiment with the basic stock market data, and finally used LSTM model to predict the closing price of the next stock exchange, and verified the effectiveness of the model on the real data sets of three Chinese listed companies [18].

Chaojie Wang et al. used the deep learning framework Transformer proposed in recent years to predict the stock market index, and conducted many back-testing experiments on major stock market indexes in the world, including the Shanghai and Shenzhen 300, the S&P 500, the Hang Seng Index and the Nikkei 225. The experimental results show that the Transformer model is significantly better than the traditional model in stock prediction [19]. However, Transformer still has deficiencies in capturing the timing information of historical data [20].

2.1 RNN Model

RNN is a special type of artificial neural network. The nodes between the hidden layers are no longer connected but connected, and the input of the hidden layer includes not only the output of the input layer but also the output of the hidden layer at the previous time. This allows it to show the dynamic time behavior of the time series. RNN has the concept of "memory", which can help them store the previously entered state or information to generate the next output of the sequence [21].

When processing stock data, the RNN uses state h to accumulate the information that has been read. h_0 includes the information of the first data, h_1 includes the information of the first two data, and so on. The state h_t contains the information of the first t + 1 data, which can be regarded as the characteristic vector extracted from this group of stock data. After the state input returns to the output layer, we can get our final result. The whole RNN has only one parameter matrix A, A is initialized randomly, and then A is learned through training data.

The DWNN model [13] proposed by Ruixun Zhang et al. and uses m sets of Seq2seq models to simultaneously process m sets of time data series. We put the CNN layer into the encoder part, While the decoder part remains independent and unaffected. Combining the advantages of CNN in processing spatially related data and the characteristics of

Seq2seq model to automatically learn the features from training datas, and then realizes the prediction of the future stock series data after the variable length by using the variable length historical stock series data.

In the code, Ruixun Zhang et al. serialized the neural network, used Seq2seq model to build a single RNN chain, and proposed the design of interval CNN layer, that is, add one CNN layer per k RNN cycle time steps. Each layer in the stacked RNN layer contains multiple RNN units and a dropout mask, which can be used to eliminate the potential strong dependence on one dimension to prevent overfitting is applied to the output of each RNN unit.

Each layer of circular memory layer builds diverse memories, and the generated states $h_0 \sim h_t$ are all output to the next layer as the input of the next layer. The Dropout parameter is set to reduce over-fitting, and according to this, the closing price of the forecast day can be obtained after the final state h_t of the multi-layer RNN passes through the regression output layer of the sigmoid activation function.

2.2 Bi-LSTM Model

In order to overcome the long-term dependence of RNN, the structure of LSTM proposed is similar to that of RNN. The change is that different functions are used to calculate the hidden state. In stock prediction, LSTM solves the problem of memory of stock historical data by presenting gate cells and storage cells in the design of neural network. In data transmission, it selectively deletes or adds the previous stock data information to the current prediction state, only learns to keep the stock data information related to the current prediction, and forgets the irrelevant data information, so as to make the data prediction more accurate. LSTM uses different gates to transfer the recently experienced data from one unit to another. These gates are called update gates, forgetting gates and output gates. The logic diagram is shown in Fig. 1:

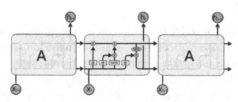

Fig. 1. Classic LSTM logic diagram (picture from http://t.csdn.cn/s3cgV)

In LSTM, the forgetting gate reads the output of the previous moment and the current input, and removes the useless data information in the current prediction through the function nonlinear mapping; The input gate is filtered through the function to determine the useful data information into the current prediction state; The output gate determines the output of a single repetitive module of the neural network by integrating the outgoing data and the cell update status value.

The structure model of Bi-LSTM neural network is composed of two independent LSTM networks. The input sequence of stock prediction data is input into two LSTM neural networks from the direction of positive and reverse order respectively for update training. The two output vectors of positive and reverse order are integrated as the output of the final prediction data. The logic diagram is shown in Fig. 2:

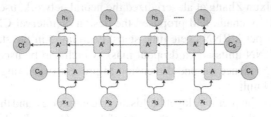

Fig. 2. Bi-LSTM logic diagram

In Bi-LSTM, the data features extracted in the positive and reverse order of the model are sufficient, and the positive and reverse data features are integrated, and then make a state that look forward and backward. Md. Arif Istiake Sunny et al. through comparing the performance of LSTM and Bi-LSTM methods by changing different parameters [22]. Finally, through systematic research, Bi-LSTM can capture the time evolution of information [23], and achieve the best stock prediction performance compared with one-way. However, it is still unable to completely solve the problem of the forgetting of long sequence data. Meanwhile, due to the limitations of the LSTM framework itself, the Bi-LSTM model is still serial extraction when extracting features, and the training speed is slow.

The above RNN and LSTM have derived a variety of morphs for stock prediction, and the prediction accuracy is improving with time, but there are still problems of gradient disappearance and gradient explosion caused by multi-stage reverse propagation [6].

2.3 Why EDA-LSTM

Because the total gradient of RNN is dominated by the gradient in the short range, the gradient disappears due to the neglect of the gradient in the long range. LSTM alleviates the gradient disappearance and gradient explosion problems in the long sequence training process by designing the conveyor belt. Traditionally, the forgetting problem can be further alleviated by training two independent LSTMs, but it will still be difficult to deal with longer sequences. In addition, the process of LSTM feature processing is serial input. Each LSTM cell means four full connection layers (MLPs). If the time span of LSTM is large and the network is deep, it will lead to excessive time consumption and average effect. Transformer makes it possible to focus on key historical data when carrying out forecasting tasks. The multiple levels of LSTM can be utilized for multiple levels of features, which can represent abstract features derived from previous levels, and thus, the level of abstraction is increased. Compared to typical networks with one hidden layer, LSTM can achieve a higher level of feature extraction by adding extra hidden layers [24]. In this paper, the feature extraction that based on the self-attention mechanism in the Transformer model, and then the LSTM is input for time series training. The self-attention mechanism is fully used to show good training effect on the periodic data of stocks. At the same time, the residual network is used to solve the gradient disappearance and improve the parallelism of calculation.

2.4   Network Architecture

The feature extraction module of the model follows the coder-decoder structure in Transformer proposed by Ashish Vaswani et al. [8], where the encoder maps an input sequence $(x_1, \ldots, x_n)$ represented by a symbol to a continuous representation sequence $z = (z_1, \ldots, z_n)$. Given z, the decoder generates an output sequence $(y_1, \ldots, y_m)$ of symbols. The output of Decoder is used as the input of LSTM. In each step, the model is automatically regressed. When generating the next step, the previously generated symbols are used as additional inputs. The specific network structure is shown in Fig. 3:

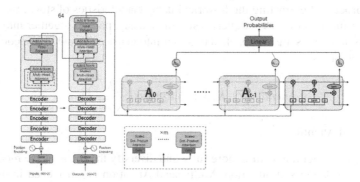

Fig. 3. EDA-LSTM network structure diagram. EDA-LSTM inputs static metadata, integrates information at any time step based on multi-header attention, and time-dependent processing is based on LSTM module.

2.5   Encoder and Decoder Stacks

Encoder: The Encoder part is consists of N = 6 stacks of the same layer. Each layer has two sub-layers. The first layer is the multi-head self-attention mechanism, and the second layer is the fully connected feedforward network. We use a residual neural network around each two sub-layers, and then carry out layer normalization. In order to facilitate these residual connections, all sub-layers and embedded layers in the model will generate outputs with dimension $d_{model} = 2$.

Decoder: Decoder is also composed of N = 6 stacks of the same layer. In addition to the original mask layer, interaction layer and sense layer, the output of the decoder layer is used as the input of the LSTM.

2.6   Gating Mechanisms for Random Mask Stock Data

Before executing the attention function, we randomly initialize a 01 matrix. The dimension of the matrix is the dimension of the input vector. A matrix element of 0 means that the corresponding position data does not participate in the calculation of the attention mechanism.

2.7 Attention Used to Calculate the Correlation Between Stock Data

- Scaled Dot-Product Attention

The attention function is used to reconstruct the input feature vector, and to get the most relevant days between the current stock data and the historical data. Query represents the information to be queried, key represents the vector to be queried, and value represents the value obtained from the query.

For each q, we calculate the dot product with all k, then divide by $\sqrt{d_k}$, and apply a softmax function to obtain the weight of the value.

In the process of extracting the historical data characteristics of stock, we compute the attention function on a set of queries simultaneously, packed together into a matrix Q. The keys and values are also packed together into matrices K and V. We compute the matrix of outputs as:

$$Attention(Q, K, V) = softmax\left(\frac{QK^T}{\sqrt{d_k}}\right)V \tag{1}$$

- Multi-Head Attention

In order to extract the feature more fully and diversify the focus of the model in the process of calculating self-attention, Multi-head Attention is introduced in the model establishment, and multiple Q, K, V matrices are updated to obtain the features extracted from multiple angles of the model, and then the data features extracted from each head are fused to make the data features better.

$$MultiHead(Q, K, V) = Concat(head_1, \ldots, head_h)W^o$$

$$where\ head_i = Attention(QW_i^Q, KW_i^K, VW_i^V) \tag{2}$$

2.8 LSTM Part for Extracting Temporal Features

The feedforward neural network of N-layer Decoder outputs 64 * 1 vector and inputs it into the LSTM of 64-layer at the same time. The self-focus mechanism of Transformer can effectively extract the important features of the historical data for the current forecast day data, which solves the problem that LSTM cannot predict the long series.

2.9 Positional Encoding

The feature extraction part of this model does not include recursion and convolution. In order to make the model use the sequence order, we must inject some information about the relative or absolute position of symbols in the sequence. We embed position coding after transforming the data into standard normal distribution. The position code and the input have the same dimension d_{model}, so they can be added. In this work, we use sine and cosine functions of different frequencies:

$$PE(pos, 2i) = sin(pos/10000^{2i/d_{model}}) \tag{3}$$

$$PE(pos, 2i + 1) = cos(pos/10000^{2i/d_{model}}) \tag{4}$$

where pos is the location, i is the dimension. Which means, each dimension of position coding corresponds to a sine wave.

2.10 Loss Function

The mean square error MSE (L2 LOSS) is used: the mean square error (MSE) is the average of the square of the difference between the predicted value f (x) of the model and the true value y of the sample. The formula is as follows:

$$loss(x, y) = \frac{1}{n} \sum_{i=1}^{n} (y_i - f(x_i))^2 \tag{5}$$

where y_i is the real value, $f(x_i)$ is the predict value and n is the number of samples. Each point of L2 LOSS function is continuous and smooth, which is convenient for derivation and has relatively stable solution.

3 Experiment

3.1 Experimental Environment

The experimental environment for training and testing EDA-LSTM model in this paper is shown in Table 1.

Table 1. Experimental environment based on Python 3.8.5

Open source third-party Python library	Version	Open source third-party Python library	Version
torch	1.12.1+cpu	tqdm	4.61.2
matplotlib	3.4.2	pandas	1.3.0
numpy	1.23.3	argparse	1.1

3.2   Data Preparation

The stock data we used for training and testing is the data of ENGRO stock and HBL stock (hereinafter referred to as stock 1 and stock 2) in the Korean Stock Exchange during the period from September 1, 2003 to December 12, 2018 (3691 days).

Stock 1 data source:

https://github.com/ZainUlMustafa/Stock-Prediction-RNN-LSTM/blob/master/Data/KSE/ENGRO_01092003_12102018.csv

Stock 2 data source:

https://github.com/ZainUlMustafa/Stock-Prediction-RNN-LSTM/blob/master/Data/KSE/HBL_01092003_12102018.csv

We use the closing price of the first 90% (3321 days) of the stock 1 for training, and the remaining data (370 days) for testing, and convert the training data into tenor format. In the training data, every 64 days form a time series, and the 65th day is used as the corresponding training tag to form a group of training data. All the data used for training (3321 days) can generate multiple groups of training data. The data of the input characteristics in each group does not overlap, which disrupts the order of the training data, calculates the loss through the prediction tag and the training tag, and updates the corresponding network parameters, all groups of training data are trained together once to complete an epoch; In the test data, every 64-day data is used as the test input feature, and the 65th day is used as the corresponding test tag to form a test sequence. Each day's closing price is predicted, and then a new set of test data is formed by the data predicted on the day and the 63 days before the day for the next round of prediction.

3.3   Experimental Parameters

To highlight the superiority of the EDA-LSTM algorithm, the RNN and Bilstm models were set with a time series length of 64, a learning rate of 0.001, an Adam optimiser, a 6-layer structure for the RNN network and a bidirectional layer number of 2 chosen for the Bilstm network.

3.4   Results

We choose MSE loss function when predicting the target series, and the loss propagates back to the whole model from the output. We have shown in Table 2 that RNN, Bi-LSTM and EDA-LSTM models under 10, 30 and 50 epochs have been trained respectively for the loss comparison of two open-source stock data forecasts.

From the simulation results in Table 2, it can be seen that EDA-LSTM has achieved a very good training effect when epochs are small. Compared with the traditional two models, EDA-LSTM has an order of magnitude improvement in accuracy - for stock 1, when Epochs = 10, the loss rate of this model is reduced by 98.59% compared with RNN model, and 97.99% compared with Bi-LSTM, which proves the advantage of replacing LSTM with self-attention mechanism for feature extraction.

Figure 4 shows the output of expected value and actual value of stock 1 from RNN model. The actual value is shown in blue, and the predicted value is shown in red.

Table 2. Univariate long series time series prediction results of RNN, Bi-LSTM and EDA-LSTM models

Epochs	10		30		50	
Stock	Stock 1	Stock 2	Stock 1	Stock 2	Stock 1	Stock 2
RNN	0.568	0.632	0.282	0.326	0.142	0.183
Bi-LSTM	0.398	0.424	0.186	0.202	0.094	0.105
Transform-LSTM	0.008	0.010	0.005	0.006	0.003	0.005

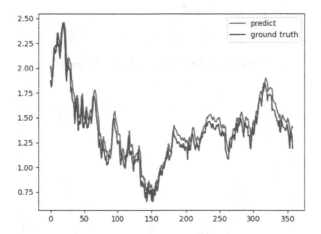

Fig. 4. Use the RNN model for stock forecasting

It can be seen from Fig. 4 that the stock trend predicted by RNN is consistent with the actual trend, but there is a delay error in time between the predicted value and the actual value.

Similarly, Fig. 5 and Fig. 6 show the results of Bi-LSTM model and EDA-LSTM model on the same data set and input, respectively.

In the end, we will combine the training loss values of the three models shown in the Table 2 and the visual gap between the predicted curves shown in figures and the actual curves to judge the prediction effects of the three models.

From the visual inspection, Bi-LSTM and EDA-LSTM predicted the actual value very closely. It is obvious that the EDA-LSTM model is closer to the actual stock price time series, and the prediction curve of the model almost coincides with the actual curve.

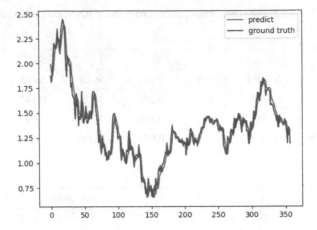

Fig. 5. Use the Bi-LSTM model for stock forecasting

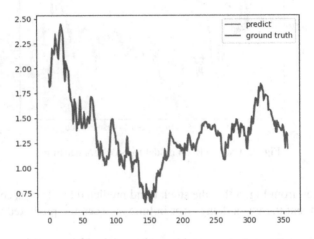

Fig. 6. Use the EDA-LSTM model for stock forecasting

4   Conclusion and Future Work

In this paper, we propose the EDA-LSTM model by combining the parallelism of Transformer in feature extraction and the accuracy of long time series learning with the accurate capture of local features by LSTM and the effective learning of time series information. For the feature extraction of historical stock data, Transformer is far better than the framework based on circular layer or convolution layer in terms of extraction efficiency and accuracy, and provides interpretability for the prediction model. LSTM with special implicit unit is an excellent variant model of RNN. It can memorize the previous information and apply it to the current output calculation, and has strong learning ability for the timing of data. In addition, in this paper, we use the same batch of stock data, and use RNN, Bi-LSTM, and EDA-LSTM models to predict, respectively. The results show that the training effect of our proposed EDA-LSTM model is much better than the other two models, and verify the potential value of the self-attention mechanism in capturing the individual dependency between the output and input of the long series of time series. Our model can be used as a starting point for the future research and implementation.

The EDA-LSTM model proposed in this paper is a further improvement of the LSTM model. Compared with the traditional methods of constructing two-way networks and multi-layer networks, Transformer's multi-head attention mechanism shows better results. In the future, we can replace the random initialization in Gating mechanisms with the gating linear unit GLU for feature selection to improve the generalization ability of the model; In addition to the known data in the past and the data to be predicted in the future, some known prior knowledge in the future may also have an impact on the data predicted by the model, so we can use these known prior information in the future as a part of the input through the gated residual network (GRN).

Acknowledgment. This work is supported by the National Nature Science Foundation of China through project 51979048.

References

1. Liu, J., Lin, C.M.M., Chao, F.: Gradient boost with convolution neural network for stock forecast. In: Ju, Z., Yang, L., Yang, C., Gegov, A., Zhou, D. (eds.) UKCI 2019. AISC, vol. 1043, pp. 155–165. Springer, Cham (2020). https://doi.org/10.1007/978-3-030-29933-0_13
2. Gong, Y., Ming-Tai Wu, J., Li, Z., Liu, S., Sun, L., Chen, C.M.: A CNN-based method for AAPL stock price trend prediction using historical data and technical indicators. In: Zhang, J.F., Chen, C.M., Chu, S.C., Kountchev, R. (eds.) Advances in Intelligent Systems and Computing. SIST, vol. 268, pp. 25–33. Springer, Singapore (2022). https://doi.org/10.1007/978-981-16-8048-9_3
3. Zheng, Y., Si, Y.W., Wong, R.: Feature extraction for chart pattern classification in financial time series. Knowl. Inf. Syst. **63**, 1807–1848 (2021). https://doi.org/10.1007/s10115-021-01569-1
4. Sakhare, N.N., Shaik, I.S., Saha, S.: Prediction of stock market movement via technical analysis of stock data stored on blockchain using novel history bits based machine learning algorithm. IET Soft., 1– 12 (2023). https://doi.org/10.1049/sfw2.12092
5. Jin, Z., Jin, Y., Chen, Z.: Empirical mode decomposition using deep learning model for financial market forecasting. PeerJ Comput. Sci. **8**, e1076 (2022). https://doi.org/10.7717/peerj-cs.1076
6. Hao, H., Wang, Y., Xia, Y., et al.: Temporal convolutional attention-based network for sequence modeling. arXiv preprint arXiv:2002.12530 (2020)
7. Wu, N., Green, B., Ben, X., et al.: Deep transformer models for time series forecasting: the influenza prevalence case. arXiv preprint arXiv:2001.08317 (2020)
8. Vaswani, A., et al.: Attention is all you need. In: Proceedings of the 31st International Conference on Neural Information Processing Systems (NIPS 2017), pp. 6000–6010. Curran Associates Inc., Red Hook (2017)
9. Lim, B., Arık, S.Ö., Loeff, N., et al.: Temporal fusion transformers for interpretable multi-horizon time series forecasting. Int. J. Forecast. **37**(4), 1748–1764 (2021)
10. Shah, D., Campbell, W., Zulkernine, F.H.: A comparative study of LSTM and DNN for stock market forecasting. In: 2018 IEEE International Conference on Big Data (Big Data), pp. 4148–4155. IEEE (2018)

11. Vanguri, N.Y., Pazhanirajan, S., Kumar, T.A.: Tversky-RideNN based feature fusion and optimized deep RNN for stock market prediction. In: 2022 4th International Conference on Inventive Research in Computing Applications (ICIRCA), Coimbatore, India, pp. 1056–1063 (2022). https://doi.org/10.1109/ICIRCA54612.2022.9985572

12. Chen, W., Yeo, C.K., Lau, C.T., Lee, B.S.: Leveraging social media news to predict stock index movement using RNN-boost. Data Knowl. Eng. **118**, 14–24 (2018)

13. Zhang, R., Yuan, Z., Shao, X.: A new combined CNN-RNN model for sector stock price analysis. In: 2018 IEEE 42nd Annual Computer Software and Applications Conference (COMPSAC), Tokyo, Japan, pp. 546–551 (2018). https://doi.org/10.1109/COMPSAC.2018.10292

14. Chung, J., Jang, B.: Hybrid CNN-LSTM model with multivariate data to increase the forecast accuracy of electricity consumption. SSRN. https://ssrn.com/abstract=4097479 or https://doi.org/10.2139/ssrn.4097479

15. Liu, S., Chen, Y.: Comparison of variant principal component analysis using new RNN-based framework for stock prediction. In: 2021 International Conference on Data Mining Workshops (ICDMW), Auckland, New Zealand, pp. 1047–1056 (2021). https://doi.org/10.1109/ICDMW53433.2021.00136

16. Zhang, X., Gu, N., Chang, J., Ye, H.: Predicting stock price movement using a DBN-RNN. Appl. Artif. Intell. **35**(12), 876–892 (2021). https://doi.org/10.1080/08839514.2021.1942520

17. Singh, N., Mohan, B.R., Naik, N.: Hybrid model of multifactor analysis with RNN-LSTM to predict stock price. In: Gupta, D., Sambyo, K., Prasad, M., Agarwal, S. (eds.) Advanced Machine Intelligence and Signal Processing. LNEE, vol. 858, pp. 107–122. Springer, Singapore (2022). https://doi.org/10.1007/978-981-19-0840-8_8

18. Weng, X., Lin, X., Zhao, S.: Stock price prediction based on LSTM and bert. In: 2022 International Conference on Machine Learning and Cybernetics (ICMLC), Japan, pp. 12–17 (2022). https://doi.org/10.1109/ICMLC56445.2022.9941293

19. Wang, C., Chen, Y., Zhang, S., et al.: Stock market index prediction using deep transformer model. Expert Syst. Appl. **208**, 118128 (2022)

20. Wen, Q., Zhou, T., Zhang, C., et al.: Transformers in time series: a survey. arXiv preprint arXiv:2202.07125 (2022)

21. Salinas, D., Flunkert, V., Gasthaus, J., et al.: DeepAR: probabilistic forecasting with autoregressive recurrent networks. Int. J. Forecast. **36**(3), 1181–1191 (2020)

22. Cho, K., Merrienboer, B.V., Gulcehre, C., et al.: Learning phrase representations using RNN encoder-decoder for statistical machine translation. Comput. Sci. (2014)

23. Shah, J., Jain, R., Jolly, V., Godbole, A.: Stock market prediction using bi-directional LSTM. In: 2021 International Conference on Communication information and Computing Technology (ICCICT), Mumbai, India, pp. 1–5 (2021). https://doi.org/10.1109/ICCICT50803.2021.9510147

24. Dey, P., et al.: Comparative analysis of recurrent neural networks in stock price prediction for different frequency domains. Algorithms **14**, 251 (2021). https://doi.org/10.3390/a14080251

CAD-Based Research on the Design of a Standard Unit Cabinet for Custom Furniture of the Cabinet Type

Guo Lan[✉] and Fan Weiquan

Jingchu University of Technology, Jingmen 448200, China
201507005@jcut.edu.cn

Abstract. Mass customization (MC) cabinet furniture has rapidly developed into the mainstream mode of furniture manufacturing in the information age with customized products and services, mass production mode and complete and unified space joint form through the design system of "standard unit cabinet + nonstandardized customized cabinet". Based on this, taking the design of a standard unit cabinet as the center, the modular design method is used to study the design of functional modules in a standard unit cabinet based on a CAD drawing form. In this paper, the relationship between the function and structure of a standard unit cabinet is analysed according to the principles of serialization, generalization and combination in the standardized design principles of a custom wardrobe, combined with the modular design method. Finally, the functional and structural design of two standardized series of wardrobes from the Korean furniture brand "Hanssen" is studied in terms of functional and structural dimensions. Conclusion: MC wardrobe can be designed using a combination of configurations of functional modules to achieve a standardized design that reduces the internal diversity of the product and increases the external diversity of the product. In addition, functional modules and structural design have an adaptive relationship in the design of a standard unit cabinet for a custom wardrobe. It is hoped that this study can have some reference value for the practice of MC furniture and the development of functional modules for custom wardrobes.

Keywords: CAD standard unit cabinet · modularity · function · structure

1 Introduction

With the rapid development of advanced manufacturing technology, computer software technology, QR code information technology and other sciences, the mass customization (MC) wardrobe not only combines the cost and speed of mass production but can also provide consumers with the advantages of personalized products, receiving increasing attention in the furniture industry. However, due to the arbitrary feature of customization, this has led to low efficiency and long delivery cycles in factory production. Therefore, in the face of the huge market demand, MC wardrobe should be standardized under the conditions of meeting mass production.

© The Author(s), under exclusive license to Springer Nature Singapore Pte Ltd. 2023
Z. Yu et al. (Eds.): ICPCSEE 2023, CCIS 1879, pp. 25–36, 2023.
https://doi.org/10.1007/978-981-99-5968-6_3

According to data released by the National Statistical Centre, Sogal, Hanssem and IKEA are the world's top three custom furniture retailers, with Sogal as the leading Chinese company in the custom wardrobe category with a market share of 13.7% and sales of 3.636 trillion yuan in 2018. Hanssem is the long-standing No. 1 brand in the Korean furniture industry, with its share of the Korean furniture market at approximately 28% by 2018. IKEA is the world's largest furniture retailer by market share, with sales of 506.582 trillion yuan in 2018. From the high market share and sales of the three major brands, it can be concluded that the custom furniture produced by these three major companies is representative of the custom furniture industry, as shown in Table 1. Of these, the products of Hanssem in Korea are mainly focused on cabinet-type custom furniture; therefore, Hanssem was taken as the research sample in this paper. Based on this, this study intends to analyse the relationship between the function, structure and standardized design of a standard unit cabinet on the basis of the modular design approach. Based on the relationship between the three, the functional and structural design of a standard unit cabinet in a custom cabinet is analysed using Hanssem as an example, and a functional module optimization method for a standard unit cabinet is proposed.

Table 1. Sales of the Top Three Brands, 2014–2018 (Unit: billion yuan).

Brand	Category	2014	2015	2016	2017	2018	2019
Hanssem	Custom furniture	1.325	1.71	1.934	2.063	1.93	2.01
Hanssem	Custom wardrobe	0.46	0.56	0.65	0.67	0.58	0.61
Sogal	Custom furniture	23.61	31.95	45.29	61.61	73.1	76.8
Sogal	Custom wardrobe	1.384	1.842	2.36	3.09	60.6	61.75
IKEA	furniture	227.178	416.494	429.55	445.217	506.582	

2   Theoretical Background

2.1   Concept and Characteristics of the MC Cabinet and Modular Design

First, we discuss the concept and design features of MC cabinet furniture. MC cabinet furniture is the integrated cabinet furniture assembled on site by professional installers after the design is customized according to the specific indoor space location and size on the basis of mass production by furniture enterprises, in accordance with the individual needs of the user. The design of the MC cabinet is a combination of a "standard unit cabinet + nonstandardized custom cabinet", which is characterized by the coexistence of standard modules and individual customization.

A standard unit cabinet is a unit cabinet that meets the common needs of users under the conditions of mass production. Basically, a custom wardrobe can be divided into four parts: doors, cabinets, hardware components and functional accessories. The unit cabinet is an important part of the overall wardrobe, including the cabinet frame and internal functional parts. The functional requirements of the custom wardrobe mainly lie in the design of the cabinet. Therefore, the functional design is mainly to meet the user's needs for storage, finishing and replacement of clothes, and its specific design tasks are manifested in the reasonableness of the cabinet size setting and the reasonableness of the product space location arrangement.

Second, modular design for MC furniture. The standardized modules of custom furniture are generally divided into three parts: surface decoration modules, functional modules and interface modules. Modular design refers to the division and design of a series of functional modules on the basis of functional analysis of a certain range of products with different functions or the same function with different performance and different specifications. Through the selection and combination of modules, different products can be formed to meet the different needs of the market design method.

2.2 Preliminary Studies

First, an overview of domestic and international research on the application of modular design in custom furniture design is presented. Economic development in foreign countries, especially in Europe and the United States, started earlier, and the design and even the furniture market has been updated very quickly, which has allowed the custom furniture industry to accelerate progress, and custom furniture has become widely popular. Research on the design of MC furniture has been carried out by national and international scholars from different perspectives. In the book "Das Baukastensystem in der Technik", published in 1961, Karl-Heinz Borowski recorded that the earliest example of modular design in furniture products was the "Ideal-Bücherschranke" designed and produced by the German company "Soennechen" approximately 1900. The book designed nine different standard modules based on three basic modules (Klein Bausteine, Grob Bausteine, Nicht Bausteine) as prototypes, which had a profound influence on subsequent research into modular design methods for products. In their 2016 publication "Semantic Segmentation of Modular Furniture", Tobias Pohlen et al. proposed a method for semantic segmentation and structural parsing of modular furniture items such as cabinet, wardrobe and bookshelf, the so-called interactive elements. In the book "Research on Modular Design of Furniture for MC", Li Bing et al. studied and analysed the modular design method of furniture for MC and the key technologies (module coding, division, interface, configuration and maintenance technology) lurking in it, starting from the concept of MC of furniture and combining it with the current situation of the furniture industry, and built a furniture modular design database platform to realize resource sharing.

Second, product status study. Through field research of Hanson's five furniture stores in Busan, Korea and Shanghai and Wuhan, China, Hanssen's brand concept of a customized wardrobe is as follows: a customized wardrobe is a system whole wardrobe that maximizes space in the form of a unit cabinet-type wall-to-wall approach. Its custom wardrobe design form is a combination of a standard unit cabinet + nonstandardized custom cabinet, of which the standard unit cabinet is divided into two series: luxury quality standard cabinet and simple and affordable standard cabinet, for a total of 16 standard unit cabinets. The details are shown in Table 2.

Table 2. Table captions should be placed above the tables.

Name	Features		Long and short clothing storage, basic type		Valuables storage, high-end type		Couple's Clothing Storage, combination type	
Luxury & quality type Standard cabinet	Made of E0 grade environmental protection material and imported PP finishing material from RENOLIT, Germany.	Standardized combination						
		Standardized unit cabinet	Long wardrobe	Long wardrobe with drawers	Short wardrobe	Right lower partition type high T-cabinet	Low T-Cabinet	Right upper partition type high T-cabinet	Open dressing cabinet
Simple and affordable type Standard cabinet	Made of E0-grade environmentally friendly materials	Standardized combination	Long and short clothes storage, basic type		Daily Clothes Storage, Common type		Clothes Storage for Couples, combination type	

(continued)

Table 2. (*continued*)

			High T-c abinet with r ight PP draw er	Low T-ca binet with ri ght partition	Short T-c abinet with l eft partition	High T -c abinet with l eft drawer	Low T-ca binet with P P drawer
		St anda rdize d uni t cabi net					
			Short wa rdrobe	Long war drobe	Long war drobe	Long war drobe with drawers	High T-ca binet

3 Basis for the Design of Standard Unit Cabinet MC Wardrobe

3.1 Design Basis

The terms "custom" and "modular" are, from a certain point of view, opposites, with custom emphasizing variability and specificity and modular emphasizing fixed constancy. However, in furniture production, for the current furniture market demand and production situation, customization serves personalization and modules serve rapid production, and they are complementary to each other.

The design of the MC wardrobe is based on standardization and the similarity and versatility of the components and product structures of the product families. Methods such as standardization and modularization are used to reduce the internal diversity of the product and increase the external diversity perceived by the customer, transforming or partially transforming the customized production of the product into the mass production of components through restructuring, thus providing the customer with low-cost, high-quality furniture with short delivery times quickly.

3.2   Design Core

The furniture module is generally divided into three parts: the surface decoration module, the functional module and the interface module. From the enterprise point of view, the increasingly competitive market environment and the increasingly rational consumer concept make homogeneous and excessively decorative products no longer favored by consumers. Therefore, product design returns to the issue of people, furniture product design should be based on functional design, and the design of functional modules is the core of custom furniture design. In addition to meeting the relationship between people and products, custom furniture also focuses on the combination of the product itself and the integrity of the combination of the product and the environment; thus, the structural design of the product becomes another design core of custom furniture.

3.3   Modular Design in Relation to the Function and Structure of the Standard Unit Cabinet

After the design core of the standardized unit cabinet has been established, it is designed and studied using modular design methods for the functions and connection structures of the standardized unit cabinet. First, the functional modules are divided and established in terms of functional differentiation, and the way in which they interface with each other is established according to the form of the functional modules. In the specific design, the functional and structural design of the standardized unit cabinet is carried out using the modular design method to form functional modules and connection structure modules. The process is based on the principles of serialization and generalization, and the design aims to reduce the internal diversity of the product through the "simplification and interchangeability" of modules. Then, based on the principle of combination, the functional modules are combined and configured according to the common needs of the user group, and the corresponding standard unit cabinet is designed using the corresponding connection structure. The process uses a certain number of modules to increase the perceived diversity of the external form of the product through design to meet the needs of the individual customization market.

Therefore, the modular design form of the standard unit cabinet can be summarized as follows: standard unit cabinet = functional module + connection structure + functional module + connection structure + functional module +..... An examination of the connection structure of the standard unit cabinet reveals that the connection structure used for the functional modules changes depending on the requirements. The connection structure consists of two types: explicit connections and implicit connection structure modules, where the explicit connections include screws, round bar tenons, concealed hinges and other hardware connections. The implicit connection structure modules refer to specially designed connection structure modules. They are combined with functional modules in the way shown in Fig. 1 and Fig. 2.

In Fig. 1, the connectors used in the combination method of explicit joining + functional modules are standardized hardware. In the integral jointing method, the functional modules are connected by means of universal hardware (eccentric connectors and wood tenon connections). The hardware used in this combination is already a standardized

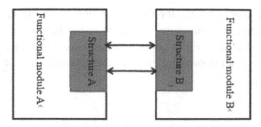

Fig. 1. Combination of explicit connections + functional modules.

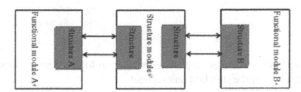

Fig. 2. Combination of implicit connections + functional modules

product and does not require a modular design. Therefore, in the constitutive relationship between the modular design and the standardization, functionality and structure of the standardized unit cabinet, the standardized hardware can be used directly to connect the functional modules to achieve standardization.

As shown in Fig. 2, in the invisible connection module + functional module combination method, the invisible connection module used is a dedicated connection structure. In contrast to the explicit connection pieces, the invisible connection pieces do not leave visible holes on the surface of the side panels and shelves. However, the different interfaces between the different panels can lead to variations in the design of the connection structure modules. As currently used in the connection of side panels to shelves and the connection of side panels to skirting boards, the complicit joiners need to be combined into special connection structure modules by forming standardized parts through modular design methods. In Fig. 2, the panels can lead to variations in the design of the connection structure modules. As currently used in the connection of side panels to shelves and the connection of side panels to skirting boards, the complicit joiners need to be combined into special connection structure modules by forming standardized parts through modular design methods. In Fig. 2, the combination is joined in a holistic way by connecting functional modules by means of implicit connection modules. Thus, in the constitutive relationship between the modular design and the function and structure of the standard unit cabinet, the standard unit cabinet, to be standardized, needs to act on the implicit connection structure through the modular design method so that it forms a standardized dedicated connection structure module.

In summary, the relationship between the modular design and the function and structure of the standard unit cabinet is based on the principle of standardization in the design of the standard unit cabinet. Through modular design acting on function and structure, functional modules and connection structure modules are formed to standardize the unit cabinet. However, there are standardized hardware connections in the connection structure that can be used directly for the connection of functional modules to form a standardized unit cabinet. Therefore, the relationship between the three of them is shown in Fig. 3. Functional modules and connection structures are diverse and need to be selected from each other depending on form and function in the course of their use.

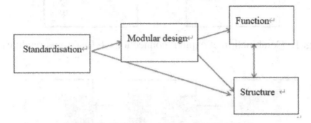

Fig. 3. Combination of the structural diagram of the relationship between the modular design and the function and structure of the standard unit cabinet

4   Analysis of the Functional Modules and Structural Design of Standard Unit Cabinets

In this paper, taking Hanssem's customized wardrobe as the center and based on the findings of a study of the current product situation, the 16 existing standardized unit cabinets of Hanssem were analysed and summarized in terms of their functional modules and connection structures using CAD software based on the relationship between modular design and the function and structure of the standardized unit cabinet. There are three steps:

- the structural forms of the existing standardized unit cabinet are counted, and the standardized unit cabinet with different functions are defined and classified in conjunction with the internal functional components;
- the modules of each type of standardized cabinet are split and combined to derive the basic functional modules and connection structures that make up the standardized unit cabinet;
- The basic function modules and connection structures are designed and applied, and the basic standardized unit cabinets are combined according to the modular design method of standardized unit cabinets.

4.1   Functional Modules for Standard Unit Cabinet

The functional module is a direct feedback of the user's needs in the design of a custom wardrobe and is one of the cores of the standard unit cabinet design. In Sect. 2, the first study of Hanson's standard unit cabinet is presented in two series of 16 models. The main difference between the two ranges of standard units is the use of different veneer panels. However, the cabinet body function is the same and similar. As the main focus of this paper is on the function and structure of the cabinet, the two series can be

Table 3. Classification of Standardized Unit Cabinet.

Functional type	Example
Hanging and stacking of long clothing	Long wardrobe, drawer-type long wardrobe
Hanging and stacking of short clothing	Short wardrobe
Hanging and stacking of long clothing, storage of small clothing	High T-cabinet with lower right partition, High T-cabinet with upper right partition, High T-cabinet with right PP drawer, High T-cabinet with left drawer, High T-cabinet with right partition
Hanging and stacking of short clothing, storage of small clothing	Low T-Cabinet, Low T-Cabinet with right partition, Low T-Cabinet with left partition, Low T-Cabinet with right PP drawer

combined and then classified and defined according to the functional characteristics of the standardized unit cabinet. The details are shown in Table 3.

The 16 standardized unit cabinets are designed and summarized according to the principles of serialization and generalization from the perspective of functional similarities and differences, resulting in four basic types, namely, shorT-cabinet, long cabinet, short T-cabinet and high T-cabinet. The sample cabinet model is brought into CAD, and the four basic types of cabinets are extracted according to the decomposition diagram of custom wardrobe products. In the process of extraction, functional parts such as drawers, shelves and clothes passages are removed. Eight functional modules can be derived to form CAD graphics, as shown in Table 4.

By comparing the eight functional modules, it can be found that four functional modules are the same, and they can be merged. The final result is six functional modules. This is shown in Fig. 4.

4.2 Standard Unit Cabinet Construction Design

At present, the standard unit cabinet of the Hanssen custom wardrobe is mostly connected by standard universal hardware in the connection structure. The connections are made with three-in-one or two-in-one connectors for the cabinet panels and concealed hinges for the door panels. The cabinet openings are designed on the basis of the 32 mm system. Rows of holes can be punched in the panels to increase the flexibility of the space for storage and to serve a decorative purpose. The choice of structural hardware mostly uses connections with the same hole size to better serve the interchangeability of the product modules.

In contrast to the three-in-one connector, the combination of invisible connectors is still used in the cabinet connection of the standardized unit cabinet, avoiding the exposure of hardware and ensuring aesthetics.

Table 4. Extraction of Standardized Unit Xtraction of Functional Modules for Standardized Unit Cabinet.

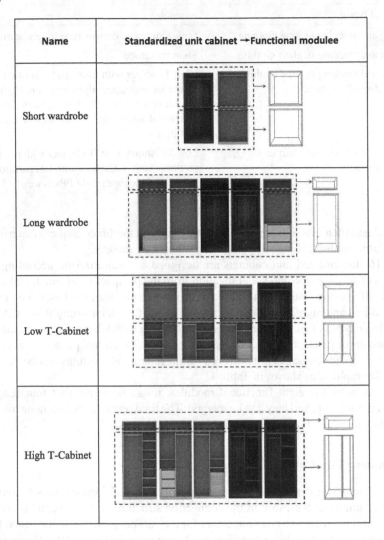

Name	Standardized unit cabinet →Functional modulee
Short wardrobe	
Long wardrobe	
Low T-Cabinet	
High T-Cabinet	

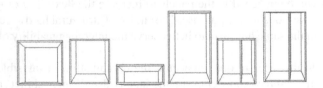

Fig. 4. Standardized unit cabinet base function module

4.3 CAD-Based Design of Standardized Unit Cabinet with Diverse Combinations

After the design analysis of the functional modules and connection structures of the standard unit cabinet in the Hanssen custom wardrobe, the six basic functional modules summarized can be used in the actual design for individual users according to the modular design method and configured in combination according to the standardization principle. Four basic standard unit cabinets are combined in the paper, and the specific results are shown in Fig. 5.

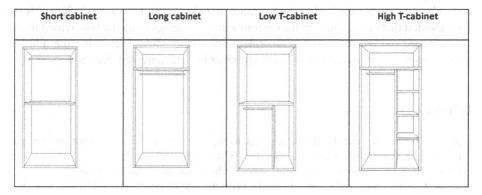

Short cabinet	Long cabinet	Low T-cabinet	High T-cabinet

Fig. 5. Base standard unit cabinet for Hanssen Custom Wardrobe.

The four basic standardized unit cabinets are a combination of functional modules based on the existing standardized unit cabinet from Hanssen Custom wardrobe and meet the requirements of current customized wardrobe production processes. In theory, each of the six basic functional modules is relatively independent. Their use in design enables the use of a small number of modules to combine a large number of forms of wardrobe. However, if the feasibility and rationality are taken into account, as well as the actual needs of the customer for the furniture, the number of combination methods is limited.

5 Conclusion

MC wardrobe design has its own unique form of custom design under standardization. How to further optimize the design of standard cabinets is a common concern for both enterprises and scholars. Through research and analysis of Hanssen's standard unit cabinet, modular design methods are combined to determine the standardized functional modules. Thus, when designing custom wardrobes for individual users, it is possible to use component-level functional modules to configure and combine designs to meet the diverse needs of unit cabinets. In this process, based on the consideration of product spatial unity, the relationship between structural design and the choice of functional module adaptability is analysed. The specific conclusions are as follows.

- When using modular design methods for functional modules in custom wardrobe design, the functional modules can be optimized through the principles of serialization, generalization and combination of standardization principles to achieve a reduction in product-oriented internal diversification.
- Through the analysis of the standardization, function and structure of a standardized unit cabinet using modular design, it is found that functional modules and connection structures could be designed using a modular approach, which helped to standardize the unit cabinet.
- Through the design application of the extracted functional modules according to the standardization principle, four basic standardized unit cabinets were designed. It is concluded that the modules of the custom wardrobe design could be subdivided from a standardized unit cabinet to component-level submodules, achieving the aim of increasing the external diversity of the product in a standardized design.

References

1. Xiong, X., Wu, W.: The development status and application technology of mass customized furniture. J. Nanjing For. Univ. (Nat. Sci. Ed.) **37**(04), 156–162 (2013)
2. Borowski, K.H.: Das Baukastensystem in der Technik. Springer, Heidelberg (2013)
3. Pohlen, T., Badami, I., Mathias, M., et al.: Semantic segmentation of modular furniture. In: 2016 IEEE Winter Conference on Applications of Computer Vision (WACV), pp. 1–9. IEEE (2016)
4. Li, B., Guan, H., Wu, W.: Research on modular design of furniture for MC. Packag. Eng. **32**(04), 66–69 (2011)
5. Anderson, P.: The competitive frontier of the 21st world enterprise: agile product development under the model of mass customization
6. 이현정: 모듈(Module) 가구디자인에 대한 연구-반복(Reiteration) 과확장(Expansion). 을 중심으로. 한국가구학회지**19**(1), 72–81 (2008)
7. 조남주: 모듈요소(modular elements)와 가구의 구조적 특성에 관한 연구한국가구학회지**14**(1), 31–39 (2003)
8. 김상돈: 한국 맞춤가구산업의 현황과 발전방안 연구/A Study on the Status and Development Plan of Korean Customized Furniture Industry. 조선대학교 대학원석사학위논문(2016)

An Improved War Strategy Optimization Algorithm for Big Data Analytics

Longjie Han, Hui Xu[✉], and Yalin Hu

School of Computer Science, Hubei University of Technology, Wuhan 430068, China
xuhui@hbut.edu.cn

Abstract. Big data analysis is confronted with the obstacle of high dimensionality in data samples. To address this issue, researchers have devised a multitude of intelligent optimization algorithms aimed at enhancing big data analysis techniques. Among these algorithms is the War Strategy Optimization (WSO) proposed in 2022, which distinguishes itself from other intelligence algorithms through its potent optimization capabilities. Nevertheless, the WSO exhibits limitations in its global search capacity and is susceptible to becoming trapped in local optima when dealing with high-dimensional problems. To surmount these shortcomings and improve the performance of WSO in handling the challenges posed by high dimensionality in big data, this paper introduces an enhanced version of the WSO based on the carnivorous plant algorithm (CPA) and shared niche. The grouping concept and update strategy of CPA are incorporated into WSO, and its update strategy is modified through the introduction of a shared small habitat approach combined with an elite strategy to create a novel improved algorithm. Simulation experiments were conducted to compare this new War Strategy Optimization (CSWSO) with WSO, RKWSO, I-GWO, NCHHO and FDB-SDO using 16 test functions. Experimental results demonstrate that the proposed enhanced algorithm exhibits superior optimization accuracy and stability, providing a novel approach to addressing the challenges posed by high dimensionality in big data.

Keywords: big data analytics · war strategy optimization · carnivorous plant algorithm · shared niche

1 Introduction

As the world enters the era of big data and increasingly complex data sets are generated, the significance of big data analysis becomes increasingly prominent. The discovery of underlying patterns and values through big data analysis has become a ubiquitous necessity for advancing various fields [1]. To this end, a multitude of intelligent optimization techniques have been devised to improve big data analysis methods [2–4].

Among these is war strategy optimization (WSO), a novel optimization technique proposed by Ayyarao et al. in 2022 [5]. WSO distinguishes itself from other swarm intelligence algorithms through its potent optimization capabilities and has been successfully applied to practical problems by numerous researchers [6–8]. However, the

WSO exhibits limitations when dealing with multipeaked problems, including insufficient convergence accuracy and susceptibility to becoming trapped in local optima. To address these issues, Mohamed M. Refaat proposed an enhancement to WSO through the incorporation of second- and third-order Longo-Kuta optimization techniques to refine the solutions obtained by the WSO [9].

Despite these advancements, big data analysis continues to face challenges posed by high dimensionality in samples [10], and current optimization methods exhibit weaknesses in their global search capabilities and remain prone to becoming trapped in local optima when dealing with high-dimensional problems. To surmount these obstacles and improve upon the original WSO search strategy, this paper introduces a novel enhanced version of the WSO based on the carnivorous plant algorithm (CPA) and shared niche, termed CSWSO.

CSWSO incorporates CPA's grouping concept to increase population diversity and modifies WSO's update strategy by guiding individuals through a combination of the best individuals within their respective groups, global best individuals and their search inertia to expand the search range. The update strategy of the carnivorous plant algorithm (CPA) is incorporated to enhance the global search capacity and continuous optimization of the population, facilitating mutual guidance among individuals within groups and information sharing among the best individuals in each group. To prevent information exchange from leading the algorithm to become trapped in local optima, a shared minor habitat approach is introduced to evolve lagging individuals in conjunction with an elite strategy to guide group evolution.

The remainder of this paper is structured as follows. Section 2 introduces the fundamental algorithms, including WSO and CPA. Section 3 presents the CSWSO proposed in this paper. Section 4 conducts an experimental analysis of CSWSO using benchmark function tests. Section 5 summarizes the work.

2 Basic Algorithms

2.1 War Strategy Optimization (WSO)

The WSO exhibits elegance in its simplicity, robustness and rapid convergence in optimization performance. Its operation proceeds as follows. Soldiers are randomly deployed on the battlefield according to Eq. (1).

$$X = LB + rand \times (UB - LB) \tag{1}$$

where UB and LB represent the upper and lower bounds of the search space, respectively, while $rand$ corresponds to a random number within the range of 0 to 1.

The soldier exhibiting the highest performance is designated as the king, while the second-best becomes the commander. When the random value exceeds the predefined threshold R, the offensive strategy is enacted through the utilization of Eqs. (2)–(4).

$$X_i(t + 1) = X_i(t) + 2 \times rand \times (C - King) + rand \times (W_i \times King - X_i(t)) \tag{2}$$

$$W_i = W_i \times (1 - \frac{R_i}{T})^\alpha \tag{3}$$

$$R_i = (R_i + 1) \times (F_n \geq F_p) + R_i \times (F_n < F_p) \tag{4}$$

where C and $King$ represent the positions of the commander and king, respectively. W_i denotes the weight assigned to King's position. α represents the exponential decay factor. T indicates the current iteration number. R_i denotes the rank of soldier i. F_n represents soldier i's fitness value. F_p denotes its previous position's fitness value. When rand $>$ R, a defensive strategy is adopted using Eq. (5).

$$X_i(t + 1) = X_i(t) + 2 \times rand \times (King - X_r(t)) + rand \times W_i \times (C - X_i(t)) \tag{5}$$

where $X_r(t)$ represents soldier i's random position in iteration t. Soldiers exhibiting low combat effectiveness may be replaced with new soldiers using Eq. (6) if rand >0.5 or Eq. (7) otherwise.

$$X_w(t + 1) = LB + \text{rand} \times (UB - LB) \tag{6}$$

$$X_w(t + 1) = -(1 - randn) \times (X_w(t) - median(X)) + King \tag{7}$$

where $X_w(t+1)$ signifies the position of the feeble soldier subsequent to $t+1$ iterations, $randn$ designates a random number drawn from a uniform distribution between 0 and 1, and the term "$median$" denotes the function yielding the median value.

2.2 Carnivorous Plant Algorithm (CPA)

The carnivorous plant algorithm (CPA) is an innovative approach proposed by Ong Kok Meng et al. in 2020 [11]. During initialization, individuals are grouped by equal differences, and an attraction rate γ of 0.8 is established during the growth phase. If rand $<\gamma$, then growth phase Eqs. (8)–(9) is executed.

$$NewCP_{i,j} = growth \times CP_{i,j} + (1 - growth) \times Prey_{v,j} \tag{8}$$

$$growth = growth\_rate \times rand_{i,j} \tag{9}$$

where $CP_{i,j}$ denotes the carnivorous plant at level i and $Prey_{v,j}$ represents a randomly selected prey. $growth\_rate$ is a constant value. If rand $> \gamma$, the prey successfully evades the trap according to Eqs. (10)–(11).

$$NewPrey_{i,j} = growth \times Prey_{u,j} + (1 - growth) \times Prey_{v,j}, u \neq v \tag{10}$$

$$growth = \begin{cases} growth\_rate \times rand_{i,j}, & f(prey_v) > f(prey_u) \\ 1 - growth\_rate \times rand_{i,j}, & f(prey_v) < f(prey_u) \end{cases} \tag{11}$$

where $Prey_{u,j}$ signifies a distinct prey selected at random. During reproduction, Eqs. (12)–(13) is employed.

$$NewCP_{i,j} = CP_{1,j} + Reproduction\_rate \times rand_{i,j} \times mate_{i,j} \tag{12}$$

$$mate_{i,j} = \begin{cases} CP_{v,j} - CP_{i,j}, & f(CP_i) > f(CP_v) \\ CP_{i,j} - CP_{v,j}, & f(CP_i) < f(CP_v) \end{cases} \tag{13}$$

where $CP_{1,j}$ denotes the optimal carnivorous plant, $CP_{v,j}$ represents a randomly selected carnivorous plant, and *Reproduction_rate* is a constant value. This process is repeated CP times.

3 Improved War Strategy Optimization Based on the Carnivorous Plant Algorithm and Shared Niche (CSWSO)

3.1 Integration with the Carnivorous Plant Algorithm

The innate limitations of the original WSO algorithm, namely, its susceptibility to local optima and its diminished global search capabilities arising from an excessive dependence on the global best solution. To address this, we introduce a novel framework that integrates the grouping strategy inspired by the carnivorous plant algorithm. By reducing the individual's undue influence on the collective population, the proposed framework redefines the fundamental structure of warfare behavior logic and search methods. As a result, the improved algorithm delineates six pivotal rules.

- The king directs the army and obtains the strategic intelligence of the entire force.
- The soldiers form units to engage in combat, and each unit is assigned a commander.
- The most advantageous position (fitness value) among all soldiers will become the king's new strategic objective (global optimal position), guiding the whole army to assault it.
- The most advantageous position among the soldiers within each unit will become the new operational objective of each unit's commander, guiding the unit to assault it.
- The soldiers have a rank parameter, and the higher the rank, the greater their ability to occupy better areas.
- Soldiers cannot become commanders or kings based on fitness values but only need to attack, report, and obey orders.

Figure 1 presents a schematic diagram of the improved search strategy for the CSWSO.

The grouping idea of the carnivorous plant algorithm was adopted to group the population. The best CP individual among the soldiers as the commander. The remaining soldiers will be merged into the corresponding group based on an arithmetic progression. The population is grouped using the grouping concept of the carnivorous plant algorithm. The best CP individual among the soldiers is designated as the commander. The remaining soldiers are merged into corresponding groups based on an arithmetic progression. Each iteration selects one update strategy from CPA and WSO. Unlike the original CPA strategy, grouping in the new CPA phase is performed only once, and individuals continue to be updated on an individual basis after grouping. Similarly, the rank weights of individuals are updated after individual updates in WSO. The reproduction phase of the original CPA is modified using Eq. (18).

$$NewCP_{i,j} = CP_{i,j} + Reproduction\_rate \times rand_{i,j} \times mate_{i,j} \tag{14}$$

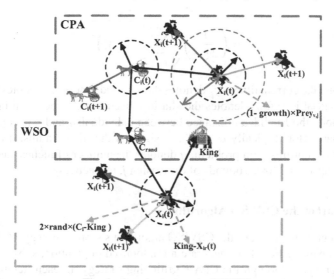

Fig. 1. Search strategy in CSWSO.

where $CP_{i,j}$ denotes the commander of group i. *Reproduction_rate* is a predefined value. Unlike the original WSO strategy, individuals in the new WSO phase are guided by different commanders, and attack and defense strategies are modified according to Eqs. (19)–(20), respectively.

$$X_i(t+1) = X_i(t) + 2 \times rand \times (C_i - King) + rand \times (W_i \times King - X_i(t)) \quad (15)$$

$$X_i(t+1) = X_i(t) + 2 \times rand \times (C_i - King) + rand \times (W_i \times King - X_i(t)) \quad (16)$$

where $X_{ir}(t)$ denotes the position of a random individual within group i, excluding themselves, at iteration t.

3.2 Shared Niche

The niche strategy, predicated on a sharing mechanism, adeptly addresses multipeak function optimization challenges [12]. This approach effectively preserves population diversity through the penalization of similar individuals and is implemented via Eqs. (14)–(17).

$$f_i' = \frac{f_i}{m_i} \quad (17)$$

$$m_i = \sum_{j=1}^{N} Sh(d_{ij}) \quad (18)$$

$$Sh(d_{ij}) = \left\{ \begin{array}{ll} 1 - \left(\frac{d_{ij}}{\sigma}\right)^{\alpha}, & d_{ij} < \sigma \\ 0, & other \end{array} \right\} \quad (19)$$

$$\sigma = \left(\frac{1}{2\sqrt[n]{q}}\right)\sqrt{\sum_{j=1}^{n}\left(x_j^u - x_j^l\right)^2} \tag{20}$$

where f_i denotes the current fitness, f_i' denotes the adjusted fitness, m_i denotes the sharing value of individual i, $Sh(d_{ij})$ denotes the sharing function value between individuals i and j, d_{ij} denotes the Euclidean distance between individuals i and j, α constitutes the adjustment parameter (typically $\alpha = 1$), σ signifies the niche radius, n denotes the dimension of the optimization problem, q denotes the number of niches, and x_j^u and x_j^l represent the upper and lower bounds of dimension j, respectively.

3.3 Flowchart of the CSWSO Algorithm

Figure 2 presents a flowchart of the CSWSO algorithm's search strategy. CSWSO commences by employing Eq. (1) to generate a randomized population, subsequently subjecting it to grouping through the CPA partitioning strategy. In each iteration, diverse groups exercise random selection between the WSO and CPA update strategies. Following this, individuals within each group update their positions and weights utilizing the corresponding formulae (8) and (10) or (19) and (20). Subsequently, the shared niche strategy (14) is employed to impose penalties and selectively identify individuals within the group who exhibit similarity and inferior quality. Subsequently, the groups implementing the CPA strategy update their most exceptional individual within the group, employing Eq. (18), whereas the groups employing the WSO strategy replace the least optimal individual utilizing either Eq. (6) or (7). Finally, the global best individual undergoes an update to ensure continuous refinement.

4   Experiments on Benchmark Test Functions

To verify the effectiveness of the CSWSO algorithm, experiments were conducted to compare it with existing WSO improvement algorithms (RKWSO), the original algorithm (WSO), and other well-known improved algorithms from recent years, including the Improved Grey Wolf Optimizer (I-GWO) [13], Nonlinear Based Chaotic Harris Hawks Optimization (NCHHO) [14], and Designing the SDO Algorithm with FDB Method (FDB-SDO) [15].

4.1   Testing Functions and Parameter Configuration

This section selected 16 classic standard test functions for experimentation, described in Table 1. These functions can be divided into four groups based on their characteristics. $f1$–$f7$ are single-modal high-dimensional functions. $f8$–$f9$ are single-modal low-dimensional functions. $f10$–$f14$ are multimodal high-dimensional functions. $f15$–$f16$ are multimodal low-dimensional functions. All functions used in this paper aim to find the minimum value.

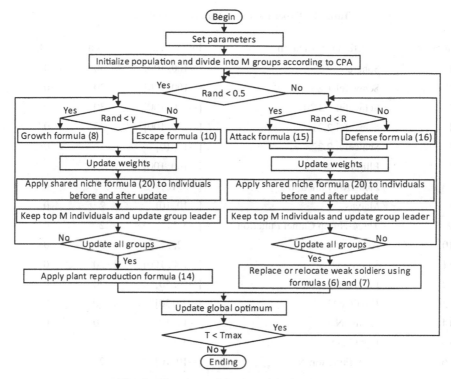

Fig. 2. Flowchart of the CSWSO algorithm

4.2 Quality Comparison Analysis of Solutions

For each comparative algorithm applied to each test function, this investigation employed the subsequent parameters. The experiments were carried out independently 30 times, with a population size of N = 30 and a dimensionality of Dim = 30 or 2. The maximum number of iterations was established as Max = 500. The average absolute error (MAE) and standard deviation (Std) were computed utilizing formula (21). It is important to note that the parameter R in the WSO algorithm was adjusted manually according to the specific test function, which may lead to overfitting. Therefore, in this study, the values of R for WSO, RKWSO and CSWSO were uniformly set at 0.5.

$$MAE = \frac{1}{n} \sum_{i=1}^{n} |f(x) - f(x^*)| \tag{21}$$

where $f(x)$ denotes the predicted value, and $f(x^*)$ denotes the true value.

Table 2 showcases the comprehensive outcomes attained from the experimental evaluations conducted on standard test functions utilizing the proposed algorithm alongside its comparative counterparts. The resulting experimental data are expressed in the form of the mean absolute error (MAE) value, supplemented by its corresponding standard

Table 1. Experimental test functions in the paper

Function	Function Information	Initial Range	Dim	Min
$f1$	Sphere	$[-100, 100]$	30	0
$f2$	Schwefel 1.2	$[-100, 100]$	30	0
$f3$	Schwefel's 2.20	$[-100, 100]$	30	0
$f4$	Schwefel 2.21	$[-100, 100]$	30	0
$f5$	Schwefel 2.22	$[-10, 10]$	30	0
$f6$	Elliptic	$[-100, 100]$	30	0
$f7$	Zakharov Function	$[-5, 10]$	30	0
$f8$	Matyas	$[-10, 10]$	2	0
$f9$	Three-Hump Camel Function	$[-5, 5]$	2	0
$f10$	Ackley	$[-32, 32]$	30	0
$f11$	Salomon	$[-100, 100]$	30	0
$f12$	Penalized	$[-50, 50]$	30	0
$f13$	Penalized2	$[-50, 50]$	30	0
$f14$	Alpine N.1	$[-10, 10]$	30	0
$f15$	Egg Crate	$[-5, 5]$	2	0
$f16$	Levy Function N. 13	$[-10, 10]$	2	0

Table 2. MAE values and standard deviations of the six algorithms (Dim $= 30,2$)

Function	MAE±Std		
	I-GWO	NCHHO	FDB-SDO
f_1	2.25E-28±2.76E-28	2.41E-235±0.00E+00	4.64E-32±2.50E-31
f_2	5.69E-03±2.07E-02	1.67E-219±0.00E+00	2.30E-33±9.58E-33
f_3	4.75E-17±4.41E-17	8.61E-122±3.76E-121	2.62E-16±6.29E-16
f_4	1.40E-05±1.61E-05	4.05E-120±1.46E-119	1.00E-16±5.06E-16
f_5	6.89E-18±5.37E-18	1.07E-120±5.78E-120	1.88E-17±9.48E-17
f_6	4.86E-25±6.05E-25	4.80E-241±0.00E+00	4.56E-32±2.06E-31
f_7	1.34E-07±2.28E-07	4.71E-195±0.00E+00	2.97E-29±1.60E-28
f_8	6.28E-150±3.38E-149	4.24E-264±0.00E+00	1.07E-106±3.07E-106
f_9	7.91E-229±0.00E+00	3.92E-238±0.00E+00	3.34E-120±1.63E-119
f_{10}	6.19E-14±1.08E-14	**8.88E-16±0.00E+00**	**8.88E-16±0.00E+00**
f_{11}	2.10E-01±3.00E-02	1.13E-90±6.08E-90	6.11E-05±1.79E-04
f_{12}	3.84E-03±1.86E-02	9.70E-07±1.26E-06	2.04E-04±2.53E-04
f_{13}	1.08E-01±1.33E-01	1.11E-05±1.30E-05	2.27E-02±3.59E-02
f_{14}	3.07E-04±2.73E-04	7.94E-125±2.46E-124	1.67E-18±7.58E-18
f_{15}	4.67E-245±0.00E+00	2.84E-249±0.00E+00	8.29E-125±3.55E-124
f_{16}	1.66E-26±4.32E-26	4.05E-05±9.85E-05	**1.35E-31±6.57E-47**

(*continued*)

Table 2. (*continued*)

Function Name	MAE±Std		
	WSO	RKWSO	CSWSO
f_1	4.76E-71±2.27E-70	**0.00E+00±0.00E+00**	**0.00E+00±0.00E+00**
f_2	2.12E-71±1.06E-70	1.02E-293±0.00E+00	**0.00E+00±0.00E+00**
f_3	8.98E-35±4.28E-34	6.14E-165±0.00E+00	**4.69E-181±0.00E+00**
f_4	3.16E-35±7.32E-35	2.60E-162±1.11E-161	**2.00E-178±0.00E+00**
f_5	4.64E-35±2.48E-34	8.38E-174±0.00E+00	**4.15E-182±0.00E+00**
f_6	3.45E-64±1.85E-63	**0.00E+00±0.00E+00**	**0.00E+00±0.00E+00**
f_7	1.83E-71±9.02E-71	1.68E-284±0.00E+00	**0.00E+00±0.00E+00**
f_8	1.34E-149±3.25E-149	**0.00E+00±0.00E+00**	**0.00E+00±0.00E+00**
f_9	1.16E-144±6.25E-144	**0.00E+00±0.00E+00**	**0.00E+00±0.00E+00**
f_{10}	**8.88E-16±0.00E+00**	**8.88E-16±0.00E+00**	8.88E-16±0.00E+00
f_{11}	5.18E-02±4.87E-02	1.27E-139±4.93E-139	**4.98E-149±1.77E-148**
f_{12}	3.30E-03±8.92E-03	2.54E-11±7.25E-11	**4.38E-13±1.40E-12**
f_{13}	1.79E+00±1.45E+00	8.68E-09±3.80E-08	**7.01E-12±2.22E-11**
f_{14}	4.52E-37±1.19E-36	4.15E-176±0.00E+00	**1.43E-180±0.00E+00**
f_{15}	4.61E-141±1.87E-140	**0.00E+00±0.00E+00**	**0.00E+00±0.00E+00**
f_{16}	4.15E-31±6.13E-31	**1.35E-31±6.57E-47**	**1.35E-31±6.57E-47**

deviation. The data that are presented in bold type denote the algorithm's superior performance in relation to others, manifesting the attainment of optimal computational results for the specific test function under consideration.

By examining Table 2, it becomes evident that the CSWSO algorithm profoundly enhances the solution quality for unimodal high-dimensional functions f1–F7 and unimodal low-dimensional functions F8–F9, surpassing the performance of the comparative algorithms. Notably, it achieves the theoretical optimal solution for six out of nine functions. Moreover, when dealing with multimodal high-dimensional functions F10–F14, although the CSWSO algorithm falls short of discovering the theoretical optimal solution, it still exhibits superior quality compared to all other algorithms. In the case of multimodal low-dimensional functions f15–F16, CSWSO yields comparable results to the RKWSO algorithm while notably improving solution quality when contrasted with the original WSO algorithm. To provide further elucidation and visually demonstrate the superiority of CSWSO in terms of stability and convergence speed, Fig. 3 illustrates the convergence trajectory of the selection function.

Drawing upon the outcomes depicted in Fig. 3, CSWSO exhibits remarkable velocity and precision of convergence in contrast to the original WSO algorithm and alternative evaluation algorithms employed across the designated four functions. It is imperative to acknowledge that while the CSWSO algorithm initially trails behind other algorithms in functions $f1$ and $f13$, it ultimately surpasses them, corroborating the discoveries outlined in Table 2 and affirming the efficacy of our enhancement strategy in augmenting the optimization prowess of WSO.

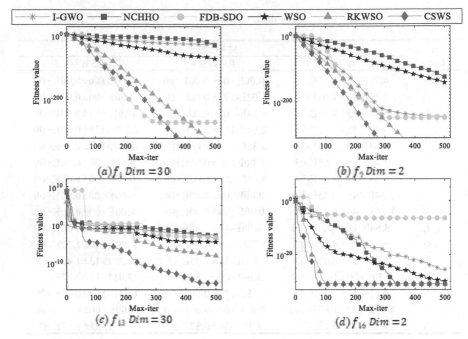

Fig. 3. Convergence curve of fitness value for six algorithms on partial functions

4.3 Friedman and Wilcoxon Tests

To objectively appraise our algorithm's optimization performance, we employed Friedman and Wilcoxon tests to conduct a comprehensive comparative analysis of its convergence capability.

Table 3. Average rankings achieved by the Friedman test for the six algorithms (Dim = 30,2)

Algorithm	Rankings
I-GWO	5.50
NCHHO	4.88
FDB-SDO	3.19
WSO	4.19
RKWSO	1.91
CSWSO	**1.34**

Table 3 illustrates the outcomes of the Friedman test performed on six algorithms across sixteen functions. The table shows that the mean rank of the CSWSO algorithm stands at 1.34, demonstrating notable statistical significance with respect to the average ranks of the WSO and RKWSO algorithms, which differ by 2.85 and 0.57, respectively.

Notably, the CSWSO algorithm demonstrates the most diminished mean rank among all algorithms examined, indicating its consistent attainment of superior precision in comparison to the other algorithms under scrutiny.

Table 4. Average rankings achieved by Wilcoxon Test for the six algorithms (Dim = 30,2)

Algorithm	p_value
I-GWO	3.81E−04
NCHHO	8.82E−04
FDB-SDO	5.17E−02
WSO	1.89E−03
RKWSO	2.73E−01

Table 4 presents the results of Wilcoxon tests conducted on six algorithms across 16 functions (Dim = 30 or 2). Although the significant differences between the CSWSO and both the FDB-SDO and RKWSO fall below the critical value of $\alpha = 0.05$, the rank mean of our CSWSO surpasses those of both aforementioned algorithms. The significant differences between our CSWSO and the remaining three algorithms are even more pronounced.

5 Conclusions

This paper introduces a war strategy optimization algorithm predicated on an improvement to the carnivorous plant algorithm, referred to as CSWSO. In comparison to its counterparts, this algorithm significantly enhances the global optimization capacity and solution quality. Nevertheless, there remains room for further improvement in our proposed algorithm. This is due to our decision to fix parameter R to circumvent overfitting issues. However, this fixed R may not fully exploit our algorithm's search capacity across different optimization problems. Consequently, future research will concentrate on mitigating the impact of parameter R on our algorithm's optimization capacity and employing our novel algorithm in resolving practical engineering optimization problems to expand its range of applications.

References

1. Gokulkumari, G.: An overview of big data management and its applications. J. Netw. Commun. Syst. **3**(3), 11–20 (2020)
2. Fong, S., Wong, R., Vasilakos, A.V.: Accelerated PSO swarm search feature selection for data stream mining big data. IEEE Trans. Serv. Comput. **9**(1), 33–45 (2015)
3. Aslan, S.: A comparative study between artificial bee colony (ABC) algorithm and its variants on big data optimization. Memet. Comput. **12**(2), 129–150 (2020). https://doi.org/10.1007/s12293-020-00298-2

4. Abualigah, L., Gandomi, A.H., Elaziz, M.A., et al.: Advances in meta-heuristic optimization algorithms in big data text clustering. Electronics **10**(2), 101 (2021)
5. Ayyarao, T.S.L.V., Ramakrishna, N.S.S., Elavarasan, R.M., et al.: War strategy optimization algorithm: a new effective metaheuristic algorithm for global optimization. IEEE Access **10**, 25073–25105 (2022)
6. Ayyarao, T.S.L.V., Kumar, P.P.: Parameter estimation of solar PV models with a new proposed war strategy optimization algorithm. Int. J. Energy Res. **46**(6), 7215–7238 (2022)
7. Xu, J., Cui, D.: Time series prediction of sediment discharge by optimizing extreme learning machine with war strategy. J. Hydroelectr. Eng. **48**(11), 36–42 (2022)
8. Kumar, V.T.R.P., Arulselvi, M., Sastry, K.B.S.: War strategy optimization-enabled Alex Net for classification of colon cancer. In: 2022 1st International Conference on Computational Science and Technology (ICCST), pp. 402–407. IEEE (2022)
9. Refaat, M.M., Aleem, S.H.E.A., Atia, Y., et al.: A new decision-making strategy for techno-economic assessment of generation and transmission expansion planning for modern power systems. Systems **11**(23), 1–42 (2023)
10. Fan, J., Han, F., Liu, H.: Challenges of big data analysis. Natl. Sci. Rev. **1**(2), 293–314 (2014)
11. Ong, K.M., Ong, P., Sia, C.K.: A carnivorous plant algorithm for solving global optimization problems. Appl. Soft Comput. **7** , 30710 (2020)
12. Miller, B., Miller, B.L., Shaw, M.J., et al.: Genetic algorithms with dynamic niche sharing for multimodal function optimization. In: Proceedings of IEEE International Conference on Evolutionary Computation, pp. 786–791. IEEE (1996)
13. Nadimi-Shahraki, M.H., Taghian, S., Mirjalili, S.: An improved grey wolf optimizer for solving engineering problems. Expert Syst. Appl. **166**, 1–25 (2020)
14. Dehkordi, A.A., Sadiq, A.S., Mirjalili, S., et al.: Nonlinear-based chaotic Harris Hawks optimizer: algorithm and internet of vehicles application. Appl. Soft Comput. **109**(2), 1–32 (2021)
15. Kati, M., Kahraman, H.T.: Improving supply-demand-based optimization algorithm with FDB method: a comprehensive research on engineering design problems. J. Eng. Sci. Des. **8**(5), 156–172 (2020)

Research on Path Planning of Mobile Robots Based on Dyna-RQ

Ziying Zhang[1(✉)], Xian Li[2], and Yuhua Wang[2]

[1] College of Computer Science, Jiaying University, Meizhou 514015, China
zhangziying@jyu.edu.cn
[2] College of Computer Science and Technology, Harbin Engineering University,
Harbin 150001, China

Abstract. The mobile robot path planning problem is one of the main contents of reinforcement learning research. In traditional reinforcement learning, the agent obtains the cumulative reward value in the process of interacting with the environment and finally converges to the optimal strategy. The Dyna learning framework in reinforcement learning obtains an estimation model in the real environment. The virtual samples generated by the estimation model are updated together with the empirical samples obtained in the real environment to update the value function or strategy function to improve the convergence efficiency. At present, when reinforcement learning is used for path planning tasks, continuous motion cannot be solved in a large-scale continuous environment, and the convergence is poor. In this paper, we use RBFNN to approximate the Q-value table in the Dyna-Q algorithm to solve the drawbacks in traditional algorithms. The experimental results show that the convergence speed of the improved Dyna-RQ algorithm is significantly faster, which improves the efficiency of mobile robot path planning.

Keywords: Reinforcement · Path planning · Dyna-Q · RBFNN

1 Introduction

Path planning is one of the key technologies in the field of mobile robot research. At present, fuzzy logic [1], artificial potential field methods [2], genetic algorithms [3], neural networks [4], Deep neural networks [5], etc., are all successful robot path planning methods, but these methods usually need to assume complete environmental configuration information. However, in a large number of practical applications, the agent needs to have the ability to adapt to an uncertain environment. Therefore, improving the self-learning ability and adaptability of robot path planning has become a key technology for scholars to study.

As an important machine learning method, reinforcement learning emphasizes the use of the 'trial and error improvement' method to realize online learning without a tutor in the process of interaction with the environment, which is an effective method to solve the sequential optimization problem. In [6], a deep Q network algorithm based on a multilayer perceptron was proposed. The algorithm improved the reward function according

© The Author(s), under exclusive license to Springer Nature Singapore Pte Ltd. 2023
Z. Yu et al. (Eds.): ICPCSEE 2023, CCIS 1879, pp. 49–59, 2023.
https://doi.org/10.1007/978-981-99-5968-6_5

to the information related to the target and distinguished the value of different actions according to the dynamic reward value to improve the efficiency of the neural network action selection. The Dyna learning framework proposed by Sutton et al. improves the efficiency by establishing an environmental model instead of the real environment and using simulation samples for learning [7]. In [8], a heuristic planning strategy is proposed to incorporate into a Dyna agent the ability of heuristic search in path planning. The proposed Dyna-H algorithm selects branches more likely to produce outcomes than other branches, just as A* does. In [9], a modified version of Dyna-learning and prioritized sweeping is proposed. The modification exploits the breadth-first search (BFS) to conduct additional modifications of the policy in the epoch mode. Liu et al. proposed an improved Sarsa-based Dyna-Sa algorithm using a Dyna framework that combines model-based and model-independent algorithms to improve planning speed [10].

A summary of the existing research findings reveals that the current research on mobile robot path planning based on reinforcement learning faces the following key issues that need to be addressed:

(1) The existing algorithm is almost a modification of the traditional reinforcement learning model, and the convergence speed is slow when the obstacle information is dense or the state space is large.
(2) Existing path planning algorithms based on reinforcement learning are in tabular form and are not applicable to the case where the dimensionality of the state space is large or continuous.

2 Reinforcement Learning

The problem to be solved in reinforcement learning is to let the agent select the optimal action through interaction with the environment and then obtain the optimal strategy to solve the problem. Reinforcement learning is different from supervised learning. There is no tutor signal in the process of learning, and the evaluation measure for intensive learning is the enhanced signal obtained from the environment. The agent must learn by itself to know how to choose the action instead of being directly given by the instructor signal. Therefore, the agent needs to continuously acquire knowledge from the environment and improve the choice of actions to adapt to the environment. The process of intensive learning is a process of continuous testing with the environment. The key elements are the environment, reward value, action, and state. These elements can establish a reinforcement learning model, as shown in Fig. 1.

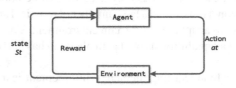

Fig. 1. Reinforcement learning model

The agent accepts the state input S of the environment and outputs the action a through an internal corresponding mechanism. The agent performs action a, arrives at state $s\prime$ and generates the corresponding enhanced signal, that is, the bonus value. The agent selects the corresponding action according to the current environmental state information and the enhanced signal, and the agent tries to maximize the return within a certain period of time. The instant rewards generated by the agent selection action will affect not only the current reward value but also the next moment and the final return. The principle that the agent follows in the process of learning is to make the positive returns in the environment continually increase and to reduce the tendency of negative returns. As an effective method to solve sequential optimization decision problems, reinforcement learning is based on the Markov decision process (MDP).

The Markov decision process is a discrete-time stochastic process, described by tuples (S, A, P, r, γ), which is a finite set of state spaces in a quintuple; A is for a limited set of action spaces, $P(s, a, s\prime) \in [0, 1]$ is the state transition probability, $r(s, a, s\prime)$: $S \times A \times S \rightarrow R$ is instant rewards for the system to select the action arrival status in the state; γ is used as a discount factor to calculate the cumulative return. The goal of reinforcement learning is to find the optimal strategy through learning in a certain Markov decision process. A policy refers to a state-to-action mapping, usually expressed as π, a distribution on a set of actions in under state s, namely:

$$\pi(a|s) = P(A_t = a|S_t = s) \tag{1}$$

In Eq. (1), the strategy refers to the probability of performing an action a in a given state s. If the action is deterministic, then the policy π can specify a certain action in each state. When the strategy is determined, the cumulative return is defined as:

$$G_t = R_{t+1} + \gamma R_{t+2} + \cdots = \sum_{k=0}^{\infty} \gamma^k R_{t+k+1} \tag{2}$$

If you arrive at S_5 the state from the state S_1, the resulting state transition sequence might be:

$$s_1 \rightarrow s_2 \rightarrow s_3 \rightarrow s_4 \rightarrow s_5$$
$$s_1 \rightarrow s_2 \rightarrow s_4 \rightarrow s_5 \tag{3}$$
$$\vdots$$

Under the strategy, using Eq. (2), you can calculate multiple cumulative return values. Since the randomness of the strategy π determines the randomness of the cumulative return, the expectation of the cumulative return is a certain value, so define its expectation as a state value function $V^{\pi}(s)$ to describe the value of the state, namely:

$$V^{\pi}(s) = E^{\pi}\left[\sum_{k=0}^{\infty} \gamma^k R_{t+k+1}|S_t = s\right] \tag{4}$$

The corresponding state-action value function (action value function) is defined as:

$$Q^{\pi}(s, a) = E^{\pi}\left[\sum_{K=0}^{\infty} R_{t+k+1}|S_t = s, A_t = a\right] \tag{5}$$

According to the Bellman equation, the state action value function can be derived as:

$$Q^{\pi}(s, a) = E^{\pi}\left[R_{t+1} + \gamma Q(S_{t+1}, A_{t+1})|S_t = s, A_t = a\right] \tag{6}$$

The temporal difference (TD) is the core idea in reinforcement learning, combining the Monte Carlo algorithm with the idea of dynamic programming. As with the Monte Carlo method, the environment model is not required to learn directly from the training samples. Similar to the dynamic programming method, the current state value is updated by the estimation of the value function of the subsequent state. The simplest time difference method is updated to a one-step TD (0) algorithm, also called a one-step TD method. Its iterative formula is as shown in Eq. (7).

$$V(s_t) = V(s_t) + \alpha \left[r_t + V(s_{t+1}) - V(s_t) \right] \qquad (7)$$

where $\alpha \in 0, 1$ is the learning rate, and the square brackets are the time difference terms, that is, the difference between the updated estimate $r_t + V(s_{t+1})$ of the state value function of the current state s and the current estimate $V(s_t)$. The estimate in the TD method is a sample of the expected value, which uses the current value $V(s_t)$ for calculation instead of $V^{\pi}(s_t)$. . The difference between the brackets in the TD algorithm update is used to evaluate the error between the current state value function and the better estimate (TD error), which exists in various forms of reinforcement learning. The most basic algorithm of the time difference method is a different strategy and model-independent Q learning algorithm proposed by Watkins et al. The Q learning algorithm adopts the off-policy method, that is, the behavior strategy and evaluation strategy are different, and the evaluation formula is different. As in Eq. (8).

$$Q(s_t, a_t) = Q(s_t, a_t) + \alpha \left(r_{t+1} + \gamma \max_a Q(s_{t+1}, a) - Q(s_t, a_t) \right) \qquad (8)$$

In Eq. (8), $\alpha \in 0, 1$ is the learning rate, and $r_{t+1} + \gamma \max_a Q(s_{t+1}, a) - Q(s_t, a_t)$ is the time difference term, which represents the difference between the optimal Q value and the current evaluation $Q(s_t, a_t)$.

3 Mobile Robot Path Planning Algorithm Based on Dyna-RQ

3.1 Dyna Learning Framework

The agent interacts with the environment, assuming that the current state is s_t, taking action a_t, and the agent enters the next state s_{t+1}, and at the same time obtains the immediate reward value r_t, that is, a real sample (s_t, a_t, s_{t+1}, r_t) is obtained, which can be used to update the value function or strategy π. At the same time, an environment model can be established, which can be an instant update or an all stored experience sample, which can be in the form of a state transition matrix. The estimated environment model is updated with real samples to obtain virtual samples, which can also update value functions and policies π. In the Dyna learning framework, the real sample and the virtual sample are used together to update the value function and strategy. The most widely used algorithm in the Dyna learning framework is the Dyna-Q algorithm. The reinforcement learning-based model and Dyna learning framework are shown in Fig. 2 and Fig. 3.

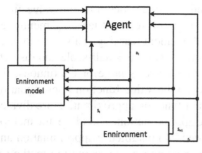

Fig. 2. Model-based reinforcement structure

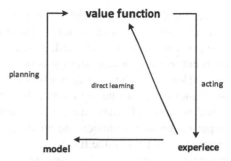

Fig. 3. Dyna learning framework

The specific process of the algorithm is **Algorithm 1**.

Algorithm 1 Dyna-Q

Initialize $Q(s, a)$ and $Model(s, a)$ for all $s \in S$ and $a \in A(s)$

Repeat（for each step of episode）:

    (a) $s \leftarrow$ current state;

    (b) $A \leftarrow \epsilon - greedy(s, Q)$;

    (c) Execute action a, observe s' and reward r;

    (d) $Q(s, a) \leftarrow Q(s, a) + \alpha \left(r + \gamma \max_{a' \in A(s')} Q(s', a') - Q(s_t, a_t) \right)$;

    (e) $Model(s, a) \leftarrow s', r$;

    (f) **for** i=1 **to** N **do**

        $s \leftarrow$ random previously observed state;

        $a \leftarrow$ random action previously taken in s;

                $s', r \leftarrow Model(s, a)$

  End

Until (s' is the terminal state);

In Algorithm 1, an environment model $Model(s, a)$ is used to record instant reward r and the next state $s\prime$ when the agent takes action a. . Step (d) is an update formula in the Q learning algorithm, and steps (e) and (f) are, respectively a learning and planning process in the Dyna-Q algorithm, if step (e) and step (f) are omitted, or parameter N = 0 in the algorithm becomes the traditional Q learning algorithm.

3.2 RBFNN Value Function Approximation Method

From the theory in the previous section, it can be concluded that the reinforcement learning method requires a clear representation of its value function. The value function is stored in the form of a table. The update and iterative process of the value function is also the process of the table. Therefore, the reinforcement learning mentioned in the previous chapter is also called tabular reinforcement learning. This form of reinforcement learning in table storage is only suitable for solving small-scale and discrete spaces. However, when the dimension of the state space is large or continuous, the value function cannot be represented by a table, so the value function approximation method is derived. In mathematics, the methods of function approximation are divided into parameter approximation and nonparametric approximation. Therefore, the method of approximating the value function is also divided into parametric approximation and nonparametric approximation. Among them, the parametric approximation method is divided into linear parametric approximation and nonlinear parametric approximation.

Unlike tabular reinforcement learning, a function approximation method is used to predict the value function. The value function at time t is no longer stored in the form of a table but is represented by a parameterized function containing the parameter vector ω. In the process of interactive learning between the agent and the environment, only the parameter ω is changed, and the final value function is determined only by the parameter ω. The parameter of a n-dimensional vector ω, the Q-valued function approximator, can have a mapping $F : R^n \rightarrow Q$. R^n is the parameter space of the n dimension, and Q is the state action value function. A parameter vector corresponds to an approximate value function, such as Eq. (9).

$$\hat{Q}(s, a) = [F(\omega)](s, a) \tag{9}$$

where $[F(\omega)](s, a)$ represents the value function estimate for the state-action pair (s, a) and stores an n-dimensional vector to replace the value function table for tabular reinforcement learning.

Because the linear function approximator is easy to implement and the algorithm is simple, the parameterized linear approximator is the most widely used. The parameterized linear function approximator consists of a n basis function (Basic Function, BF). The linear approximation of the approximate state action Q-valued function is expressed as Eq. (10).

$$\hat{Q}(s, a) = [F(\omega)](s, a) = \sum_{i=1}^{n} \phi_i(s, a)\omega_i = \phi^T(s, a)\omega \tag{10}$$

In Eq. (10), $\phi(s, a) = [\phi_1(s, a), \phi_2(s, a), \ldots, \phi_n(s, a),]^T$ is an n dimensional vector composed of basic functions.

The nonlinear approximation method is mainly a neural network. However, when the approximation model of the value function is determined, that is, the linear approximation method selects the basis function, the neural network determines the structure. Then, the approximation of the value function is equivalent to the approximation of the parameters.

The Dyna-RQ algorithm proposed in this chapter uses RBFNN to approximate the value function in the Dyna-Q algorithm. The most basic form of RBFNN is a three-layer feedforward neural network, as shown in Fig. 4. The input layer is composed of source nodes, the number of nodes is equal to the dimension of the input vector, and the input signal is passed to the input layer node. The second layer is the only hidden layer in the network, and its role is to achieve a nonlinear transformation of the input space to the hidden space. This nonlinear transformation is achieved by the radial basis function of the implicit node, which produces a local response to the input signal. When the input signal is near the center of the function, the hidden layer node will produce a larger output. In most cases, the implicit space has a higher dimension. The third layer is the output layer, and the output layer neurons are linear, providing a response to the input mode acting on the input layer.

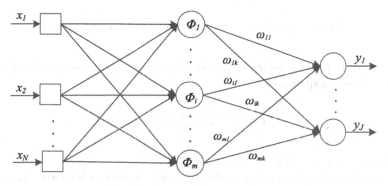

Fig. 4. RBFNN network structure

Let the output be $y_k = \left[y_{k1}, y_{k2}, y_{k3}, \ldots y_{kJ}\right]$ and j be the number of output layer vectors, representing the resulting output of the k-th vector. When RBFNN inputs training sample $x_k \in R^n$, the result of the j-th neuron output of the network is:

$$y_{kj} = \sum_{i=1}^{m} \omega_{ij}\phi(x_k, c_i), j = 1, 2, \ldots, J \tag{11}$$

The basis function of RBFNN is generally a Gaussian kernel function; then, $\phi(x_k, c_i)$ can be expressed as:

$$\phi(x_k, c_i) = \exp\left(-\frac{1}{2\sigma^2}\|x_k - c_i\|^2\right) \tag{12}$$

where the output of the first hidden layer node is $\phi\|x_k - X_i\|$ and $X_i = [x_{i1}, x_{i2}, \ldots, x_{im}]$ is the center of the basis function. Compared with other feedforward networks, RBFNN has good approximation characteristics. The single hidden layer RBFNN can approximate any nonlinear function defined on R^n with arbitrary precision.

3.3 Dyna-RQ Path Planning Algorithm

Using RBFNN to approximate the value function of the Dyna-Q algorithm, a structure diagram is shown in Fig. 5.

The first layer is the input layer, and the state information of the environment and the actions performed are input to the RBFNN network. The dimension of the input vector $D = n + 1$ where n is the amount of status information, and $n + 1$ is the action to be selected.

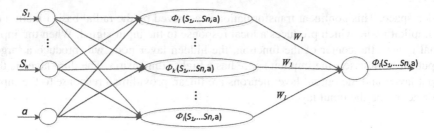

Fig. 5. Dyna-Q approximation algorithm structure diagram

The second layer is the hidden layer, where each node represents a $n+1$ dimensional Gaussian function, expressed as:

$$\varphi_k(x) = \exp\left(\sum_{i=1}^{n+1} \frac{\|x_i - c_{ik}\|^2}{2\sigma_{ik}^2}\right), k = 1, 2, ..., K \tag{13}$$

where N is the number of hidden layer nodes.

The third layer is the output layer, which has only one node, which is used to indicate the Q value output. The expression for the Q value is:

$$Q(s, a) = \sum_{i=1}^{n+1} \omega_i \varphi_i \tag{14}$$

For traditional table-type reinforcement learning, the Q-value table update method updates each $Q(s, a)$ according to Eq. (8) with continuous trial and error and iteration. The table of Q values will gradually converge to a more stable state. Similarly, for the initial sample of the RBF network, we can also initialize it to a random number. The fitting process of the RBF network is to adjust the parameters of the network by the error between the network output of the sample and the output of the target. Therefore, during the update process of the Dyna-RQ algorithm, each time an action is performed, a corrected Q value is obtained, and one sample is added. In the Q-learning algorithm, we can obtain the updated value of Q based on Eq. (8). Each time an action is performed, a corrected Q value is obtained, and one sample is added. A new Q update value is obtained, which is equivalent to obtaining a new sample, adding it to the original sample to participate in the training of the RBF network, and updating the network parameters.

In this paper, the supervised center-selection method in RBFNN is used to learn and adjust the center vector X_i of the basis function, the standard deviation σ_i of the basis function, and the connection weight ω of the hidden layer and the output layer in the three-parameter hidden layers of the RBF network.

The function approximation procedure of the Dyna-RQ algorithm is as follows:

Step 1: Initialize the Dyna-RQ algorithm parameters, learning factors, $0 \le \gamma \le 1$ attenuation factors $0 \le \gamma \le 1$, initialize the RBF network, and determine the network structure and parameters.
Step 2: For the status, select the action according to the $\varepsilon - greedy$ policy.
Step 3: Execute $s\prime$, , obtain the new environment status and immediately return r(t + 1).
Step 4: Update the Q function distribution with Eq. (8).

Step 5: Detect the convergence condition, such as satisfying, and end the algorithm. Otherwise, go to step 2.

4   Simulation Experiment

This chapter will verify the effectiveness of the proposed algorithm. The experiment uses V-REP simulation software to build a real environment with obstacle scenes. By constructing the communication interface between ROS and V-REP simulation software 3.4.0, it is possible to use the ROS terminal to send control commands to V-REP and see the algorithm running effect on the visual simulation platform.

4.1   Reward Function Settings

The reward function is an indicator that quantifies the effect of the action on the target during the interaction between the mobile robot and the environment and evaluates a score for each action in a given state. In this paper, the state set is divided into four categories: safety state SS (the possibility of robot collision with obstacles is very low), nonsafe state NSS (the possibility of a robot colliding with an obstacle is high, since the environment in this article is a raster map, that is, the state corresponding to the grid around the obstacle is defined as an unsafe state), winning state WS (the robot reaches the target), and failure state. FS (the robot collides with an obstacle). The setting of the bonus value is related to the state transition. The bonus value from the nonsecure state to the safe state is set to 1, the nonsafe state from the safe state, and the bonus value from the nonsecure state to the nonsafe state are set to -1. The bonus function is set as in formula (2).

$$r = \begin{cases} 2, SS \to WS \\ 1, NSS \to SS \\ 0, NSS \to NSS \\ -1, SS \to NSS : NSS \to NSS \\ -2, NSS \to FS \end{cases} \tag{15}$$

4.2   Parameter Setting and Result Analysis

As shown in the figure, a complex obstacle environment is established, and as a simulation background of the path planning experiment of the robot, G is the target point. During the simulation experiment, the robot is used as the research object, and the irregular cubic object acts as an obstacle. Figure 6 shows the path planning of the robot from the starting point in the simple environment and complex environment through the obstacle area until reaching the target point.

Using the Dyna-RQ algorithm and Dyna-Q algorithm, path planning is carried out 20 times, data processing is performed by MATLAB, and the convergence of the average iteration step and the average cumulative reward value of different algorithms is obtained.

According to Fig. 7, the Dyna-Q algorithm starts to approximate convergence after approximately 150 learning iterations, and the Dyna-RQ algorithm averages

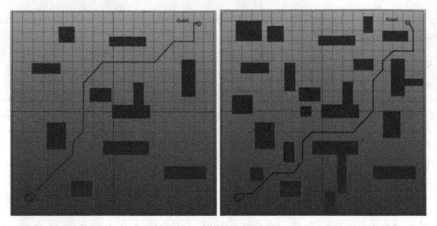

Fig. 6. Effect of path planning in a simple and complex environment

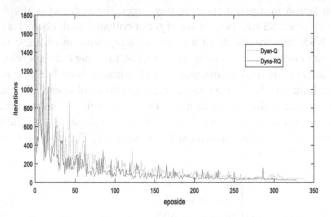

Fig. 7. Comparison of convergence between Dyna-RQ algorithm and Dyna-Q algorithm

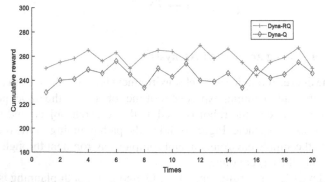

Fig. 8. Comparison of convergence between average cumulative reward value Dyna-RQ algorithm and Dyna-Q algorithm

approximately 120 iterations. From this, Dyna-RQ converges faster than Dyna-Q convergence.

At the same time, this paper has performed experiments 20 times. The experimental results are shown in Fig. 8. It can be seen from the figure that the average cumulative reward value of the Dyna-RQ algorithm proposed in this paper is higher than that of the traditional Dyna-Q.

In summary, the Dyna-RQ algorithm has better convergence than the Dyna-Q algorithm when applied to path planning tasks in complex obstacle environments. The experimental results show that the Dyna-RQ algorithm proposed in this paper has a good path planning effect in a simple environment and a complex environment, and the convergence speed is higher than that of Dyna-Q.

5 Conclusion

This paper solves the problem of mobile robot path planning by using the Dyna learning framework in reinforcement learning. To solve the problem that the traditional algorithm converges slowly and is not suitable for continuous and large-scale state spaces, this paper uses RBFNN to approximate the value function table in the Dyna-Q learning algorithm. The RBFNN can directly obtain the corresponding Q value function through the corresponding state input. Applicable to situations with a large state space.

References

1. Guo, N., Li, C., Wang, D., Zhang, N., Liu, G.: Path planning of mobile robot based on prediction and fuzzy control. Comput. Eng. Appl. **56**(8), 104–109 (2020)
2. Wang, Y., Chirikjian, G.S.: A new potential field method for robot path planning. In: IEEE International Conference on Robotics and Automation, Proceedings. ICRA. IEEE, vol. 2, pp. 977–982 (2000)
3. Roberge, V., Tarbouchi, M., Labonte, G.: Comparison of parallel genetic algorithm and particle swarm optimization for real-time UAV path planning. IEEE Trans. Ind. Inform. **9**(1), 132–141 (2013)
4. Zhang, Y., Li, S., Guo, H.: A type of biased consensus-based distributed neural network for path planning. Nonlinear Dyn. 1–13 (2017)
5. Hu, W., Lei, B.: Application research of computer vision technology based on deep neural network in path. Planning **6**(13), 113–116 (2022)
6. Cao, J., Liu, Q.: Research on path planning algorithm based on deep reinforcement learning. **39**(11), 231–237 (2022)
7. Sutton, R.S., Barto, A.G.: Reinforcement learning: an introduction. Mach. Learn. **8**(3–4), 225–227 (1992)
8. Santos, M., Botella, G.: Dyna-H: a heuristic planning reinforcement learning algorithm applied to role-playing game strategy decision systems. Knowl.-Based Syst. **32**(8), 28–36 (2012)
9. Zajdel, R.: Epoch-incremental dyna learning and prioritized sweeping algorithms. Neurocomputing **319**, 13–20 (2018)
10. Liu, S., Tong, X.: Urban transportation path planning based on reinforcement learning. J. Comput. Appl. **41**(1), 185–190 (2021)

Small Target Helmet Wearing Detection Algorithm Based on Improved YOLO V5

Jiajing Hu, Junqiu Li$^{(\boxtimes)}$, and Qinghui Zhang

Southwest Forestry University, Kunming 650224, China
`li_junqiu@sohu.com`

Abstract. To solve problems such as the low detection accuracy of helmet wearing, missing detection and poor real-time performance of embedded equipment in the scene of remote and small targets at the construction site, the text proposes an improved YOLO v5 for small target helmet wearing detection. Based on YOLO v5, the self-attention transformer mechanism and swin transformer module are introduced in the feature fusion step to increase the receptive field of the convolution kernel and globally model the high-level semantic feature information extracted from the backbone network to make the model more focused on helmet feature learning. Replace some convolution operators with lighter and more efficient Involution operators to reduce the number of parameters. The connection mode of the Concat is improved, and 1×1 convolution is added. The experimental results compared with YOLO v5 show that the size of the improved helmet detection model is reduced by 17.8% occupying only 33. 2 MB, FPS increased by 5%, and mAP@0.5 reached 94.9%. This approach effectively improves the accuracy of small target helmet wear detection, and meets the deployment requirements for low computational power embedded devices.

Keywords: Helmet wearing detection · YOLO V5 · Small object detection · Transformer · Swin Transformer · Involution

1 Summary

1.1 Subsection Sample

The construction [1] industry is in a period of rapid development, and the annual output value of the construction industry has been increasing. Wearing safety can timely and efficiently protect the heads of construction personnel and prevent and reduce head injuries caused by the collapse of construction sites or falling objects. The manual supervision method [2] is often costly, time-consuming and error-prone, but does not meet the requirements of modern construction safety management. The quantity of pixels and texture information of remote target images are small, which is difficult to detect. Therefore, the detection of small target helmet wearing is of great significance to the safety of construction personnel.

Z. Yu et al. (Eds.): ICPCSEE 2023, CCIS 1879, pp. 60–77, 2023.
https://doi.org/10.1007/978-981-99-5968-6_6

Fund projects : Yunnan Agricultural Joint Special Project(202301BD70001-127), Key Project of Key Laboratory of National Forestry and Grassland Administration for Forest Ecological Big Data at Southwest Forestry University(2022-BDK-05)
Author profile: First author Hu Jiajing(2000-), Woman, Undergraduate, Research direction: machine learning , E-mail:masterHuj@163.com; Corresponding author Li Junqiu(1978-) , Woman, associate professor, Research direction: information processing and intelligent control, E-mail: li_junqiu@sohu.com

In recent years, with the continuous development of deep learning object detection technology, domestic and foreign researchers have gradually applied deep learning technology to the detection of safety helmets. At present, methods based on deep learning are the mainstream of target detection and can be divided into two categories: one-stage target detection algorithms and two-stage target detection algorithms. YOLO [3, 4], SSD [5], anchor-free and other series are representative of the one-stage target detection algorithm, which proposes the solution of target detection based on regression. Although the speed is fast, the accuracy is lower and the detection effect for small targets is not ideal. R-CNN [6], Fast-RCNN [7], Faster-RCNN [8] and other series represent two-stage target detection algorithms, which put forward have begun that has high accuracy but lower speed. Many scholars have begun to apply deep learning algorithms to helmet wearing detection tasks and have achieved good results. Sun LiCheng [9] and Wang Liang proposed the improved YOLO v5 with R-UNet by combining it with explicit information to enhance the network's context-aware ability and the modification of the training loss. Xu Xianfeng [10] improved the detection speed by introducing the lightweight MobileNet network and building the MobileNet-SSD, which adopted the migration learning strategy to overcome the difficulty of model training. Zhang Jin [11] introduced a multispectral channel attention module into the feature extraction network based on YOLO v5, enabling the network to learn the weight of each channel independently.

Based on key frame detection under mobile terminal equipment, this paper proposes an improved YOLO v5, self-attention mechanism and hierarchical swin transformer based on a sliding window. This mechanism integrates the features extracted from the backbone network more efficiently, retains more abundant feature information, and enhances the learning of multiscale features of the mode. It also replaces some convolution operations with lighter Involution inner convolution operators, reducing the number of parameters and computation while still maintaining high accuracy. At the same time, it improves the Concat connection mode, adds a convolution operation, and iterates the optimal weight value output. The improved model is deployed on the embedded device to verify the small target objects in the monitoring screen in real time.

2 Related Work

2.1 Introduction of YOLO V5

The YOLO (You Only look Once) algorithm is a single-stage target detection algorithm proposed by Joseph Redmon et al. in 2015. The model uses regression to solve the target end to end and divides the input image into an S × S grid. If the center point of the detection object is in the grid, the grid is responsible for predicting the detection object. Each grid outputs the information of B bounding boxes (rectangular areas containing objects) and the conditional probability information of C objects belonging to this category.

The YOLO v5 network module is mainly divided into an input terminal, backbone network, neck and output terminal. On the basis of YOLO v4, the input terminal adds mosaic data enhancement, adaptive anchor box calculation and adaptive image scaling. The backbone network is mainly composed of a focus structure, convolution module, C3 and spatial pyramid pooling SPP [12] (spatial pyramid pooling) module. The neck part adopts the FPN + PANet [13] structure for multiscale fusion of features. This structure can transfer the high-level features and supplement the low-level semantics to obtain high-resolution and strong semantic features, which is conducive to small target detection. The binary cross entropy loss function is used at the output end to calculate the classification and confidence loss. The bounding box uses GIOU loss [14] as the loss function, and non-maximum suppression (NMS) is used to filter the multitarget box to improve the accuracy of target recognition.

2.2 Self-attention

The attention mechanism [15] mimics the internal process of biological observation behavior and increases the observation precision of the interest area. Attention mechanisms can quickly extract important features of sparse data, so they are widely used in machine translation [16], speech recognition [17], image processing [18] and other fields.

Transformer [19] is a self-attention mechanism proposed by Ashish Vaswani on the basis of an attention mechanism in 2015, which is used for natural language processing tasks. The Transformer adopts an encoder -decoder architecture. In computer vision tasks, the self-attention mechanism converts the input feature map into feature vectors, calculates the attention weight between each pair of feature vectors, and generates the updated feature map. Each pixel has global feature information. This mechanism can capture the global information of the image, reduce the dependence of the model on external information and is better at capturing the internal correlation of data or features. It also improves the shortcomings of traditional convolutional neural networks, such as slow training and convergence, and has a larger receptive field, so that it can obtain the dependency relationship between features with longer distance intervals in space. Due to the fixed length of the input sequence, the number of calculation steps, calculation amount and calculation complexity of Transformer are reduced compared with RNN.

The research data source is a helmet wearing target with a viewing distance of 100 m to 200 m, and the size is generally 20×20 pixels. According to the definition of the MS COCO dataset [14], an absolute value of pixels less than 32×32 can be called a small target. The experiment of YOLO v5 is mainly based on the MS COCO dataset with a relatively broad target. Many experiments have been carried out on the helmet wearing detection algorithm, and it is found that the imaging pixel information and texture information of remote small targets are difficult to extract,, resulting in unsatisfactory detection results. Therefore, in the neck feature fusion step of YOLO V5, this paper introduces the transformer's self-attention mechanism and hierarchical swin transformer module based on a sliding window to replace the partial convolution operation, improve the ability of extracting shallow contour information of the model, and reduce the calculation amount of the model.

3 Innovation Method

3.1 Swin Transformer

In the Transformer model, for the words given in NLP (Natural Language Processing), the length of the input sequence is certain, and the amount of computation is small. However, in the field of computer vision, the proportion of visual elements is different. There will be objects of different sizes in the same scene, and the computational complexity will increase in squares, which puts forward higher requirements for resolution and algorithm performance. Swin Transformer [20] aims to solve the problems caused by the differences between natural language processing and computer vision. Swin Transformer proposed two major improvements. First, a hierarchical transformer is drawn, and the patch merging operation is introduced to reduce the resolution, adjust the number of channels, form a hierarchical design and save a certain amount of computation.

The algorithm uses technology such as a feature pyramid network (FPN) or U-Net to predict small-scale targets. Similar to a convolutional neural network, the Transformer can be divided into several stages for processing. Based on the hierarchical feature map, the self-attention calculation is limited to nonoverlapping local windows. At the same time, a cross-window connection is allowed to make the model focus on the information of the neighborhood window, increase the receptive field of the model, and thus bring higher efficiency. Second, this paper proposes a self-attention mechanism based on a sliding window, as shown in Fig. 1. In layer l, the conventional window partition mode is used to calculate self-attention in each window. In the next layer, the sliding window moves to the lower right corner to generate a new window layer $(l + 1)$. The self-attention calculation of layer $(l + 1)$ in the new window crosses the boundary of the upper middle window and provides new correlation information.

The basic component unit of the Swin Transformer is the Swin Transformer Block, as shown in Fig. 2. Each swin transformer block consists of two blocks. The first block is composed of computing units such as layer norm, multihead self-attention, and multi-layer perceptron through residual connections. The only difference between the second

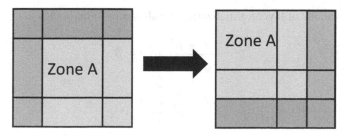

Fig. 1. Sliding window operation

block and the first block is that the feature data are divided into windows before the multi-head self-attention mechanism. The second block calculates self-attention and slides the window to realize window movement. Compared with Transformer's global attention calculation, it reduces a certain calculation quantity.

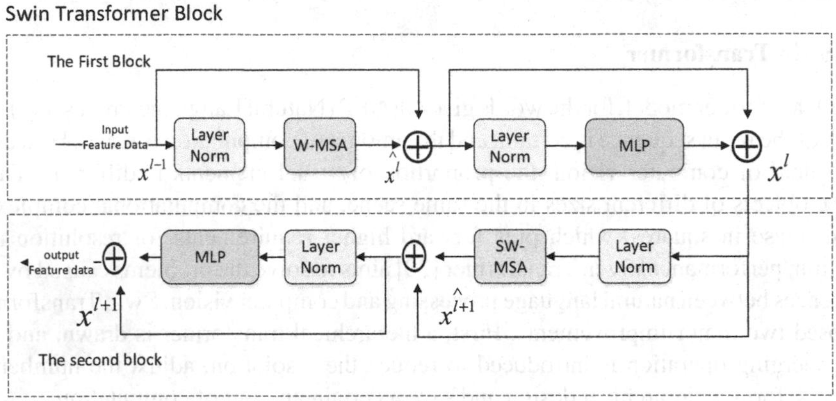

Fig. 2. Swin Transformer block

The calculation of input characteristic data X from layer 1 to layer $(1 + 1)$ is shown in formula (1–4), where $\hat{x}^l$ and x^l represent the output of the SW-MSA module and the MLP module of the first layer.

$$\hat{x}^l = W\text{ - }MSA(LN(x^{l-1}))+x^{l-1} \tag{1}$$

$$x^l = MLP(LN(\hat{x}^l)) + \hat{x}^l \tag{2}$$

$$\hat{x}^{l+1} = SW\text{ - }MSA(LN(x^l))+\hat{x}^l \tag{3}$$

$$x^{l+1} = MLP(LN(\hat{x}^{l+1})) + \hat{x}^{l+1} \tag{4}$$

The calculations of layer normalization are shown in formula (5–7).

$$\hat{z}^l = \frac{z^l - \mu^l}{\sqrt{\sigma^l + \varepsilon}} \odot \gamma + \beta \tag{5}$$

$$\mu^l = \frac{1}{n^l} \sum_i^{n^l} z_i^l \tag{6}$$

$$\sigma^l = \frac{1}{n^l} \sum_i^{n^l} (z_i^l - \mu^l)^2 \tag{7}$$

In the above formulas, z_i^l is the net output of layer l, is the mean and variance of the output of the first layer; γ $\beta$and are parameters that can be learned and changed; it is a point multiplication operation; and ε is any small number.

For the helmet dataset of small targets, it is easy to lose fine-grained information in the small pixel image during convolution sampling. Therefore, this paper replaces the bottleneck convolution operation in feature fusion section C3 with the swin transformer block structure to increase the feature information of the small helmet wearing target. This improvement reduces the impact of too little learning of fine-grained features, and greatly reduces the number of network parameters, making the model lighter.

3.2 Involution

Involution [21] is the framework drawn by Li Duo and Hu Jie in 2021. Traditional convolution [22] has two characteristics: spatial invariance. The convolution kernel weights in the convolution neural network are shared, and he size of the convolution kernel is generally 1×1. Small-sized static convolution kernels have difficulty capturing features in spatial dimensions. Channel specificity is that different channels have different weights, resulting in redundancy in convolution. The main structure of convolution is shown in Fig. 3.

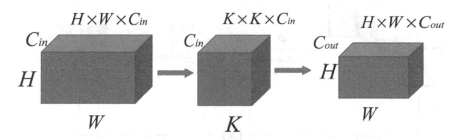

Fig. 3. Main structure of convolution

The length and width of the input image and output image are H × W. The size of the convolution kernel is K × K. The numbers of input channels and output channels are C_{in} and C_{out}, respectively. Parameter quantity P_1 and calculation quantity Q_1 are shown in formula (8) and formula (9), respectively.

$$P_1 = K^2 CinCout \tag{8}$$

$$Q_1 = HWK^2 CinCout \tag{9}$$

Consequently, the article proffers an Involution operator with spatial specificity and channel invariance. For the helmet data graph with an input dimension of, select the feature vector of one of the coordinate points, first expand the feature map into the shape of Kneel through $\varnothing$ (FC-BN-ReLU-FC) and reshape (Channel-to-Space) operations, obtain the corresponding evolution kernel of the coordinate point, and then perform multiply with the neighborhood feature vectors of the coordinate points on the input feature map to obtain the output feature map. The complete process is shown in Fig. 4. Where $\Omega_{i,j}$ is the K × K neighborhood near the coordinates (i, j). r is the channel reduction ratio, which can greatly reduce mAP.

$$X_{i,j}\ 1\times1\times C\ \xrightarrow{FC}1\times1\times\frac{C}{r}\xrightarrow{FC}1\times1\times K^2\text{—Reshape}\blacktriangleright K\times K\times1\text{—Multi-Add}\blacktriangleright Y_{i,j}:1\times1\times C$$

$$X_{\Omega_{i,j}}:K\times K\times C\ \text{———Multi-Add}$$

Fig. 4. Complete process of involution

The input dimension is 1 × 1 × C. The size of the convolution kernel is K × K. Through linear transformation, compress the channel, r is the channel reduction ratio, and then divide the c channels into G groups, and transform them into K × K × G. G is taken as 1. The whole implementation process is shown in Fig. 5.

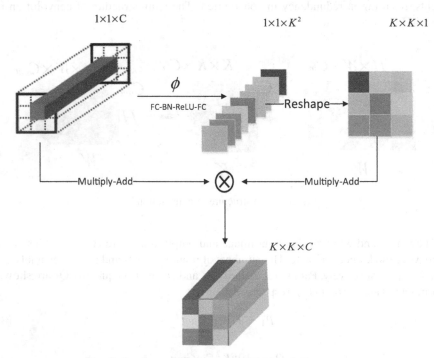

Fig. 5. Implementation process of the involution operator

For the Involution operator, its parameter quantity P_2 and calculation quantity Q_2. As shown in formula (10) and formula (11).

$$P_2 = \frac{C^2 + CGK^2}{r} \tag{10}$$

$$Q_2 = HWK^2C \tag{11}$$

It is not difficult to see that the calculation amount of the solution is linear with the number of channels. Due to the channel reduction factor r, compared with the convolution of the solution, the solution kernel simplifies the operation and complexity of the model, and accelerates the convergence of the model.

In this paper, the convolution in the bottleneck in C3 is replaced by a lighter and more efficient convolution operator. Applying it to the model reduces the size of the model, improves the speed of the model, and increases the accuracy, which is more conducive to deployment in terminal devices.

3.3 Vector Splicing Concat

In the lightweight network ShuffleNet [23, 24], it is proposed that operations such as adding element-by-element, ReLU activation function, etc., will increase the network computation. Consequently, YOLO v5 replaces the original add with Concat vector splicing. In vector stitching, the weight of each feature vector is the same, and the extracted features are relatively single, which is not conducive to flexibly adjusting parameters according to different inputs.

This paper proposes the MyConcat operation to replace the Concat operation, as shown in Fig. 6. First, the output feature weight value is normalized, and the gradient parameter w_i is introduced on the basis of Concat and learning rate e. The feature data vector is spliced and transferred to the ReLU activation function, and then sent to a 1×1 convolution.

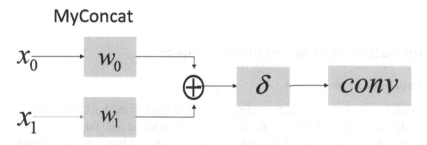

Fig. 6. Schematic diagram of Myconcat

A 1×1 convolution has two functions. First, realize the linear combination change of information between channels, which is equivalent to reducing the dimension of the feature vector to the number of original channels to reduce the parameters and deepen the number of network layers. The second is for the parameter w_i Gradient descent, searching for the best weight value of the objective function through continuous iteration. These two functions improve the weak learning ability of Concat.

The parameters in formula (12) are the output result of the entire MyConcat operation, δ is the activation function, this paper selects the ReLU activation function, w_o and w_i

are the feature weights, ε is the minimum value, and formulas (13) and (14) are the characteristic weights w_o and w_i of the normalized processing formula.

$$y = conv(\delta(\sum_{i=0}^{1} w_i x_i)) \tag{12}$$

$$w_0 = \frac{w'_0}{w'_0 + w'_1 + \varepsilon} \tag{13}$$

$$w_1 = \frac{w'_1}{w'_0 + w'_1 + \varepsilon} \tag{14}$$

Formula (15) is the prediction function. Assume that the feature components are x_1, $x_2, \ldots x_n$, and the weight of each special feature is $\theta_1, \theta_2, \ldots, \theta_n$. The estimated value is calculated using h_θ (w) for representation. Formula (16) is the loss function formula of gradient decline, in which m is the number of feature vectors and y^j is the output value of the j-th feature vector.

The purpose of introducing convolution is to optimize the error function, that is, how to integrate the function $J(\theta)$ convergence to the minimum, which is essentially a constantly changing θ value. Here, gradient search is used to find the optimal solution of the parameter.

$$h_\theta(w) = \theta_0 + \theta_1 w_1 + \theta_1 w_1 + \ldots + \theta_n w_n \tag{15}$$

$$J(\theta) = \frac{1}{2m} \sum_{j=1}^{m} [h_\theta(w^j) - y^j] \tag{16}$$

4 Innovation and Improvement of Papers

4.1 Backbone Improvement

The effective features of the helmet dataset are extracted in the backbone network and fused with the features of the neck network after C3.To better feed the features extracted by the backbone into the neck module, the improved backbone network structure is shown in Fig. 7.

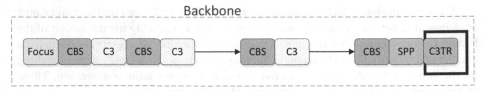

Fig. 7. Improved neck network structure

4.2 Neck Improvement

YOLOv5 fuses the features extracted from the backbone network in the neck framework. However, for dense targets, small targets, and multiscale helmet datasets, it is easy to lose fine-grained information in small pixel images during convolutional sampling, resulting in incomplete feature fusion and unsatisfactory final results. Therefore, this article introduces the transformer's self-attention mechanism and a sliding window-based hierarchical swin transformer module to replace partial convolution operations in the neck feature fusion step of the YOLO V5 algorithm, improving the model's ability to extract shallow contour information while reducing the computational complexity of the model. Replacing some C3 operations with more concise and efficient C3Inv operations can effectively reduce the size of the model. The improved neck network structure is shown in Fig. 8.

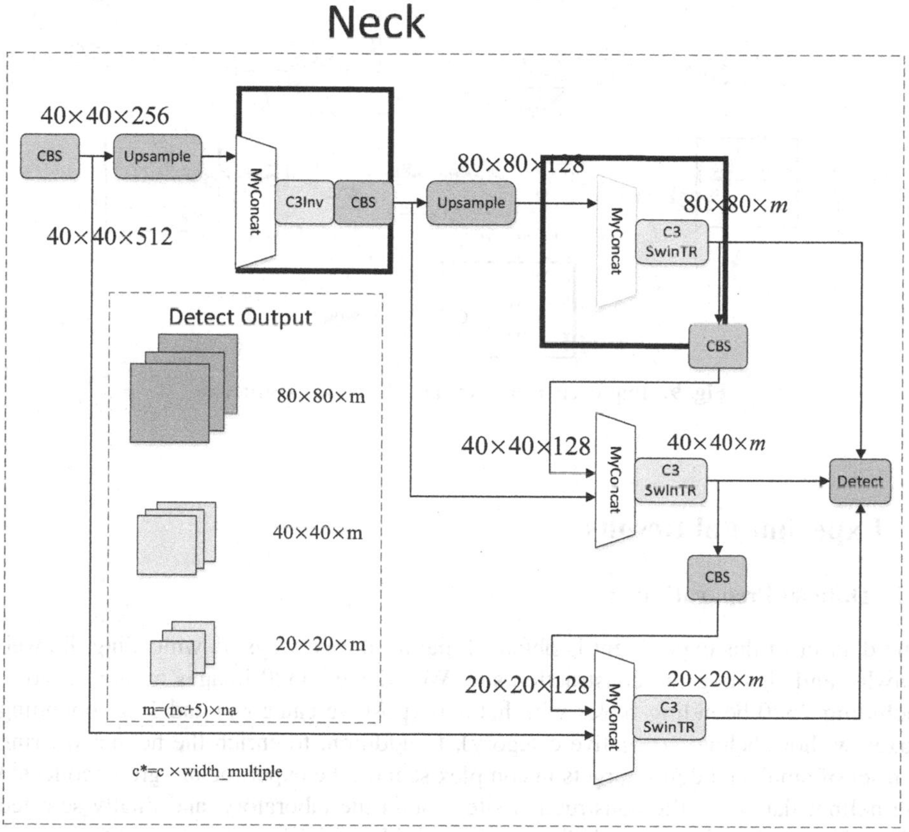

Fig. 8. Improved neck network structure

4.3 Overall Improvement

The improved YOLO v5 network structure is shown in Fig. 9. The solid line boxes, and in the figure correspond to the introduction of the Swin Transformer structure, the introduction of the Involution structure and the improvement of the Concat operation.

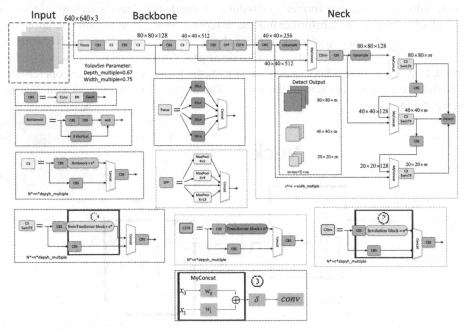

Fig. 9. Improved YOLO v5m model network structure

5   Experimental Results

5.1   Dataset Preparation

The dataset of this experiment is obtained mainly in two ways: downloading the web crawler and shooting the construction site. We crawled 4320 images on the network, including 2570 bounding boxes with helmets (positive category) and 1750 bounding boxes without helmets (negative category). In addition, to enrich the helmet wearing dataset of small and dense targets in complex scenes, the experimental group collected the helmet dataset at the construction site around the laboratory, and finally selected 1500 construction pictures and 80 construction videos as the verification set to evaluate the performance of the improved YOLO v5 model and verify the performance of the algorithm in different scenarios.

Dataset Annotation

Dataset annotation was performed using LabelImg annotation software to annotate images and visual images. LabelImg is a graphic image annotation tool. It is written in Python and uses Qt as its graphical interface. Annotations are saved as XML files in PASCAL VOC format, and support YOLO.

The helmet dataset is randomly divided into 80% for model training and 20% for model validation. Finally, the labelled dataset is stored in the VOC standard dataset format for training. The LabelImg operation and each bounding box label are shown in Fig. 10. The left image (a) shows the LabelImg labelling operation, and the right image (b) shows the bounding box label.

(a) LabelImg tag operation   (b) Bounding box label

Fig. 10. LabelImg operation diagram

5.2 Experimental Environment

To verify that the improved YOLO v5 model has good performance, the computer hardware configuration used in this experiment is shown in Table 1.

Table 1. Hardware condition configuration

Name	Related configurations
CPU	Intel(R)core(TM)5-10210U
RAM	8 GB
GPU	NVDIA GeForce GTX 1060 6 GB
Operating system	Window10
data handling	Python3.7

5.3 Evaluating Indicator

The performance indicators of the target detection model are divided into speed indicators and accuracy indicators.

(1) Precision refers to the correct proportion of predictions in all prediction boxes.

$$Precision = \frac{TP}{TP + FP} \times 100\% \tag{17}$$

(2) Recall refers to the proportion correctly predicted in all label boxes.

$$Recall = \frac{TP}{TP + FN} \times 100\% \tag{18}$$

(3) Average precision (AP) sets the confidence threshold IOU_{tru} changes from $[0,1]$, calculates the precision and recall corresponding to each threshold, and draws the PR curve, whose surrounding area is the AP of this category.

$$AP = \int_0^1 P_{trh}(r)dr \tag{19}$$

(4) mAP@0.5 Is the intersection and combination ratio threshold IOU_{tru} When thetruth is taken as 0.5, the average value of AP of each category.

$$mAP@0.5 = \frac{1}{N} \sum_{i=1}^N AP_i(IOU_{tru} = 0.5) \tag{20}$$

5.4 Improvement Results

Ablation Experiment

To further verify the optimization effect of each module of the improved model, the experimental group set up an ablation experiment, and the experimental results are shown in Table 2. In the table, "1" means replacing Bottleneck in C3 with Swin Transformer block; "2" indicates the introduction of the Involution inner roll operator; "3" means replacing Concat with MyConcat. The final experimental results show that the AP value of the optimized network model is increased by 3.37%, the model is only 33.2 M, and the performance is improved compared with the original YOLO v5m.

Comparative Experiment

Table 3 shows the comparative experimental data between the improved YOLOv5 algorithm and other mainstream algorithms. Through comparison, the performance improvement of the improved YOLO v5 algorithm is further verified.

Table 2. Ablation Experiment

Network	1	2	3	AP%	Size
YOLOv5m	×	×	×	89.1%	39.4 M
modify 1	✓	×	×	91.5%	36.4 M
modify 2	×	✓	×	92.0%	37.3 M
modify 3	×	×	✓	91.2%	34.2 M
Final modify	✓	✓	✓	**92.1%**	**33.2 M**

Table 3. Comparative Experiment

Module	mAP%	Rcall%	Prec%	Size	FPS
Fast-RCNN	91.2%	90.3%	92.1%	132 M	2.7
SSD	80.3%	77.1%	75.7%	103 M	9.5
YOLO v3	83.2%	81.7%	84.1%	230 M	21.7
YOLO v4	87.3%	88.1%	89.1%	247 M	30.0
YOLO v5	91.0%	87.8%	91.4%	40.4 M	39.4
YOLO v5 modify	**92.1%**	**88.3%**	**93.6%**	**33.2 M**	**42.3**

Although the improved YOLO v5m model is slightly inferior to Fast-RCNN in terms of recall and precision, the FPS of the improved YOLO v5m model is 15.7 times that of Fast-RCNN. The improved YOLO v5 model is smaller and faster, and the improvement in real-time performance is beneficial to deployment in terminal devices. It is slightly superior in terms of mAP, recall, precession, model size and reasoning speed. The experimental results show that the algorithm takes into account the accuracy and speed of detection at the same time, and meets the requirements of real-time detection.

Recall and Precision Comparison Chart

The comparison of recall and average accuracy between YOLOv5m and improved YOLOv5m in the training results of the helmet wearing dataset is shown in Figs. 11 and 12. The left figure (a) shows the experimental results of YOLOv5m, and the right figure (b) shows the improved YOLOv5m experimental results. The improved network has significant improvements in Recall and Precision.

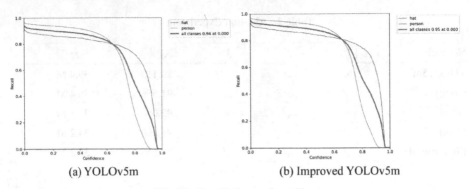

(a) YOLOv5m   (b) Improved YOLOv5m

Fig. 11. Recall Comparison Chart

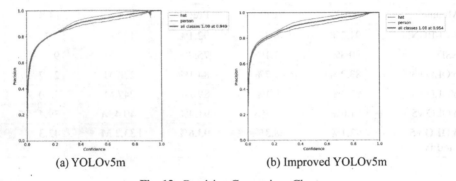

(a) YOLOv5m   (b) Improved YOLOv5m

Fig. 12. Precision Comparison Chart

Analysis of Test Results

As shown in Fig. 13, the left (a) is the original YOLO v5 model renderings, and the right (b) is the improved YOLO v5 model renderings. It can be seen from the result chart that in small-scale target detection, the YOLO v5 algorithm can accurately locate the position of the helmet, but to some extent, it has missed the detection and inaccurate positioning of small targets in the distance. Compared with the left figure, the improved model has fewer cases of missing detection, and can locate more accurately. It has a high degree of recognition for dense small scale targets. From the experimental results, it can be seen that the method in this paper is applicable to small target helmet wearing detection in different scenarios, and has achieved significant results.

(1)Small target detection 1

(2)Small target detection 2

(3)Small target detection 3

(4)Dense target detection

(a)YOLO v5   (b)Improved YOLO v5

Fig. 13. Recognition effect of YOLOv5 algorithm and improved YOLOv5 algorithm

6 Conclusion

To solve the problems of low helmet wearing detection accuracy in the scene of remote small targets at the construction site, missing detection and poor real-time performance of embedded equipment, this paper proposes an improved YOLO v5 small target helmet wearing detection algorithm. Based on the single-stage target detection algorithm YOLO v5 m, this paper proposes a helmet wearing detection algorithm that introduces a transformer. It replaces some convolution operations in YOLO v5 with algorithms such as Transformer, Swing Transformer, and Revolution inner roll operator, and improves the Concat connection mode. The advantage of this model is that it increases the learning of fine-grained and multiscale features of the network, and effectively reduces the number of parameters and computation of the network. The model in this paper is deployed on mobile devices, and its timeliness and performance are relatively stable. However, the improved algorithm does not work well in the detection of safety helmets in dark scenes, which needs to be further strengthened in the future.

References

1. Feng, P., Sai, Y.X.: Common safety hazards and corresponding countermeasures on construction. Constr. Saf. **6**, 33–35 (2014)
2. Yang, Y.B., Li, D.: Improved lightweight helmet wearing detection algorithm of YOLOv5. Comput. Eng. Appl. 1–8, 2021–01–19
3. Redmon, J., Divvala, S., Girshick, R., et al.: You only look once: unified, real-time object detection. In: Proceedings of the IEEE Conference on Computer Vision and Pattern Recognition, pp. 779–788 (2016)
4. Redmon, J., Farhadi, A.: YOLO:9000: better, faster, stronger. In: Hawaii, USA: Proceedings of the IEEE Conference on Computer Vision and Pattern Recognition (2017)
5. Girshick, R., Donahue, D., et al.: Region-based convolutional networks for accurate object detection and segmentation. IEEE Trans. Pattern Anal. Mach. Intell. **38**(1), 142–158 (2015)
6. Girshick, R.: Fast R-CNN. In: Proceedings of the IEEE International Conference on Computer Vision. IEEE, Piscataway, pp. 1440–1448 (2015)
7. Rensq, S., Hekm, K., Girshick R, et al.: Faster R-CNN: towards real time object detection with region proposal networks. IEEE Trans. Pattern Anal. Mach. Intell. **39**(6), 1137–1149 (2017)
8. Liu, W., Anguelov, D., Erhan, D., Szegedy, C., Reed, S., Fu, C.-Y., Berg, A.C.: SSD: Single shot multibox detector. In: Leibe, B., Matas, J., Sebe, N., Welling, M. (eds.) ECCV 2016. LNCS, vol. 9905, pp. 21–37. Springer, Cham (2016). https://doi.org/10.1007/978-3-319-464 48-0_2
9. Sun, L.C., Wang, L.: An improved YOLO V5-based algorithm of safety helmet wearing detection. In: 2022 34th Chinese Control and Decision Conference, pp. 2031–2035 (2022)
10. Xu, X.F., Zhao, W.F., Zou, H.Q., et al.: Detection algorithm of safety helmet wear based on MobileNet-SSD. Comput. Eng. **47**(10), 298–305, 313 (2021)
11. Zhang, J., Qu, P.Q., Sun, C., et al.: Helmet wearing detection algorithm based on improved yolov5. Comput. Appl. **42**(4), 1292–1300 (2022)
12. Liu, S., Qi, L., Qin, H., et al.: Path aggregation network for insurance segmentation. In: Proceedings of the IEEE Conference on Computer Vision and Pattern Recognition, pp. 8759–8768 (2018)

13. Lin, T.Y., Dollár, P., Girshick, R., et al.: Feature pyramid networks for object detection. In: Proceedings of the IEEE Conference on Computer Vision and Pattern Recognition, pp. 2117–2125 (2017)

14. Rezatofihhi, H., Tsoi, N., Gwak, J., et al.: Generalized intersection over union: a metric and a loss for bounding box regression. In: Proceedings of the IEEE/CVF Conference on Computer Vision and Pattern Recognition, pp. 658–666 (2019)

15. Liu, J.H.: Active and Semisupervised Learning based on ELM for Multiclass Image Classification. Southeast University, Nanjing (2016)

16. Wang, H., Shi, J.C., Zhang, Z.W.: Text semantic relation extraction of LSTM based on attention mechanism. Appl. Res. Comput. 35(5), 1417–1420 (2018)

17. Tang, H.T., Xue, J.B., Han, J.Q.: A method of multiscale forward attention model for speech recognition. Acta Electron. Sin. 48(7), 1255–1260 (2020)

18. Wang, W.G., Shen, J.B., Yu, Y.Z., et al.: Stereoscopic thumbnail creation via efficient stereo saliency detection. IEEE Trans. Visual Comput. Graph. 23(8), 2014–2027 (2016)

19. Vaswani, A., Shazeer, N., Parmar, N., et al.: Attention Is All You NeNeed. arXiv:1706.03762 (2017)

20. Liu, Z., Lin, Y., Cao, Y., et al.: Swin Transformer: Hierarchical vision Transformer using shifted window. arXiv:2103.14030 (2021)

21. Li, D., Hu, J., Wang, C.H., et al.: Involution: Inverting the Inherence of Convolution for Visual Recognition. arXiv:2013.062 55 (2021)

22. Zhou, F.Y., Jin, L.P., Dong, J.: A review of convolutional neural networks. J. Comput. Sci. 6(40), 1230–1251 (2017)

23. Zhang, X.Y., Zhou, X.Y., Lin, M.X., et al.: ShuffleNet: an extremely efficient convolutional neural network for mobile devices. In: Proeedings of the IEEE Conference on Computer Vision and Pattern Recognition, pp. 6848–6856 (2018)

24. Ma, N.N., Zhang, X., Zheng, H.T., et al.: ShuffleNet V2: practical guidelines for efficient CNN architecture design. European Conference on Computer Vision. Springer, Cham (2018)

Research on Dance Evaluation Technology Based on Human Posture Recognition

Yanzi Li, Yiwen Zhu, Yanqing Wang$^{(\boxtimes)}$, and Yiming Gao

Nanjing Xiaozhuang University, Nanjing 211171, Jiangsu, China
wyq0325@126.com

Abstract. In view of the increase in the number of people participating in dance rating assessments, this paper proposes a dance assessment technology based on human body posture recognition. This technique adopts the human target detection of the dance video, extracts bone key points, and then uses the video data set collected by professional dancers to conduct PoseC3D model training, enabling the model to classify the basic movements of the dance; then, the dynamic time normalization algorithm is used to evaluate the classified movements. The experimental results show that this technology can accurately identify the basic movements of various dances and accurately give the evaluation score of the corresponding movements, thus reducing the work intensity of the assessment staff.

Keywords: Body Pose Recognition · Pose Estimation · Dynamic Time Warping · Deep Learning

1 Introduction

Since the 18th National Congress of the CPC, the Party and the State have attached great importance to aesthetic education. In 2022, the General Office of the CPC Central Committee and The General Office of the State Council jointly issued for the first time the programmatic document on aesthetic education, "Opinions on Comprehensively Strengthening and Improving Aesthetic Education in Schools in the New Era", proposing to explore including art subjects in the academic level examinations of junior and senior high schools. Aesthetic education evaluation has become a lever to leverage the implementation and development of aesthetic education in schools [1]. As a part of aesthetic education, dance has also become an important content of current attention. An increasing number of people are learning dance. After learning dance to a certain extent, it is necessary to carry out dance assessment, and the workload of dance assessment personnel also increases. Moreover, the assessment standard mainly depends on the professional ability of dance assessors and the subjective evaluation of teachers. Although many times attention is given to the beauty of dance, the main part of the dance assessment is basic skills assessment and body standards, which can be achieved through technology. Moreover, it costs considerable time and manpower to evaluate dance movements by personal visual observation. It is of great application value and practical significance to propose a kind of quantitative and automatic dance evaluation technology.

Z. Yu et al. (Eds.): ICPCSEE 2023, CCIS 1879, pp. 78–88, 2023.
https://doi.org/10.1007/978-981-99-5968-6_7

With the development of computer vision technology, many studies have tried to classify and evaluate human motion based on computer vision. The current methods of action classification and behavior recognition mainly include action recognition models, temporal motion detection models, temporal motion monitoring models and action recognition models based on bone key points. Yan Guoliang [2] used the method of motion capture to identify dance video movements. Although motion capture has a high recognition accuracy for overall dance movements, this method requires video motion capture equipment, which is complicated to operate. Bi Xuechao [3] used a method based on multifeature fusion to identify the dance movements of dancers. Although multifeature fusion can identify key points of human posture in complex environments and thus identify dance movements, it is unable to classify dance movements or make real-time evaluations of dance movements.

Therefore, this paper proposes to use the posture estimation model based on ResNet-50 to detect human objects in dance videos extract bone key points, and then input the obtained sequence of human bone key points into PoseC3D. The basic dance movements are classified, and the corresponding movements completed after classification are evaluated by the DynamicTime Warping algorithm, which greatly reduces the work intensity of the dance assessor, effectively improves the assessment speed, and makes the assessment more rapid and efficient.

2 Theory and Method

This section mainly introduces the human posture recognition method, posture estimation method and dynamic time regulation adopted by the Institute of Dance Evaluation Technology.

2.1 Attitude Estimation

The human posture estimation method [4, 5] processes RGB images or videos through deep learning and estimates the corresponding positions of key points of the human skeleton [6, 7], thus depicting the shape of the human body. According to the different starting points of prediction, human posture estimation methods are mainly divided into top-down and bottom-up methods: high-level abstract or low-level pixel. The top-down approach starts with high-level abstractions, first detecting the human body, then generating the human position in a boundary box, and finally estimating the human posture. The bottom-up approach begins by predicting all body parts of the human body in an input image and grouping them and connecting key points of the human skeleton through mannequin fitting or other algorithms. Depending on the method, the body parts may be joints, limbs or small template patches.

When the deep learning network layer is deeper, in theory, the expression ability will be stronger, but only by increasing the number of network layers is enhancing the network learning method not always feasible because the CNN network reaches a certain depth and then deepens, and classification performance will not improve but will lead to slower network convergence lower accuracy, and disappearance of the stochastic gradient. Even if the data set increases, solving the problem of overfitting, classification performance and accuracy will not improve, and the ResNet model can solve all these problems and thus introduce the BN layer and residuals.

The ResNet-50 model contains two basic speeds, called Conv Block and Identity Block, where the Conv block input and output dimensions are different, so they cannot be continuously connected, which is used to change the dimension of the network; the Identity block input dimension and output dimension are the same, which can be connected to deepen the network. The purpose of the BN layer (batch normalization) is to preprocess the feature map of a batch (batch) to meet the distribution law of mean 0 and variance 1 to accelerate the convergence of the network. In addition, the ResNet-50 pose estimation model [8] will have higher accuracy when using the COCO data set. Therefore, this paper uses the ResNet-50 pose estimation model trained based on the COCO data set to detect the human body in the dance video and then identify the key points of human bones and connect them.

The full name of the COCO data set is Microsoft Common Objects in Context, or MS COCO for short. It is a large image data set that is used for object detection and segmentation, key point detection of people, filling segmentation and subtitle generation in the field of machine vision and contains 330,000 images. The key points of human bone based on the COCO data set are shown in Fig. 1, and the corresponding relationship between key points of human bone and various parts of the human body is shown in Table 1. The sequence of key points of human bone is processed to obtain the final classification and evaluation of dance movements.

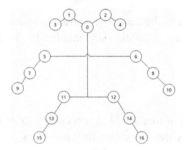

Fig. 1. The key bones of the human body.

Table 1. The corresponding relationship between the key points of the human skeleton and various parts of the human body.

Human body parts	Serial number of key points in human bones
nose	0
Left eye	1
Right eye	2
Left ear	3

(*continued*)

Table 1. (*continued*)

Human body parts	Serial number of key points in human bones
Right ear	4
Left shoulder	5
Right shoulder	6
Left elbow	7
Right elbow	8
Left wrist	9
Right wrist	10
Left hip	11
Right hip	12
Left knee	13
Right knee	14
Left ankle	15
Right ankle	16

2.2 Human Posture Recognition

Human posture recognition [9] detects key points of the human body through images or videos and classifies the behaviors of targets in images or video segments. However, many skeleton-based action recognition methods adopt graph convolutional networks (GCNs) [10] to extract features from the human skeleton. Although GCN is widely used, this method still has some defects in robustness, compatibility and scalability. Therefore, the PoseC3D model is adopted in this paper [11].

PoseC3D is a skeletal motion recognition method based on 3D-CNN. In this method, two-dimensional human posture is used as input. Based on the extracted two-dimensional human posture, it is stacked into T sheets of a two-dimensional key point heatmap with a shape of $K \times H \times W$ to generate a 3D heatmap stack with a shape of $K \times T \times H \times W$, which is used as the input and output of the 3D-CNN and finally outputs the classification category of actions, which has a better recognition effect. The PoseC3D network structure is shown in Fig. 2, where C is the number of key points, T is the timing dimension, namely, the number of continuous frame heatmaps, and H and W are the height and width of the frame, respectively.

First, a 3D heatmap with a size of $17 \times 32 \times 56 \times 56$ was input. Second, features with a size of $32 \times 32 \times 56 \times 56$ were output through Conv1, and then features with a size of $512 \times 32 \times 7 \times 7$ were output through three layers of ResNetLayer. After global average pooling, 512-dimensional feature vectors are obtained and finally input into the classification of actions performed at the fully connected layer.

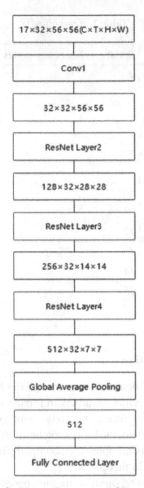

Fig. 2. Network structure of PoseC3D.

2.3 Dynamic Time is Structured

In the time series, the length of the two time series that need to be similar may not be the same. In the field of speech recognition, different people have different speeds and intonations; in the field of action recognition, different people have different durations when performing the same action. In addition, different time series may only have displacement on the time axis, that is, in the case of reduction displacement, two time series are consistent. In these complex cases, the similarity between two time series cannot be obtained effectively by using the traditional Euclidean distance. Based on the dynamic programming strategy, the DTW algorithm makes nonlinear time-domain alignment adjustments for two time sequences to correctly calculate the similarity between them, to find the best matching result. In recent years, it has mainly been applied in the matching of time series.

The dynamic time warping method in the similarity measurement of time series can avoid the influence of time series displacement and dislocation. Dynamic time warping uses a dynamic programming method that can align and match two action sequences of different lengths to obtain action scores.

As shown in Fig. 3, for two time series of the same length $1 = \{a1, a2..., a16\}$ and time series $2 = \{b1, b2..., b16\}$, the length of time series 1 is equal to that of time series 2, but the length of the two time series may not be equal in actual matching. The dotted line between time series represents the similar points between the two time series. The dynamic time warping algorithm calculates the similarity between two time series through the sum between the similar points.

Time series 1

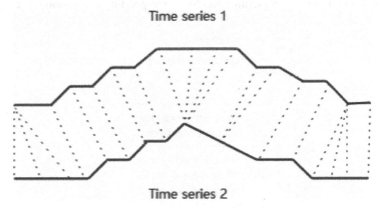

Time series 2

Fig. 3. The regularity of the two time series matches.

In the similarity calculation, the similarity calculation between time series 1 and time series 2 is shown in Eq. (1).

$$d_{ij} = \sum_{k=1}^{17} \sqrt{(a_i - b_j)^2 (x_k^2 + y_k^2)} \tag{1}$$

17 is the key points of human bones obtained by the ResNet-50 attitude estimation model, and the key points are recorded by two-dimensional coordinates. d_{ij} represents the total distance between frame i in sequence 1 and 17 key points of the human skeleton in frame j in sequence 2. According to Formula (1), the distance between two sequences of key points in human bones can be obtained, as shown in Formula (2).

$$DTW(X, Y) = \min \sum_{i=1}^{n} \sum_{j=1}^{m} d_{ij} \tag{2}$$

In Formula (2), the length of time series 1 is n, and the length of time series B2 is m. The distance between time series 1 and time series 2 can be calculated by Formula (2). At the same time, the mapping relationship between the DTW distance and the score of the dance evaluator was established to give the final evaluation of the dance movements of the dancers under evaluation.

2.4 Dance Evaluation Model

The dance recognition model and evaluation model are composed of three parts: human posture recognition model, PoseC3D model and action scoring algorithm. The specific workflow is shown in Fig. 4.

The dance assessment need to input has recorded the dance video, after using the human posture recognition model of the video generation frame, and the detection box in the human body using PoseC3D human bone key identification, identification, after the completion of the human bone key sequence for a series of processing, after the ResNet-50 model to identify the action of the classification, finally, according to the category of dance movements through the DTW algorithm to evaluate the kind of dance action, the final output action pictures and the corresponding evaluation score.

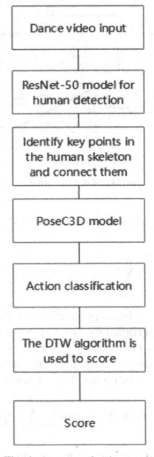

Fig. 4. Dance evaluation model

3   Analysis of Experimental Results

3.1   Production of Data Sets

The data set adopted in this paper is the video collected in the dance classroom. National second-level dancers are taken as shooting objects to collect standard movements, and dance trainees are taken as shooting objects to collect nonstandard movements.

3.2   Classification and Treatment of Dance Movements

According to the basic skills of dance, the basic skills of dance movements are divided into two categories: the upper part and the lower part. The training demonstration diagram is shown in Table 2. The dance posture emphasizes "hand, eye, body and Dharma" and is accomplished by coordinating the body's torso, legs, arms, head and eyes. Basic dance postures include basic shapes and postures of hands and feet, basic postures, training of body methods, combining waist and leg movements and skills to form a group of dance movements with a strong sense of sculpture and expression.

Table 2. Basic skills training demonstration map.

Stretch	Hook	Chest-waist push	Kneel down	Kneel down and grab the foot	Stand at the waist	Frog span	Sit by the moving legs
Cross fork	Vertical fork	Vertical fork hind feet	Ground press side leg	Ground press behind crotch	Lossy fork	hipSide leg	Front leg

Based on the dynamic programming strategy, the DTW algorithm makes nonlinear time domain alignment adjustments for two time sequences to correctly calculate the similarity between them for matching. However, the start point and end point of the time series have a great influence on the accuracy of the dynamic time warping algorithm. To ensure the standardization of dance movements, fine-grained division is carried out, and rules are formulated for the starting point and ending point of dance movements according to the standard requirements of dance movements.

The original dance movement video was input into the ResNet-50 human pose estimation model, and the acquired key points of human bones were used to calculate the start point and end point of fine-grained movements according to the dance movement division rules to obtain a more accurate dance movement data set.

3.3 Experimental Results

Every 2 s, a frame of dancing video of dancers is randomly collected for posture recognition. According to the main parts of the human body, the key points of human bones are extracted and the human skeleton is detected to realize action recognition. The results of some experiments are shown in Fig. 5.

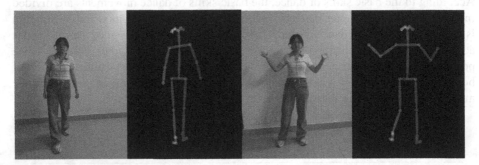

Fig. 5. Body skeleton recognition diagram of dance movements.

4   Experimental Analysis

I n the introduction, theory and method, this paper mentions some shortcomings of the traditional gesture recognition algorithm, such as the inability to classify and evaluate the dance movements and the complex operation. Therefore, in this paper, we propose corresponding methods to improve these problems and use the human posture model based on the ResNet-50 method and PoseC3D method to extract and process the key points of human bones to realize the behavior detection of the human body.

Most traditional bone action recognition work uses GCN to extract bone features, but the GCN method still has some defects in robustness, compatibility and scalability. Therefore, this paper uses the PoseC3D model framework. PoseC3D is a skeleton behavior identification framework based on 3D-CNN with good identification accuracy and efficiency. Compared with GCN-based methods, PoseC3D can more efficiently extract spatiotemporal features in the human skeleton sequence, be more robust to noise in the skeleton sequence and have better generalization. Moreover, unlike the traditional GCN method based on a human 3-dimensional skeleton, PoseC3D can achieve a better identification effect by using only 2-dimensional human skeleton heatmap stacking as input. The experiments show that the PoseC3D model framework is better and the identification accuracy is higher. To show a more intuitive comparison of the PoseC3D method and GCN method, this paper lists their performance in different situations, as shown in Table 3.

Table 3. Compare the table between GCN and PoseC3D.

method	GCN	PoseC3D
Posture recognition test	The recognition accuracy is general	High recognition accuracy
robustness	Easy to be affected	Less impact
extendibility	Group identification is not efficient	High group identification accuracy
compatibility	It is difficult to feature-fuse with other 3D-CNN-based modes	Easy to feature fusion with other modes

As shown in Table 2, the PoseC3D model is more suitable for the human posture recognition module. Compared with the traditional GCN, the recognition accuracy is higher, and the recognized actions can be classified by combining with the ResNet-50 method.

5 Conclusion

In this paper, a hybrid model of pose estimation, human body pose recognition and scoring modules is proposed. First, the human body pose estimation model based on ResNet-50 is used to extract the key points of human bones from dance videos, and then the dance movements are classified by the PoseC3D model. Finally, the movements are scored by the DTW algorithm. Through multidimensional visualization analysis of the basic data, this technology realizes the objective and accurate classification and evaluation of dance movements, reduces the workload of dance assessment personnel, and can effectively improve the assessment speed.

References

1. Zheng, L., et al.: studied the construction of students' dance literacy evaluation index system and curriculum. teaching material. Coach. Meth. **42**(02), 131–137 (2022)
2. Yan, G.: a research on dance video motion recognition technology based on motion capture. J. Chifeng Univ. (Natl. Sci. Ed.) **38**(09), 48–52 (2022)
3. Xuechao, B.: Research on dance movement recognition technology based on multifeature fusion. Electr. Des. Eng. **28**(18), 189–193 (2020)
4. Pranjal, K., Siddhartha, C., Kumar, A.L.: Human pose estimation using deep learning: review, methodologies, progress and future research directions. Int. J. Multimedia Inf. Retrieval **11**(4), 489-521 (2022)
5. Yinong, D., Luo, J., Jin, F.: An overview of human pose estimation methods based on deep learning. Comput. Eng. Appl. **55**(19), 22–42 (2019)
6. Alaoui, A.Y., Fkihi, S.E., Thami, R.O.H.: Fall detection for elderly people using the variation of key points of human skeleton. IEEE Access **7**, 154786-154795 (2019)
7. Zeng, W., Ma, Y., Li, W.: Research on the classification of soldier training movements based on bone key points. J. Hebei Acad. Sci. **39**(01), 7–14 (2022)

8. Jinzi, L., et al.: Rock image intelligent classification and recognition based on resnet-50 model. J. Phys. Conf. Ser. **2076**(1), 012011 (2021)
9. Jurjiu, N., et al.: A systematic review of integrated machine learning in posture recognition. Timisoara Phys. Educ. Rehabil. J. **14**(27), 15–20 (2021)
10. Gao, M., et al.: 3D bone point action recognition based on the convolution of multiresidue maps. Small Microcomput. Syst. **43**(12), 570–2574 (2022)
11. Zhou, S., Chen, Z., Deng, Y.: Based on PoseC3D tennis motion recognition and evaluation method. Comput. Eng. Sci. **45**(01), 95–103 (2023)

Multiple-Channel Weight-Based CNN Fault Diagnosis Method

Peng Xu[1], Xinyu Liu[1], Junyu Lin[2(⊠)], Zhongyu Lu[3], Fengming Li[3], and Husheng Gou[4]

[1] Jiangsu JARI Technology Group Co. Ltd, Lianyungang, China
[2] Institute of Information Engineering, Chinese Academy of Sciences, Beijing, China
linjunyu@iie.ac.cn
[3] Harbin Engineering University, Harbin, China
[4] China International Engineering Consulting Corporation, Beijing, China

Abstract. It is difficult to comprehensively extract device status information for CNNs under a single source high-frequency timing signal, and CNNs cannot effectively achieve precise identification and classification based on the importance of multichannel features. This article proposes a CNN fault diagnosis method based on multi -channel weight adaptation. This method first normalizes different data sources as input as different channels of CNN, and uses the characteristics of convolutional networks to achieve the characteristics of different data sources. Fusion and extraction. Then, the SNET module is embedded into the CNN network, adapted to the weight of each channel, and the accuracy of classification is improved. Finally, through comparative experiments, this method can further improve the accuracy of fault recognition.

Keywords: Convolutional Neural Network · Timing Features · Weight Adaptation

1 Introduction

With continuous innovation in production and manufacturing, the components of equipment are more closely related to each other, and the failure of equipment during operation can bring huge economic losses and safety problems to industrial production. Traditional fault diagnosis methods use statistical analysis to extract signal features, which are often limited by expert experience. CNN has a unique structure with powerful data mining capabilities as well as data fusion capabilities to achieve better performance at a much lower cost [1]. The original signal input or image information can be directly used as model input, and the results are directly output after network operations, which are adaptive and do not require feature extraction work, realizing an adaptive input-to-output recognition process. At present, using CNN to diagnose equipment faults has become a more mainstream fault diagnosis method: Zhao [2]. Proposed a deep network model combining long short-term memory (LSTM) and a convolutional neural network to detect and diagnose the wear degree of tools, and the model was proven to have a

© The Author(s), under exclusive license to Springer Nature Singapore Pte Ltd. 2023
Z. Yu et al. (Eds.): ICPCSEE 2023, CCIS 1879, pp. 89–105, 2023.
https://doi.org/10.1007/978-981-99-5968-6_8

very good detection Han Tao [3] et al. used a wavelet transform to process the vibration signal of the bearing to obtain the corresponding coefficient matrix and form a feature map, and then used CNN to identify and diagnose, which was experimentally verified to achieve a good fault identification rate. Wang [4] et al. used a particle swarm optimization algorithm to design the hyperparameters of CNN, including the CNN learning rate and the size of the convolutional kernel, to achieve adaptive selection of parameters in different application scenarios, increasing the model adaptation capability and achieving more than 94% diagnostic accuracy [5].

Most of the current convolutional neural network (CNN)-based fault diagnosis is to converts the one-dimensional vibration signal into a two-dimensional matrix by time-frequency domain transformation or segmentation combination, and obtains the fault information through the powerful feature extraction ability of CNN [6] to segments the one-dimensional vibration signal into several fixed-length sequence segments by a sliding window. The timing segment signals are directly fed into the 1D CNN, and the fault information is obtained by using the 1D CNN [7]. For high-frequency timing signals, a large amount of signal trend feature information and signal fluctuation feature information will be implied in a longer segment of the timing signal, and these features can reflect the operation status of the equipment at a deeper level [8]. However, because the signal needs to be divided into timing segments of certain width as the input of CNN, it will cause the CNN to lose the feature information of signal timing change during the training and learning process, so it is difficult to combine the fault information in the signal with time change in fault diagnosis. Most of the current research methods of fault diagnosis are based on single-source signals as the input information of fault diagnosis, which can achieve high recognition accuracy of fault diagnosis, but it is difficult to reflect the state information of the equipment in a comprehensive and multidimensional way, and cannot reflect the state information of the equipment comprehensively [9].

In this paper, we take bearing equipment as the research object and adopt a one-dimensional CNN network to conduct relevant research on bearing fault diagnosis, which not only takes the vibration signal as the original input, but also considers the multi-dimensional signal as the signal input for fault diagnosis to improve the accuracy of diagnosis and extract the equipment condition information more comprehensively.

2   MCCNN-1D Model Based on Multichannel Weight Adaptation

In this section, the fault diagnosis method based on multichannel weight adaptive CNN is described. First, a CNN-based data fusion strategy and adaptive network module structure are introduced, then a multichannel weighted adaptive data fusion method is proposed, and finally, the MCCNN-1D model is proposed based on this method for solving the problem of multidimensional signals as input sources in fault diagnosis.

2.1   Fault Diagnosis Strategy Based on CNN Data Fusion

As equipment monitoring becomes more difficult, the detected equipment signals show nonlinear, uncertain and irregular changes. A single signal source as a basis for diagnosis difficulty reflecting the status of the equipment in a comprehensive and multiangle manner, and data fusion technology can be a good solution to this problem, which can effectively use the signals collected by multiple sensors for fusion. The three types of data fusion are shown in Fig. 1, and the fault diagnosis strategy based on CNN data fusion is explained below.

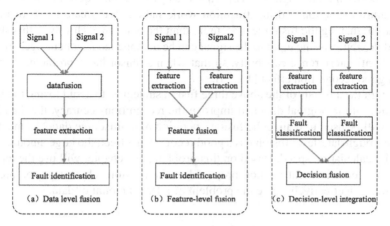

Fig. 1. Three fusion strategies.

Data-Level Fusion. Data-level fusion means directly fusing the signals acquired by sensors in some way and then using the fused information as input to a CNN for training and fault identification using CNN networks. Data-level fusion is the processing of signals acquired by multiple sensors by intercepting signal segments of the same length and then combining the information by connecting the signal segments in series or in parallel. The parallel method usually transforms the signals acquired by multiple sensors into single image information, and then performs image feature extraction based on a two-dimensional convolutional neural network to finally achieve diagnosis; the series method usually combines the signals from multiple sensors into a one-dimensional signal and then uses a one-dimensional convolutional neural network for feature extraction and fault diagnosis. Data-level fusion can lose the least amount of data so that the finer features in the fused data can be extracted and have a higher recognition rate, but due to the large amount of data, it will lead to a great increase in computation.

Feature-Level Fusion. Feature-level fusion is the fusion of the extracted features of the signals under multiple sensors. Different signal feature extraction methods are adopted according to different application areas, and the respective feature vectors are extracted from the signals of multiple sensors. Then, feature fusion is completed using CNN [10]. Commonly used signal feature analysis methods are used to extract features under multiple dimensions, including the time domain, frequency domain, and time-frequency domain, and then integrated into one-dimensional feature information or two-dimensional image feature information and sent to CNN for training and fault identification. Feature-level fusion realizes the dimensionality reduction of the signal, which can reduce the computational effort. However, due to the loss of some information of the original signal, the fusion performance will be degraded compared to data-level fusion [11].

Decision-Level Integration. Decision-level fusion extracts the signals under each sensor by CNN for features, performs fault identification, and then fuses the individual results to perform fault diagnosis [12]. This strategy has high requirements for the previous feature extraction and identification work and has the greatest loss of original information, making the fusion less accurate than the other two. However, this strategy has good flexibility as the input to the signal can be homogeneous or heterogeneous, and has good anti-interference capability, so that when a sensor has a problem, it can still obtain a good recognition effect [13, 14].

Through the analysis of the above three fusion strategies, to avoid losing the feature information of the original data and improve the recognition accuracy, the data fusion strategy will be used to realize the fusion operation of multisource data. As mentioned above, although data-level fusion has a good recognition effect, the large amount of data affects the calculation speed. Based on this problem, this chapter will use the temporal compression enhancement method proposed in the previous chapter to compress the data, which can effectively solve the problem of a large amount of data.

2.2 SENet Block Network Module Structure

SENet (Squeeze-and-Excitation Networks) is a network structure module proposed by Hu et al. in 2017, embedded in CNN and applied in the field of image recognition, and it can also be flexibly embedded into other network models, whose structure is shown in Fig. 2. The advantage of SENet is that it can learn adaptively through the training of the network to learn The advantage of SENet is that it can learn feature weights adaptively through the training of the network to differentiate the weight of the features and make the network more accurate. The four core steps of the SENet embedded CNN are described below.

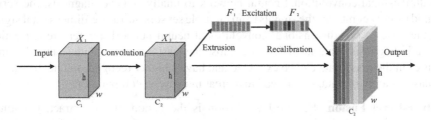

Fig. 2. SENET module structure.

Feature Extraction. The process of X_1 to X_2 in Fig. 2 is the feature extraction process, and the convolution operation in the CNN is implemented here. The formula is shown in Eq. (1).

Squeeze Operation. In Fig. 2, X_2 to F_1 is the squeezing process, the squeezing operation is extracted to C_2 feature channels, and then the global average pooling operation is used to calculate the feature values on each channel to get the global features under that channel, so that the feature map of $w * h * C_1$ can be mapped to the weighted features of $1 * 1 * C_2$ size, and the perceptual field of the CNN network can be increased at the same time, the formula is shown in Eq. (1).

$$Z_c = F_{sq}(u_c) = \frac{1}{W_{k*H}} \sum_{i=1}^{W} \sum_{j=1}^{H} u_c(i,j) \tag{1}$$

where represents each feature in the feature, and represent the 2D feature length and width, respectively, and represents the summation and averaging operation on the 2D feature values.

Incentive Operation. The extrusion process is shown in Fig. 2 from F_1 to F_2. After obtaining the global features, it is necessary to obtain the dependency relationships among the channels. Two points need to be satisfied, one is that the channel relations can be acquired by self-learning, and the second is that the learned relations are not mutually exclusive, because there are many channel features, so its structure adopts the sigmoid form of the gating mechanism. The important weights of each channel can be obtained after the excitation operation, and the nonlinear relationship between the learned channels is shown in formula (2).

$$s = F_{ex}(z, W) = \sigma(g(z, W)) = \sigma(W_2 \delta(W_1, z)) \tag{2}$$

where $W_1 \in R^{\frac{c}{r}*c}$, $W_2 \in R^{\frac{c}{r}*c}$. To enhance the generalization ability of the network and reduce the complexity of the network, its structure is a bottleneck structure with two fully connected layers, where one fully connected layer acts as a dimensionality reduction with a dimensionality reduction factor of r and is then activated using ReLU. The role of the second fully connected layer is to recover the original dimensionality.

Recalibration. After the excitation operation, the weight value of each channel is obtained, and Eq. (3) is the recalibration process based on the channel feature weights. That is, the above learned weight parameters are multiplied by the original features of each channel to calculate the new channel features. The SENet module essentially takes the input as the condition, adaptively learns the channel weight parameters during the learning process, and recalibrates the original feature weights to finally make each channel have different weights to improve the discriminative ability of the model.

$$x_c = F_{scale}(u^c, s) = s_c * u_c \tag{3}$$

where u_c indicates the features of each channel, s_c indicates the weight value of the acquired channels, and finally, the original features of each channel are multiplied by the weight of each channel to obtain the output.

2.3 Multichannel Weighted Adaptive Data Fusion Method

Based on the data-level feature fusion strategy, a multichannel weighted adaptive data fusion method is proposed in this section. First, this section explains the method in terms of the general idea. The method is considered from two aspects of data fusion methods and weights between channels, in which a data fusion method based on one-dimensional CNN multichannel superposition is used for data adaptive fusion, followed by a multichannel weight adaptive method for adaptive weight learning of each channel to more accurately and effectively fuse data information while extracting more critical data features.

General Overview of the Method. Convolutional neural networks were first applied in the field of image processing and have achieved great results in the field of recognizing RGB images. CNNs use RGB three pixel features as the input of three channels when processing image information. Inspired by CNNs in image recognition, this section proposes a method to apply multi-channel data superposition to the fusion of multi-source signals based on the characteristics of multichannel CNNs through multichannel superposition of a convolutional neural network for multisource feature fusion and feature extraction to increase the multiplicity of data, so that the CNN network can more comprehensively use the data under multiple features to learn and more comprehensively extract the state information of the equipment and better identify the fault information. Different data sources have different weights for feedback of device status. If we can make the network learn the feature weights under each channel adaptively in the learning process and consider the weights under different channels in fault diagnosis, it will greatly increase the accuracy of identification. Therefore, a multichannel weight adaptive method is also proposed, which embeds SENet into CNN and can establish the interdependence between each channel by using the characteristics of SENet, which can enhance the features that are beneficial to the network, suppress the features that are not important to the network, and adaptively learn the weights of each channel, Through the training of the model, the convolutional kernel adaptively learns to the multichannel By training the model, the convolutional kernel can adaptively learn the fused weight values of data sources under multiple channels, realize parameter weighting for the input features of each channel, and adaptively sense the importance of multisource timing data for device state recognition. Achieving more accurate classification.

Multichannel Superposition of the Number Fusion Method. The data fusion method is based on the multichannel feature of a convolutional neural network, which normalizes different data sources as the input of different channels, and can realize the feature extraction and fusion of different data sources by using the own characteristics of the convolutional network [15]. The multichannel convolution operation is shown in Fig. 3. Each channel corresponds to the input of a one-dimensional timing signal, each channel corresponds to a convolutional kernel, and the feature extraction and feature fusion of each channel is realized by a convolutional operation.

Based on the powerful feature extraction ability of the convolutional neural network, the input of each channel is first extracted to the corresponding features after the respective convolutional kernel completes the convolutional computation, and then the features extracted from each channel are accumulated to achieve data fusion, multiple fusion features are obtained, and multiple signal features continue to be used as the next convolutional layer of the CNN [16–18]. Then, the features extracted from each channel are accumulated to obtain multiple fused features, and the multiple signal features are used as the multichannel input of the next convolutional layer of the CNN to continue

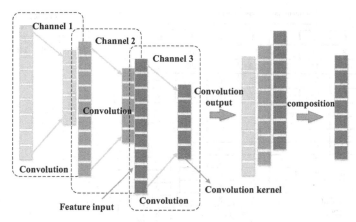

Fig. 3. Multichannel superposition data fusion schematic diagram.

the convolutional operation under each channel to finally achieve the purpose of feature extraction and fusion to realize fault diagnosis under signal fusion.

The information fusion process of multiple channels is shown in Eq. (4).

$$X_l = \sum_{j=1}^{m} f_i \left(\sum_{i=1}^{k} \left(x_{ij}^{l-1} * w_{ij}^{l-1} \right) + b_j^{l-1} \right) \tag{4}$$

where x_l denotes the output of the l-th convolutional layer; x_{ij}^{l-1} denotes the i-th feature input under the j-th channel in the $l-1$ convolutional layer, with k feature inputs; w_{ij}^{l-1} denotes the convolution kernel under the j-th channel in the $l-1$ convolutional layer; b_{ij}^{l-1} represents the bias of the $l-1$ channel j; $f_i(x)$ indicates the activation function of the $l-1$ th convolutional layer channel j; the number of channels is m; and the convolution operation of the convolutional layer effectively realizes the fusion of multichannel information and the extraction of key features.

It is assumed that five sensors are used to monitor the device, where the sensor types and sensor deployment locations are different. The acquired timing sequences are under the same time and there is an intrinsic connection between the timing sequences, so a five-channel CNN is used as the input of the initial signal. The computational process of multichannel superposition fusion is shown in Fig. 4. Each channel has a corresponding convolutional kernel, which is 2 * 1 with a step size of 2, and a complementary zero operation is used. The convolution operation is performed using m (m = 1, 2, 3....n) groups of convolution kernels (only the convolution kernels under one group are illustrated in the figure), and the fusion features are extracted to m groups of numbers using the convolution kernels of m groups, and the fusion features of m groups will continue the convolution fusion of the above multichannel superposition.

Multichannel Weight Adaptive Method. The SENet network module is embedded in the CNN network structure, and the SENet module is introduced after each convolutional layer [19]. That is, it can be realized that after each convolution operation to obtain channel features, the corresponding feature weights are calculated by the SENet network

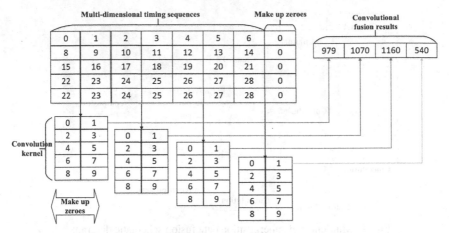

Fig. 4. Schematic diagram of the convolution process with multiple source inputs

module, and the features are rescaled and calculated to obtain the features with weights as the input of the next convolution layer [20, 21]. Figure 5 illustrates the flow structure of SENet embedded in a CNN network by embedding SENet into one layer of the CNN. Suppose the number of channels is c, the width is w, and the height is h. After the convolution operation of the convolution layer, the number of channels with features of c is obtained, and the weight values of the channels are obtained after the abovementioned squeezing and excitation operations. Finally, after the rescaling operation, the weights are weighted and multiplied channel by channel, and the new features are adopted as the input of the next layer, thus making multiple channels with different weights.

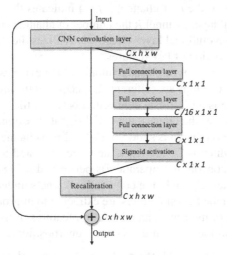

Fig. 5. Three common incentive functions.

where the feature value obtained after setting a certain layer of convolution calculation is SENet learning under multiple channels is divided into the following three steps:

Step 1: Forming the feature description of each channel by using the global features of each channel as the global feature representation under that channel;

Step 2: Generating a weight value for each feature channel pass, representing the importance of the feature channel;

Step 3: Multiply the weight value with the original feature to complete the rescaling of the original feature channel, and output the rescaled feature as the new feature value as the lower level feature. The network structure of the embedded SENet is shown in Fig. 6.

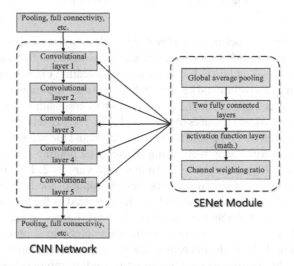

Fig. 6. Schematic diagram of SENet embedded CNN structure

Generally, the output of the convolutional layer does not consider the weight of each channel and the dependency relationship between channels, but only convolves each channel. By embedding SENet after each convolutional layer of the CNN, the channels in each layer have their own weights, so that the CNN can learn the dependencies of each channel during the learning process by using the SENet module, and extract features from the channel data under different data sources in a selective manner, thus increasing the accuracy of feature fusion in the network.

2.4   MCCNN-1D Model Structure

Based on the method proposed in Sect. 2.2, this section proposes the MCCNN-1D (Convolution Neural Network with Multichannel, MCCNN-1D) network model, which adaptively fuses and extracts more comprehensive features of multisource signals through a multichannel one-dimensional CNN network, combined with the powerful feature extraction capability of CNN. The structure of the MCCNN-1D model is shown in Fig. 7.

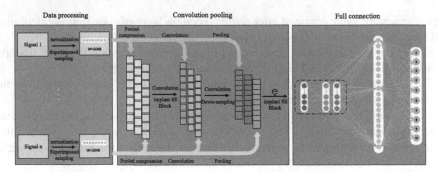

Fig. 7. MCCNN-1D network structure.

The MCCNN-1D network uses two levels of pooling layers to compress and down-scale the signals, and the MCCNN-1D network is connected with four convolutional layers after the two pooling layers for data fusion and feature extraction [22]. The first layer still uses a large convolutional kernel (16 * 1) to extract the temporal features of the signal, while the other convolutional layers use a small convolutional kernel (3 * 1) to extract the high-dimensional features of the signal [23]. The SENet module is embedded between each convolutional layer and the pooling layer, and the weights of each channel of the network are learned adaptively using SENet. The pooling layers are connected after the SENet module, and the pooling window widths are all 2. The last pooling layer is connected to the fully connected network, and the input of the fully connected layer is 100, as shown in Fig. 7. Finally, the Softmax layer is used as the last layer of the fully connected layer as the probabilistic output of the final diagnostic recognition results. Since multiple channels lead to more network parameters, to suppress overfitting, a dopout function is added, and the neurons in the network are set to fail with a probability of 0.40 to reduce the number of parameters, improve the training speed, and prevent overfitting. All the convolutional layers are edge-zeroed with the SAME function, and the hyperparameters of the MCCNN-1D model are set as shown in Table 1.

3 Experiment and Analysis

In this section, the abovementioned model is validated and three sets of comparison experiments are used to validate the analysis of the MCCNN-1D model, thus verifying the feasibility of the method proposed in this chapter.

3.1 Purpose of the Experiment

The performance of the MCCNN-1D model under multichannel input fusion is analysed by comparing experiments with multisource signals as inputs with single-source signals for inputs. The MCCNN-1D is also compared with and without the SENet module to verify the effect of SENet under multichannel weight adaptation, and the feasibility of the proposed method in this chapter is verified by the above experimental analysis.

Table 1. .

Serial number	Network Level	Convolution kernel size/step size	Number of input channels	Number of output channels
1	Level 1 average pooling layer	2 * 1 * 3/2 * 1	3	3
2	Secondary maximum pooling layer	2 * 1 * 3/2 * 1	3	3
3	Convolutional layer 1	16 * 1 * 3/8 * 1	3	8
4	Pooling layer 1	2 * 1 * 8/2 * 1	8	16
5	Convolution layer 2	3 * 1 * 16/1 * 1	16	32
6	Pooling layer 2	2 * 1 * 32/2 * 1	32	64
7	Convolution layer 3	3 * 1 * 64/1 * 1	64	128
8	Pooling layer 3	2 * 1 * 128/2 * 1	128	128
9	Convolution layer 4	3 * 1 * 128/1 * 1	128	64
10	Pooling layer 4	3 * 1 * 64/2 * 1	64	64
11	Fully connected layer	100	1	100 * 1
12	Softmax	10	1	10

3.2 Experimental Data Set

Data set 1: The CWRU bearing data set is still used, the acceleration sensors are deployed to different positions, and the vibration signals under different orientations are collected, divided into base end (BA), drive end (DE), and fan end (FE), The data set at 48 kHz is used, and 1000 sample data are collected for each state type, of which 70% of the samples are used for training and 30% of the samples are used for testing, The signals at the three positions constitute a multisource signal. As shown in Table 2. And all four loads are sampled, and the datasets under 0 hp, 1 hp, 2 hp and 3 hp are coded as A, B, C and D.

Data set 2: Vibration signals collected by acceleration sensors from the planetary gearbox equipment at SE, where the vibration signals include three directions (x, y and z). There are three fault states pitting, broken teeth and wear. The experiment simulates 10 different conditions at three speeds with 1000 samples of each fault type, including 500 vibration signals and 500 sound signals, as shown in Table 3. Seventy percent of the samples are used for training and 30% for testing.

Table 2. .

Injury site		Normal	Rolling body			Inner Circle			Outer ring		
Tags		1	2	3	4	5	6	7	8	9	10
Damage diameter(inch)		0	0.007	0.014	0.021	0.007	0.014	0.021	0.007	0.014	0.021
De (A/B/C/D)	Training	700	700	700	700	700	700	700	700	700	700
	Testing	300	300	300	300	300	300	300	300	300	300
BA (A/B/C/D)	Training	700	700	700	700	700	700	700	700	700	700
	Testing	300	300	300	300	300	300	300	300	300	300
FE (A/B/C/D)	Training	700	700	700	700	700	700	700	700	700	700
	Testing	300	300	300	300	300	300	300	300	300	300

Table 3. .

Injury site		Normal	Pitting			Inner Circle			Outer ring		
Tags		1	2	3	4	5	6	7	8	9	10
Damage diameter(inch)		0	0.007	0.014	0.021	0.007	0.014	0.021	0.007	0.014	0.021
Vibration - x-axis	Training	350	350	350	350	350	350	350	350	350	350
	Testing	150	150	150	150	150	150	150	150	150	150
Vibration - y-axis	Training	350	350	350	350	350	350	350	350	350	350
	Testing	150	150	150	150	150	150	150	150	150	150
Vibration-z-axis	Training	350	350	350	350	350	350	350	350	350	350
	Testing	150	150	150	150	150	150	150	150	150	150

3.3 Experimental Protocol and Analysis of Results

In this section, we will first analyse the performance of multisource signal fusion by comparing three single-source data signals with multisource signals under dataset 1, and analyse the performance of MCCNN-1D with and without the embedded SENet module under dataset 1, and analyse the generalization ability of the model by experimenting under four loads of dataset 1 and under different datasets, through the above three aspects. The performance of the MCCNN-1D model is used to verify the feasibility of the multichannel weighted adaptive data fusion method. The experimental protocols and results analysis of the three experiments are described in detail below.

Experiment 1: Performance Comparison Under Multichannel Data Source Fusion and a Single Data Source. In this experiment, the results of different data sources under a single channel are compared with those under multichannel data source fusion. Group An in dataset 1 is used for the experiments, and DE, BA, and FE single signals and three fusions (called MULs) are used as inputs for comparison experiments to analyse the performance of the MCCNN-1D model under multichannel fusion. The batch size was set to 150, the initial learning rate of Adam was set to 0.001, the input width was set to 4096, 30 rounds of training were performed, 10 sets of reexperiments were performed, and the average accuracy was obtained, as shown in Table 4.

Table 4. .

Data source	Accuracy rate (%)	Loss rate	Training time (s)
DE	0.9685	0.0390	201
BA	0.9780	0.0419	198
FE	0.9698	0.0435	210
MUL	0.9889	0.0371	361

The trend of training accuracy over time for three single-signal versus multisource signal experiments recorded simultaneously is shown in Fig. 8 below, and one of the 10 experiments is shown and illustrated.

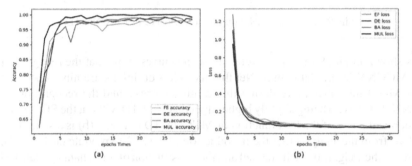

Fig. 8. Single source and fusion signal comparison experimental results.

As shown in Fig. 8 (a), the fault recognition rates of DE, FE, and BA under all three single signals are significantly lower than those of the three after multichannel fusion. The accuracy rate of MUL training in the figure rises faster, rises more smoothly and fluctuates less. From Fig. 8 (b), it can be seen that the rate of decrease of the loss rate under the MUL fusion signal is faster and the loss is smaller, which means that the global minimum can be found faster. This experiment shows that MCCNN-1D uses multichannel fusion to extract more comprehensive device state features, and multiple sources can play a complementary role, which makes the model more accurate. The average training time in Table 4 shows that the training time for MUL data is longer because the increase in the number of channels increases the model parameters, which leads to an increase in training time, and an increase in the amount of training data due to the multisource signals as input. It also leads to an increase in training time. In summary, although the complexity of the model increases and the training time grows, the overall effect achieves some improvement.

Experiment 2: SENet Module Performance Analysis on MCCNN-1D. This experiment compares MCCNN-1D with a network model that removes the SENet module from MCCNN-1D (referred to as M-1D) to analyse the impact of the SENet module on the model performance. The experiments were conducted using group An in dataset

1 and with MUL fusion data, the batch size was set to 150, the initial learning rate of Adam was set to 0.001, the input width was set to 4096, and 30 rounds of training were conducted, and the results are shown in Fig. 9.

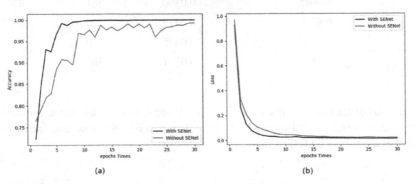

(a)   (b)

Fig. 9. Effect of SENet on the performance of MCCNN-1D.

As shown in Fig. 9 (a), the waveforms of the curves show that the recognition rate of the MCCNN-1D model with SENet fluctuates less during the training process, while that of M-1D fluctuates more during the training process, and the recognition rate of MCCNN-1D ends up being slightly higher than that of M-1D without the SENet module. The decreasing rate of the loss function of MCCNN-1D in Fig. 9 (b) is also more rapid. It is clear from the analysis that the introduction of the SENet module enables adaptive learning of the weights of each channel and the construction of interchannel relationships, thus enabling the MCCNN-1D model to quickly filter the noncritical information through the weights during the learning process and effectively focus on training and learning, resulting in less fluctuation during the training process.

Experiment 3: MCCNN-1D Model Generalization Capability Analysis. To validate the generalization ability of the MCCNN-1D model, the following experiments will be conducted: Experiment a. Experiments are made under dataset 1 for four sets of datasets under loads A, B, C and D to validate the generalization ability under different loads. Experiment b. Experimental analysis is performed on MCCNN-1D under dataset 2 to verify the generalization ability under different datasets.

Experiment a was conducted using four datasets under four loads, A, B, C and D, with DE, BA, FE single signal as input and MUL multisignal as input, with batch size set to 150, initial learning rate of Adam set to 0.001 and input width set to 4096, for 30 rounds of training and 10 sets of re-experiments, and average accuracy was obtained. The experimental results are shown in Fig. 10.

As shown in Fig. 10, the MCCNN-1D achieves a high recognition rate under four different working conditions of load and maintains a certain stable range, while all are higher than the recognition accuracy under a single signal, indicating that the MCCNN-1D has a good generalization ability and can still have a high recognition rate under multiple working conditions with a certain stability.

Experiment b uses the three reversed vibration signals from data set 2 as the input under multichannel. The signals in each of the three directions were separately used as

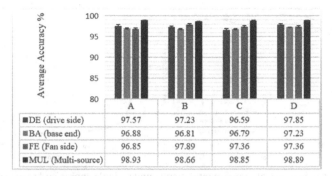

	A	B	C	D
■DE (drive side)	97.57	97.23	96.59	97.85
■BA (base end)	96.88	96.81	96.79	97.23
■FE (Fan side)	96.85	97.89	97.36	97.36
■MUL (Multi-source)	98.93	98.66	98.85	98.89

Fig. 10. Performance analysis under the single source and fusion signal under different loads.

inputs as a comparison experiment, and the experimental results are shown in Table 5. Table captions should be placed above the tables.

Table 5. .

Data source	Accuracy rate (%)	Loss rate	Training time (s)
Vibration - x-axis	0.9630	0.0518	301
Vibration - y-axis	0.9599	0.0489	318
Vibration-z-axis	0.9628	0.0535	325
MUL	0.9839	0.0471	420

The trend of training accuracy over time for three single-signal versus multisource signal experiments recorded simultaneously is shown in Fig. 11 below, and one of the 10 experiments is shown and illustrated.

From the result data in Table 5. Table captions should be placed above the tables., it can be seen that under experimental data set 1, all three single signals have a high recognition accuracy, but the recognition rate is significantly improved under the fusion of the three signals, with an average accuracy rate of 98.39%, which indicates that the fusion model can learn multiple signal features, thus making the recognition better.

As shown in Fig. 4.11, under the test of experimental data set 2, the accuracy of the single vibration signal in all three directions as a single channel input is not as high as the accuracy of the vibration signal in all three directions as a three-channel fusion input of MCCNN-1D. By proving the feasibility of the method under different datasets, it also proves that MCCNN-1D has a certain generalization ability. The high diagnostic recognition rate under different datasets still shows that the model has good generalization ability.

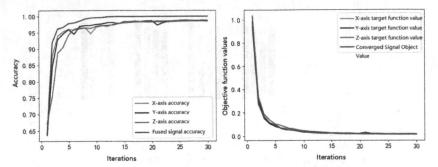

Fig. 11. Comparison of experimental results under single source and fused signals.

4   Conclusion and Future Work

In this paper, we propose a CNN fault diagnosis method based on multichannel weighting to solve the above problems. In this paper, a CNN fault diagnosis method based on multichannel weight adaptation is proposed to solve the above problems. First, based on the multichannel feature of CNN, different data sources are normalized as the input of different channels, and the features of convolutional networks can be extracted and fused with different data sources by using their own characteristics. Then the SENet module is introduced and embedded into the CNN, which adaptively learns the weights of each channel to improve the accuracy of classification. The method is then used to design the MCCNN-1D network for application in fault diagnosis. Finally, through comparison experiments, it is demonstrated that the method can further improve the fault identification accuracy.

As the fault diagnosis method proposed in this paper has some details that need to be improved and supplemented, the future research directions and expected results are summarized as follows.

As the equipment will collect considerable noisy data from the sensors under complex working conditions, the fault diagnosis of the equipment under noise needs to be the next research step to increase the generalization capability of the network model.

The fault diagnosis methods studied in this paper are based on supervised learning, and unsupervised fault identification methods based on them are the focus of future work.

Acknowledgements. This work is financially supported by: The National Key R&D Program of China (No. 2020YFB1712600); The Fundamental Research Funds for Central University (No. 3072022QBZ0601); and The National Natural Science Foundation of China (No. 62272126).

References

1. Zheng, J.D., Pan, H.Y., Yang, S.B., et al.: Generalized composite multiscale permutation entropy and Laplacian score based rolling bearing fault diagnosis. Mech. Syst. Signal Process. **99**(15), 229–243 (2018)

2. Rui, Z., Yan, R.Q., Chen, Z.H.: Deep learning and its application to machine health monitoring: a survey. IEEE Trans. Neural Netw. Learn. Syst.
3. Han, T., Yuan, J.H., Tang, J., et al.: Intelligent composite fault diagnosis method for rolling bearings based on MWT and CNN. Mech. Trans. 12(4), 139–143 (2016)
4. Wang, F., Jiang, H., Shao, H., et al.: An adaptive deep convolutional neural network for rolling bearing fault diagnosis. Meas. Sci. Technol. 28(9), 095005 (2017)
5. Yang, B., Liu, R., Chen, X.: Fault diagnosis for a wind turbine generator bearing via sparse representation and shift-invariant K-SVD. IEEE Trans. Ind. Inf. 13(3), 1321–1331 (2017)
6. Wen, L., Li, X.Y., Gao, L., et al.: A new convolutional neural network-based data-driven fault diagnosis method. IEEE Trans. Industr. Electron. 65(7), 5990–5998 (2018)
7. Wu, D.H., Ren, G.Q., Wang, H.G., Zhang, Y.Q.: The review of mechanical fault diagnosis methods based on convolutional neur. J. Mech. Strength 42(05), 1024–1032 (2020)
8. Wang, Z., Zhang, Z., Qin, J., Ji, C.: Fault diagnosis technology based on convolutional neural network. J. Comput. Appl. 42(04), 1036–1043 (2022)
9. Shi, F., Cao, H.R., Wang, Y.K., et al.: Chatter detection in high-speed milling processes based on ON-LSTM and PBT. Int. J. Adv. Manufact. Technol. 111(11–12), 3361–3378 (2020)
10. Lan, C.F., Li, S.J., Chen, H., et al.: Research on running state recognition method of hydro turbine based on FOA-PNN. Measurement 169, 108498 (2021)
11. Huang, J., Wei, L.J.: Research on gear fault diagnosis based on synchronous compression cross wave. Comput. Measurement Control 28(11), 41–44+49 (2020)
12. Zhu, J., Deng, A.D., Li J., et al.: Resonance-based sparse improved fast independent component analysis and its application to the feature extraction of planetary gearboxes. J. Mech. Sci. Technol. 34(11), 4465–4474 (2020)
13. Yang, X.Z., Zhou, J.X., Deng, J.M.: Fault diagnosis system of motor bearing based on improved Bayesian classification. Mach. Tool Hydraulics 48(20), 172–175 (2020)
14. Chen, X.C., Feng, D., Lin, S.: Mechanical fault diagnosis method of high voltage circuit breaker operating mechanism based on deep auto encoder network. High Volt. Eng. 46(09), 3080–3088 (2020)
15. Zhang, S.S., Zhang, T.: Sensor fault diagnosis method based on CGA-LSTM. In: Proceedings of the 13th Annual China Satellite Navigation Conference, pp. 1–5 (2022)
16. Gao, S.C., Li, X.P., Zhang, W.: Research on key technologies of PHM for aerospace complex system based on data driven. In: Proceedings of the 3rd Academic Conference on Systems Engineering, pp. 220–227 (2022)
17. Liu, Y., Yin, C.H., Hu, D., Zhao, T., Liang, Y.: Fault Detection of communication satellite based on cyclic neural network. Comput. Sci. 47(02), 222–232 (2020)
18. Liu, Q., Rong, L.L., Yu, K.: Public opinion evolution model of microblog network considering the influence of multilayer neighbor nodes. J. Syst. Eng. 32(06), 721–731 (2017)
19. Li, B., Chow, M.Y., et al.: Neural-network-based motor rolling bearing fault diagnosis. IEEE Trans. Industr. Electron. 47(5), 1060–1069 (2000)
20. Wu, C.Z., Jiang, P.C., Feng, F.Z., et al.: Gearbox fault diagnosis based on one-dimensional convolution neural network. J. Vibr. Shock 37(22), 51–56 (2018)
21. Quan, W., Wang, K., Yan, D., et al.: Distinguishing between natural and computer generated images using convolutional neural networks. IEEE Trans. Inf. Forensics Secur. 13(11), 2772–2787 (2018)
22. Ferreira, V.H., Zanghi, R., Fortes, M.Z., et al.: A survey on intelligent system application to fault diagnosis in electric power system transmission lines. Electr. Power Syst. Res. 136, 135–153 (2016)
23. Zhang, R.T., Chen, Z.G., Li, B.B., Jiao, B.: Research on gearbox fault diagnosis based on DCNN and XGBoost algorithm. J. Mech. Strength 42(05), 1059–1066 (2020)

Big Data Management and Applications

Design and Implementation of Key-Value Database for Ship Virtual Test Platform Based on Distributed System

Qingyu Meng[1], Kejia Zhang[1(✉)], Haiwei Pan[1], Maocai Yuan[2], and Baoying Ma[1]

[1] College of Computer Science and Technology, Harbin Engineering University, Harbin, China
kejiazhang@hrbeu.edu.cn
[2] China Ship Scientific Research Center, Wuxi, China

Abstract. The virtual test platform is a vital tool for ship simulation and testing. However, the numerical pool ship virtual test platform is a complex system that comprises multiple heterogeneous data types, such as relational data, files, text, images, and animations. The analysis, evaluation, and decision-making processes heavily depend on data, which continue to increase in size and complexity. As a result, there is an increasing need for a distributed database system to manage these data. In this paper, we propose a Key-Value database based on a distributed system that can operate on any type of data, regardless of its size or type. This database architecture supports class column storage and load balancing and optimizes the efficiency of I/O bandwidth and CPU resource utilization. Moreover, it is specifically designed to handle the storage and access of large files. Additionally, we propose a multimodal data fusion mechanism that can connect various descriptions of the same substance, enabling the fusion and retrieval of heterogeneous multimodal data to facilitate data analysis. Our approach focuses on indexing and storage, and we compare our solution with Redis, MongoDB, and MySQL through experiments. We demonstrate the performance, scalability, and reliability of our proposed database system while also analysing its architecture's defects and providing optimization solutions and future research directions. In conclusion, our database system provides an efficient and reliable solution for the data management of the virtual test platform of numerical pool ships.

Keywords: Key-Value databases · multimodal data fusion · heterogeneous data · distributed systems · columnar-like storage · indexing

1 Introduction

The ship industry is a modern comprehensive industry and a strategic industry combining military and civilian aspects, which has an inevitable demand for improving the comprehensive strength of the country, accelerating the pace of marine development, safeguarding national maritime rights and interests, guaranteeing the safety of water transportation, maintaining the growth of the national economy and ensuring the safety of national defense. However, ship research faces the problem that the cost is too high due to high test costs and variable sea environments.

Z. Yu et al. (Eds.): ICPCSEE 2023, CCIS 1879, pp. 109–123, 2023.
https://doi.org/10.1007/978-981-99-5968-6_9

To solve this problem, China Numerical Pool launched a ship virtual test system that can not only complete algorithms in various environments but also generate test reports automatically and realize automation and intelligence of virtual tests for different ship types and virtual test environments to meet the needs of industrial application accuracy. However, in the process of conducting virtual tests, the system generates a large amount of data to be managed, such as user names, hull parameters, currents, wind speed, waves, test reports, pictures, etc. A large number of heterogeneous data types, such as relational data, files, text, images, videos and animations, are generated. The storage and management of these data is a challenging problem for distributed systems [1, 2]. To solve this problem, key-value databases are ideal due to their simplicity, scalability and flexibility. However, it is still challenging to design a distributed deployment of a key-value database that can handle multiple data types and support high availability and consistency [3–5].

With the rapid development and wide application of multimedia technology, various unannotated data in multimedia applications are expressed through various modes [6, 7]. More abundant information can be obtained by clustering multi-modal data [8–12]. This pose a challenge for storing multimodal data. If we consider the storage of multimodal data at the underlying level, it would greatly facilitate data analysis.

To address this problem, this thesis presents the design of a key-value database designed to handle heterogeneous data types and meet the requirements of distributed systems, which we call heterogeneous multimodal data fusion. The database can store and retrieve any type of data, including relational data, files, text, images, videos, and animations. At the same time, the database supports multicopy consistency, high availability and ACID constraints. To reduce write amplification and optimize I/O bandwidth and CPU resources, the database uses column-like storage. Additionally, an indexing mechanism is introduced to achieve efficient data retrieval and to support the concatenation of multiple values of a single entity. We also propose a mechanism to concatenate multiple interpretations of the same substance (e.g., ship descriptions), which we call heterogenous multimodal data fusion.

In summary, the research in this thesis aims to provide an applicable database management scheme for the numerical pool virtual test system to efficiently store and manage the large amount of heterogeneous data generated by the system. In this paper, we review the related work on key-value databases and distributed systems in Sect. 2. In Sect. 3, we will describe the requirements of the key-value database and propose a design approach to meet these requirements. In Sect. 4, we will experimentally evaluate our design and demonstrate its efficiency, scalability, and reliability. Finally, in Sects. 5 and 6, we summarize our research and discuss future work.

2   Related Work

Both relational and nonrelational databases can be used to manage the data generated by virtual experiments on numerical pools. Relational databases are traditional database types that use a tabular approach to storing data and organizing it in the form of rows and columns, such as MySQL and Oracle. Nonrelational databases, on the other hand, are relatively new database types that do not use a tabular format but instead use a variety of different data models to store data, such as key-value pairs, documents, graphs, etc., such as MongoDB and Redis [13]. Therefore, which database type is more suitable for managing the data generated by virtual experiments on numerical pools? We have performed the following research.

In ship virtual trials, data types include files, images, text and animations, and the data size may be very large, reaching several GB. Such data size and type can pose challenges for processing and storage in relational databases, which store data in tabular form and are less suitable for storing large unstructured data [13, 15].

In contrast, nonrelational databases (NoSQL) have greater advantages in storing and processing large unstructured data because they were originally designed to store unstructured or semistructured data. Both document-based databases and columnar databases can handle various types of data well and do not require predefined table structures such as relational databases.

In addition, with the continuous development of big data technology, the processing speed of nonrelational databases is increasing, and there are good solutions for the needs of large-scale data storage and processing capabilities, such as distributed storage and computing.

Therefore, in combination with data type and size considerations, nonrelational databases are more suitable for managing and processing large unstructured data in ship virtual tests. The data generated by numerical pool virtual tests are diverse, including files, images, text, animations, etc. Key-value databases are widely used in various applications due to their simplicity, scalability and flexibility. Many key-value databases, such as Redis, Cassandra, and Riak, have been developed to handle large amounts of data and provide high availability and fault tolerance. However, these databases are designed for similar data types, such as strings, integers, and lists. They do not support multiple data types, such as files, images, and videos.

To handle heterogeneous data types, several key-value databases have been developed, such as MongoDB and Couchbase. These databases support a variety of data types, including relational data, files, text, and images. However, they are not designed for distributed deployments with multicopy consistency and high availability. They also do not support ACID constraints [13].

In the field of distributed systems, a number of approaches have been proposed to ensure consistency and high availability in distributed databases. Two-phase commit (2PC) and three-phase commit (3PC) are widely used protocols for distributed transaction management. However, they suffer from high latency and low availability. Paxos and Raft are consensus protocols that ensure the consistency and high availability of distributed systems. However, they are complex to implement and require a high degree of coordination between nodes.

MongoDB is suitable for storing larger files but does not support transaction operations, while Redis is suitable for storing small data and transaction operations, but there are limitations for larger files. Therefore, in the application of the storage of virtual experimental data of numerical pools, whether it is a relational database or a nonrelational database that is more widely used in the market, it cannot meet the unified storage of heterogeneous data [13, 14]. In this regard, we dissect the database from both indexing and storage aspects and propose an MDF (multimodal data fusion) architecture to solve the problem of unified operation of such multimodal data generated by numerical pool virtual experiments. We also propose a mechanism to link multiple interpretations of the same substance (e.g., ship descriptions) together, e.g., a ship document, which may have text, table, file, picture, etc., representations, needs to conveniently invoke other interpretations based on one of them, and we call this mechanism heterogenous multimodal data fusion.

3   Methodology

As shown in 错误!未找到引用源。, MDF uses a hash table as the basic unit for storing data and uses a hash table to store and index key-value data. A hash table is a data structure that stores data in the form of key-value pairs, which maps keys to a bucket B_i by a hash function, and each bucket holds a chain table for storing conflicting key-value pairs. The advantage of a hash table is that it can achieve constant level $O(1)$ lookup and insertion operations, but it also has the problem of hash conflict, i.e., multiple different key values are mapped into the same bucket, and we resolve the conflict through a chain table (Fig. 1).

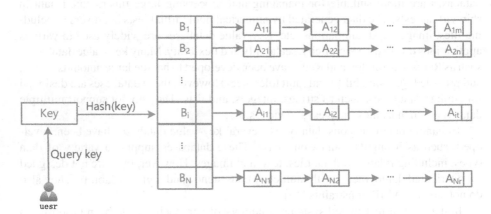

Fig. 1. The basic unit hash table of MDF

In MDF, the hash table is an important data structure that is widely used to store key-value data and is also the basis for MDF to implement other data structures. MDF uses multiple hash tables to implement a database instance, and each hash table can hold multiple key-value pairs. Different from key-value databases on the market, the MDF hash table does not store simple key-value pairs but judges based on the space occupied by the value. When the space occupied by the value is less than a certain threshold ∂, the stored A_{ij} is the offset address of the disk block and the block, and the address points to the value we want to find; however, when the space occupied by the value exceeds ∂, information such as the file name or path is stored, and then the distributed file is accessed by the system. At this time, the client sends the stored information to the metadata service of the file system. After receiving the query request, the metadata service will locate the data node storing the file according to the file name or path. The metadata service sends a query request to the data node, requesting the data node to return the metadata information of the file. After receiving the query request, the data node will search for the metadata information of the file and return it to the metadata service. After the metadata service receives the metadata information returned by the data node, it returns the result to the client.

As shown in 错误!未找到引用源。. In MDF, a corresponding hash table is created for each type of data, and the hash table is used as the basic unit. For example, strings, pictures, files, and animations all have their own hash tables. As long as the space occupied by the value is less than ∂, the value A_{ij} stored in the linked list corresponding to the hash table is the disk block and block offset address. When the occupied space reaches a certain ∂, information such as file name or path is stored, which is used to access the distributed file system. The hash table composed of all the data of each user constitutes a hash table bucket. A user's hash table buckets should be as close as possible on disk so that they can be read into memory at one time (Fig. 2).

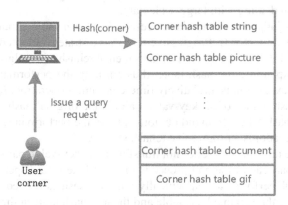

Fig. 2. Hash table bucket of user corner

In MDF, the size of the hash table is dynamically adjusted. When the load factor of the hash table ($If = n/m$, where n represents the number of elements in the table and m represents the size of the hash table) exceeds the threshold, MDF will automatically double the size of the hash table and restart Remap all key-value pairs into new buckets. Similarly, when the load factor is too small, MDF will automatically reduce the size of the hash table to reduce memory waste.

In the implementation, the hash table in MDF consists of an array and several linked lists. Each element in the array is called a bucket, and each bucket stores the head pointer of the linked list. When multiple keys have the same hash value, they will be put into the same bucket and stored through a linked list.

When a key-value pair needs to be queried, the MDF will first calculate the hash value of the key and then use the hash value to locate the corresponding bucket, traverse the linked list in the bucket, and find the corresponding A_{ij}. When A_{ij} is the disk address, we call the corresponding disk block to fetch the value; otherwise, we send it to the distributed file system to query the data.

To improve the efficiency of the hash table, MDF uses some optimization techniques in the hash table, such as:

- When expanding the hash table, MDF will preallocate a certain amount of space to avoid frequent memory allocation and fragmentation.
- MDF uses the MurmurHash2 algorithm as the hash function, which has a lower hash collision rate and faster hash calculation speed.
- MDF maintains a time round of the LRU approximation algorithm for each hash table, which is used to clear key-value pairs that have not been used for a long time to reduce memory usage.
- For each hash table, MDF maintains a meta-information structure of the hash table, including information such as the size of the hash table and load factor, to facilitate the operation and management of the hash table.

In short, MDF uses a hash table to store and index key-value data. It has efficient lookup and insertion performance and supports optimization techniques such as dynamic resizing and preallocation of space. The use of these technologies makes MDF have better performance when dealing with large-scale data.

In MDF, it is judged whether the corresponding hash table should be expanded according to the load factor If. Progressive rehash is a strategy adopted by MDF when expanding the hash table, also known as incremental rehash. This algorithm can avoid blocking when expanding the hash table, thus ensuring the performance of MDF. In MDF, hash table expansion is a relatively time-consuming operation. It is necessary to recalculate the hash values of all key-value pairs in the original hash table and assign them to the new hash table. To avoid obvious jitter in the performance of MDF during this process, MDF adopts a progressive rehash strategy.

In progressive rehash, MDF does not transfer all the key-value pairs at one time but only transfers a part each time and proceeds in several steps to disperse the impact of operations on MDF performance. Specifically, when the hash table needs to be expanded, the MDF will store the original hash table and the new hash table at the same time, and then each time a request is processed, a part of the key-value pairs in the original hash table will be moved to the new hash table. This process may take a certain amount of time, but it can be gradually completed during the running of MDF, avoiding the performance impact of large-scale hash table expansion at one time.

The main idea of progressive rehash is to decompose the expansion process of the hash table into multiple small steps and only migrate a small part of the data in the hash table each time. Specifically, incremental rehash includes the following steps:

1. Apply for memory space for the new hash table, and set the expansion flag to 1, indicating that the hash table is being expanded.
2. Sequentially migrate some key-value pairs from the old hash table to the new hash table. This step will not affect the client operation because the key-value pairs in the new hash table will not be exposed to the client.
3. The new hash table is used as the current hash table, and the expansion flag is set to 0, indicating that the expansion is complete. At this time, if the client operates, it will process the old hash table and the new hash table at the same time to ensure data consistency.

4. In the subsequent time, MDF will continue to migrate the data in the old hash table to the new hash table in the same way until all the key-value pairs in the old hash table have been migrated. During this period, the old hash table and the new hash table coexist.

It can be seen that the progressive rehash decomposes the hash table expansion process into multiple small steps, migrating only a portion of data at a time, which can avoid blocking when performing hash table expansion. This algorithm can effectively improve the performance and stability of MDF.

MDF contains a large amount of heterogeneous source multimodal data; for example, for a ship file, the test results may contain data in various forms, such as text, images, and animations. All these data are used to describe the same ship file from different perspectives. To facilitate future data analysis, we want to link these different forms of data together like a column store, as shown in 错误!未找到引用源。and can be retrieved together. We refer to this process as heterogeneous multimodal data fusion (Fig. 3).

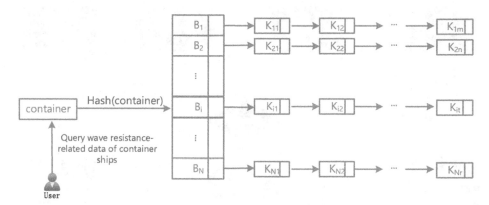

Fig. 3. Heterogeneous Multimodal Data Fusion

To integrate heterogeneous multimodal data, we still use hash tables for storage. Different from the previous representation of A_{ij}, we use the keyword Key, which is represented as K_{ij} in the figure. When the user needs to inquire about wave resistance-related data about container ships, the system will query the hash table to obtain the corresponding name, calculate and locate a bucket in the hash table through hash function calculation, and then in the linked list of the bucket to obtain the key of the heterogeneous multimodal data, the system queries the hash table again to obtain the values of various modalities.

Since all indexes of a user are stored in a hash table bucket, they can be loaded into the main memory at once. When users frequently operate data, the hash table bucket corresponding to the user will be stored in the main memory, which means that no matter how many queries are performed, only one I/O operation is required to retrieve the index. For the value corresponding to the key, if the space occupied is less than ∂, then its A_{ij} is the disk address pointed to. The data of the same user are stored close to the disk, which can reduce the seek time of the disk, and here, RAID can be used for parallel operation.

When the space occupied by value is greater than ∂, its A_{ij} is connected to the distributed file systems. The general flow chart is shown in 错误!未找到引用源。(Fig. 4).

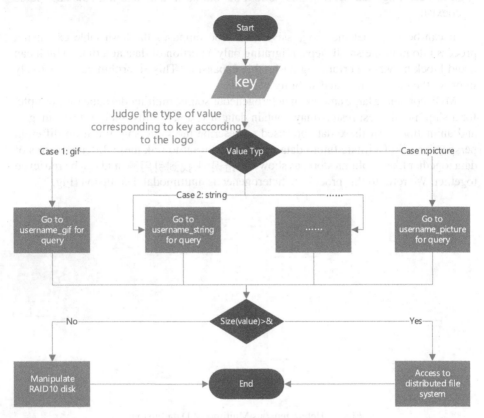

Fig. 4. The general flow chart

In short, for this kind of heterogeneous multimodal data, we can quickly obtain data through parallel operations and adopt certain strategies on the disk to make its performance close to that of column storage.

MDF also provides fault recovery and persistence functions. In this paper, MDF mainly implements this function in the following ways.

- The disk snapshot will periodically store the data snapshot in memory to the disk to form an MDB file. After the MDB mechanism is enabled, the MDF will perform a snapshot operation at regular intervals and write the data in the memory into a temporary file. After the snapshot is executed, the MDF will replace the temporary file with the previous snapshot file. If the server shuts down unexpectedly, we can use the latest MDB file to restore the state of the MDF server.
- Write append, which is to write all the write commands executed by the MDF server into a log file on the disk in an appended way. This file saves all the write commands

executed by the MDF server and can be reconstructed by playing back the commands in the file out the dataset.

- RAID disk array, we will use RAID10 to ensure the efficiency and reliability of storing small data.
- Distributed file system. Large files will be connected to the distributed file system. The experiment in this paper is Hadoop; of course, other distributed file systems can also be selected. To ensure data security and reliability. What kind of files will be connected to the distributed file system depends on the threshold ∂ we set.

4 Experiments

4.1 Lab Environment

CPU model and quantity: Intel(R) Xeon(R) CPU E5-2620 v2 @ 2.10.
Memory: 32 GB.
Storage: 1 TB.
Network interface card: 9.8 GB.
Operating System: Ubuntu 18.04.6 LTS.
Compiler: GCC (Ubuntu 7.5.0-3ubuntu1 ~ 18.04) 7.5.0, Python 3.6.9

4.2 Experimental Content

We compare the performance of all databases by inserting and deleting four types of data. These are character strings, pictures, videos, and large files. All the data are inserted each time, and then 20% of the data are randomly selected for deletion. All experiments are performed five times, and the average value is taken.

4.3 Time Comparison of Storing 100,000 Key-Value Strings and Deleting 20% of the Data

As shown in 错误!未找到引用源。. The time spent by Redis on storing data and deleting 20% of the data is 14.8547 and 2.1910; MongoDB is 121.4117 and 822.3114; MySQL is 587.3374 and 1536.9407; MDF is 0.3230 and 0.3050.

From the experimental data, it can be concluded that under the string data type, the MDF database performs better than Redis, MongoDB and MySQL in storing word data and deleting data. Specifically, in the case of storing 100,000 key-value strings, the time taken by MDF is only 0.3230 s, while the times required by Redis, MongoDB and MySQL are 14.8547 s, 121.4117 s and 587.3374 s, respectively, so we can consider that MDF is more time efficient under the string data type (Fig. 5).

4.4 Time Comparison for Storing 810 Photos and Deleting 20% of the Data

As shown in 错误!未找到引用源。. The experimental results show that the time spent by Redis on storing data and deleting 20% of the data is 14.6821 and 0.0945; MongoDB is 26.4531 and 0.6076; MySQL is 109.5634 and 322.7612; and MDF is 204.6715 and 1.0126.

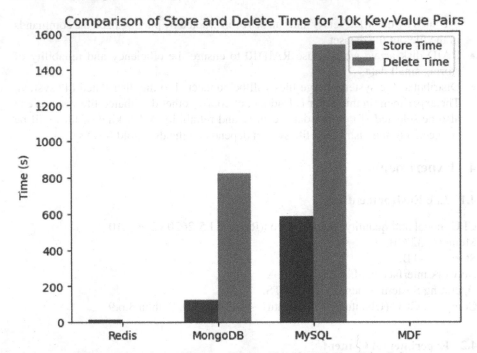

Fig. 5. Comparison of Store and Delete Time for 10k Key-Value Pairs

We selected 810 photos, totaling 6.5 Redis still excels at storing and deleting data, including photos, while MDF lags behind. Specifically, Redis has a storage time of 14.6821 s and a deletion time of only 0.0945 s. MongoDB and MySQL are also slightly faster than MDF for storage operations. However, MDF is still very fast in terms of deletion operations (Fig. 6).

4.5   Time Comparison of Storing 40 Videos and Deleting 20% of the Data

As shown in 错误!未找到引用源。. The time spent by Redis on storing data and deleting 20% of the data is 22.3474 and 0.3886; MongoDB is 31.6924 and 0.4249; MySQL is 130.6361 and 75.5450; MDF is 26.1780 and 0.0564.

We selected 40 videos, totaling 2 GB. Redis still performs better for storing and deleting data, including videos, but MDF performs better than MongoDB and MySQL. Specifically, Redis has a storage time of 22.3474 s and a deletion time of 0.3886 s. MySQL takes longer to store and delete than MongoDB and MDF, probably because it requires more processing and transformation time, as well as scalability and capacity limitations when dealing with large data (Fig. 7).

4.6   Time Comparison for Storing and Deleting Files Larger than 4.5

As shown in 错误!未找到引用源。. The result of the experiment was that Redis and MySQL could not do the job and ended in failure. The time spent by MongoDB on storing and deleting data is 103.6615 and 43.6481, and the MDF is 33.8952 and 0.0764.

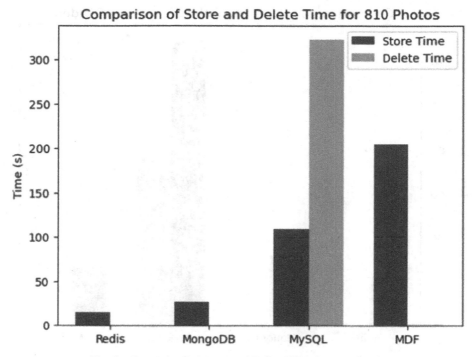

Fig. 6. Comparison of Store and Delete Time for 810 Photos

We prepared a 4.6 GB file. Neither Redis nor MySQL can handle large files, probably because they are in-memory databases, which are not suitable for large files. While both MongoDB and MDF are capable of handling large files, MongoDB takes longer to store and delete than MDF (Fig. 8).

The following is our analysis and summary of the experimental data:

- Experiment of storing 100,000 key-value strings

In this experiment, the MDF database clearly performed well, and its storage and deletion operations were much faster than those of the other databases. Redis also performed very well, but in comparison, MongoDB and MySQL performed poorly. This may be due to MongoDB and MySQL requiring more disk I/O operations and index maintenance.

- Experiments on storing and deleting photos

In this experiment, Redis still performed very well. MongoDB and MySQL also perform much better than when storing string experiments, but their performance still lags behind Redis and MDF. This may be because photo data are larger than string data and require more I/O operations and index maintenance.

- Experiments on storing and deleting videos

In this experiment, Redis did not perform as well as in the previous two experiments but still performed best among all databases. MongoDB also performed very well, but

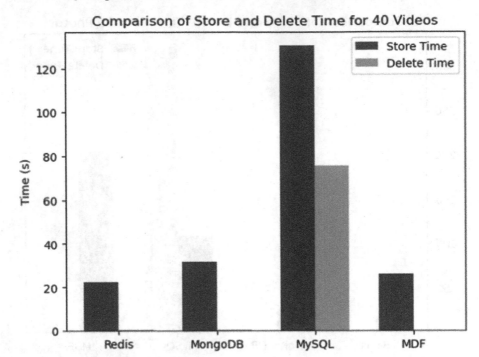

Fig. 7. Comparison of Store and Delete Time for 40 videos

MySQL performed poorly. Similar to the photo experiment, this may be because video data are larger than string data and require more I/O operations and index maintenance.

- Experiments on storing and deleting large files

In this experiment, neither Redis nor MySQL was up to the task, MongoDB performed well, but MDF performed best. This may be because MDF is designed to store multimodal data and employs some special storage and indexing strategies that make it perform well when dealing with large files.

In general, according to the above experimental results. We can draw the following conclusions:

Redis performs best when storing and reading small data.

MongoDB performs well with moderately sized data.

MySQL is suitable for handling medium-sized structured data.

MDF excels at handling multimodal data, especially when storing large files.

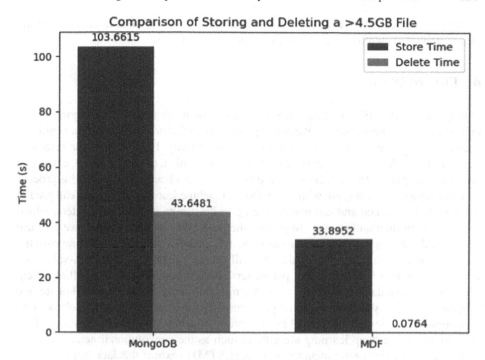

Fig. 8. Comparison of Storing and Deleting a > 4.5 GB File

5   Conclusion and Discussion

From the experimental results, it can be concluded that MDF does not store data for the shortest time among all types of data, but it is the best database that can best take into account multimodal data fusion and speed. In terms of data storage, MDF is not as efficient as Redis because Redis is an in-memory database that has the advantages of fast reading and writing, but its data capacity is limited. In MDF, only small data are processed in memory, and data larger than the set threshold ∂ are connected to the distributed file system Hadoop. To ensure that Hadoop manages more data, we set the value of ∂ as small as 256 KB. That is, part of the speed is given up to ensure that more data can be managed by the distributed file system Hadoop. If the scene involves a large amount of small data and frequent access, then the size of ∂ can be increased so that more data can be processed in memory, thereby improving our speed.

We also conducted a separate test on Hadoop and found that there is almost no difference in the time to import large files directly into Hadoop and the time to import files into Hadoop through the database. This means that the processing efficiency of large files depends on the efficiency of the distributed file system. This experiment only uses one server and does not give full play to the advantages of the distributed file system. If clusters are used for deployment, the speed will be greatly improved. It is different for Redis and MySQL. For large files, if the storage limit is not exceeded, they also need to convert the file into binary data first and then process it as a string, which increases the time cost of converting data.

In conclusion, MDF databases have advantages in multimodal data fusion, are time efficient and are able to store data of different types and sizes.

6 Future Work

Our future work will investigate the application of machine learning algorithms to enhance database performance. We aim to predict nodes that are likely to fail and replicate data to other nodes proactively to ensure high availability. Furthermore, for heterogeneous multimodal data, we can use the attention mechanism to train a model, extract its features, and predict the results of the ship virtual test in advance. Typically, this process can take up to two weeks, but with the proposed machine learning model, we can predict the results in advance and use reverse tuning to determine optimal parameters, which can save both time and money. To predict the possibility of node failure, we will use supervised learning methods such as decision trees and neural networks. To improve the accuracy and reliability of the model, we will use techniques such as cross-validation and grid search to tune the model parameters. Data mining techniques will also help identify potential data correlations to use for model training and prediction. For storage purposes, similarity detection can be performed to eliminate redundancy, followed by compression for storage efficiency [20]. For processing heterogeneous multimodal data, we propose using deep learning algorithms such as the convolutional neural network (CNN) and long short-term memory network (LSTM) to extract the data characteristics. The attention mechanism will further improve the model's accuracy and interpretability. When it comes to the correlation of massive data, a common approach is to embed textual values into high-dimensional vectors and use similarity algorithms to establish associations on these high-dimensional vectors [16–19]. Ultimately, the model will predict the virtual test results of the ship, and we will determine the optimal parameters through reverse tuning to save test costs and time. To validate the effectiveness and reliability of these methods, we recommend conducting experiments and tests on a large amount of real data. We will also work with industry experts to understand practical applications' constraints and needs and optimize our methods and algorithms based on their feedback continuously. In conclusion, our future research aims to explore machine learning algorithms' application in database performance improvement and heterogeneous multimodal data processing. We aim to provide reliable and efficient solutions for practical applications.

References

1. Xu, W., Li, S., Zhao, S., et al.: Analysis of big data platform based on hadoop framework. China Secur. **2020**(4), 38–45 (2020)
2. Yang, J., Liu, Z.: Construction and management of safety production information sharing platform based on big data analysis. Shanxi Coal **2020**(3), 123–127 (2020)
3. Wang, L.: Design and implementation of big data platform for smart campus in universities. J. Hebei Univ. Natl. (Nat. Sci. Edn.) **40**(2), 88–93 (2020)
4. Peng, S.: Research on the construction of smart education cloud platform under the background of big data. Software **41**(4), 229–231, 243 (2020)

5. Liu, C., Xu, W.: Ideas for the construction of municipal government big data platform. Inform. Technol. Inform. **2020**(3), 173–176 (2020)
6. Li, Y., Yang, M., Zhang, Z.: A survey of multi-view representation learning. IEEE Trans. Knowl. Data Eng. **31**(10), 1863–1883 (2019)
7. Lu, R., Jin, X., Zhang, S., Qiu, M., Wu, X.: A study on big knowledge and its engineering issues. IEEE Trans. Knowl. Data Eng. **31**(9), 1630–1644 (2019)
8. Chao, G., Sun, S., Bi, J.: A survey on multi-view clustering. ArXiv, vol. abs/1712.06246 (2017)
9. Wang, H., Yang, Y., Liu, B.: GMC: graph-based multiview clustering. IEEE Trans. Knowl. Data Eng. **32**(6), 1116–1129 (2020)
10. Luong, K., Nayak, R.: A novel approach to learning consensus and complementary information for multiview data clustering. In: 2020 IEEE 36th International Conference on Data Engineering (ICDE), pp. 865– 876 (2020)
11. Wu, J., Lin, Z., Zha, H.: Essential tensor learning for multi-view spectral clustering. IEEE Trans. Image Process. **28**(12), 5910–5922 (2019)
12. Kang, Z., et al.: Multi-graph fusion for multi-view spectral clustering. Knowl. Based Syst. **189**, 105102 (2020)
13. Li, L.: Application scenarios of relational databases and NoSQL databases. Electron. Technol. Softw. Eng. **16**, 184–187 (2022)
14. Shi, X., Yang, H.: Big data processing and analysis based on the integration of hadoop and MongoDB. Comput. Knowl. Technol. (29), 1–2+10 (2019). https://doi.org/10.14004/j.cnki. ckt.2019.3401
15. Marathe, A.P., Lin, S., Yu, W., El Gebaly, K., Per- ˚Ake Larson, Sun, C.: Integrating the Orca optimizer into MySQL. In: Proceedings of the 25th International Conference on Extending Database Technology, EDBT 2022, Edinburgh, UK, March 29 - April 1, 2022. OpenProceedings.org, pp. 511–523 (2022)
16. Wang, F., Yiu, M.L., Shao, Z.: Accelerating similarity-based mining tasks on high-dimensional data by processing-in-memory. In: 2021 IEEE 37th International Conference on Data Engineering (ICDE), pp. 1859–1864. IEEE (2021)
17. Dong, Y., Takeoka, K., Xiao, C., Oyamada, M.: Efficient joinable table discovery in data lakes: a high-dimensional similarity-based approach. In: 2021 IEEE 37th International Conference on Data Engineering (ICDE), pp. 456–467. IEEE (2021)
18. Peng, B.: Data series indexing gone parallel. In: 2020 IEEE 36th International Conference on Data Engineering (ICDE), pp. 2059–2063. IEEE (2020)
19. Wang, Y., et al.: HowSim: a general and effective similarity measure on heterogeneous information networks. In: 2020 IEEE 36th International Conference on Data Engineering (ICDE), pp. 1954–1957. IEEE (2020)
20. Richly, K., Schlosser, R., Boissier, M.: Joint index, sorting, and compression optimization for memory-efficient spatio-temporal data management. In: 2021 IEEE 37th International Conference on Data Engineering (ICDE), pp. 1901–1906. IEEE (2021)

Big Data Mining and Knowledge Management

Research on Multi-Modal Time Series Data Prediction Method Based on Dual-Stage Attention Mechanism

Xinyu Liu[1], Yulong Meng[2,3](✉), Fangwei Liu[2], Lingyu Chen[2], Xinfeng Zhang[2], Junyu Lin[4], and Husheng Gou[5]

[1] Jiangsu JARI Technology Group Co. Ltd., Lianyungang 222006, China
[2] Harbin Engineering University, Harbin 150001, China
mengyulong@hrbeu.edu.cn
[3] Modeling and Emulation in E-Government National Engineering Laboratory, Beijing, China
[4] Institute of Information Engineering, Chinese Academy of Sciences, Beijing, China
[5] China International Engineering Consulting Corporation, Beijing, China

Abstract. The production data in the industrial field have the characteristics of multimodality, high dimensionality and large correlation differences between attributes. Existing data prediction methods cannot effectively capture time series and modal features, which leads to prediction hysteresis and poor prediction stability. Aiming at the above problems, this paper proposes a time-series and modal feature enhancement method based on a dual-stage self-attention mechanism (DATT), and a time series prediction method based on a gated feedforward recurrent unit (GFRU). On this basis, the DATT-GFRU neural network with a gated feedforward recurrent neural network and dual-stage self-attention mechanism is designed and implemented. Experiments show that the prediction effect of the neural network prediction model based on DATT is significantly improved. Compared with the traditional prediction model, the DATT-GFRU neural network has a smaller average error of model prediction results, stable prediction performance, and strong generalization ability on the three datasets with different numbers of attributes and different training sample sizes.

Keywords: Multi-modal time series data · Recurrent neural network · Self-attention mechanism

1 Introduction

Time series data prediction is one of the most popular research directions in time series data mining [1–3]. Time series data prediction methods have developed from methods completely based on mathematical statistics to methods based on deep learning [4, 5]. The target data of the research have also developed from simple low-dimensional single-modal time series data at the beginning to complex high-dimensional multi-modal time series data [6].

Z. Yu et al. (Eds.): ICPCSEE 2023, CCIS 1879, pp. 127–144, 2023.
https://doi.org/10.1007/978-981-99-5968-6_10

In recent years, algorithms based on neural networks have emerged in an endless stream, and many time series prediction models based on deep learning have been developed. To make full use of the correlation features between previous and subsequent moments in time series data, Jeffrey Elman et al. proposed a recurrent neural network (RNN) [7]. Heimes et al. applied RNNs to the task of predicting the remaining life of an engine [8] and achieved good performance. However, RNNs are prone to gradient disappearance and gradient explosion during backpropagation [9].

Based on the problems existing in RNNs, Jurgen Schmidhuber and his student Cummins proposed a long short-term memory neural network (LSTM) in 1997 [10], which can avoid gradient explosion and gradient disappearance to the greatest extent. It became the preferred classic model in natural language processing and time series prediction tasks. Schmidhuber tried to apply LSTM to music composition, speech recognition and other fields [11, 12], and achieved preliminary results. Shi et al. used convolution to replace the matrix product in the traditional LSTM model and proposed a convolutional LSTM neural network [13]. Alahi used LSTM to predict human trajectories in crowded spaces [14]. LSTM greatly alleviates the gradient problem by adding gating units, but at the same time brings the problems of larger parameter size and slower training speed.

In 2014, Kyunghyun Cho et al. proposed an LSTM-based variant-gated recurrent unit (GRU) network for the first time [15]. They simplified the three gated units of LSTM neurons into two, which made the structure simpler and the training speed faster. Chung, Gulcehre and Cho et al. compared the effects of LSTM and GRU models on tasks based on speech signal modelling, and found that the two models had similar effects when the training data size was not large [16]. Zhang et al. used a GRU network to predict the traffic flow of urban expressways based on weather conditions and other data [17]. Emily L. Aiken used GRU networks for flu prediction in US cities [18]. Sun et al. used a neural network based on GRU hybrid stacking to predict traffic flow [19]. The improved GRU based on LSTM simplifies the gating unit by abandoning the cell state, which improves the efficiency of the model to a certain extent, but at the same time leads to the fact that the prediction effect of GRU is not as good as LSTM in the face of large-scale time series data [20].

When dealing with multi-modal time series prediction tasks, the existing time series prediction algorithms are strongly dependent on the dataset, do not capture the correlation features of the time series context well, and lack targeted learning of the correlation features between modalities of multi-modal time series data. By studying the problems of insufficient prediction accuracy and prediction hysteresis in the existing multi-modal time series data prediction algorithms, starting from the characteristics of the data and the prediction model itself, this paper first proposes a feature enhancement method based on dual-stage self-attention mechanism to improve the prediction hysteresis and accuracy. Then, aiming at the problem that the time series correlation of the prediction algorithm itself is not strong, and the time series prediction ability and training speed cannot be achieved simultaneously, an improved gated feedforward recurrent unit based on LSTM is proposed, which not only improves the training speed of the algorithm, but also improves the accuracy of time series prediction to a certain extent.

2 Feature Enhancement Method Based on a Dual-Stage Self-attention Mechanism

2.1 Self-attention Mechanism

Adding the self-attention mechanism to the time series prediction model can weight the features in the sequence, and the important information will receive higher weights. This mechanism gives the neural network the ability to selectively learn the data [21, 22]. The self-attention mechanism in the time series task is shown in Fig. 1.

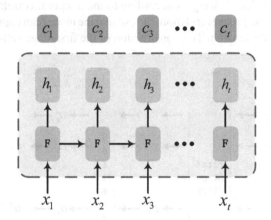

Fig. 1. Self-attention mechanism in time series tasks

In Fig. 1, x_t represents the input at time t, h_t represents the state of the hidden layer corresponding to x_t, and c_t represents the corresponding context vector. F can be an activation function or a neuron such as an RNN. The workflow of the self-attention mechanism is as follows: First, the hidden layer state h_t at the current time is calculated according to the context vector c_{t-1} at the previous time and the input x_t, at time t. Then the correlation degree $\widetilde{\alpha}_{it}$ is found between h_t at time t and the hidden layer state at each past time. Next, Softmax normalization is applied to $\widetilde{\alpha}_{it}$ to obtain the attention weight α_{ti}. Finally, the context vector c_t at time t is obtained by taking the weighted average of all hidden layer states h according to α_{ti}.

2.2 Dual-stage Self-attention Mechanism

In this section, by analysing the characteristics of multi-modal time series data, a multi-modal time series data feature enhancement method based on a dual-stage self-attention mechanism is proposed. First, the first-stage self-attention layer is added after the input layer, and the correlation features between each time step are adaptively extracted by referring to the previous hidden layer state to solve the problem that the model is not sensitive to the target features. Second, by fusing the second-stage self-attention mechanism after the prediction unit, the correlation feature weights between different attribute columns of multi-modal time series data are extracted, so that the neural network can selectively train according to the importance of each attribute of the data.

The First-stage Attention Layer. When performing regression tasks such as time series prediction, it is usually necessary to feed a large number of time series, which contain the target features as well as a large number of common features of the data. As a result, the neural network is more prone to fit the common features that are absolutely dominant in number, while ignoring the few target features we need. To solve this problem, this section proposes a time series feature enhancement method based on a self-attention mechanism. This method uses the self-attention mechanism to extract the features of the time dimension, and assigns weights according to the temporal correlation between the current moment and the past several moments, so that the model can capture the important features in the time dimension. The construction of the first-stage self-attention layer is shown in Fig. 2:

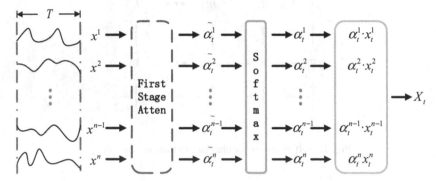

Fig. 2. The structure of the first-stage attention layer

Step 1: Feed the time series data to the input layer as a matrix [batch_size, windows, features], where batch_size is the number of training data in each batch, windows represent the prediction of the current value using the data of the previous window time, and features represent the number of attribute columns of the multi-modal time series data.

Step 2: The attention layer is used to fully connect with the input layer, and the activation function is tanh.

Step 3: The activation layer is added after the attention layer, and the activation function selects softmax, and the output is the weight matrix on the time series.

Step 4: Finally, the output of activation layer is multiplied by the input of the input layer to complete the weighted average calculation of the time series input sequence.

The Second-stage Attention Layer. In multi-modal time series prediction tasks, it often occurs that no matter how the model is tuned, the prediction curve cannot fit the original data well, especially when the correlation between attributes of multi-modal time series datasets is very different, and the phenomenon is more obvious. In practical applications, it is difficult to have a fixed standard for the correlation between attributes, and it is too difficult to eliminate the attribute columns with little correlation with the target feature manually. To solve this problem, this section proposes a method of attribute feature enhancement based on a self-attention mechanism. By adding a self-attention layer after the prediction unit, this method assigns a weight attribute to each attribute column of multi-modal time series data, so that the model has the ability to independently and selectively learn about the dimension of the attribute column. The construction of the second-stage attention mechanism is shown in Fig. 3:

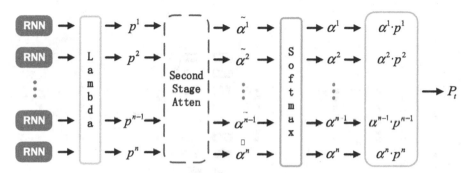

Fig. 3. The structure of the second-stage attention layer

Step 1: First, the matrix with [batch_size, lstm_units] from the LSTM output is transposed to obtain the matrix with [batch_size, lstm_units, features]. Here, lstm_units represents the output dimension of the LSTM hidden layer.

Step 2: The lambda expression takes the mean of the second dimension of the matrix obtained in Step 1 and produces the matrix of [batch_size, features].

Step 3: The matrix in Step 2 is fed into the attention layer, and the feature weights are calculated for each data point in each batch.

Step 4: The weights of the attention layer are input to the activation layer, and the activation function is selected as softmax; that is, the normalized weight matrix is obtained.

Step 5: Finally, matrix multiplication is performed on the output of the activation layer and the input of the input layer to complete the weighted average calculation of each attribute of the time series input sequence.

2.3 Prediction Model Combining a Dual-stage Self-attention Mechanism

In the previous section, the dual-stage self-attention mechanism is introduced in detail, and this section introduces the workflow of the whole time series prediction model and the structure of the neural network. Since the prediction mechanisms studied in this paper are all based on variants of recurrent neural networks, the prediction units are all represented using RNN units for ease of presentation.

Prediction Model for Multi-modal Time Series Data based on DATT. The workflow of the prediction model and the structure of the neural network are shown in Fig. 4. First, for multi-modal time series data with more than 15 attributes, the denoising autoencoder is used to reduce the dimension and extract the key attributes. Then, the extracted data are used as the input of the time series prediction model. After the time series weighting of the first-stage self-attention mechanism, the time series data of each modality are input to the time series prediction unit. Then, the output data of the time series prediction unit is averaged in the time series dimension, and a one-dimensional vector with the same number of elements and modal features is obtained. Finally, the one-dimensional vector is input into the second-stage self-attention layer for modal feature weighting, and the prediction result is output through the fully connected layer.

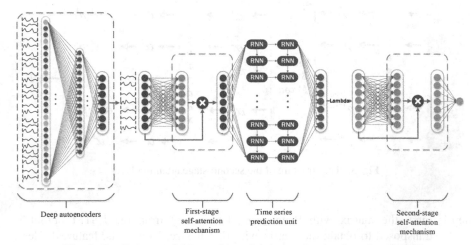

Fig. 4. Prediction model for multi-modal time series data based on DATT

3 Gated Feedforward Recurrent Neural Network

Aiming at the problems existing in the current time series data prediction algorithms, this paper proposes gated feedforward recurrent unit (GFRU).

3.1 Time Series Data Prediction Methods Based on a Recurrent Neural Network

To overcome the gradient problem, various excellent variants based on RNN have been proposed in succession, among which LSTM and GRU neural networks are the most widely used [23].

Long Short-Term Memory Neural Network (LSTM). The long short-term memory neural network (LSTM) is different from the traditional recurrent unit that covers the information of the previous time step at each time step. The LSTM unit decides whether to retain the information of the previous time step through the introduced gating unit. Therefore, LSTM can capture and remember long-distance information [24]. However, at the same time, because LSTM introduces three gating units, the scale of the training parameters of the whole network model is greatly increased, which leads to problems such as difficult model training, slow iteration of network weight parameters, and easy overfitting.

Gated Recurrent Neural Network (GRU). Gated recurrent neural networks consist of gated recurrent neural units, which are an excellent LSTM-based variant. By removing the cell state in LSTM neurons, GRU uses the update gate to replace the forget gate and input gate of LSTM, which reduces the parameter scale by 25% and has a certain improvement in training speed. Since the input of GRU in calculating the gating matrix is the information from the previous time step, the cell output is used instead of the original way of calculating the cell state and cell output respectively, which leads to its performance inferior to LSTM in large-scale data prediction tasks [25].

3.2 GFRU Neuron and Network Structure

Based on the problems of LSTM and GRU, this section proposes a variant-GFRU based on LSTM improvement, introduces the structure and forward propagation process of GFRU in detail, and illustrates the advantages of GFRU from the perspective of mathematical derivation.

GFRU Neurons. GFRU neurons are derived from the optimization and improvement of LSTM neurons, and are variants of RNNs, similar to LSTM and GRU. The GFRU neuron is shown in Fig. 5. C_{t-1} represents the cell state at the previous time, h_{t-1} is the output at the previous time, x_t is the input at the current time, C_t is the cell state at the current time, and h_t is the output at the current time.

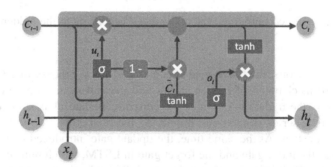

Fig. 5. GFRU neuron structure

The cell state C_{t-1} at the previous moment, through a feedforward mechanism, is involved in deciding the weight of the update gate. The update gate u_t of GFRU is determined by the input x_t at the current time, the output h_{t-1} at the last time and the cell state C_{t-1} at the last time. The update gate weight is multiplied with these three parameters respectively, and then the respective bias matrices are added and summed. A value between 0 and 1 is output through the sigmoid activation function; the larger the value is, the more information is retained at the previous time; otherwise, the information is retained less. The update gate replaces the role of the forget gate and the input gate, and its formula is as follows:

$$u_t = \sigma \left(W_u \cdot [C_{t-1}, h_{t-1}, x_t] + b_u \right) \tag{1}$$

The formula for the input intermediate state at the current time is as follows:

$$T_t = \tanh \left(W_i \cdot [h_{t-1}, x_t] + b_i \right) \tag{2}$$

The cell state at the current time is determined by the update gate, u_t determines how much information from the past time is kept, and $1 - u_t$ determines how much input from the current time is received by the cell state. The formula for the cell state at the current time is as follows:

$$C_t = u_t * C_{t-1} + (1 - u_t) * T_t \tag{3}$$

The output gate is similar to the input gate, which is determined by the input x_t at the current time, the output h_{t-1} at the previous time and the cell state C_{t-1} at the previous time. The output gate controls how much feature information of the cell state at the current time flows into the next time. The output gate formula is as follows:

$$o_t = \sigma \left(W_o \cdot [h_{t-1}, x_t] + b_o \right) \tag{4}$$

The cell state at the current time goes through the *tanh* activation function, under the control of the output gate o_t and outputs h_t at the next time. The output of the hidden layer at time t is as follows:

$$h_t = \tanh(C_t) * o_t \tag{5}$$

Finally, the predicted output at time t is updated according to the hidden layer:

$$\widehat{y_t} = V h_t + b_y \tag{6}$$

It can be seen from the above mathematical derivation process that GFRU adds the feedforward mechanism of hidden states, and the cell state at the previous time can participate in the update iteration of the current cell state, thus strengthening the connection of time series context and improving the prediction ability of time series data to a certain extent. At the same time, the update gate introduced by GFRU plays a similar role as the input gate and the forget gate in LSTM, which reduces the weight parameters by 12.5%. It improves the prediction ability and reduces the parameter scale of the entire neural network at the same time, which helps to improve the iteration speed of model parameters.

Construction of the GFRU Network Model. In this subsection, a neural network model with multiple hidden layers based on GFRU units is constructed. Figure 6 is a deep GFRU neural network with n hidden layers with input x and output y. Assuming that the multi-modal time series data have k columns, the first t data are used to predict the data at the next moment each time, and the number of GFRU units in each layer is m, and the number of parameters to be trained of the whole network is:

$$num_{para} = n * ((3k + 2m) * m + m) \tag{7}$$

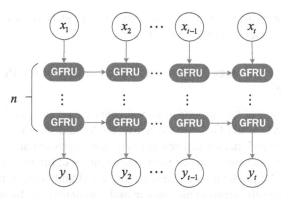

Fig. 6. GFRU Neural Network

Taking the update gate matrix u_t^n at time t of the nth layer as an example, the lower corner label denotes time t, and the upper corner label denotes the nth hidden layer. h_t^{n-1} is the output of the previous neural network and the input to the nth layer, $x_t^n = h_t^{n-1}$. The output at the current moment is o_t, W represents the weight parameters of each gate in the current hidden layer, and b is the corresponding bias matrix. Here, σ represents the sigmoid activation function.

$$\begin{cases} u_t^n = \sigma\left(W_u^n h_t^{n-1} + U_u^n h_{t-1}^n + Y_u^n C_{t-1}^n + b_u^n\right) \\ T_t^n = \tanh\left(W_T^n h_t^{n-1} + U_T^n h_{t-1}^n + b_T^n\right) \\ o_t^n = \sigma\left(W_o^n h_t^{n-1} + U_o^n h_{t-1}^n + b_o^n\right) \\ C_t^n = C_{t-1}^n \odot u_t^n + T_t^n \odot \left(1 - u_t^n\right) \\ m_t^n = \tanh\left(C_t^n\right) \\ h_t^n = m_t^n \odot o_t^n \\ \widehat{y_t^n} = \sigma\left(V h_t + b_y\right) \end{cases} \qquad (8)$$

Except at time 0, the neurons at each subsequent time receive the hidden layer state h_{t-1}^n and cell state C_{t-1}^n of the neurons at the previous time in the same layer and the output h_t^{n-1} of the neurons at the same time in the previous layer. They are respectively multiplied by the weight matrix of the update gate, summed up and added to the bias matrix, and the update gate u_t^n is obtained after activation by the sigmoid function. Then the update gate determines how many cell states of the previous time and the input of the current time are kept in the cell state of the current time, and finally the hidden state h_t^n of the cell of the current time and the output of the cell of the current time are updated.

4 Experiments and Analysis

To visually compare the performance between models, it is necessary to quantify the predictive metrics of the models. In this paper, we choose MAE (mean absolute error), RMSE (root mean square error) and MAPE (mean absolute percentage error) from the common evaluation indicators of regression prediction models as the evaluation criteria of the time series prediction model in this paper.

4.1 Effect of the Dual-stage Self-attention Mechanism on the Prediction Performance

In this section, a total of two groups of experiments are set up: (1) By comparing the performance of LSTM and GRU in the native model and the model fused with the first-stage self-attention mechanism respectively, the improvement effect of the first-stage self-attention mechanism on the prediction accuracy of time series data is verified. (2) By comparing the prediction performance of LSTM and GRU in the cases of only fusing the first-stage self-attention mechanism and simultaneously fusing the dual-stage self-attention mechanism, it verifies the stable performance of selective learning ability given to the model by the second-stage self-attention mechanism when dealing with multi-modal time-series data with large differences in attribute correlation.

The experimental data are from the Beijing Multi-Site Air-Quality Data Set (BMAQ), which is the UCI machine learning standard dataset publicly available from the University of California Irvine.

Effect of the First-stage Self-attention Mechanism on Prediction Performance. First, the native LSTM prediction model, the LSTM prediction model fused with the first-stage self-attention mechanism (FATT-LSTM), the native GRU prediction model, and the GRU prediction model fused with the first-stage self-attention mechanism (FATT-GRU) are implemented. The attribute data such as CO, NO2, SO2, PM10, wind speed and PM2.5 monitored at No. 1035 in the BMAQ dataset were used to predict PM2.5, and the training and prediction data ratio was 8:2. The prediction unit was composed of two-layer LSTM or GRU units, and the output of the hidden layer was 16. Each of the above four models is trained for 100 rounds, and each group of experiments is repeated 20 times. The final results are averaged and two floating point numbers are retained. The experimental results are as follows:

Table 1. The errors of the four prediction models

Model	MAE	RMSE	MAPE
LSTM	20.08	34.02	28.22
FATT-LSTM	16.56	24.03	23.02
GRU	21.05	33.96	30.24
FATT-GRU	17.23	24.5	25.45

From Table 1 and Fig. 7, on the BMAQ dataset, the three errors of FATT-LSTM are reduced by 17.53%, 29.37% and 18.43% compared with LSTM, and the three errors of FATT-GRU are reduced by 18.15%, 27.86% and 15.84% compared with GRU. Experiments show that the first-stage self-attention mechanism has a stable improvement on the prediction performance of both LSTM and GRU prediction models.

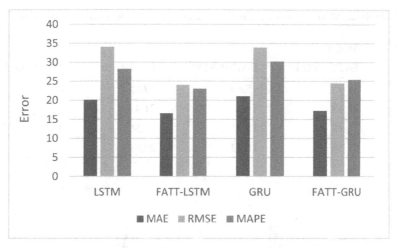

Fig. 7. Error comparison of the four models under the same dataset

Effect of the Second-Stage Self-Attention Mechanism on Prediction Performance. First, the LSTM and GRU prediction models of the first-stage self-attention mechanism (FATT-LSTM, FATT-GRU) and the LSTM and GRU prediction models of the dual-stage self-attention mechanism on the basis of FATT (DATT-LSTM, DATT-GRU) are implemented. The experiment used CO, NO2, SO2, PM10, wind speed and PM2.5 data from monitoring stations 1001 and 1002 in the BMAQ dataset (hereinafter referred to as the 1001-1 and 1002-1 datasets, respectively), as well as CO, NO2, SO2, PM10, wind direction, PM2.5 data from monitoring stations 1001 and 1002 (hereafter referred to as the 1001-2 and 1002-2 datasets, respectively) to predict PM2.5. The prediction step size is 5; that is, the data of the previous 5 days are used to predict the data of the next day. Compared to the 1001-1 and 1002-1 datasets, the 1001-2 and 1002-2 datasets only replace the wind speed column with the wind direction attribute column. Intuitively, PM2.5 is more correlated with wind speed than with wind direction, and the attribute correlation difference between the 1001-2 and 1002-2 datasets is larger than that between 1001-1 and 1002-1. Each of the above four models was trained for 100 rounds, and each group of experiments was repeated 20 times. The final results were averaged, two floating point numbers were retained, and the experimental results are shown in Table 2.

It can be seen from Table 2 and Fig. 8 that under the above four different datasets, the MAE of DATT-LSTM is reduced by 17.88% on average compared with FATT-LSTM, and the MAE of DATT-GRU is reduced by 15.51% on average compared with FATT-GRU. Experiments show that the second-stage self-attention mechanism can effectively capture the correlation characteristics between data attributes, solve the performance of the prediction model when the correlation between time series data attributes is very different, and has a more significant role in improving the prediction effect.

Table 2. MAE errors of the four prediction models under different datasets

Dataset	Model			
	FATT-LSTM	DATT-LSTM	FATT-GRU	DATT-GRU
1001-1	18.75	15.84	19.55	16.24
1001-2	20.49	16.56	21.43	17.43
1002-1	18.10	15.75	17.56	15.21
1002-2	21.56	16.42	20.01	17.40

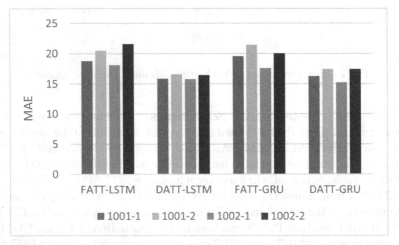

Fig. 8. Error comparison of four prediction models under different datasets

4.2 Validation of Time Series Prediction Performance of GFRU and DATT-GFRU

This section first studies the prediction accuracy performance of GFRU under different datasets by using datasets with different characteristics and sample sizes, and compares it with LSTM, GRU, TSANet (a fusion network model based on TCN and self-attention) and LSTM-DANN to verify the time series prediction performance of GFRU. Then, the DATT-GFRU (dual-stage self-attention based gated feedforward recurrent unit) neural network fused with the dual-stage self-attention mechanism is implemented, and its model structure is shown in Fig. 4. The RNN units are replaced by GFRU units, which is the DATT-GFRU neural network. Training and testing are carried out on different datasets, the regression fitting performance of the model under different datasets is studied, and the generalization performance of DATT-GFRU for multi-modal time series data prediction is verified.

In this experiment, in the TensorFlow1.14 environment, the open source deep learning framework keras is used to realize GRU, LSTM, TSANet, LSTM-DANN and GFRU neural networks, and the performance of the GFRU neural network is analysed through comparative experiments. The experimental data were used in the UCI machine learning standard datasets BPM (Beijing PM2.5 Data Set), PPG (PPG-DaLiA Data Set) and AEP (Appliances energy prediction Data Set) three datasets published by the University of California Irvine. The information of the four datasets is shown in Table 3.

Table 3. Multi-modal time-series datasets

Dataset	Attribute Columns Number	Dataset Size
AEP	29	19735
BPM2.5	13	43824
PPG	11	8300000

By setting up comparative experiments, the performance of GFRU and DATT-GFRU on multi-modal time series prediction tasks is analysed from the perspective of test error and training time, and its advantages over LSTM, GRU, TSANet, LSTM-DANN and DATT-LSTM neurons are analysed. The power consumption, temperature, humidity, rain and snow state, outdoor temperature and other data of each room in the family recorded by the AEP dataset were used to predict the power consumption, and the data of the past 144 time steps were used to predict the energy consumption of the next 6 units of time. The above five models were trained for 20 rounds and repeated 20 times. The attribute columns of PM2.5, temperature, pressure, wind, wind direction, rain and snow in the BPM2.5 dataset were used to predict PM2.5, and the prediction step was 5, that is, the PM2.5 data of the next day were predicted by using the data of the previous 5 days. The five models were trained 40 times and repeated 20 times. The attribute data of ECG, skin potential, body temperature, acceleration, respiratory rate and other attributes sampled at 700 Hz by RespiBAN, a chest wearable device in the PPG dataset, were used to predict heart rate. The data of 3500 time steps were used to predict the next 700 data points, and each set of experiments for the five models was repeated 20 times with 50 rounds of training each time. The final results are averaged, two floating point numbers are kept, and the experimental results are shown in Table 4.

Validation of Time Series Prediction Performance of GFRU. To show the data in Table 4.2 more intuitively, the MAE error and training time are selected and plotted in bar charts, as shown in Fig. 9 and Fig. 10.

As seen in Table 4, Fig. 9 and Fig. 10, in general, the three models have GRU > GFRU > LSTM > TSANet > LSTM-DANN in training speed and GFRU > TSANet > LSTM-DANN > LSTM > GRU in prediction effect. Since GFRU adds the feedforward mechanism of the cell state at the last time and strengthens the relevance of the temporal context, the overall linear regression fitting ability is stronger than that of LSTM. At the same time, compared with LSTM, GFRU reduces the gating weight parameters by 12.5%, and improves the overall training speed of the neural network by approximately 6–10%.

Experimental results show that by using the update gate to replace the input gate and the forget gate in LSTM, GFRU reduces the scale of model parameters and improves the training speed. At the same time, the model is easier to train and not easy to overfit.

Table 4. Performance of seven models on different datasets

Model	Index	Multi-modal time-series datasets		
		AEP	BPM2.5	PPG
LSTM	MAE	20.12	22.34	26.54
	RMSE	26.42	31.51	38.9
	Training Time (s)	182	397	1562
GRU	MAE	20.25	25.46	34.43
	RMSE	28.32	34.81	53.78
	Training Time (s)	156	332	1289
GFRU	MAE	18.05	21.57	24.94
	RMSE	25.83	30.85	36.45
	Training Time (s)	165	358	1423
TSANet	MAE	19.03	21.76	25.54
	RMSE	25.94	30.98	37.36
	Training Time (s)	202	431	1745
LSTM-DANN	MAE	19.27	21.96	26.01
	RMSE	26.20	31.02	37.74
	Training Time (s)	225	484	1905
DATT-LSTM	MAE	14.48	17.25	22.12
	RMSE	16.85	21.42	28.76
	Training Time (s)	230	491	1982
DATT-GFRU	MAE	13.76	15.73	20.46
	RMSE	14.41	18.02	23.61
	Training Time (s)	212	468	1823

In addition, by adding the feedforward mechanism of cell state, the correlation of the time series context is strengthened, and the time series prediction ability of the model is improved to a certain extent.

Validation of the Prediction Effect and Generalization Ability of DATT-GFRU. As seen from Table 4, on three time-series datasets with different training samples and different numbers of attributes, AEP, BPM2.5 and PPG, compared with the DATT-LSTM neural network, the training time of the DATT-GFRU neural network is reduced by 6.84% on average, the MAE error is reduced by 7.09% on average, and the RMSE is reduced by 16.09% on average. To intuitively show the prediction performance of the two models, the BPM2.5 dataset is taken as an example, part of the test data is taken, and the prediction effect diagram of the DATT-GFRU and DATT-LSTM models is drawn. Figure 11 and Fig. 12 show the prediction performance of the DATT-LSTM and DATT-GFRU neural networks, respectively, on the BPM2.5 dataset. On the whole,

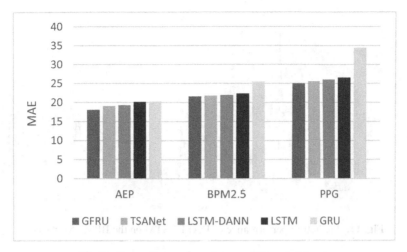

Fig. 9. Error of several models on different datasets

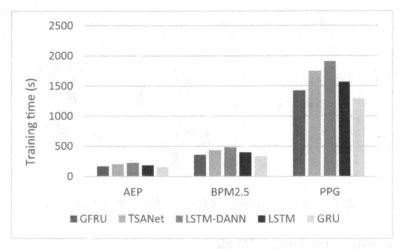

Fig. 10. Training time of several models on different datasets

the two have a good fitting effect on the test data, and there is no obvious hysteresis of the prediction curve. However, in some peaks and troughs, the latter performs better, which also indicates that the GFRU neuron is stronger than the LSTM neuron in the overall prediction effect.

Experiments show that, due to the cell state feedforward mechanism and dual-stage self-attention mechanism of GFRU, the DATT-GFRU neural network has a smaller pre-error than the traditional prediction model on three datasets with different numbers of attributes and different training sample sizes. The prediction performance of the model is stable and shows strong generalization ability.

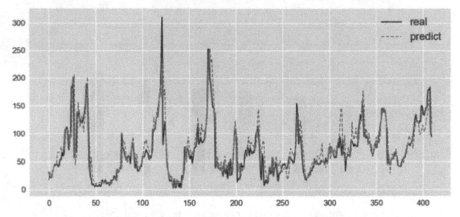

Fig. 11. Prediction performance of DATT-LSTM on the BPM2.5 dataset

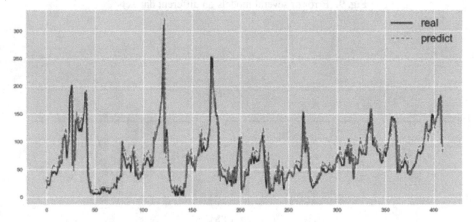

Fig. 12. Prediction performance of DATT-GFRU on the BPM2.5 dataset

5   Conclusion and Future Work

Aiming at the problem that the existing multi-modal time series data prediction methods cannot effectively capture the time series and modal characteristics, resulting in prediction hysteresis and poor prediction stability, this paper first proposes a dual-stage self-attention mechanism and an improved GFRU neural network and then designs and implements the DATT-GFRU prediction model.

Experiments show that the dual-stage self-attention mechanism can fully capture the characteristics of time series data, and has the ability to selectively learn the column attributes of the data, which greatly reduces the phenomenon of prediction data hysteresis and improves the prediction ability of the model under multi-modal time series data. The improved GFRU improves the problem of slow training and easy overfitting of LSTM by introducing the update gate, and improves the problem that LSTM is not sensitive to time series changes by introducing the feedforward mechanism of the cell state. The prediction performance of the GFRU model has unique advantages compared with the traditional GRU and LSTM models. The DATT-GFRU prediction model shows strong generalization ability.

The dual-stage self-attention mechanism proposed in this paper enhances the data characteristics, but the time series prediction hysteresis problem always exists, and further optimizing the algorithm for prediction hysteresis is the focus of future work.

Acknowledgements. This work is financially supported by: The National Key R&D Program of China (No. 2020YFB1712600); The Fundamental Research Funds for Central University (No. 3072022QBZ0601); The National Natural Science Foundation of China (No. 62272126); and The National Natural Science Foundation of China (No. 61872104).

References

1. Sun, Q.S., Zhang, J.X., Cheng, H.Y., Zhang, Q., Wei, X.P.: Financial time series data prediction by attention-based convolutional neural network. J. Comput. Appl. **42**(S2), 290–295 (2022)
2. Wang, T., Wang, M.: Communication network time series prediction algorithm based on big data method. Wireless Pers. Commun. **102**(2), 1041–1056 (2017). https://doi.org/10.1007/s11277-017-5138-7
3. Li, H.L.: Dynamic time warping based on time weighting for time series data mining. Inform. Sci. **547** (2021)
4. Li, J., Zhu, L., Zhang, Y., Guo, D., Xia, X.: Attention-based multi-scale prediction network for time-series data. China Commun. **19**(5), 286–301 (2022)
5. Wang, H., Zhang, Z.: TATCN: Time series prediction model based on time attention mechanism and TCN. In: 2022 IEEE 2nd International Conference on Computer Communication and Artificial Intelligence (CCAI), pp. 26–31. Beijing, China (2022)
6. Xing, Z.K., He, Y.G.: Multi-modal information analysis for fault diagnosis with time-series data from power transformer. Int. J. Electric. Power Energy Syst. **144** (2023)
7. Elman, J.L.: Finding structure in time. Cogn. Sci. 14(2), 179–211 (1990)
8. Heimes, F.O.: Recurrent neural networks for remaining useful life estimation. Prognostics and Health Management, 2008. PHM 2008. International Conference on IEEE (2008)
9. Zhao, J., Zeng, D., Liang, S., et al.: Prediction model for stock price trend based on recurrent neural network. J. Ambient Intell. Hum. Comput. **12**, 745–753 (2021)
10. Hochreiter, S., Schmidhuber, J.: Long short-term memory. Neural Comput. 9(8), 1735–1780 (1997)
11. Graves, A., Schmidhuber, J.: Framewise phoneme classification with bidirectional LSTM networks. In: 2005 IEEE International Joint Conference on Neural Networks, 2005. IEEE 4, 2047–2052 (2005)
12. Eck, D., Graves, A., Schmidhuber, J.: A new approach to continuous speech recognition using LSTM recurrent neural networks. Technical Report (2003)
13. Shi, X., Chen, Z., Wang, H., et al.: Convolutional LSTM Network. A Machine Learning Approach for Precipitation Nowcasting (2015)
14. Alahi, A., Goel, K., Ramanathan, V., et al.: Social LSTM: human trajectory prediction in crowded spaces. In: Proceedings of the IEEE Conference on Computer Vision and Pattern Recognition, pp. 961–971 (2016)
15. Tong, Y., Tien, I.: Time-series prediction in nodal networks using recurrent neural networks and a pairwise-gated recurrent unit approach. ASCE-ASME J. Risk Uncert. Eng. Syst. Part A. Civil Eng. **8**(2), 04022002 (2022)

16. Cho, K., Van Merriënboer, B., Gulcehre, C., et al.: Learning phrase representations using RNN encoder-decoder for statistical machine translation. arXiv preprint arXiv: 1406.1078 (2014)

17. Zhang, D., Kabuka, M.R.: Combining weather condition data to predict traffic flow: a GRU-based deep learning approach. IET Intel. Transport Syst. **12**(7), 578–585 (2018)

18. Aiken, E. L., Nguyen, A. T., Santillana, M.: Towards the Use of Neural Networks for Influenza Prediction at Multiple Spatial Resolutions. arXiv preprint arXiv:1911.02673 (2019)

19. Sun, P., Boukerche, A., Tao, Y.: SSGRU: a novel hybrid stacked GRU-based traffic volume prediction approach in a road network. Comput. Commun. **160**, 502–511 (2020)

20. Ji, S.P., Meng, Y.L., Yan, L., et al.: GRU-corr neural network optimized by improved PSO algorithm for time series prediction. Int. J. Artific. Intell. Tools **29**(07n08), 2040010 (2020)

21. Gao, C., Zhang, N., Li, Y., et al.: Self-attention-based time-variant neural networks for multi-step time series forecasting. Neural Comput. Appl. **34**, 8737–8754 (2022)

22. Su, Y., Cui, C., Qu, H.: Self-attentive moving average for time series prediction. Appl. Sci. **12**(7), 3602 (2022)

23. Noh, S.H.: Analysis of gradient vanishing of RNNs and performance comparison. Information **12**(11), 442 (2021)

24. Frame, J.M., Kratzert, F., Raney, A., et al.: Post-processing the national water model with long short-term memory networks for streamflow predictions and model diagnostics. JAWRA J. Am. Water Resourc. Assoc. **57**(6), 885–905 (2021)

25. Zeng, C., Ma, C., Wang, K., Cui, Z.: Parking occupancy prediction method based on multi factors and stacked GRU-LSTM. IEEE Access **10**, 47361–47370 (2022)

Prediction of Time Series Data with Low Latitude Features

Haoran Zhang[1], Haifeng Guo[1], Donghua Yang[1(✉)], Mengmeng Li[1], Bo Zheng[2], and Hongzhi Wang[1]

[1] Harbin Institute of Technology, Harbin, China
`yang.dh@hit.edu.cn`
[2] ConDB, Beijing, China

Abstract. The main purpose of this paper is to study the key technology for the prediction of time series data. It has a very wide range of applications, such as forecasting sales. Forecasting sales can be said to play an important role in company operations. Whether for saving costs or inventory scheduling, accurate prediction can save unnecessary waste. From this aspect, this paper uses a neural network to achieve the purpose of the prediction.

The application of neural networks in prediction has been a long time. However, most of them have not performed much research on the structure and input of neural networks, and it is not easy to process time series data. Usually, there will be many features. However, the features of data in some scenarios are small. In this paper, we determined how to predict through low-latitude features. At first, among all the ways of preprocessing data, the paper selects a mathematical method. After that, this paper builds three models in two aspects: the input and the network structure. To improve the accuracy of the results, this paper proposes two means. One is based on the seasonal characteristics of commodities. The other is based on the prediction error, called exponential smoothing. Finally, according to the results of the experiment, we come to some conclusions.

Keywords: Data processing · Neural network · Prediction model

1 Introduction

1.1 Background

Data carry information through pictures, text, voice and other carriers to record objective phenomena. Businesses will use the data to predict certain decision making, as will companies. Companies will use historical data to predict the next sales period and use the profits of the same industry or financial data of related data or marketing feedback data to make production decisions [4,8].

Supported by The National Key Research and Development Program of China (2020YFB1006104).

Z. Yu et al. (Eds.): ICPCSEE 2023, CCIS 1879, pp. 145–164, 2023.
https://doi.org/10.1007/978-981-99-5968-6_11

It makes everything possible because the time of big data is coming. Social networking, e-commerce and mobile communications, as the lead of the Internet era, bring people into an era of big data. The big data era, with copious amounts of various types of information and data, leads companies to face a new competitive environment. In particular, business leaders need to meet in the operation and management of information resources for the era of big data, customer experience, marketing innovation effects requirements and to promote development and improve business management and market competitiveness through marketing efficiency. Managing business in the era of big data, it stressed that companies should have strong management of large data processing technology, play well with use of data onto the wealth of information, highlight big data analysis and application of market and consumer-related. Meanwhile, companies should adapt to meet the characteristics of the new era of big data to improve their competitiveness in the market [12].

Big data marketing is the product of the rapid development of Internet technology products. Faced with rich, complex information and data, companies must improve the efficiency of marketing campaigns all the time. Therefore, marketing analysis-based big data come into being.

1.2   Related Work

Traditional predictions are based on complex mathematical models, and different prediction models have different ranges of applicability. Through advanced analytics, companies can perceive consumer behavior patterns associated with demand signals and use predictive analytics and data mining technology to shape future needs. Forty years ago, sales forecasting began to be implemented by computers. It combines the traditional statistical methods of computer information processing and uses a combination of quantitative methods such as the moving average method, exponential smoothing, and regression analysis. It focuses on trends and changes, relies on historical data and finds a linear or nonlinear function completion prediction. The development of artificial neural networks has experienced roughly three climaxes: cybernetics from the 1940s to the 1960s, mid-term syndication from the 1980s to the 1990s, and recent deep learning. People have never stopped studying and exploring neural networks. After the neural network was trained by back propagation of error, research on neural networks has become increasingly deep with increasing computing power. When people find that neural networks can fit the function well, especially nonlinear functions, the application of neural networks is very extensive. People tend to train the network with known data and then predict the unknown data to achieve a certain purpose. Because of its strong fitting ability, it is widely used in all fields.

Artificial neural networks (ANNs) have been found to be useful for sales or demand forecasting [2,10,15]. [16] compares in detail the effects of neural networks and other models. Support vector machine dominated the position for a long time, but the design of the function compared to the neural network is too complicated and professional. Similar to environmental science, similar research on prediction is reflected [9]. In the mechanical engineering field, [11] used a neural network as a classifier. Additionally, to adapt to a more complex scenario, some other models need to be fused, but the neural network is still the main body [1]. There are also some neural networks that are ingeniously designed for use in forecasting [5, 7]. Regarding time series prediction, there are long-term research results [3]. With the rise of machine learning, researchers have tried to use a variety of methods for prediction [13, 14]. Studies have been performed to study the performance of artificial neural networks of prediction and even compare them with other prediction models using time series data [6, 17].

1.3 Main Research Content

According to previous research results, they usually build a model based on the artificial neural network or its optimization and come up with results from this model. However, they rarely discuss the impact of the input and the structure of the neural network on the results. Most importantly, their data may have many features. Currently, there are few studies on the prediction of low latitude data. The data used in this paper are clothing sales data. Time series data on clothing sales are irregular but also affected by many factors, so this prediction is very difficult. In addition, there are very few ways to get data because the shop regarded this as a trade secret. Even if data are available, we cannot guarantee the correctness of the data. Nonoptimized neural networks are usually error-prone for prediction. There are always different levels of optimization. However, how to optimize is usually difficult. A proper and easy method to optimize is more difficult. Finally, considering the reason why the volume of data is not so large, we abandon the recent hot and deep learning model. In this paper, we use an artificial neural network for training. First, this paper uses a traditional mathematical method for smoothing data. Second, we have developed novel ways of using low-latitude feature data to make the prediction and determine which is the best. The average percentage error and absolute error are used to measure the results. Finally, to adapt to the characteristics of the data, this paper combines traditional models and a novel method for optimization.

1.4 Structure Introduction

The system can be abstracted as shown in Fig. 1:

The data process module is used to reduce data noise. Then, 'clean' data are transferred to the next module. The prediction model uses data to make forecasts. In addition, we discuss two ways of improving means. Finally, we obtain the prediction result.

Sample Heading (Third Level). Only two levels of headings should be numbered. Lower level headings remain unnumbered; they are formatted as run-in headings.

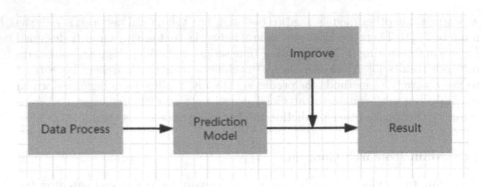

Fig. 1. System structure.

2   Data Acquisition and Processing

2.1   Data Acquisition

This part is to obtain the sales data of the store. What is needed is a large amount of accurate, time series sales data. The first thought of obtaining data is web crawler. However, www.taobao.com shut down the URL interface, so we cannot acquire the data we want. Second, I tried to find proper data from the famous data set like UCI, but there was not suitable data either. Until I found a platform which is professional selling data about E-commerce. The data of an example store are shown in Fig. 2.

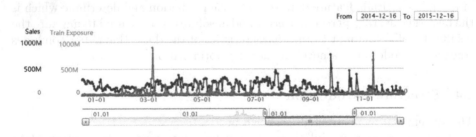

Fig. 2. Data of store.

However, the data is only in the figure! Thus, viewing the page source code and determining the data format is shown in Fig. 4 (Fig. fig:fig3):

The date and salesamount fields are what we needed. Write a simple program and obtain the data.

{"date": "2013-05-23", "order_amount": 289, "sales": 53391, "sales_amount": 290},

{"date": "2013-05-24", "order_amount": 246, "sales": 44293, "sales_amount": 246}

Fig. 3. Data format for instance.

2.2 Data Processing

Now we have the time series data with a number of approximately 900. However, when observing data, some data points will be much higher than the value of the adjacent point or much lower. There exists error in these data. Analysing and correcting possible errors in the accuracy of the results becomes critical. This paper selects a data point smoothing algorithm. First, we need to determine which dots are 'bad'. Therefore, we need to define some rules to determine which dot is bad:

1. When the data have exponential growth;
2. When the data differ by one magnitude compared with the next dot.

Judgments need three dots: current dot n, previous dot n-1 and n-2.

1. If $f_n > f_{n-1} * 10$ or $f_n * 10 < f_{n-1}$, then dot n is 'bad';
2. If $f_n * f_{n-2} > f_{n-1}^2$ and $f_n > f_{n-1} > f_{n-2}$, then dot n is 'bad'.

To minimize the impact, limit to three dots and select a three-dot linear smoothing algorithm, as shown in Algorithm 1.

Algorithm 1. Three Point Linear Smoothing Algorithm

Require: $in_{n-3}, in_{n-2}, in_{n-1}, in_n$
Ensure: $out_{n-2}, out_{n-1}, out_n$
1: **function** LINEARSMOOTHING$(in_{n-3}, in_{n-2}, in_{n-1}, in_n)$
2:   $out_{n-2} \leftarrow (0.5 * in_{n-2} + 2.0 * in_{n-1} - in_n)/6.0$
3:   $out_{n-1} \leftarrow (in_{n-2} + in_{n-1} + in_n)/3.0$
4:   $out_n \leftarrow (5.0 * in_n + 2.0 * in_{n-2} - in_{n-3})/6.0$
5: **return** out

2.3 Conclusion

This chapter introduces the means of obtaining data. In addition, after data was downloaded, it was found that some data was not realistic. First, define a rule and write an algorithm to define 'bad' data dots. Then, in order to achieve the purpose of improving data and make the impact minimized, select data point smoothing algorithm.

3 Implement a Neural Network and Establish Forecasting Models

3.1 Neural Network Overview

One of the most prominent feature of neural network is learning, and a lot of problems can be solved because of its feature. Learning algorithm has a very important role in the development process of the neural network. Currently, the neural network model people proposed are based on the learning algorithm. Since the learning rule was proposed by Hebb, people have proposed a variety of learning algorithms and the error back-propagation method (BP) is most widely affected. Until today, BP method is still the most widely used algorithm in the field of automatic control.

Neural network can be seen as composed of interconnected neurons. Each neuron has the ability to simply reflect the essential characteristics of the nonlinear. It is precisely because of these basic units interconnected so that the neural network can build any nonlinear continuous function. Through this inductive process, it allows the network to obtain the inherent law of sequence, which can be predicted the sequence. BP algorithm itself because of its simple, small amount of calculation, and other advantages, is now one of the preferred algorithm for multilayer feedforward network training and it has been widely used in a variety of practical problems.

3.2 Backpropagation Learning Algorithm

The basic idea of the BP algorithm is to learn from the signal forward propagation and the error back propagation of the two processes. In the forward propagation phase, the learning signal enters through the input layer, calculated by each hidden layer, and transmitted to the output layer. If the actual output of the output layer does not match the desired output, it goes to the error back propagation phase. Error back-propagation puts the output error although each hidden layer neuron by the error back propagation algorithm, and all the neurons of layers share the error to obtain all the error signal of layers. Then, the neural network updates weights depending on the error. Therefore, with the process of neural network weight adjustment, the learning procedure continues. The learning process of the signal forward propagation and the error back propagation has been carried out to the output error of the network of an acceptable level or to be a predetermined number of times.

BP Algorithm Principle. The BP algorithm is a multilayer feed-forward network learning algorithm. The structure is shown in Fig. 4.

It has an input layer and output layer and is hidden between the output and input layers. The middle layer is named the hidden layer because it does not connect directly to the outside world. Its properties affect the relationship between the input and output layers. It says that you can the performance of the entire multilayer neural network by changing weights of hidden layer. Suppose there is an m-layer neural network, and put the sample X into the input layer. Suppose the total income of i neuron of the kth layer is U_i^k and its output is X_i^k. The weight coefficient from i neuron of the $k - 1_{th}$ layer to j neuron of kth layer is W_{ij}. Excitation function of each neuron is f. Therefore, the relationship between variables can be expressed by the following mathematical formula:

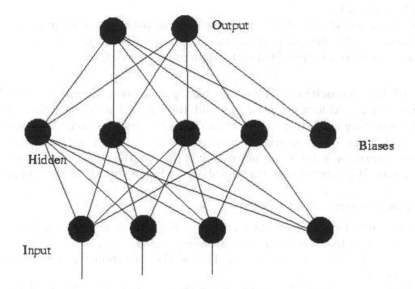

Fig. 4. Data format for instance.

It has an input layer and output layer and is hidden between the output and input layers. The middle layer is named the hidden layer because it does not connect directly to the outside world. Its properties affect the relationship between the input and output layers. It says that you can the performance of the entire multilayer neural network by changing weights of hidden layer.

Suppose there is an m-layer neural network, and put the sample X into the input layer. Suppose the total income of i neuron of the k_{th} layer is U_i^k and its output is X_i^k. The weight coefficient from i neuron of $k - 1_{th}$ layer to j neuron of k_{th} layer is W_{ij}. The excitation function of each neuron is f. Therefore, the relationship between variables can be expressed by the following mathematical formula:

$$X_i^k = f(U_i^k) \tag{1}$$

$$U_i^k = \sum_j W_{ij} X_j^{k-1} \tag{2}$$

As we know now, the back-propagation algorithm has two steps, and these two processes are outlined below:

1. Forward propagation:
   Sample input is processed from the input layer although the hidden layer by layer. After passing through all the hidden layers, the data are transmitted to the output layer. In the process of layer by layer, the status of each layer of neurons only affects the state of the next layer of neurons. Compare the output and the desired output in the output layer; if the current outputs are not equal to the expected output, then enter the back-propagation process.
2. back propagation:
   The error signal spreads through the reverse path that the signal spreads although in the forward propagation. The respective weights of each hidden layer neuron can be modified to the minimum error signal.

Step of BP Algorithm. When the back-propagation algorithm is used in the multilayer feed-forward networks and the excitation function is sigmoid, use the following steps to recursively calculate the network weights. Note that when each layer has n neurons, there is $i = 1, 2, \ldots, n, j = 1, 2, \ldots, n$. For the ith neuron of k layer, it has n weight factors, $W_{i1}, W_{i2}, \ldots, W_{in}$, on the other hand, W_{in+1} represents the threshold θ_i. When inputting the sample X, $x = (x_1, x_2, \ldots, x_n, 1)$.
The algorithm steps are as follows:

1. Set the initial value of weight factor W_{ij}. Set a small nonzero random value to the weight factor of each layer. However, among them $W_{in+1} = -\theta$
2. Input a sample $x = (x_1, x_2, \ldots, x_n, 1)$, and the corresponding desired output $Y = (Y_1, Y_2, \ldots, Y_n)$.
3. Calculate the output of each layer: For the i_{th} input X_i^k of the k_{th} layer,

$$U_i^k = \sum_{j=1}^{n+1} W_{ij} X_j^{k-1}, X_{n+1}^{k-1} = 1, W_{i,n+1} = -\theta \tag{3}$$

$$X_i^k = f(u_i^k) \tag{4}$$

4. Calculate learning error d_i^k of each layer
   For the output layer $k = m$,

$$d_i^m = X_i^m(1 - X_i^m)(X_i^m - y_i) \tag{5}$$

   For the other layer,

$$d_i^k = X_i^k(1 - X_i^k) \sum_i W_{ij} d_i^{k+1} \tag{6}$$

5. Adjust weight W_{ij} and threshold θ

$$\Delta W_{ij}(t + 1) = \alpha \Delta W_{ij}(t) - \eta \cdot d_i^k \cdot X_j^{k-1} \tag{7}$$

Among them, η is the leaning rate, $\eta = 0.1\ 0.4$, α weight correction constant, $alpha = 0.7\ 0.9$.

6. After calculating the weights of each layer, it determines that the output can meet the requirements according to a given index. If the output is qualified, the algorithm ends; if not, return to step 3 to continue.

3.3 Prediction Model

This paper selects the neural network with one hidden layer. Most of the forecasting work is to continually select and construct features through the existing features until the best result is obtained. Most do not have much research on the effects of neural network structure and input, especially for time series data. This paper provides a novel way of thinking about how to construct features using neural networks. We explored how BP neural network parameters affect the result. Combined with the traditional prediction method, we simplified the prediction model.

Ideas of Modeling. At the beginning, introduce some standard statistical error term.

The basic calculation prediction model can be written as:

$$Error = e_t = A_t - F_t \tag{8}$$

e_t is the error of time period t; A_t is the actual sales of time period t; F is the predictive value of time period t.

This paper selects two standards to measure the results. One is the average error. The actual formula is as follows:

$$ME = \frac{1}{n} \sum_{t=1}^{n} (A_t - F_t) \tag{9}$$

t is the time period; n is the number of time periods; A_t is the actual sales of time period t; F_t is the predictive value of time period t.

The other is the mean absolute percentage error, MAPE. It is most commonly used as the fit for an accurate prediction method. MAPE is obtained by calculating the mean absolute percentage error for each period. The actual formula is as follows:

$$MAPE = \frac{1}{n} \sum_{t=1}^{n} \frac{|A_t - F_t|}{A_t} \times 100 \tag{10}$$

This mean, as a percentage, is a relative indicator, and it contributes to comparison with the prediction results in different models.

Regarding comprehension of the forecasting model fitting process, it is a model designed to fit the needs of history (train neural network). Then, compare the model generated, new historical demand estimates (fitting) value with actual (known) needs. Therefore, associate the error calculation and fitting demand with the known requirements. Now that we can compare the actual demand with the forecast value, we can also do the same for the future forecast. It is important to note that a good (or low) error fitting model does not necessarily mean that it will predict a good (precise) future demand. It only confirms that the model can well predict the past history of demand. However, if a model can forecast the demand of the past, it is likely to predict the future demand with a similar error. Ideas as follows:

1. Determine the time sequence or data set (product historical data).
2. Divide the data set into two parts: Within the sample data and outside the sample data.
3. Select a quantitative prediction method (this paper uses a neural network).
4. Use the forecasting method and create predictions.
5. Operate model to obtain fitting results using sample data.
6. Compare the predicted values with the outside sample data set.
7. Evaluate the results to predict how good the future forecasting model matches with historical demand.

Prediction Model. Because of the time sequence data, time is also a factor that must be considered, especially for this data source for Taobao women's clothing stores. Therefore, the paper considers the prediction model to be abstract as follows:

$$L = Sys(L', t) \tag{11}$$

where L' is the historical sales data, t is the time factor, and L is the prediction sales.

The prediction model is established next. The starting point of the prediction model is established on the basis of the characteristics of the BP neural network. By changing the structure of the neural network, different neural networks are established to predict the result, and then the models are compared. In this paper, by changing the number of neural network inputs, changing the output data and changing the neural network algorithm to set up different neural networks, the goal of establishing different models is achieved.

Model1: Time forecasts 10-days total sales
At first glance, you may not understand the meaning of title. The thinking of this model is through the implicit periodic sales to forecast. Because observations every month can be divided into three parts, it is necessary to build a model based on periodic sales to predict future sales. Therefore, the input data must first be processed. Because the input is time, month can be the integer part. The first ten days is 0.1, the middle ten days is 0.2, and the last ten days is 0.3. For example, the first ten days in April is 4.1, and the middle ten days in March is 3.2. However, the last ten days in different months are different, and the difference will influence the result. Therefore, we need to deal with the data.

This paper uses a simple method of average weight. The following algorithm can reduce the impact:

If there are 31 days in a month:

Then, add the last 11 days of sales data and then multiply by a weighting factor of 10/11. The results obtained are trained as input data of the neural network.

If the month is February:

If the year is leap year:

Then, add the last 9 days of sales data and then multiply by a weighting factor of 10/9. The results obtained are trained as input data of the neural network.

Else

Then, add the last 8 days of sales data and then multiply by a weighting factor of 10/8. The results obtained are trained as input data of the neural network.

To date, we have determined that the input is 10-day sales. However, it is necessary to normalize the data so that the range of all the data is uniform and kept in [0,1] to fit the function in the neural network. If x and y are used to represent the values before and after the conversion, the normalization method is calculated as follows:

$$y = (x - MinValue)(MaxValue - MinValue) \qquad (12)$$

MaxValue and MinValue represent the maximum and minimum of the sample, respectively. In this model, the MaxValue is 12.3, and the MinValue is 1.1. This model can be abstracted as follows:

$$L = Sys(t) \qquad (13)$$

L represents sales for 10 days, and t is the time factor. Calculate ME and MAPE by comparison with predicted results and the real value.

Model 2: Ten-consecutive-day sales data to predict the next day
The first model is the abstract based on periodic observation on sale. Therefore, the thinking of this model is based on the analysis of continuous sales and makes the next step of predictions. Then, we have two essential inputs: sales and time. Therefore, we naturally obtain our output: prediction of result. In this model, ignore the consideration of the total sales and consider a single date of sales. This model may be more appropriate for predicting judgment. In this model, we need to input sales of 10 consecutive days. Because of the nature of the time of the sales, we also need to input the time factor. The input format is the month for the integer part and the day for the decimal part. For example, March 23 is 3.23, and December 29 is 12.29. Then, we can start training.

First, the data should be normalized. The method is the same as the previous model. However, MaxValue is the maximum sales sample, and MinValue is the minimum sales sample. In addition, this model has 11 inputs: the top 10 for 10 consecutive days of sales data and the 11th for the date of the predicted day. Then, this model can be abstracted as follows:

$$L = Sys(L_1, L_2, \ldots, L_{10}, t) \tag{14}$$

L for predicting sales for one day, t for the time factor. By comparison with predicted results and real values, calculate ME and MAPE. Here is the point, the result L is only the sales for one day, but only one day is meaningless. We determine an approximate estimation method by taking the output we obtain as a new round of input of predictions. Therefore, we can obtain a series of predictions. In this process, we obtained 7 days of forecast sales and compared the sum of the prediction with the real value.

Model 3: Twenty-consecutive-day sales data to predict the next day
In fact, the third model extends the second model. This model explores how much the number of continuous days affects the result. By comparison with the results of the different models, we can draw a conclusion about this assumption.

First, the data must be normalized. The method is the same as the second model. The format of the time factor is also the same as the second model. The difference is that this model needs to input 20 continuous days of sales. It also needs to input the date of the predicted day. Then, this model can be abstracted as follows:

$$L = Sys(L_1, L_2, \ldots, L_{20}, t) \tag{15}$$

The method used to obtain the result is the same as mentioned above. By comparison with predicted results and real values, calculate ME and MAPE.

3.4   Model to Improve

After the models were set up, to make the results more precise, this paper makes some improvements. Most previous work uses many complicated methods to improve the result. The two approaches presented in this paper are closely related to traditional methods.

Simple Classifier. Because of considering the cyclical nature of sales. We can achieve a simple classifier to classify the off-season and peak season. However, why can a simple classifier do this?

The simple classifier can complete the goal because the neural network has a great degree of adaptive nature itself. It can accept a certain range of changes but cannot accept too wide a range of change. Therefore, we establish a model with a simple classifier to distinguish between the off-season and peak season and train two neural networks. In line with this idea, we make some improvements.

First, we consider a problem: why can we classify the off-season and peak season? The peak season sales are obviously higher than the off-season sales. Next, we should describe the 'high' in the algorithm and determine it. In this paper, the ratio is used to describe 'high'. According to the past sales amount of this shop, when the sales of this month are larger than 1.3 times the sales of one month before this month, we generally think that this month is in the peak season. This is only a simple judgement. A complete algorithm is as follows:

Algorithm 2. Judge Peak Month

Input: *Monthplus*.
Output: *Flag*.
 1: **function** JUDGEPEAKMONTH(*Month*)
 2: $i \leftarrow 2$
 3: $beginMonth \leftarrow Month_1$
 4: **while** $i \leq 12$ **do**
 5:    **if** $beginMonth \times plus \leq Month_i$ **then**
 6:      $Flag_i \leftarrow 1$
 7:    **else**
 8:      $beginMonth \leftarrow Month_i$
 9:    **end if**
10:    $i \leftarrow i + 1$
11: **end while**
12:
13: **return** *Flag*

Single Exponential Smoothing. At this point, what this article discussed is not to deal with the error. As we discussed earlier, if F_t represents the future demand forecast and when a new demand period becomes reality and can be observed, the actual sales Y_t becomes available and allows us to calculate the prediction error, $Y_t - F_t$. SES (single exponential smoothing) evaluates the previous demand in essence, using the prediction error of the result to adjust the result and generate the forecast of the next period:

$$F_{t+1} = F_t + \alpha(Y_t - F_t) \tag{16}$$

α is constant in a range of 0 to 1.

3.5 Conclusion

This chapter implements neural network prediction, identifies three models, and compares the results from the aspects of input and neural network structure. We find that model 1 has poor performance compared with model 2 and model 3. Model 3 has a stable structure compared with model 2. However, in a certain range, model 2 has the best performance. In the experiment, MAPE is volatile, but ME is stable. Therefore, the model fit for continuous prediction. To improve the model, this paper proposes two kinds of improved schemes. The classifier has a good performance in ME, and single exponential smoothing has a better performance in MAPE. Therefore, we can select different tools according to different purposes.

4   Construct and Analysis of Prediction Results

By default, $\alpha = 0.3, \eta = 0.9$, the number of hidden layers is 6, and the results of the contrast under different models are shown in Table 1.

Table 1. ME and MAPE of different models

Category	ME	MAPE
Model1	9896.03	0.39
Model2	80.60	0.21
Model3	396.50	0.28

The result of the first model presents 10-day total sales. It is normal that ME is large, so we compare the MAPE. Overall, model 2 has better performance.

Next, we explore how the same model of different neural network structures influences the results. First, we consider different numbers of neurons in the hidden layer effect on the prediction results. We select two models that have better performance, model 2 and model 3. The results are shown in Table 2, Fig. 5 and Fig. 6.

Table 2. Different numbers of neurons in the hidden layer of model 2

Number of neurons	ME	MAPE
3	−40.80	0.32
4	−70.5	0.35
5	−48.11	0.32
6	80.86	0.21
7	33.86	0.24
8	3.94	0.27
10	22.39	0.23
12	43.54	0.21
14	−4.60	0.25
16	119.65	0.20
20	334.03	0.21
24	222.15	0.21

We can see from Table 2, Fig. 5 and Fig. 6 that when we change the hidden layer, the ME of model 2 shows a large fluctuation, but the MAPE is stable in a certain range.

Next, we explore model 3. The results are shown in Table 3, Fig. 7 and Fig. 8:

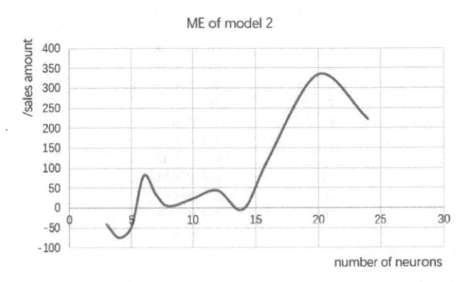

Fig. 5. The ME results for different numbers of neurons in hidden layer model 2.

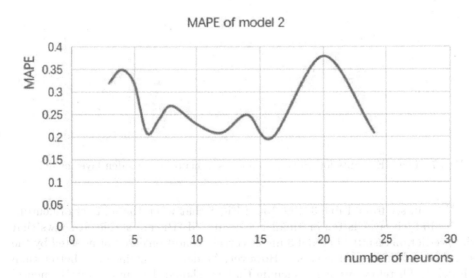

Fig. 6. MAPE results for different numbers of neurons in hidden layer model 2.

Table 3. Different numbers of neurons in the hidden layer of model 2

Number of neurons	ME	MAPE
3	335.40	0.29
4	363.72	0.29
5	341.75	0.28
6	396.50	0.28
7	468.80	0.20
8	417.75	0.28
10	442.27	0.27
12	637.17	0.47
14	516.78	0.34
16	818.74	0.66
20	644.46	0.48
24	608.81	0.44

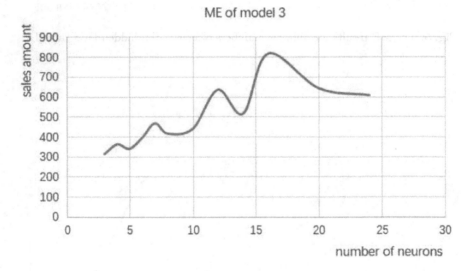

Fig. 7. The ME results for different numbers of neurons in hidden layer model 2.

We can see from Table 3, Fig. 7 and Fig. 8 that when the number of neurons in the hidden layer is ten or smaller, ME and MAPE are stable. It shows that the prediction results of model 3 under certain conditions are not affected by the structure of the neural network. However, Model 2 is significantly better than Model 3. Therefore, we can conclude that we should determine suitable neural network parameters to obtain better results. We can see that even if the MAPE in model 2 is still not low, the ME is low enough. We hypothesize that this may be because in the seven-day prediction, some data are higher than the real value,

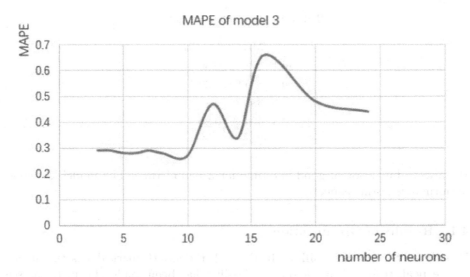

Fig. 8. MAPE results for different numbers of neurons in hidden layer model 2.

and some are lower. Therefore, when the prediction value is summed, there will be a lower ME.

Next, we explore how the neural network algorithm affects the results. We explore the influence of alpha and *alpha* and η of different values on the result.

First, when $\alpha = 0.3$, different η influences on the results are shown in Table 4:

Table 4. Different η of model 3

η	ME	MAPE
0.9	396.5	0.28
0.8	276.63	0.30
0.7	203.41	0.31

Next when $\eta = 0.9$, different α influences on the results are shown in Table 5:

We can see that this has little effect on the results. It also shows that in the neural network algorithm, the learning step size has a larger influence on the result.

We can conclude that model 1 has poor performance compared with models 2 and 3. Compared with model 2, model 3 is less affected by the neural network structure in a certain range. However, model 2 has the best performance. Comparing the results through different methods, we can see that MAPE is volatile, but ME is stable and ideal. Therefore, these models fit for continuous prediction so that the total error is small. In fact, we seldom predict the sales of one day.

Table 5. Different α of model 3

α	ME	MAPE
0.1	223.62	0.30
0.2	355.69	0.29
0.3	381.75	0.28
0.4	406.98	0.28

We always buy a lot of good one-time for a period time. The model has good performance in this aspect.

4.1   Results of Optimization

Result of Simple Classifier. In the end, the month marked 1 is the month in the peak season. Now, a simple classifier has been made. Then, let us see how the results are optimized. We select model 2, which has better performance (Table 6).

Table 6. The result of model in possession of classifier or not

Type	ME	MAPE
Have classifier	43.36	0.23
Don't have classifier	80.60	0.21

Result of Single Exponential Smoothing. In this paper, $\alpha = 0.9$ (distinguish with weights fixed constant), model 2, the number of neurons in hidden layers is 6, $\alpha = 0.3, \eta = 0.9$ and the results are shown in Table 7:

Table 7. The comparison of result optimized or not

Type	ME	MAPE
Normal	43.36	0.23
Optimized	72.22	0.15

We can find out that the MAPE is decreased a lot!

5   Conclusions

In this paper, three sales forecasting models are established by using neural network technology, and two methods are used to improve them.

First, we perform sales data acquisition. However, the data has some 'bad points'. Data processing is performed by using the three-point-one smoothing algorithm for the purpose of minimizing data impact and improving the data.

The prediction model is composed of four parts. First, we establish a BP network. Then, three models are established according to different thoughts: time forecasts 10-day total sales, ten-consecutive-day sales data to predict the next day, and twenty-consecutive-day sales data to predict the next day. Then, the advantages and disadvantages of the model are discussed from two aspects: the input and the structure of the neural network. These models are all designed to solve the problem of low data dimensions. We also propose a novel method of prediction in which we make the current forecast as a new round of input. We conclude that suitable parameters are important. Finally, two optimization schemes are proposed. Which are, respectively the predictive model with classifier and the predictive model with one exponential smoothing method. It is concluded that the performance of the second model is optimal under certain conditions and is suitable for continuous prediction.

References

1. Amiri, M., Amnieh, H.B., Hasanipanah, M., Khanli, L.M.: A new combination of artificial neural network and k -nearest neighbors models to predict blast-induced ground vibration and air-overpressure. Eng. Comput. **32**(4), 631–644 (2016)
2. Atiya, A.F., Elshoura, S.M., Shaheen, S.I., Elsherif, M.S.: A comparison between neural-network forecasting techniques-case study: river flow forecasting. IEEE Trans. Neural Netw. **10**(2), 402–9 (1999)
3. Chatfield, C., Weigend, A.S.: Time series prediction: forecasting the future and understanding the past: Neil A. Gershenfeld and Andreas S. Weigend, 1994, 'the future of time series'. In: Weigend, A.S., Gershenfeld, N.A. (eds.) International Journal of Forecasting, vol. 10, no. 1, pp. 161–163. Addison-Wesley, Reading (1994). 1–70
4. Chen, C.H.: Neural networks for financial market prediction. In: IEEE International Conference on Neural Networks, 1994. IEEE World Congress on Computational Intelligence, vol. 2, pp. 1199–1202 (2002)
5. Faraggi, E., Kloczkowski, A.: GENN: a general neural network for learning tabulated data with examples from protein structure prediction. Methods Mol. Biol. **1260**(1260), 165 (2015)
6. Huarng, K., Yu, H.K.: The application of neural networks to forecast fuzzy time series. Phys. A **363**(2), 481–491 (2006)
7. Hussain, A.J., Fergus, P., Al-Askar, H., Al-Jumeily, D., Jager, F.: Dynamic neural network architecture inspired by the immune algorithm to predict preterm deliveries in pregnant women. Neurocomputing **151**(3), 963–974 (2015)
8. Kaastra, I., Boyd, M.: Designing a neural network for forecasting financial and economic time series. Neurocomputing **10**(3), 215–236 (1996)

9. Keskin, T.E., Düğenci, M., Kaçaroğlu, F.: Prediction of water pollution sources using artificial neural networks in the study areas of sivas, karabük and bartın (turkey). Environ. Earth Sci. **73**(9), 5333–5347 (2015)

10. Lu, C.J., Lee, T.S., Lian, C.M.: Sales forecasting for computer wholesalers: a comparison of multivariate adaptive regression splines and artificial neural networks. Decis. Support Syst. **54**(1), 584–596 (2012)

11. Lyon, E., Dearden, G., Cheng, H., Shenton, T., Page, V., Kuang, Z.: Neural network prediction of engine performance for second pulse fire/no fire decision making in dual pulse laser ignited engines. Plant Cell **16**(6), 1365–77 (2015)

12. Pindoriya, N.M., Singh, S.N., Singh, S.K.: Application of adaptive wavelet neural network to forecast operating reserve requirements in forward ancillary services market. Appl. Soft Comput. **1**, 1811–1819 (2011)

13. Scott, S.L., Varian, H.R.: Bayesian variable selection for nowcasting economic time series. In: NBER Working Papers (2012)

14. Smith, C., Wunsch, D.: Time series prediction via two-step clustering. In: International Joint Conference on Neural Networks, pp. 1–4 (2015)

15. Taylor, J.W., Buizza, R.: Neural network load forecasting with weather ensemble predictions. IEEE Power Eng. Rev. **22**(7), 59–59 (2007)

16. Were, K., Bui, D.T., Dick, Ø.B., Singh, B.R.: A comparative assessment of support vector regression, artificial neural networks, and random forests for predicting and mapping soil organic carbon stocks across an afromontane landscape. Ecol. Indicators **52**, 394–403 (2015)

17. Zhang, G.P., Qi, M.: Neural network forecasting for seasonal and trend time series. Eur. J. Oper. Res. **160**(2), 501–514 (2005)

Lightweight and Efficient Attention-Based Superresolution Generative Adversarial Networks

Shushu Yin[1], Hefan Li[2], Yu Sang[1(✉)], Tianjiao Ma[1], Tie Li[1], and Mei Jia[1]

[1] School of Electronic and Information Engineering, Liaoning Technical University,
Huludao 125105, China
sangyu2008bj@sina.com

[2] The First Affiliated Hospital, Dalian Medical University, Dalian 116011, China

Abstract. To address the problems of lack of high-frequency information and texture details and unstable training in superresolution generative adversarial networks, this paper optimizes the generator and discriminator based on the SRGAN model. First, the residual dense block is used as the basic structural unit of the generator to improve the network's feature extraction capability. Second, enhanced lightweight coordinate attention is incorporated to help the network more precisely concentrate on high-frequency location information, thereby allowing the generator to produce more realistic image reconstruction results. Then, we propose a symmetric and efficient pyramidal segmentation attention discriminator network in which the attention mechanism is capable of deriving finer-grained multiscale spatial information and creating long-term dependencies between multiscale channel attentions, thus enhancing the discriminative ability of the network. Finally, a Charbonnier loss function and a gradient variance loss function with improved robustness are used to better realize the image's texture structure and enhance the model's stability. The findings from the experiments reveal that the reconstructed image quality enhances the average peak signal-to-noise ratio (PSNR) by 1.59 dB and the structural similarity index (SSIM) by 0.045 when compared to SRGAN on the three test sets. Compared with the state-of-the-art methods, the reconstructed images have a clearer texture structure, richer high-frequency details, and better visual effects.

Keywords: Superresolution · Generative adversarial networks · Attention mechanism · Texture structure · Residual dense blocks

1 Introduction

Superresolution (SR) creates a high-resolution (HR) image from a single low-resolution (LR) picture, thereby enhancing the image's clarity and quality. Superresolution reconstruction at the algorithmic level is one of the most important areas of study in computer vision and image processing. Numerous real-world applications include healthcare imaging [1], security [2], satellite remote sensing [3], facial recognition [4], and radar imaging [5]. SR has the discomfort problem of outputting multiple HR images for input LR

© The Author(s), under exclusive license to Springer Nature Singapore Pte Ltd. 2023
Z. Yu et al. (Eds.): ICPCSEE 2023, CCIS 1879, pp. 165–181, 2023.
https://doi.org/10.1007/978-981-99-5968-6_12

images. As a result, many SR techniques, such as interpolation-based [6] approaches, reconstruction-based [7] approaches, shallow learning [8]-based approaches, and the more recent deep learning [9]-based methods, have been suggested.

Deep learning algorithms have become increasingly prevalent in superresolution image reconstruction in recent years [10–16] due to their benefits, such as quick computing speed and outstanding performance. Deep learning methods aim at hierarchical representation using neural networks [17] to achieve high-level feature extraction of the input image. Although the deep learning-based SR algorithm can recover lost details from LR observations, it employs a pixel loss function during training, which can lead to overly smooth edges and the loss of high-frequency details in the generated images. This algorithm also maximizes the peak signal-to-noise ratio (PSNR) measurement. However, having a greater PSNR metric cannot guarantee an aesthetically appealing SR image. To solve these problems, generative adversarial networks (GANs) were introduced to the SR task, which uses a multilayer neural network combining generative and discriminative networks to continuously improve the image generation quality and obtain realistic image texture features using adversarial learning, thus making the reconstructed image present better visual realism. The close combination of generative adversarial networks and superresolution further promotes the development of image superresolution reconstruction towards high-quality perception.

In the GAN-based SR method [18–22], the discriminator network collaborates with the generator network to generate SR photos with enhanced high-frequency characteristics that more closely resemble HR images taken in the real world. One of the first applications of generative adversarial networks in superresolution practice was implemented in the SRGAN model [18], in which the generator network takes LR pictures as input and generates HR pictures through a deep residual network, and the discriminator network is responsible for determining the truth of the images that are produced from the real images and reconstructing the SR images using perceptual loss and adversarial loss. The ESRGAN model [19] boosts the SRGAN by calculating the loss using the map of features before activation and adapting the standard discriminative network to a relative discriminative network, i.e., instead of judging whether the output picture is true or false, more true or more false probability is evaluated, highlighting the picture's detail enhancement. Nevertheless, images reconstructed using the SRGAN model frequently lose a substantial quantity of high-frequency details, resulting in tessellation and ring effects. This article improves the SRGAN model based on this information. The following is a summary of the key contributions:

- We design a residual dense coordinate attention generation network (RRDB-CARB) consisting of residual dense blocks (RRDB) and coordinated attention (CA) to collect the depth characteristics of a picture and subsequently concentrate on the high-frequency location information.
- We propose a symmetric pyramid split attention discriminant network (SPSAD) discriminator network that can better manage the spatial information of multiscale input feature maps and enhance the discriminator model's capacity to distinguish between true and false.
- We replace the original MSE loss with a Charbonnier loss with better robustness. In light of this, since pixel loss will make the generated image contours blurred and

the network training of GAN is often unstable, which may produce redundant and inaccurate texture information, this paper introduces Gradient Variance (GV) loss, which improves the produced texture's structure via gradient mapping using SR and HR pictures, and the activation functions in this model all use Aconc.

- The outcomes of the experiments show that in all three test sets, the model presented in this study outperformed the SRGAN model, and the rebuilt images exhibit better visual results and texture structure.

2 Related Work

2.1 Superresolution of Images Based on Deep Learning

Deep learning-based methods use multilayer nonlinear transformations to describe abstract features, and their nonlinear fitting performance is superior to that of traditional methods, showing great potential in superresolution reconstruction. Recently, deep-learning reconstruction methodologies have attracted widespread interest because of the increasing demand for their implementation in superresolution reconstruction.

Dong et al. [10] implemented the superresolution method (SRCNN) using a full-forward neural network, which provides an enormous boost in precision compared with conventional algorithms but makes the network computationally intensive and inefficient due to presampling the LR to the required size. Shi et al. [11] used subpixel convolutional layers (ESPCN) to convert LR photos to HR photos for this purpose, allowing for fast and efficient end-to-end learning. Inspired by the VGG network model, Kim et al. [12] proposed a VDSR network with 20 weight layers that uses residual learning and gradient cropping to achieve excellent convergence. Subsequently, DRCN was proposed by Kim et al. [13] to achieve parameter sharing among recursive units by repeatedly feeding data into a recursive layer containing 16 recursive units. Lim et al. [14] suggested the EDSR, which eliminates the meaningless batch normalization (BN) layer in traditional residual networks and adds a constant scaling layer to keep shallow image information and decrease the number of parameters. Several lightweight model structures have been developed recently. Bhardwaj et al. [15] proposed the SESR algorithm using a linear overparameterization technique to achieve a good balance in terms of superresolution photograph quality and computation. Zhang et al. [16] suggested the ECBSR algorithm, which solves the superresolution problem for arbitrary multiples using structural reparameterization techniques.

2.2 Generative Adversarial Network-Based Image Superresolution

Ian Goodfellow et al. [23] put forward the GAN in 2014; the network provides a strong fitting capability and is significant in the realm of computer vision. It is made up of a generator (G) and a discriminator (D), and the network is trained so that G is able to become fuzzy for D, while D can differentiate well between images produced by generator G and original images. GAN-based approaches include LGGAN [24], UCTGAN [25], and House-GAN [26], which aim to progressively extend the application domain and functionality of generative adversarial networks. Simultaneously, many GAN-based SR methods [18–22] have emerged, yielding excellent outcomes and boosting the efficiency of network models. SRGAN was suggested by Ledig et al. [18] and employs adversarial learning to improve image sharpness. Wang et al. [19] developed ESRGAN, which uses residual density blocks instead of the original residual blocks to reconstruct images with better texture information and sharper visual effects. Chen et al. [20] suggested that HSRGAN obtains image features by integrating various modules to gradually generate more natural texture images. Lei et al. [21] proposed HFF-SRGAN to reduce image noise and improve the learning capacity of the model by introducing high-frequency information to make the generated SR photos clear and natural. Jia et al. [22] suggested an improved SRGAN algorithm to improve image reconstruction quality by incorporating an attention mechanism in the generator and employing a double discriminator.

3   Methodology

In this article, the SRGAN algorithm is enhanced. We use a generative adversarial network learning approach to jointly train a mutually adversarial generator along with a discriminator. After iterative training, the network reaches Nash equilibrium, making the reconstructed picture indistinguishable from the original picture to the human eye. Figure 1 depicts the model's action procedure. First, the LR picture is input to the generator network (RRDB-CARB), composed of residual dense blocks and improved lightweight coordinate attention, in which the residual dense blocks extract richer image feature information in a densely connected manner and the coordinated attention mechanism gives priority to high-frequency detail so that the generator obtains more realistic image reconstruction results. The generator network's SR photos and the actual HR photos are then fed into a symmetric pyramidal segmented attention discriminator network (SPSAD), where the PSA seeks to learn attention weights with low model complexity and efficiently combine local and global attention to create long-term channel dependencies, thereby improving the discriminator's potential to differentiate true from false. In this article, the model adopts a novel combination of perceptual loss functions during the training procedure, incorporating Charbonnier loss with better noise reduction, gradient variance loss with enhanced texture details, higher-level semantic-based perceptual loss, and adversarial loss to achieve optimization of the total loss function, make the training of the model more stable, and generate clearer image texture details.

3.1   RRDB-CARB

The generator network is a structure based on residual dense blocks and improved coordinate attention (RRDB-CARB); Fig. 2 depicts its composition. The network is composed of three sections: shallow feature extraction, consisting of a single convolutional layer; deep feature extraction, consisting of multiple cascaded RRDB modules; and the image reconstruction module, consisting of an upsampling module and a coordinate attention module. Where each dashed box represents a submodule. Following the entry of the low-resolution picture I_{LR} into the generator, the superresolution picture I_{SR} is the outcome via the generator network.

The specific process is as follows: first, the network extracts shallow features from LR images by 3×3 convolutional layers F_0.

$$F_0 = H_{Conv}(I_{LR}), \tag{1}$$

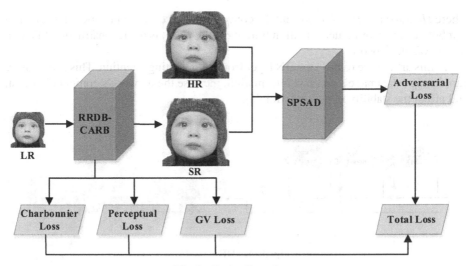

Fig. 1. Entire flow chart

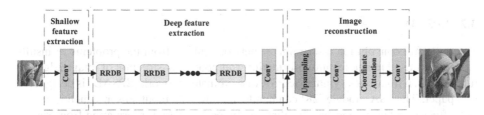

Fig. 2. Residual Dense-Coordinated Attention Generation Network (RRDB-CARB)

where $H_{Conv}(\cdot)$ means convolution and the outcome F_0 is input to the depth feature extraction module. The process is represented as:

$$F_t = H_{RRDB}(F_{t-1}), t = 1, 2, ..., M \tag{2}$$

where F_{t-1} and F_t denote the input and output of the t-th RRDB block, respectively, and $H_{RRDB}(\cdot)$ represents the RRDB deep feature extraction function. The RRDB block can learn deep semantic features while improving the underlying feature extraction. Its structure is shown in Fig. 3. Each RRDB block is made up of three cascaded dense blocks, each of which consists of four closely connected convolutions.

In the image reconstruction process, the upsampling module consists of subpixel convolutional layers, and since the channel relationships in the feature map after upsampling are important for the image reconstruction quality, the coordinate attention after upsampling is introduced to the generator. Coordinative attention [27] emphasizes high-frequency location information and has the advantage of being lightweight and efficient by adding almost no computational overhead. The reconstruction process is as follows:

$$I_{SR} = H_{Conv}(F_{CA}(H_{Conv}(H_{up}(F_t + F_0)))) = R_{rec}(I_{LR}), \tag{3}$$

where $H_{Conv}(\cdot)$ and $H_{up}(\cdot)$ represent the convolution and upsampling operations, $F_{CA}(\cdot)$ symbolizes the coordinated attention function, and $R_{rec}(\cdot)$ is the reconstruction function for the whole network.

In this article, we remove the BN layer from coordinating attention. This is because it can damage the image's contrast information, increase the network's computation cost, and affect the stability of network training.

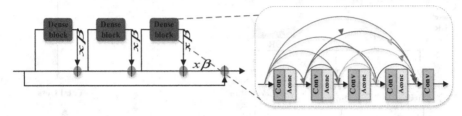

Fig. 3. RRDB module

3.2 SPSAD

Since the full convolution discriminator network suffers from the problems of insufficiently extracted image feature information and poor reconstruction of image edge texture details, a front-back symmetric PSA discriminator network (SPSAD) is designed in this paper. The PSA module [28] can handle a multiscale input tensor, which is a novel and effective attention mechanism that combines contextual information from various scales and successfully integrates local and global attention. This improves information interaction between local and global channels and creates long-term channel dependencies. As shown in Fig. 4, the network keeps the BN layer used to stabilize the training model. It takes the genuine HR photograph and the rebuilt SR picture as inputs, with its goals being to identify the error likelihood between the recreated SR picture and the genuine HR picture, as well as to gain insight into the data distribution of the input images. The network first passes through the convolution and activation layers after receiving the images, followed by three residual blocks and a PSA module, and finally outputs the reconstruction results through the convolution layer and the softplus function. Each of these residual blocks includes a level of convolution, the BN level, and a nonlinear activation function, Aconc, which can effectively learn the input image feature representation and thus boost the discriminative network's capacity. Aconc learns whether to activate each neuron or not, effectively improving the efficiency of network transmission. The overall training process is performed until the generator obtains the desired image and the discriminator is unable to discriminate between the two different inputs.

3.3 Loss Function

Content Loss. Since the MSE loss function is more sensitive to outliers by optimizing the squared difference between HR and LR pixels, the resulting pictures lack high-frequency

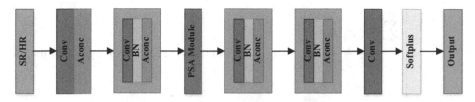

Fig. 4. Symmetric PSA Discriminator Network (SPSAD)

information and will degrade the model's performance. Therefore, the Charbonnier loss, which has better robustness, is used in the paper instead of the MSE loss to better handle the outliers in the network and save the time of network training, as in Eq. (4).

$$L_{Charb}\left(I_{HR}, \hat{I}_{HR}\right) = \lambda E_{LR,HR\ P_{data}(LR,HR)}\varphi(I_{HR} - G_\theta(I_{LR})), \tag{4}$$

where λ represents the hyperparameter and $\varphi(x) = \sqrt{x^2 + \varepsilon^2}$ is the penalty function for the Charbonnier loss. $\varepsilon = 10^{-7}$.

Since using only pixel loss tends to make the generated images appear blurry and artifactual, the characteristic loss values are acquired by applying a pretrained VGG-16 model. As in Eq. (5).

$$L_{VGG} = \frac{1}{W_{i,j}H_{i,j}} \sum_{x=1}^{W_{i,j}} \sum_{y=1}^{H_{i,j}} \left(\phi_{i,j}(I_{HR})_{x,y} - \phi_{i,j}\big(G_{\theta_G}(I_{LR})\big)_{x,y}\right)^2, \tag{5}$$

where $W_{i,j}$ and $H_{i,j}$ denote the dimensionality of the picture, and $\phi_{i,j}(\cdot)$ denotes the VGG model function.

Adversarial Loss. We employ the same adversarial loss function that SRGAN does, which measures the likelihood that the reconstructed high-resolution picture given by the discriminator network is the true image, as in Eq. (6).

$$L_{adv} = \sum_{n=1}^{N} -\lg D_{\theta_D}\big(G_{\theta_G}(I_{LR})\big), \tag{6}$$

where $G_{\theta_G}(I_{LR})$ indicates the rebuilt picture and $D_{\theta_D}\big(G_{\theta_G}(I_{LR})\big)$ stands for the likelihood that the discriminator reconstructs the graphic created via the generator as a natural photo.

Gradient Variance Loss. Since the variance of the gradient map generated by the pixel loss generation networks is vastly shorter than that of the original image and the rebuilt SR picture is often oversmoothed, this paper introduces the texture structure enhancement loss function [29], which deploys the gradient mapping of SR and HR photographs to boost the texture details and edge information of the generated images. Given the variance mappings v_X^{SR}, v_y^{SR}, v_x^{HR}, and v_y^{HR} corresponding to the I_{SR} and I_{HR} images, as shown in Eq. (7):

$$L_{GV} = E_{SR} \left\| v_X^{SR} - v_X^{HR} \right\|_2 + E_{SR} \left\| v_y^{SR} - v_y^{HR} \right\|_2, \tag{7}$$

where $V_i = \left[\dfrac{\sum\limits_{j=1}^{n^2}(\check{G}_{i,j}-\mu_i)^2}{n^2-1}\right], i = 1, ..., \frac{w \cdot h}{n^2}$, denotes the matrix variogram for the i-th

element, $\check{G}_{i,j}$ denotes the expanded gradient plot, and μ_i reflects the mean of the i-th picture block.

Total Loss Function. In conclusion, Eq. (8) summarizes the loss function.

$$L_{total} = L_{Charb} + \alpha L_{VGG} + \beta L_{adv} + \gamma L_{GV}, \tag{8}$$

4 Experiments

4.1 Experimental Setup

The hardware equipment of this experiment is an Intel® Xeon(R) E5-2678v3 CPU @2.5 GHz, 32 GB of running memory, and an NVIDIA GeForce RTX 3090 24-GB graphics card. The virtual environment is Anaconda 3, Pycharm (Python 3.9), Cuda version 11.4, Pytorch is the deep learning engine, and Ubuntu is the operating system. A total of 800 training sets and 100 validation sets from the DIV2K [30] dataset are used for the experiments. The test sets were tested using three generic benchmark sets: Set5 [31], Set14 [32], and BSD 100 [33]. The pictures are reduced to 88 × 88 size and input into the model for training, and the LR photos are constructed by four times double-triple downsampling of the HR photograph. The batch size is 64, the learning rate is 2×10^{-4}, and the adaptive matrix (Adam) is selected as the optimizer, with $\alpha = 6 \times 10^{-3}$, $\beta = 10^{-3}$, and $\gamma = 10^{-5}$ in Eq. (8), for a total of 400 epochs trained.

4.2 Evaluation Indicators

The peak signal-to-noise ratio (PSNR) and structural similarity (SSIM) are two objective quality assessment metrics that are frequently employed. PSNR is a popular way to compare pixel differences, and the greater the number, the more accurate the picture that comes out. SSIM is a brightness, contrast, and structure-based measurement method that determines the similarity of two graphs. The value varies from 0 to 1, with a value near 1 demonstrating that there is less error between the rebuilt and initial pictures. PSNR is computed by first calculating the mean squared error (MSE), and the full mathematical expression is:

$$MSE = \frac{1}{H \times W}\sum_{i=1}^{H}\sum_{j=1}^{W}[X(i,j) - Y(i,j)]^2 \tag{9}$$

$$PSNR = 10 \cdot \log_{10}[\frac{(2^n-1)^2}{MSE}] \tag{10}$$

where X and Y stand for the initial and produced photograph, respectively, and H and W are the sizes of pictures X and Y, respectively.

$$SSIM(X, Y) = \frac{(2u_X u_Y + C_1)(2\sigma_{XY} + C_2)}{(u_X^2 + u_Y^2 + C_1)(\sigma_X^2 + \sigma_Y^2 + C_2)}, \tag{11}$$

where u_X and u_Y reflect the averages of X as well as Y, σ_X^2 and σ_Y^2 indicate the variances of X along with Y, and σ_{XY} represents the covariances of X together with Y.

4.3 Analysis of Results

Experiments were run on the datasets Set5, Set14, and BSD100 to investigate the generalization of the model in this article, and the reconstructed findings were compared with Bicubic, SRCNN [10], ESPCN [11], SESR [15], ECBSR [16], SRGAN [18], ESRGAN [19], HFF-SRGAN [21], and literature [22]. These classical algorithms are compared subjectively and objectively. Therefore, the experiments are fair, and all the training sets are DIV2K and are performed at a magnification of 4. The stability of the network during training is observed on a TensorBoard and represented by a graph. Figure 5 depicts the initial SRGAN's generator loss function variation curve, while Fig. 6 depicts the model in this paper's generator loss function variation curve. When comparing Figs. 5 and 6, the loss convergence and reconstruction effects of this article's approach are superior.

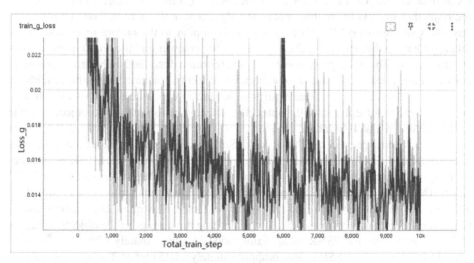

Fig. 5. SRGAN generator loss function variation curve

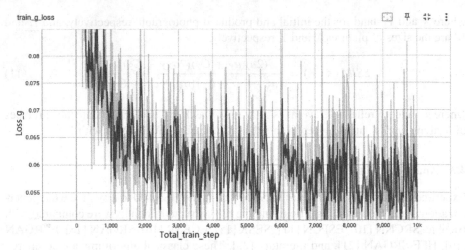

Fig. 6. The variation curve of the loss function of the method generator in this article

Figures 7, 8 and 9 present a subjective comparison of the approach used in this article and the classical algorithm, and the images of "woman" in the Set5 dataset, "pepper" in the Set14 dataset, and "floor" in the BSD100 dataset are selected for visualization. The reconstructed images from SRGAN basically overcome the ringing effect, but the reconstructed images are too smooth. ESRGAN generates the clearest effect map, but the objective value is lower than that of the algorithm in this article. The literature [22] describes an improved SRGAN that is visually similar to the picture reconstructed by this approach in this study, but the objective value is lower and the output picture lacks texture details. The photograph rebuilt by the algorithm described in the article has more distinct texture details, similar color vibrancy to the initial picture, and the highest PSNR and SSIM values. It has been demonstrated that the approach described in this article is an effective one for enhancing the subjective experience as well as the objective value of the picture in a well-balanced manner.

Tables 1 and 2 present the PSNR and SSIM values of various SR restoration techniques, respectively. The table demonstrates that the deep learning-based SR reconstruction approach outperforms the conventional bicubic interpolation approach. The PSNR and SSIM values of the approach in this article are higher than those of other models in all three test sets under the 4-fold magnification factor, illustrating the superior reconstruction performance of this model. By calculating the average of the objective metrics in the three test sets, the PSNR value of this model is approximately 1.59 dB higher than that of SRGAN, and the SSIM value is approximately 0.045 higher. The network training stability of this method is better than that of the initial SRGAN approach, and the level of accuracy of the images that were reconstructed is better than that of other network models, according to comparisons between the loss function, reconstructed images, and assessment metrics.

(a) original   (b) Bicubic   (c) SRCNN   (d) ESPCN   (e) SRGAN
PSNR=26.460   PSNR=29.007   PSNR=27.855   PSNR=27.938
SSIM=0.832   SSIM=0.885   SSIM=0.858   SSIM=0.859

(f) ESRGAN   (g) Reference[22]   (h) Ours
PSNR=29.730   PSNR=29.833   PSNR=33.710
SSIM=0.895   SSIM=0.902   SSIM=0.931

Fig. 7. Results of contrast "woman" picture rebuilding on test set 5

(a) original   (b) Bicubic   (c) SRCNN   (d) ESPCN   (e) SRGAN
PSNR=30.586   PSNR=32.827   PSNR=31.231   PSNR=30.925
SSIM=0.837   SSIM=0.864   SSIM=0.843   SSIM=0.836

(f) ESRGAN   (g) Reference[22]   (h) Ours
PSNR=32.615   PSNR=33.520   PSNR=33.773
SSIM=0.833   SSIM=0.873   SSIM=0.875

Fig. 8. Results of contrast "pepper" picture rebuilding on test set 14

(a) original   (b) Bicubic   (c) SRCNN   (d) ESPCN   (e) SRGAN
PSNR=24.133   PSNR=24.928   PSNR=24.577   PSNR=24.499
SSIM=0.692   SSIM=0.743   SSIM=0.727   SSIM=0.717

(f) ESRGAN   (g) Reference[22]   (h) Ours
PSNR=25.086   PSNR=25.443   PSNR=39.122
SSIM=0.773   SSIM=0.771   SSIM=0.978

Fig. 9. Results of contrast "floor" picture rebuilding on test BSD 100

Table 1. The PSNR values of various SR approaches were compared on three test sets.

Dataset	Scale	Bicubic	SRCNN [10]	ESPCN [11]	SESR-M11 [15]	ECBSR-M4C16 [16]	SRGAN [18]	ESRGAN[19]	HFF-SRGAN [21]	Reference [22]	Ours
Set5	4	28.416	30.445	29.140	31.270	31.040	29.182	30.459	29.278	31.182	**31.536**
Set14	4	26.086	27.554	26.338	27.940	27.780	26.906	26.606	27.738	27.811	**28.341**
BSD 100	4	25.955	26.856	26.278	27.200	27.090	26.301	25.315	26.477	27.155	**27.270**

Table 2. The SSIM values of various SR approaches were compared on three test sets.

Dataset	Scale	Bicubic	SRCNN [10]	ESPCN [11]	SESR-M11 [15]	ECBSR-M4C16 [16]	SRGAN[18]	ESRGAN[19]	HFF-SRGAN [21]	Reference [22]	Ours
Set5	4	0.810	0.864	0.832	0.881	0.881	0.834	0.851	0.853	0.879	**0.886**
Set14	4	0.704	0.754	0.740	0.766	0.769	0.740	0.713	0.779	0.776	**0.781**
BSD 100	4	0.667	0.711	0.697	0.723	0.728	0.688	0.650	0.729	0.723	**0.730**

5 Ablation Study

Ablation studies were carried out for every component in this article to assess the achievement of the RRDB module, Charbannier module, and GV module, as well as the CA module and PSA module. Based on the SRGAN model, the same number of rounds are trained with a 4-fold amplification factor for all the models under the same parameters. Tables 3 and 4 illustrate the PSNR and SSIM values obtained through the experimental results. According to the tables, adding any of the improvement modules to the initial SRGAN approach may boost the rebuilding outcome values of the algorithm, and adding five modules at the same time improves the objective values the most. The subjective effect plots are shown in Figs. 10 and 11, and it can be observed that adding different modules separately achieves better objective assessment values than the original model. The improvement in the visual effect of a single module is not very obvious, while when five modules are used at the same time, the improvement in the visual effect is very significant. In conclusion, adding each module improves image reconstruction, and ablation tests further show the viability and efficacy of the algorithm presented in this article.

Table 3. PSNR values across three sets of tests with various section combinations

Method	Set5	Set14	BSD100
SRGAN(Baseline)	29.182	26.906	26.302
+RRDB	30.411	27.349	26.861
+Charbannier	31.103	27.802	27.124
+Gradient Variance	30.781	27.617	27.022
+Coordinate Attention	30.550	27.456	26.845
+PSA	30.734	27.602	26.997
Ours	**31.536**	**28.341**	**27.270**

Table 4. SSIM values across three sets of tests with various section combinations

Method	Set5	Set14	BSD100
SRGAN(Baseline)	0.834	0.740	0.690
+RRDB	0.860	0.761	0.712
+Charbannier	0.879	0.776	0.722
+Gradient Variance	0.869	0.769	0.715
+Coordinate Attention	0.867	0.767	0.714
+PSA	0.866	0.767	0.715
Ours	**0.886**	**0.781**	**0.730**

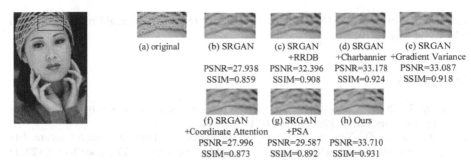

(a) original (b) SRGAN (c) SRGAN (d) SRGAN (e) SRGAN
+RRDB +Charbannier +Gradient Variance
PSNR=27.938 PSNR=32.396 PSNR=33.178 PSNR=33.087
SSIM=0.859 SSIM=0.908 SSIM=0.924 SSIM=0.918

(f) SRGAN (g) SRGAN (h) Ours
+Coordinate Attention +PSA
PSNR=27.996 PSNR=29.587 PSNR=33.710
SSIM=0.873 SSIM=0.892 SSIM=0.931

Fig. 10. Contrast of Set 5's "woman" image reconstruction findings

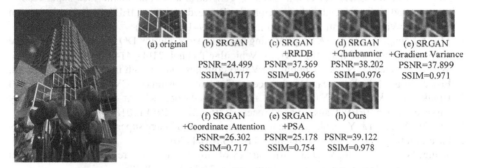

(a) original (b) SRGAN (c) SRGAN (d) SRGAN (e) SRGAN
+RRDB +Charbannier +Gradient Variance
PSNR=24.499 PSNR=37.369 PSNR=38.202 PSNR=37.899
SSIM=0.717 SSIM=0.966 SSIM=0.976 SSIM=0.971

(f) SRGAN (e) SRGAN (h) Ours
+Coordinate Attention +PSA
PSNR=26.302 PSNR=25.178 PSNR=39.122
SSIM=0.717 SSIM=0.754 SSIM=0.978

Fig. 11. Contrast of Set 14's "floor" image reconstruction findings

6 Conclusion

In this paper, we propose a lightweight and efficient attentional superresolution-based generative adversarial approach. First, the generator adopts a residual dense structure with a coordinated attention mechanism; the residual dense block extracts deep semantic features of the picture to ease the training difficulty; an effective Aconc activation function adaptively selects activation neurons to enhance network transmission ability; and coordinated attention centering emphasizes high-frequency content. Then, a symmetric PSA discriminator network is designed, which lets the discriminator acquire wealthier picture characteristic detail due to the capacity of the PSA module to manage the multiscale input tensor. Finally, the Charbonnier loss and gradient variance loss are used to guide the network to generate clearer, more realistic images with richer texture structures. The experimental results indicate that the approach suggested in this article can achieve optimal outcomes in objective assessment indices compared to other techniques. In regard to visual excellence, the suggested approach can handle blurring artifacts well and recover more high-frequency details, which improves the visual perception quality.

Acknowledgements. This work was supported in part by the Basic Scientific Research Project of Liaoning Provincial Department of Education under Grant Nos. LJKQZ2021152 and LJ2020JCL007; in part by the National Science Foundation of China (NSFC) under Grant No.

61602226; and in part by the PhD Startup Foundation of Liaoning Technical University of China under Grant Nos. 18-1021.

References

1. Shang, J., Zhang, X., Zhang, G., et al.: Gated multi-attention feedback network for medical image super-resolution. Electronics **11**(21), 3554 (2022)
2. Li, H., Zheng, Q., Yan, W., et al.: Image superresolution reconstruction for secure data transmission in Internet of Things environment. Math. Biosci. Eng. **18**(5), 6652–6672 (2021)
3. Jia, S., Wang, Z., Li, Q., et al.: Multiattention generative adversarial network for remote sensing image superresolution. IEEE Trans. Geosci. Remote Sens. **60**, 1–15 (2022)
4. He, J., Shi, W., Chen, K., et al.: Gcfsr: a generative and controllable face super resolution method without facial and gan priors. In: Proceedings of the IEEE/CVF Conference on Computer Vision and Pattern Recognition, pp. 1889–1898 (2022)
5. Ma, Y., Zeng, Y., Sun, S.: A deep learning based super resolution DOA estimator with single snapshot mimo radar data. IEEE Trans. Vehicular Technol. **71**(4), 4142–4155 (2022)
6. Wu, C.Y., Singhal, N., Krahenbuhl, P.: Video compression through image interpolation. In: Proceedings of the European Conference on Computer Vision (ECCV), pp. 416–431 (2018)
7. Irmak, H., Akar, G.B., Yuksel, S.E.: A map-based approach for hyperspectral imagery superresolution. IEEE Trans. Image Process. **27**(6), 2942–2951 (2018)
8. Liu, N., Xu, X., Li, Y., et al.: Sparse representation based image superresolution on the KNN based dictionaries. Opt. Laser Technol. **110**, 135–144 (2019)
9. Wang, W., Hu, Y., Luo, Y., et al.: Brief survey of single image superresolution reconstruction based on deep learning approaches. Sens. Imag. **21**(1), 1–20 (2020)
10. Dong, C., Loy, C.C., He, K.M., et al.: Image superresolution using deep convolutional networks. IEEE Trans. Pattern Anal. Mach. Intell. **38**(2), 295–307 (2016)
11. Shi, W.Z., Caballero, J., Huszár, F., et al.: Real- time single image and video superresolution using an efficient subpixel convolutional neural network. In: Proceedings of the IEEE Conference on Computer Vision and Pattern Recognition, Las Vegas, Washington, pp. 1874–1883 (2016)
12. Kim, J., Lee, J.K., Lee, K.M.: Accurate image superresolution using very deep convolutional networks. In: Proceedings of the IEEE Conference on Computer Vision and Pattern Recognition, pp. 1646–1654 (2016)
13. Kim, J., Lee, J.K., Lee, K.M.: Deeply recursive convolutional network for image superresolution. In: Proceedings of the IEEE Conference on Computer Vision and Pattern Recognition, pp. 1637–1645 (2016)
14. Lim, B., Son, S., Kim, H., et al.: Enhanced deep residual networks for single image superresolution. In: Proceedings of the IEEE Conference on Computer Vision and Pattern Recognition Workshops, pp. 136–144 (2017)
15. Bhardwaj, K., Milosavljevic, M., O'Neil, L., et al.: Collapsible linear blocks for superefficient super resolution. arXiv:2103.09404 (2021)
16. Zhang, X., Zeng, H., Zhang, L.: Edge-oriented convolution block for real-time super resolution on mobile devices. In: Proceedings of the 29th ACM International Conference on Multimedia, pp. 4034–4043 (2021)
17. Pandey, G., Ghanekar, U.: A conspectus of deep learning techniques for single-image superresolution. Pattern Recognit. Image Anal. **32**(1), 11–32 (2022)
18. Ledig, C., Theis, L., Huszár, F., et al.: Photo-realistic single image superresolution using a generative adversarial network. In: Proceedings of the 2017 IEEE Conference on Computer Vision and Pattern Recognition, Honolulu, 21–26 July 2017, pp. 105–114. IEEE Computer Society, Washington (2017)

19. Wang, X., Ke, Y., Shixiang, W., Jinjin, G., Liu, Y., Chao Dong, Y., Qiao, C.C., Loy,: ESR-GAN: enhanced super-resolution generative adversarial networks. In: Leal-Taixé, L., Roth, S. (eds.) Computer Vision – ECCV 2018 Workshops: Munich, Germany, September 8-14, 2018, Proceedings, Part V, pp. 63–79. Springer International Publishing, Cham (2019). https://doi.org/10.1007/978-3-030-11021-5_5

20. Chen, W., Ma, Y., Liu, X., et al.: Hierarchical generative adversarial networks for single image superresolution. In: Proceedings of the IEEE/CVF Winter Conference on Applications of Computer Vision, pp. 355–364 (2021)

21. Lei, J., Xue, H., Yang, S., et al.: HFF-SRGAN: superresolution generative adversarial network based on high-frequency feature fusion. J. Electron. Imaging 31(3), 033011 (2022)

22. Jia, M., Lu, M., Sang, Y.: Advanced generative adversarial network for image superresolution. In: Wang, Y., Zhu, G., Han, Q., Wang, H., Song, X., Lu, Z. (eds.) ICPCSEE 2022. CCIS, vol. 1628, pp. 193–208. Springer, Singapore (2022). https://doi.org/10.1007/978-981-19-5194-7_15

23. Goodfellow, I.J., Pouget-Abadie, J., Mirza, M., et al.: Generative adversarial nets. In: Proceedings of the Annual Conference on Neural Information Processing Systems, pp. 2672–2680.Curran Associates, Red Hook/Montreal (2014)

24. Tang, H., Xu, D., Yan, Y., et al.: Local class-specific and global image-level generative adversarial networks for semantic-guided scene generation. In: Proceedings of the IEEE/CVF Conference on Computer Vision and Pattern Recognition, pp. 7870–7879 (2020)

25. Zhao, L., Mo, Q., Lin, S., et al.: Uctgan: Diverse image inpainting based on unsupervised cross-space translation. In: Proceedings of the IEEE/CVF Conference on Computer Vision and Pattern Recognition, pp. 5741–5750 (2020)

26. Nauata, N., Hosseini, S., Chang, K. H., et al.: House-gan++: generative adversarial layout refinement network towards intelligent computational agent for professional architects. In: Proceedings of the IEEE/CVF Conference on Computer Vision and Pattern Recognition, pp. 13632–13641 (2021)

27. Hou, Q., Zhou, D., Feng, J.: Coordinate attention for efficient mobile network design. In: Proceedings of the IEEE/CVF Conference on Computer Vision and Pattern Recognition, pp. 13713–13722 (2021)

28. Zhang, H., Zu, K., Lu, J.E.: An Efficient Pyramid Split Attention Block on Convolutional Neural Network. arXiv:2105.14447 (2021)

29. Abrahamyan, L., Truong, A.M., Philips, W., et al.: Gradient variance loss for structure-enhanced image superresolution. In: ICASSP 2022 IEEE International Conference on Acoustics, Speech and Signal Processing (ICASSP), pp. 3219–3223 (2022)

30. Timofte, R., Agustsson, E., Van Gool, L., et al.: Ntire 2017 challenge on single image superresolution: methods and results. In: Proceedings of the IEEE Conference on Computer Vision and Pattern Recognition Workshops, pp. 114–125 (2017)

31. Bevilacqua, M., Roumy, A., Guillemot, C., et al.: Low-complexity single-image superresolution based on nonnegative neighbor embedding. In: Proceedings of the 23rd British Machine Vision Conference (BMVC), pp. 135.1–135.10. BMVA Press (2012)

32. Jiang, Y.N., Li, J.H., Zhao, J.L.: A superresolution reconstruction algorithm for images based on generative adversarial networks. Comput. Eng. 47(03), 249–255 (2021)

33. Martin, D., Fowlkes, C., Tal, D., et al.: A database of human segmented natural images and its application to evaluating segmentation algorithms and measuring ecological statistics. In: Proceedings Eighth IEEE International Conference on Computer Vision, pp. 416–423 (2001)

The Multisource Time Series Data Granularity Conversion Method

Chongyang Leng, Qilong Han, and Dan Lu$^{(\boxtimes)}$

Harbin Engineering University, Harbin, China
{lengchongyang,hanqilong,ludan}@hrbeu.edu.cn

Abstract. Granular information has emerged as a potent tool for data representation and processing across various domains. However, existing time series data granulation techniques often overlook the influence of external factors. In this study, a multisource time series data granularity conversion model is proposed that achieves granularity conversion effectively while maintaining result consistency and stability. The model incorporates the impact of external source data using a multivariate linear regression model, and the entropy weighting method is employed to allocate weights and finalize the granularity conversion. Through experimental analysis using Beijing's 2022 air quality dataset, our proposed method outperforms traditional information granulation approaches, providing valuable decision-making insights for industrial system optimization and research.

Keywords: Time series data · Granular conversion · Fuzzy c-means clustering · Multiple linear regression

1 Introduction

Information granulation has emerged as a powerful approach to address uncertainty and complexity in intelligent information processing [1]. The main part of information granulation includes dividing complex problems into simpler granules, enhancing analysis and problem solving, etc. Appropriately chosen granules can improve efficiency while ensuring satisfactory solutions. Since Zadeh's seminal 1979 paper on information granularity [2], researchers have conducted in-depth research on the theory and models of granular computing.

Traditional granulation algorithms, such as interval information granulation, process time-series data by dividing equally spaced time windows, with the interval's maximum and minimum values as the binary tuple bounds. However, this method has limitations, as the resulting granules may cover all data samples but fail to represent interval data characteristics. Researchers have proposed clustering-based granulation methods to construct information granules based on data distribution characteristics. Nevertheless, hard clustering algorithms such as k-means require defining sample similarity and only yield good results under specific pattern distributions [3]. Consequently, traditional time-series data mining methods do not fully satisfy domain requirements.

Z. Yu et al. (Eds.): ICPCSEE 2023, CCIS 1879, pp. 182–191, 2023.
https://doi.org/10.1007/978-981-99-5968-6_13

To address the drawbacks of traditional granulation, clustering-based granulation has been proposed, which can construct information granules effectively based on data distribution characteristics. However, hard clustering methods such as k-means require the definition of similarity between samples and can only have good results in certain situations [4], indicating that traditional time series data mining methods are insufficient for practical needs. This paper proposes a time series granulation method based on the Fuzzy C-Means algorithm. Compared to k-means clustering, the FCM algorithm offers more flexible clustering results. The paper incorporates a multiple linear regression model and entropy weight method to optimize the granulation process, improving the interpretability and accuracy of multisource time series data granulation by considering other relevant data. After analysing the current state of research both domestically and internationally and addressing the shortcomings of existing research methods, a superior time series granulation method is proposed and applied to an air quality dataset for granulation and performance analysis.

The contributions of this paper can be summarized as follows:

(1) In information granulation research using fixed time windows, the proposed FCM-based improved algorithm effectively enhances granulation outcomes.
(2) The information granulation method proposed in this paper introduces a multiple linear regression model, taking into account the impact of other data on the target data during granulation.

In conclusion, the proposed time-series granulation method holds significant potential for numerous applications and can serve as a valuable contribution to the research domain of data mining and analysis.

2 Background

2.1 Granulation of Time Series Information

Time Series Definition. Massive amounts of data are generated across various industries, with most of these data carrying a time label. These long-term observations are commonly referred to as time series data [5]. Time series data comprise real-valued waveforms obtained by monitoring a process at a given sampling frequency, continuously collected with changing timestamps and unaffected by factors such as the system environment [6]. Each element of a time series consists of a time stamp and a numerical value, and the values of the data are dependent on changes in time, which can be denoted as Eq. (1):

$$U = \{(t_1, u_1), (t_2, u_2), ...(t_n, u_n)\} \tag{1}$$

where (t_n, u_n) represents the collection value u_n at time n.

In general, the time interval between data in the time series is consistent, so the set U can be simply denoted as $U = \{u_1, u_2, ...u_n\}$. In the analysis of time-series data, statistical models, data mining, machine learning, and other theoretical methods have been successfully applied in a variety of fields, including short-term wind speed prediction, stock forecasting, economic prediction, and healthcare monitoring. The model proposed in this paper primarily employs clustering as a means of data mining for time-series data.

2.2 Information Granulation

Information granulation is a basic problem of granular computing and serves as the initial step in problem solving within granular computing models [7]. Information granulation and information granules are prevalent in human cognitive, decision-making, and reasoning processes and are intimately connected to information granularity. For instance, when monitoring a city's traffic flow, sensors gather traffic volume data every minute. Analysing these raw data can be computationally demanding during calculations or model training. In such cases, we typically employ information abstraction based on time units such as years, months, or days to obtain various sizes of information granules, implicitly representing the information granularity levels utilized during the granulation process[8]. Furthermore, information granules generated at different granularity levels can be transformed into one another, as depicted in Fig. 1.

Fig. 1. Granularity levels

2.3 Fuzzy Clustering

The fuzzy c-means clustering algorithm is a useful and successful algorithm. By optimizing the objective function, the algorithm obtains the membership degree matrix data, which represents each sample point to all class centers. The class membership of each sample point is determined to achieve automatic classification of the sample data [9]. According to the size of its membership value, samples are assigned a membership function for each cluster and classified. The model mainly includes three key parameters: a fixed number of clusters, one centroid for each cluster, and each data point belonging to the cluster corresponding to the nearest centroid. The steps of the FCM clustering algorithm are shown in Fig. 2.

In summary, the FCM algorithm fundamentally minimizes the objective function and iteratively computes the membership degrees and cluster centers.

2.4 Multivariate Linear Regression Model

Multiple linear regression is a regression analysis technique in statistics that involves two or more independent variables and a dependent variable[10]. It assumes that there exists a relationship between the dependent variable and the independent variables.

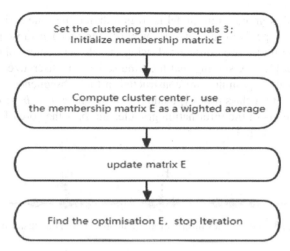

Fig. 2. Cluster flowchart

Assuming that there is a relationship between Y and $X_1, X_2, \ldots, X_n$, the multiple linear regression model can be represented mathematically as Eq. (2):

$$Y = \beta_0 + \beta_1 X_1 + \ldots + \beta_n X_n + \varepsilon \tag{2}$$

where β_0 is the regression constant and $\beta_1, \beta_2, \ldots, \beta_n$ is the global regression parameter. When $n = 1$, Eq. (2) is called the unary linear regression model; when $n \geq 2$, it is called the multiple linear regression model. ε is a random error and follows an ε $N(0, \sigma 2)$ distribution. The commonly used estimation method for parameter β is the least squares estimation method, for which the objective function is shown in Eq. (3).

$$Q(\beta) = \sum_{i=1}^{n} \|y_i - x_i \beta\|^2 \tag{3}$$

When constructing a multiple linear regression model, it is important to ensure that there is a significant impact on the independent variables to dependent variable and that they exhibit a true linear relationship.

3 The Proposed Method

3.1 Based on the Fuzzy C-Means Clustering Granulation Model

In 1996, Claudio Bettini and colleagues first introduced information granulation methods into time series mining by investigating different granularities of time series on the time axis, examining their properties and mining algorithms [11]. Over the years, information granulation methods have continuously evolved [12], including interval-based granulation, clustering-based granulation, and fuzzy set-based granulation. This paper primarily focuses on the study of granulation methods based on fuzzy clustering.

The view of this granulation model is to first divide the time series S into subsequences X and then perform fuzzy C-means clustering on each subsequence X. Then, the information granules are constructed using the cluster centers obtained from the clustering process [13]. Assuming that the time series S is discretized to form subsequence X_i information granules are constructed in the subsequence, which is a triad $\Omega_i = \{a_i, b_i, c_i\}$ with a class prototype number of 3, where $a_i < b_i < c_i$, a_i and c_i are defined as the bounds of the information granule, and b_i is the core of the information granule, as shown in Fig. 3.

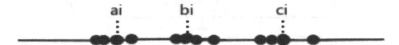

Fig. 3. Time series information granulation based on clustering

This method can construct information granules based on the data distribution characteristics of each time subsequence, which can comprehensively express the characteristics of the entire sequence.

3.2 Improved Fuzzy C-Means Granulation Model

This paper proposes an improved algorithm that incorporates a multiple linear regression model into the granulation model mentioned in Sect. 3.1. The structure is shown in Fig. 4.

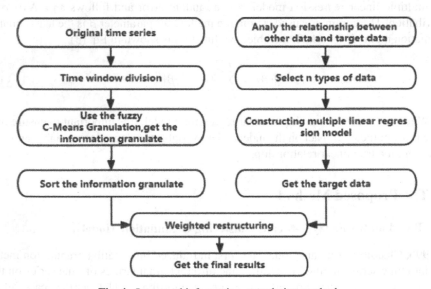

Fig. 4. Improved information granulation method

There are three steps in our proposed method:

(1) Divide the original time-series data into n subsequences $S = \{s_1, s_2, \ldots, s_n\}$ and perform fuzzy C-means clustering on each subsequence. Set the number of clusters to $m = 3$ and initialize the membership matrix u_{ij} with random values between 0 and 1. Compute the clustering centroid for each subsequence by taking a weighted average using the membership matrix as weights. Then, update the membership matrix u_{ij} using the updated clustering centroids c_i. The computation of the clustering centroids is shown in Eq. (4), and the update of the membership matrix is shown in Eq. (5):

$$c_j = \frac{\sum_{i=1}^{N} u_{ij}^m \cdot s_i}{\sum_{i=1}^{N} u_{ij}^m} \tag{4}$$

$$u_{ij} = \frac{1}{\sum_{k=1}^{C} \left(\frac{\|s_i - c_j\|}{\|s_i - c_k\|} \right)^{\frac{2}{m-1}}} \tag{5}$$

When the objective function is minimized and the membership matrix and clustering centroids reach the optimal solution, set the number of iterations to $k = 3$ based on the results. The objective function is shown in Eq. (6):

$$J_m = \sum_{i=1}^{N} \sum_{j=1}^{C} u_{ij}^m \|s_i - c_j\|^2, 1 \leq m \tag{6}$$

Therefore, information granules can be obtained and sorted internally from small to large.

(2) Train a multivariate linear regression model on the time-series data to obtain a representation of the linear relationship between other source data and the target data and obtain a target data value through this model.

(3) Use the entropy weight method to recombine the values in the information granules with the values obtained through the multivariate linear regression equation to obtain the final information granules.

The proposed model can address the issue of multisource time series. Previous studies may have only considered single granulation processing of time series data without taking into account their multisource features. This paper builds upon this and enhances granulation processing results using a linear regression model. The entropy weighting method is employed to scientifically and reasonably allocate weights to the data. Consequently, optimized information granules can be obtained.

4 Experiment

4.1 Dataset

This paper conducts model analysis and research using the measured air quality data in Beijing, China, in 2022 as sample data. This paper conducts model analysis and research using the measured air quality data in Beijing, China, in 2022 as sample data. One hour is a sampling point, and the sampling data include $PM_{2.5}$, PM_{10}, and so on. The following experiments and analyses will use these 8760 sets of data. The original data are shown in Fig. 5.

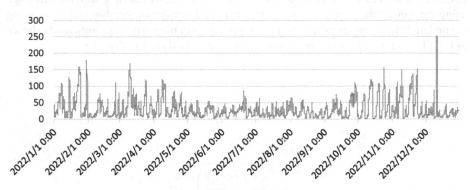

Fig. 5. $PM_{2.5}$ original data

The experiment selected $PM_{2.5}$ values from the dataset, which represent the air quality evaluation significantly, to perform information granularity transformation. The dataset is divided into training and testing sets in an 8:2 ratio.

4.2 Evaluation Metrics

Root Mean Square Error. The root mean square error can be used to evaluate the bias between the practical data and the granulated nucleus [13], as shown in Eq. (7):

$$RMSE = \sqrt{\frac{\sum_{1}^{N} (y_i - \hat{y}_i)^2}{N}} \tag{7}$$

Mean Absolute Error. The mean absolute error is used to represent the average value of the absolute error between the predicted value and the practical data [14]. The value is inversely proportional to the reliability, as shown in Eq. (8):

$$\text{MAE}(X, h) = \frac{1}{m} \sum_{i=1}^{m} |h(x_i) - y_i| \tag{8}$$

4.3 Experimental Results

During the information granulation process of the original $PM_{2.5}$ time series, a window of 6 sampling points (6 h) was taken as an operational window, and the fuzzy C-means clustering method was used to granulate each window. After clustering, $C = \{c_1, c_2, ..., c_j\}$ was obtained, where the triplets in c_j represent the interval lower limit LOW, interval upper limit UP, and interval core R (corresponding to the actual value in the interval) of each operational window. The scatter plot is shown in Fig. 6.

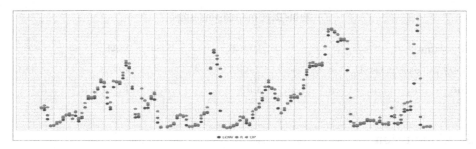

Fig. 6. The time series granulate scatter plot

The above experimental process is based on fuzzy C-means clustering for time series granulation, followed by the application of a multiple linear regression model. From the dataset, it can be observed that PM_{10} and NO_2 are positively correlated with $PM_{2.5}$. The training dataset was then inputted into a linear regression model in statsmodels(), and the model was continuously fitted to obtain a model description[15], as shown in Table 1.

Table 1. Experimental results.

Argument	coef	Std err
const	−5.6	0.312
PM_{10}	0.35	0.003
NO_2	0.71	0.013

As observed from the above figure, the model fitting is satisfactory. The PM_{10} corresponding regression parameter is 0.35, the NO_2 corresponding regression parameter is 0.71, and the constant term of the regression model is −5.6. This result can be written as Eq. (9):

$$PM_{2.5} = -5.6 + 0.35PM_{10} + 0.71NO_2 \qquad (9)$$

The obtained linear regression equation was used to predict $PM_{2.5}$ in the test dataset. The predicted values were then combined with the information granules obtained in the previous step using entropy weighting. First, the data were normalized, and the information entropy of PM_{10} and NO_2 was calculated. The weights were then calculated based on the information entropy. After training, a weight ratio of $PM_{2.5}$:$PM_{2.5}' = 0.58$:0.42 was obtained, where the former represents the clustered $PM_{2.5}$ value and the latter represents the linear regression predicted value. The data were then recombined according to this weight ratio to obtain the final set of information granules $C = \{c_1', c_2', ..., c_j'\}$, completing the time series granulation conversion.

This paper also compared with the fuzzy clustering granulation method and evaluated the model performance in terms of RMSE and MAE. The results are shown in Tab. 2:

It can be observed that the improved granulation model can better preserve the data characteristics and performs well in the two evaluation indicators.

Table 2. Comparison results.

Algorithm	RMSE	MAE
fuzzy c-means method	2.05	1.07
ours	1.75	1.52

5 Conclusion

This paper presents an improved time series granulation method based on FCM clustering, which more comprehensively and accurately extracts time series features compared to existing models. Additionally, a multiple linear regression model is introduced to integrate the target time series data with other relevant data, emphasizing the representation of multisource data correlations. Finally, the entropy weight method is used to assign weights to data generated by both models, creating information granules. Extensive experiments on the China-Beijing air quality dataset demonstrate the effectiveness of the proposed model, showing that its performance surpasses related methods.

Acknowledgements. This work was supported by the National Key R&D Program of China under Grant No. 2020YFB1710200.

References

1. Fujita, H., Gaeta, A., Loia, V., Orciuoli, F.: Resilience analysis of critical infrastructures: a cognitive approach based on granular computing. IEEE Trans. Cybern. **49**(5), 1835–1848 (2019)
2. Hailan, C.: Research on time series information granulation method for clustering and prediction. Beijing Univ. Sci. Technol. (2021). https://doi.org/10.26945/d.cnki.gbjku.2021.000062
3. Song, X.J., Huang, J.J., Song, D.W.: Air Quality Prediction based on LSTM-Kalman Model. ITAIC, pp. 695–699 (2019)
4. Wang, Y., Dou, Y., Meng, R.: A multikernel neural network short term load forecasting model based on fuzzy C-means clustering variational modal decomposition and group intelligence optimization. High Voltage Technol. **48**(04), 1308–1319 (2022). 10. 13336/j.1003–6520.hve.20210664
5. Li, X., Shifei, D.: A novel clustering ensemble model based on granular computing. Appl. Intell. **51**, 5474–5488 (2021)
6. Xu, L., Ding, S.: A novel clustering ensemble model based on granular computing. Appl. Intell. **51**, 5474–5488 (2021). https://doi.org/10.1007/s10489-020-01979-8
7. Jiye, L., Yuhua, Q., Li Deyu, H., Qinghua.: Granular theory of computation and method of big data mining. Chin. Sci. Inform. Sci. **45**(11), 1355–1369 (2015)
8. Ane, B., Angel, C., Usue, M., et al.: A review on outlier/anomaly detection in time series data. ACM Comput. Surv. **54**(3), 1–33 (2021)
9. Lu, W., Pedrycz, W., Liu, X., et al.: The modelling of time series based on fuzzy information granules. Expert Syst. Appl. **41**(8), 3799–3808 (2014)

10. Li, S.T., Cheng, Y.-C., Lin, S.-Y.: A FCM-based deterministic forecasting model for fuzzy time series. Comput. Math. Appl. **56**(12), 3052–3063 (2008). ISSN 0898-1221. https://doi.org/10.1016/j.camwa.2008.07.033

11. Bettini, C., Wang, X., Jajodia, S.: Testing complex temporal relationships involving multiple granularities and its aplpication to data mining(Extended Abstract). ACM SIGACT-SIGART Symposium on Principles of Database Systems, pp. 68–78 (1996)

12. Wang, L., Liu, X., Pedrycz, W.: Effective intervals determined by information granules to improve forecasting in fuzzy time series. Expert Syst. Appl. **40**(14), 5673–5679 (2013). ISSN0957-4174. https://doi.org/10.1016/j.eswa.2013.04.026

13. Jinghua, X., Jing, R.: Research on the combination model of exchange rate forecast based on random forest algorithm and fuzzy information granulation. Quant. Techn. Econ. **38**(01), 135–156 (2021). https://doi.org/10.13653/j.cnki.jqte.2021.01.008

14. Chen, J., Shen, Y., Lu, X., et al.: An intelligent multiobjective optimized method for wind power prediction intervals. Power Syst. Technol. **40**(10), 2758–2765 (2016)

15. Antonio, A.-F.M.:. Comparison Between Fuzzy C-means Clustering and Fuzzy Clustering Subtractive in Urban Air Pollution. In: International Conference on Electronics Communications and Computers CONIELECOMP, CONFERENCE (2010)

Outlier Detection Model Based on Autoencoder and Data Augmentation for High-Dimensional Sparse Data

Haitao Zhang, Wenhai Ma, Qilong Han[✉], and Zhiqiang Ma[✉]

College of Computer Science and Technology, Harbin Engineering University, Harbin, China
{zhanghaitao,mawenhai,hanqilong,mazhiqiang}@hrbeu.edu.cn

Abstract. This paper aims to address the problems of data imbalance, parameter adjustment complexity, and low accuracy in high-dimensional data anomaly detection. To address these issues, an autoencoder and data augmentation-based anomaly detection model for high-dimensional sparse data is proposed (SEAOD). First, the model solves the problem of imbalanced data by using the weighted SMOTE algorithm and ENN algorithm to fill in the minority class samples and generate a new dataset. Then, an attention mechanism is employed to calculate the feature similarity and determine the structure of the neural network so that the model can learn the data features. Finally, the data are dimensionally reduced based on the autoencoder, and the sparse high-dimensional data are mapped to a low-dimensional space for anomaly detection, overcoming the impact of the curse of dimensionality on detection algorithms. The experimental results show that on 15 public datasets, this model outperforms other comparison algorithms. Furthermore, it was validated on industrial air quality datasets and achieved the expected results with practicality.

Keywords: High-dimensional · data augmentation · attention mechanism · Outlier Detection

1 Introduction

In practical applications, outlier detection is widely used in fields such as fraud detection [1], industrial fault detection [2], intrusion detection [3], and medical diagnosis [4]. Especially with the advent of the industrial age, as the trend of massive and high-dimensional data patterns continues to emerge, traditional outlier detection methods based on clustering, distance, density, and angle have shown a significant decrease in effectiveness [5]. Therefore, data anomaly detection in the industrial field has become a research hotspot [6]. High-dimensional sparse data face many problems, such as uneven distribution, high dimensionality, and sparsity, as well as noise and redundancy. Therefore, it is necessary to research new efficient and rigorous outlier detection methods to improve the accuracy and reliability of data mining.

This article proposes a novel outlier detection model that combines data augmentation and autoencoders. The model first solves the data imbalance problem through

Z. Yu et al. (Eds.): ICPCSEE 2023, CCIS 1879, pp. 192–206, 2023.
https://doi.org/10.1007/978-981-99-5968-6_14

data augmentation algorithms. Then, the structure of the neural network is determined by calculating the weights of data features, allowing the model to better learn feature information during the training process. Finally, the encoding-reconstruction detection module reconstructs the data and uses the weighted KNN algorithm to detect anomalies in the reconstructed data, which solves the problem of low accuracy caused by "dimensional disaster".

- The model proposes a weighted SMOTE algorithm to fill in the minority class samples in the dataset to address the data imbalance problem and generate a high-quality dataset.
- To solve the problem of parameter tuning complexity during training, the SEAOD model utilizes attention mechanisms to calculate feature weights and sets thresholds to determine the number of features. This enables the model to better learn feature information from the dataset, further reducing the complexity of parameter tuning.
- Using an autoencoder to reconstruct the data improves the accuracy of abnormal detection in high-dimensional sparse data by using a weighted KNN algorithm. Compared to other benchmark methods, this model demonstrates improved performance, with better detection results on actual industrial datasets, conveying practical application value and practical significance.

2   Related Work

Outlier detection, also known as abnormal or novel detection, has always been a popular field of research for various research communities. Breunig et al. proposed a local outlier factor (LOF) algorithm [7] to calculate local abnormality factors and define local outlier values that are related to the density of neighbors around each data point. Yang et al. [8] used representative data point selection and the global optimal GMM fitting process to determine the number of mixture components, and detected outliers through the similarity of data samples. Sarvari. et al. proposed an autoencoder ensemble method (BAE) [9], which uses unsupervised autoencoders to build a basic abnormal detection model and constructs multiple autoencoder ensembles using reinforcement learning algorithms to better discover abnormal data.

It is easy to generate overfitting and uncertainty of neural network layers on small and medium-sized datasets, affecting detection accuracy [10, 11]. Song et al. proposed a fair anomaly detection method based on deep clustering [12], which utilizes data clustering and deep learning models to detect outliers, while taking into account sensitive attributes of each data point for balancing, so as to enhance fairness and avoid unfair impacts on particular populations. Li et al. proposed a new deep structured abnormal detection framework [13], which expresses cross-modal data as a graph structure, extracts and interacts features through graph convolutional networks, and finally applies fusion representation learning and abnormal detection modules to detect abnormal data across modalities. Lu et al. proposed using neural networks to identify patterns in time series data [14], effectively identifying abnormal data and solving the challenges of sequence anomalous value detection. Although it can detect overfitting issues in deep learning, it requires parameter tuning and model optimization.

The data collected by industrial sensors are large, incomplete, complex, and irregular, and the effectiveness of traditional detection methods is poor. Merim Dzaferagic et al. [15] used the generation countermeasures network (GAN) to generate missing sensor measurements. Liu et al. used SVDD (surface vertex detection) machine learning technology for uncertain data outlier detection [16]. Zhang et al. proposed an improved deep SVDD model [17]. Using data preprocessing techniques to process the dataset to remove noise and missing values, effective separation of normal data and abnormal data was achieved. Zhou et al. proposed a method that improves the prediction ability of the variational autoencoder (VAE) [18] by optimizing its structure on training data, and experimental verification was conducted on an industrial big dataset. In solving the issues of abnormal detection and data imbalance in the industrial field, the combination of deep learning and traditional machine learning methods is crucial.

3   Anomaly Detection Model Based on Autoencoder and Data Augmentation

3.1   Problem Description

In many cases, there is often imbalance in the dataset, resulting in low accuracy of detection algorithms, which can lead to false positives or false negatives. Industrial data often contain noise and redundant data. To address the problem of insufficient samples in minority classes within the dataset as well as optimization of parameters during neural network training, multiple modules are needed to process the data and improve the accuracy of detection results. $X = \{x^1 x^2, \ldots, x^m\}$, m is the number of data samples in spatial data, and n is denoted as the sample feature dimension. The dataset X contains normal data p and abnormal data q $p + q = m$.

3.2   Model Construction

To address the issues of data imbalance, difficult parameter tuning, and low accuracy in high-dimensional sparse data, this paper proposes a data augmentation and autoencoder outlier detection model, as shown in Fig. 1. The data augmentation module introduces computer vision data augmentation algorithms into outlier detection to resolve the issue of sample imbalance by increasing the number of minority class samples and generating higher quality data. In the attention mechanism module, feature weights are calculated to accurately determine the neural network structure, improving the efficiency and accuracy of data feature learning during the model training process. The encoding-decoding detection module maps high-dimensional sparse data into a low-dimensional space to mitigate the impact of dimensionality on outlier detection.

3.3   Data Augmentation Module

In deep learning network training, the quantity and quality of samples are crucial to the generalization performance and robustness of the learned model. In industrial sensor data collection, errors in human or machine operations may lead to missing data, which can impact the stability of neural network learning results. To address this issue, this paper applies data augmentation algorithms to the outlier detection field for missing data

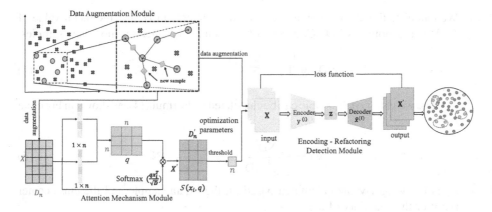

Fig. 1. SEAOD model diagram

imputation. Compared to traditional methods of missing data imputation such as median, mean, or statistical methods, our proposed data augmentation method is similar to image flipping, shifting, and other operations. By filling in a few missing points, this method processes the original dataset and increases its diversity and quantity, thus alleviating the problem of sample imbalance. This article proposes a weighted oversampling algorithm that adjusts the number of new samples generated for each minority class sample by computing the distance between these samples and assigning different weights to them. This approach increases the number of samples that are more representative of the minority class.

The derivation process of the weighted SMOTE algorithm:

Assume that dataset X contains a minority class with N samples, and each sample has C features.

(1) Calculate the Euclidean distance between each sample and other samples, as shown in Formula (1).

$$D_{ij}(x_i, x_j) = \sqrt{\sum_{k=1}^{c}(x_{i,k} - x_{j,k})^2} \tag{1}$$

where $i = 1, 2, ..., N, j = 1, 2, ..., N, i \neq j$

(2) Calculate the sum D_i distance from sample x_i to other samples. A larger D_i indicates that sample x_i is closer to the boundary, and a smaller D_i indicates that sample B is closer to the center, as shown in Formula (2):

$$D_i = \sum_{j=1, j \neq i}^{M} D_i(x_i, x_j), i = 1, 2, ..., N \tag{2}$$

(3) Normalize D_i, as shown in Formula (3):

$$D_{i'} = \frac{D_i - D_{i(min)}}{D_{i(max)} - D_{i(min)}} \tag{3}$$

(4) We calculate the absolute value between all elements in D' and the mean value of D' This is denoted as ND'. Consequently, Formula (4) generates more new samples.

$$ND' = \left| D'_i - \frac{\sum_i^M D'_i}{M} \right| i = 1, 2, \ldots, N \tag{4}$$

The weight of each sample can be calculated by Formula (4), as shown in Formula (5):

$$w_i = \frac{ND'}{\sum_i^M ND'}, i = 1, 2, \ldots, N \tag{5}$$

w_i is the weight value calculated based on the i data sample and the class center point or the boundary of the class.

Let us assume that the weighted SMOTE algorithm combines the few class samples in the original dataset to create N new samples. For the i-th sample of few classes in the original dataset x_i, the Euclidean distance between x_i and all other samples of the same class is calculated and the sum is calculated using formula (5) to obtain the weight value w_i.Find the k nearest samples as its neighbor samples, $N_W = \lfloor N \times w_i \rfloor$ sample $\{x_1, x_2, \ldots, x_n\}$ is randomly selected from its k neighbor samples. $\lfloor \rfloor$ is rounded down to generate a new sample of a few samples x_i in these samples. As shown in Formula (6):

$$x_{new} = x_i + \gamma |x_i - x_n| \tag{6}$$

where, γ is a random number between 0 and 1 (Fig. 2).

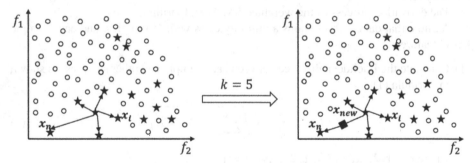

Fig. 2. Weighted SMOTE algorithm oversampling diagram

To address the problem of noisy data generation and sample boundary blurring caused by weighted SMOTE's nearest-neighbor interpolation, this paper introduces the ENN algorithm for data undersampling. The ENN algorithm traverses the dataset, calculates the frequency of each class label in k-nearest neighbor samples for each sample, and removes the sample points where the majority label appears most frequently. By combining the weighted SMOTE algorithm with the ENN algorithm, our model can integrate and complement the advantages of both algorithms, thus improving the effectiveness of data augmentation during the preprocessing stage. Figure 3 showns the comparison of results before and after data augmentation.

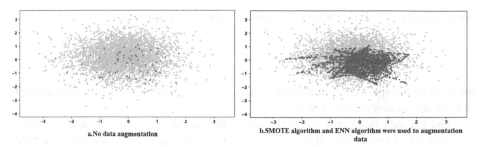

a.No data augmentation

b.SMOTE algorithm and ENN algorithm were used to augmentation data

Fig. 3. Comparison diagram of the visualization results of the data augmentation experiment

3.4 Attention Mechanism Module

As shown in Fig. 4, by visualizing the average values of each feature in the dataset, it is observed that all points are clustered around the center. Therefore, a transition matrix is constructed using the average values of the data sample features. The attention mechanism is used to compute the similarity between feature vectors of data samples, and a threshold is set to determine their relevance to each other, thus determining the neural network structure, as shown in Fig. 1.

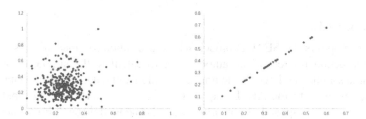

Fig. 4. Dataset WBC sample visualization

The group of query vectors q is defined by calculating the median feature values of $X = [x_1, x_2, ..., x_n]$ in the dataset matrix. The median feature values for each vector in the dataset are chosen so that they may approach the center point, as shown in Formula (7).

$$q\left(x_k^T, x_k\right) = x_k^T x_k \tag{7}$$

where, x_k represents the feature average vector of a data sample, and k represents the feature dimension. The feature average vector is multiplied by itself to obtain the feature similarity transition matrix.

Use the pointwise product model in the attention mechanism to calculate the similarity of weights, achieved by combining query vector group q and the feature vectors of the data samples through matrix multiplication operations. When the dimensionality of the input vector is high, due to the large variance, it may cause the gradient of the softmax function to decrease and affect the performance of the model. To address this issue, use the scaled pointwise product model shown in Formula (8):

$$s(x_i, q) = \frac{qx_i^T}{\sqrt{d}} \tag{8}$$

where d is the input vector dimension, $s(x_i, q)$ is the attention scoring function, and q is the feature median vector group.

The softmax function is used to convert the output values of the attention score function into a probability distribution in the range [0, 1], determining which features have high correlation between each other. These features are further filtered through the setting of threshold values to determine the structure of the neural network.

3.5 Encoding-Refactoring the Inspection Module

The SEAOD model adopts an attention mechanism to calculate weights and decide on neural network structure. Thus, it can avoid overfitting and enhance generalization ability by learning from feature information during the training process. Autopen-coders use backpropagation and optimal methods such as gradient descent to reconstruct output data that accurately imitate input data, which is why they are mostly employed for training data reconstruction.

3.5.1 Autoencoder

This article proposes the SEAOD model, which uses attention mechanisms during the data reconstruction process of an autoencoder to calculate the similarity of features between data samples and avoids the impact of redundant data. Using the data augmentation X_{new} as input to the AE. Let $x_i = (x_1, x_2, \ldots, x_n)$. After being input into the encoder of the AE composed of n neurons, X_{new} is encoded through Formula (9):

$$y^{(i)} = f\left(w_1 x^{(i)} + b_1\right) \tag{9}$$

where w_1 represents the weights between neural networks, b_1 is the hidden layer bias and f is the activation function. $y^{(i)}$ represents the latent space data.

After obtaining the encoded result $y^{(i)}$, it is input into the decoder layer to train and obtain reconstructed data $\hat{x}^{(i)} = (\hat{x}^1, \hat{x}^2, \ldots, \hat{x}^n)$, as shown in Formula (10):

$$\hat{x}^{(i)} = g\left(w_2 y^{(i)} + b_2\right) \tag{10}$$

where w_2 represents the weights between neural networks.

The goal of an autoencoder is to minimize the similarity between the input data and the output data through iterative training, thereby continuously reducing the error between the data. The calculation formula for the reconstruction error $L(x, \hat{x}')$ is shown in Formula (11):

$$L(x, \hat{x}') = \sum_{i=1}^{n} \left\| g(w_2 f(w_1 x^{(i)} + b_1) + b_2) - x_i \right\|^2 \tag{11}$$

First, the data augmentation sample X_{new} is compressed into low-dimensional features y_i using the encoder and then reconstructed using the decoder. Due to the different number of units in the hidden layer, the model may experience overfitting, which can affect the effectiveness of data dimensionality reduction. Using an attention mechanism can help the model better learn data features during training.

3.5.2 Outlier Detection

In this paper, the weighted KNN algorithm is used to perform anomaly detection on reconstructed data from the autoencoder. Unlike the basic KNN algorithm, the weighted KNN algorithm considers the weight of each neighboring point, thus avoiding the influence of distant points on the detection results. The augmented data are first reconstructed through the autoencoder and then input into the weighted KNN algorithm for anomaly detection. To determine the selection of k, a grid search is used. This algorithm calculates the distance and weight between each data point in the dataset and judges whether the data point is an outlier based on the size of its weight.

4 Experiment and Analysis

4.1 Introduction to Datasets

In this paper, 15 public outlier detection benchmark datasets from ODDS [19] of UCI and a real industrial air quality dataset were selected. Detailed information on the number of datasets and the number of outliers in each public dataset is shown in Table 1.

4.2 Evaluation Indicators

In abnormal detection models, commonly used evaluation metrics include AUC values, accuracy, recall rate, precision rate, and F1 scores to evaluate the detection effectiveness of the results.

$$\text{precision} = \frac{TP}{TP + FP} \tag{13}$$

$$\text{ACC} = \frac{TP + TN}{TP + TN + FP + FN} \tag{14}$$

$$F1 = \frac{2}{\frac{1}{\text{precision}} + \frac{1}{\text{recall}}} \tag{15}$$

4.3 Results and Analysis

The anomaly detection performance of the SEAOD model was verified by experiments and compared with five high-dimensional sparse data anomaly detection algorithms (OC-SVM [20], LOF [21], COPOD [22], SOM-DAGMM [23], and GAN-VAE). The experimental results are presented in Table 2, with the highest scores highlighted in bold.

Table 1. Outlier detection datasets (ODDS) [19]

Datasets	Points	dim	Outliers (%)
arrhythmia	452	274	66 (15%)
breastw	683	9	239 (35%)
cardio	1,831	21	176 (9.6%)
wine	129	13	10 (7.7%)
letter	1,600	32	100 (6.25%)
lympho	148	18	6 (4.1%)
mammography	11,183	6	260 (2.32%)
optdigits	5,216	64	150 (3%)
satellite	6,435	36	2036 (32%)
satimage-2	5,803	36	71 (1.2%)
Speech	3,686	400	61 (1.65%)
wbc	278	30	21 (5.6%)
Musk	3,062	166	97 (3.2%)
Vowels	1,456	12	50 (3.4%)
pendigits	6,870	16	156 (2.27%)

For experimental fairness, we used the mean accuracy rate (ACC) of 10 independent tests for model training. For the LOF algorithm, the number of neighbors k is selected by performing cross-validation within the range of 10–100 and choosing the best detection result from the 10 nearest neighbors. The OC-SVM algorithm uses a Gaussian kernel function (RBF). The network structure for the GAN-VAE algorithm based on deep learning and the autoencoder in the fusion SOM-DAGMM algorithm are the same as the SEAOD model. In the attention mechanism module of the SEAOD model, thresholds of 0.5, 0.75, and 0.9 are selected and determined through feature extraction result verification on different datasets.

From Table 2, it can be seen that the SEAOD model demonstrated high levels of abnormal detection on nine datasets (such as arrhythmia and letter) among the tested ones. For example, on the high-dimensional sparse arrhythmia dataset, the average accuracy has been improved by 9.62%, 10.34%, 12.9%, 4.67% and 2.89%. Traditional outlier detection algorithms are affected by the "curse of dimensionality" when processing high-dimensional sparse data, resulting in a decrease in anomaly detection efficiency.

The GAN-VAE algorithm based on neural networks reconstructs high-dimensional sparse data and learns classification boundaries through the generated latent anomaly points. However, the training results of the model largely depend on the quality of the input data. If there is imbalance or noise in the data, it may have a negative impact on the generated results. The fusion SOM-DAGMM algorithm relies mainly on modelling the distribution characteristics of data samples for anomaly detection, which requires high

Table 2. Average accuracy of anomaly detection performance (maximum score highlighted in bold)

Datasets	LOF	OC-SVM	COPOD	SOM-DAGMM	GAN-VAE	Ours model
arrhythmia	0.8397	0.8342	0.8153	0.8794	0.8946	**0.9205**
breastw	0.4671	0.7226	**0.9877**	0.9370	0.9210	0.9692
cardio	0.8635	0.8813	0.9236	**0.9512**	0.8755	0.8808
Wine	**0.9845**	0.9230	0.8653	0.8938	0.9357	0.9801
letter	0.9296	0.8921	0.8562	0.9312	0.9130	**0.9484**
lympho	0.933	0.7666	0.8936	0.9291	0.9580	**0.9666**
mammography	0.7615	0.9497	0.9132	0.7290	0.8906	**0.9751**
optdigits	0.9511	0.9175	0.860	0.8159	0.9294	**0.9689**
satellite	0.6107	0.6977	0.7470	0.9217	**0.9450**	0.8628
satimage-2	0.9741	0.9612	0.9130	0.8912	0.9586	**0.9940**
speech	0.8298	0.9105	0.8942	0.9710	0.9230	**0.9834**
vowels	0.9656	0.9108	0.9342	0.9333	0.9574	**0.9828**
wbc	0.921	0.9078	0.8627	**0.9647**	0.9150	0.9342
Musk	0.933	0.9216	0.9342	0.9493	**0.9693**	0.9684
pendigits	0.9559	0.9352	0.9010	0.9478	0.9105	**0.9912**
AUG	0.8613	0.8754	0.8867	0.9097	0.9264	**0.9551**

data quality. The SEAOD model is proposed in this paper, which uses data augmentation to address the issue of imbalanced data samples by filling minority class samples. Observing the results in Table 2, the average accuracy score of the SEAOD model on 15 public datasets is 0.9551, which is better than its comparative baseline methods.

4.4 Model Iteration Experiment

Taking the satellite dataset as an example, the effect of the change in the number of iterations of the model on the AUC, ACC, and F1 scores was analysed experimentally. The number of self-encoder iterations in the model is set to 200, 250, 300, 500, 750, and 1000. According to the results in Fig. 5, the changes in the model's iteration numbers have a significant impact on the evaluation metrics. Specifically, when the number of iterations of the training network was 750, the AUC value, F1 score, and accuracy were 0.9144, 0.9292, and 0.9494, respectively, reaching their maximum values for each evaluation metric.

4.5 Ablation Experiment

A group of control experiments was set up to validate the rationality of the abnormal detection model based on autoencoders and data enhancement proposed in this paper.

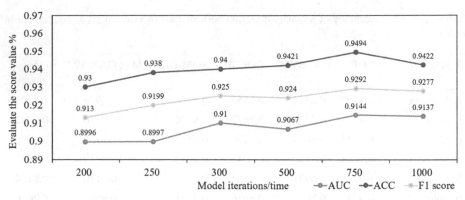

Fig. 5. Analysis of the number of iterations of the training network

Approach 1 employed a weighted KNN outlier detection algorithm, approach 2 utilized an autoencoder in combination with a weighted KNN (W-KNN) algorithm (AE-N), and approach 3 used the proposed anomaly detection model, which combines data augmentation and an autoencoder (SEAOD), to detect outliers. Experimental Plan 2 and Plan 3 also incorporate attention mechanisms to calculate the similarity of feature weights, determine the neural network structure, and address the complexity of neural network parameter adjustment.

After the experiment, the best experimental results were selected, and the performance of the three anomaly detection approaches was compared. The experimental evaluation metrics include AUC values, accuracy, and F1 scores among others. The results show that the SEAOD model in terms of AUC value, accuracy, and F1 score, fully demonstrates the effectiveness and feasibility of the model and has more significant advantages compared to the other two approaches.

Table 3. Outlier detection results between modules

Datasets	AUC			ACC			F1		
	W-KNN	AE-N	SEAOD	W-KNN	AE-N	SEAOD	W-KNN	AE-N	SEAOD
arrhythmia	**0.7965**	0.7738	0.7851	0.8527	0.8718	**0.9205**	0.7795	0.8213	**0.8426**
breasw	0.9453	0.9570	**0.9651**	0.9472	0.9498	**0.9692**	0.9135	**0.9376**	0.9227
wbc	0.7361	0.7539	**0.8018**	0.9018	0.9173	**0.9342**	0.7904	0.8261	**0.9045**
vowels	0.70	0.723	**0.849**	0.9369	0.9777	**0.9828**	0.8245	0.8503	**0.9159**
lympho	0.728	0.7649	**0.766**	0.9421	**0.9865**	0.9666	**0.8579**	0.8264	0.836
letter	0.8235	0.8461	**0.8763**	0.9468	0.9437	**0.9484**	0.8951	0.9049	**0.9123**

In Table 3, the proposed SEAOD model has shown significant advantages in detecting arrhythmia datasets. Compared to other comparison methods, this model achieved a 7.95% and 5.58% improvement in accuracy for data sample detection, and an 8.09% and 2.59% improvement in F1 score. In six public datasets, the AUC value, accuracy, and F1 score of the SEAOD model are generally higher than those of the two compared methods. The SEAOD model's advantages come from using an attention mechanism, calculating the similarity of feature weights to determine neural network structure, solving the problem of complex parameter tuning, and filling minority class anomalous samples through a data augmentation algorithm to address imbalanced data. However, traditional detection algorithms perform well for the lower-dimensional breast dataset since dimensionality limitations do not affect them.

Taking into account the experimental results of all datasets, compared to the average values of other comparative approaches, the SEAOD model shows an average improvement in AUC value, accuracy, and F1 score, as shown in Table 4.

Table 4. Average improvement of anomaly detection using the SEAOD model.

Index	AUC Score		Accuracy		F1 Score	
average growth rate	6.63%	4.65%	3.51%	1.32%	5.4%	3.24%

4.6 Industrial Application Experiment

To verify the applicability of the SEAOD model in the industrial field, an experimental dataset of air quality data from industrial emissions in a certain region was used. This dataset contains 11 features, including sulfur dioxide, atmospheric pressure, nitrogen dioxide, wind direction, PM10, temperature, humidity, ozone, carbon monoxide, PM2.5 and wind speed, which were transmitted in real-time by each national control station sensor.

To showcase the distribution of outliers in industrial data, a random sample of 5000 data points was selected from a dataset of 45000 and visualized using the t SNE algorithm. As shown in Fig. 6, the overall data exhibit a nonlinear distribution, and abnormal data points are scattered among normal data points, which can cause confusion in abnormal detection algorithms and affect the accuracy of detection.

To verify the utility of the model in industrial settings, we carried out experiments utilizing air quality data collected from industrial sources. Figure 7 shows that the distribution of minority-class abnormal samples is uneven when data augmentation is not applied. Therefore, we employed data augmentation to fill in the minority class samples, increasing outlier detection accuracy.

As shown in Table 5, traditional detection algorithms exhibit low accuracy for industrial data, as missing values or noise in the industrial data can affect the detection algorithms. To address this issue, the SEAOD model is proposed in this paper. First, the data augmentation module solves the problem of sample imbalance in industrial data. Second, the attention mechanism is used to compute feature correlations and determine the neural network structure, further solving the problem of complex parameter tuning. Compared with other existing anomaly detection algorithms, the model has a clear advantage in dealing with data anomalies in the industrial field.

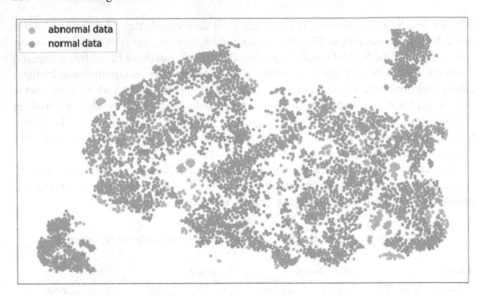

Fig. 6. Visualization of air quality data

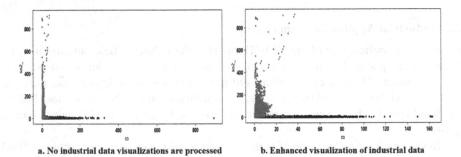

a. No industrial data visualizations are processed   b. Enhanced visualization of industrial data

Fig. 7. Comparison visualization of industrial data before and after enhancement

Table 5. Industrial dataset detection results in different algorithms

	AUC	ACC	F1
LOF	0.5581	0.5429	0.6672
OC-SVM	0.5327	0.52	0.6661
KNN	0.771	0.683	0.7534
GAN-VAE	0.8903	0.8881	0.8925
Ours model	**0.9071**	**0.9073**	**0.9042**

5 Conclusion

In this article, traditional machine learning methods are combined with deep learning methods to address the imbalance and noise problems present in high-dimensional sparse data. The SEAOD model proposes a weighted SMOTE algorithm to address sample imbalance and generates high-quality training data. At the same time, attention mechanisms are used to calculate the feature weights to determine the structure of a neural network and an autoencoder is used to reconstruct data samples. Finally, the model is combined with an abnormal detection algorithm for detection. Experimental results show that the model has high accuracy and the practicality is verified on the industrial air quality dataset. Regarding future research, several aspects could be further strengthened: (1) how to measure the deviation of outlier detection methods; (2) how to detect and ensure the fairness of the outlier detection system using deep learning models; and (3) how to make the outlier detection method more transparent.

Acknowledgement. This work is supported by the National Key R&D Program of China under Grant No. 2020YFB1710200.

References

1. Porwal, U., Mukund, S.: Credit card fraud detection in e-commerce. In: 2019 18th IEEE International Conference on Trust, Security and Privacy in Computing and Communications/13th IEEE International Conference on Big Data Science and Engineering (TrustCom/BigDataSE), pp. 280–287. IEEE (2019)
2. Zhang, L., Lin, J., Karim, R.: An angle-based subspace anomaly detection approach to high-dimensional data: With an application to industrial fault detection. Reliab. Eng. Syst. Saf. **142**, 482–497 (2015)
3. Alrawashdeh, K., Purdy, C.: Toward an online anomaly intrusion detection system based on deep learning. In: 2016 15th IEEE International Conference on Machine Learning and Applications (ICMLA), pp. 195–200. IEEE (2016)
4. Gebremeskel, G.B., Yi, C., He, Z., et al.: Combined data mining techniques based patient data outlier detection for healthcare safety. Int. J. Intell. Comput. Cybern. (2016)
5. Liu, W., Pan, R.: Outlier mining based on variance of angle technology research in high-dimensional data. In: 2015 10th International Conference on Intelligent Systems and Knowledge Engineering (ISKE), pp. 598–603. IEEE (2015)
6. Yang, Z., Ge, Z.: Rethinking the value of just-in-time learning in the era of industrial big data. IEEE Trans. Industr. Inf. **18**(2), 976–985 (2021)
7. Breunig, M.M., Kriegel, H.P., NgR, T., et al.: LOF: identifying density-based local outliers. In: Proceedings of the 2000 ACM SIGMOD International Conference on Management of Data, pp. 93–104 (2000)
8. Yang, X., Latecki, L.J., Pokrajac, D.: Outlier detection with globally optimal exemplar-based GMM. In: Proceedings of the 2009 SIAM International Conference on Data Mining, pp. 145–154. Society for Industrial and Applied Mathematics (2009)
9. Sarvari, H., Domeniconi, C., Prenkaj, B., Stilo, G.: Unsupervised boosting-based autoencoder ensembles for outlier detection. In: Karlapalem, K., et al. (eds.) PAKDD 2021. LNCS (LNAI), vol. 12712, pp. 91–103. Springer, Cham (2021). https://doi.org/10.1007/978-3-030-75762-5_8

10. Seiffert, C., Khoshgoftaar, T.M., Van Hulse, J.: Hybrid sampling for imbalanced data. Integr. Comput.-Aided Eng. **16**(3), 193–210 (2009)
11. Cheng, L., Wang, Y., Liu, X., et al.: Outlier detection ensemble with embedded feature selection. In: Proceedings of the AAAI Conference on Artificial Intelligence **34**(04), 3503–3512 (2020)
12. Song, H., Li, P., Liu, H.: Deep clustering based fair outlier detection. In: Proceedings of the 27th ACM SIGKDD Conference on Knowledge Discovery and Data Mining, pp. 1481–1489 (2021)
13. Li, Y., Liu, N., Li, J., et al.: Deep structured cross-modal anomaly detection. In: 2019 International Joint Conference on Neural Networks (IJCNN), pp. 1–8. IEEE (2019)
14. Lu, W., Cheng, Y., Xiao, C., et al.: Unsupervised sequential outlier detection with deep architectures. IEEE Trans. Image Process. **26**(9), 4321–4330 (2017)
15. Dzaferagic, M., Marchetti, N., Macaluso, I.: Fault detection and classification in Industrial IoT in case of missing sensor data. IEEE Internet Things J. **9**(11), 8892–8900 (2021)
16. Liu, B., Xiao, Y., Cao, L., et al.: SVDD-based outlier detection on uncertain data. Knowl. Inf. Syst. **34**, 597–618 (2013)
17. Zhang, Z., Deng, X.: Anomaly detection using improved deep SVDD model with data structure preservation. Pattern Recogn. Lett. **148**, 1–6 (2021)
18. Zhou, X., Hu, Y., Liang, W., et al.: Variational LSTM enhanced anomaly detection for industrial big data. IEEE Trans. Industr. Inf. **17**(5), 3469–3477 (2020)
19. Campos, G.O., Zimek, A., Sander, J., et al.: Data Min. Knowl. Discov. **30**, 891–927 (2016)
20. Anaissi, A., Braytee, A., Naji, M.: Gaussian kernel parameter optimization in one-class support vector machines. In: 2018 International Joint Conference on Neural Networks (IJCNN), pp. 1–8. IEEE (2018)
21. Xu, Z., Kakde, D., Chaudhuri, A.: Automatic hyperparameter tuning method for local outlier factor, with applications to anomaly detection. In: 2019 IEEE International Conference on Big Data (Big Data), pp. 4201–4207. IEEE (2019)
22. Li, Z., Zhao, Y., Botta, N., et al.: COPOD: copula-based outlier detection. In: 2020 IEEE International Conference on Data Mining (ICDM), pp. 1118–1123. IEEE (2020)
23. Chen, Y., Ashizawa, N., Yean, S., et al.: Self-organizing map assisted deep autoencoding Gaussian mixture model for intrusion detection. In: 2021 IEEE 18th Annual Consumer Communications and Networking Conference (CCNC), pp. 1–6. IEEE (2021)

Dimension Reduction Based on Sampling

Zhuping Li[1], Donghua Yang[1($\boxtimes$)], Mengmeng Li[1], Haifeng Guo[1],
Tiansheng Ye[2], and Hongzhi Wang[1]

[1] Harbin Institute of Technology, Harbin, China
yang.dh@hit.edu.cn
[2] ConDB, Beijing, China
daniel.ye@cnosdb.com

Abstract. Dimension reduction provides a powerful means of reducing the number of random variables under consideration. However, there were many similar tuples in large datasets, and before reducing the dimension of the dataset, we removed some similar tuples to retain the main information of the dataset while accelerating the dimension reduction. Accordingly, we propose a dimension reduction technique based on biased sampling, a new procedure that incorporates features of both dimensional reduction and biased sampling to obtain a computationally efficient means of reducing the number of random variables under consideration. In this paper, we choose *Principal Components Analysis*(PCA) as the main dimensional reduction algorithm to study, and we show how this approach works.

Keywords: PCA · dimensional reduction · biased sampling

1 Introduction

Data gathering has become easier with the rise of the Internet, which results in high-dimensional large-scale datasets. For example, the Institute of Remote Sensing and Digital Earth can gather 4 TB datasets each month, and those datasets always have high dimensions. An American health care company can gather 2.1PB datasets every year. The high-dimensional datasets with huge sample lengths have resulted in huge storage and computing time [8], while we have difficulties handling those huge datasets due to the limitations of computer performance. Hence, it is necessary to reduce the data dimension and sample length.

Some dimension reduction algorithms aim to speed up the process by reducing dataset length. Current research uses a nature-inspired *Simulated Annealing*(SA) algorithm for the purpose of data dimensionality and sample size length [7]. The drawback of such an approach is that SA must be computed in $O(\alpha N^{\beta})$, which is inappropriate for large-scale datasets. Furthermore, Szymon Lukasik

This paper was supported by The National Key Research and Development Program of China (2020YFB1006104), The Opening Project of Intelligent Policing Key Laboratory of Sichuan Province (ZNJW2023KFZD004), Sichuan Police College (CJKY202001) and NSFC grant (62232005).

Z. Yu et al. (Eds.): ICPCSEE 2023, CCIS 1879, pp. 207–220, 2023.
https://doi.org/10.1007/978-981-99-5968-6_15

et al. proposed a linear dimensionality reduction algorithm in which fast simulated annealing (FSA) is used to determine the transformation matrix [8]. Although FSA has significant computational complexity, it still does not solve the problem of disposing of very high-dimensional problems with very large sample sizes.

In addition, some algorithms aim to reduce the dimension directly but do not take into account the influence of dataset length. A sampling-based algorithm for dimension reduction is proposed for removing redundancy and noise in large datasets. One sampling-based algorithm is fast low-rank approximation, which tasks $O(M(\frac{k}{\epsilon}+k^2 log k)+(m+n)(\frac{k^2}{\epsilon^2}+\frac{k^3 log k}{\epsilon}+\frac{k^4}{log k^2}))$ [4]. Another is the bicriteria algorithm, a randomized algorithm that runs in $\tilde{O}(mnk^3(k/\epsilon)^{p+1})$ and finds an $\tilde{O}(k^2(k/\epsilon)^{p+1})$-dimensional subspace [4]. The computation time of sampling-based algorithms will grow exponentially with increasing k-dimensional linear subspace.

Principal Component Analysis. (PCA) [1] is the most well-known dimension reduction algorithm and one of the most classical algorithms in data science. PCA is simple and has wide applicability. We can obtain lower dimensional datasets with PCA. However, we can hardly process all records in datasets by PCA since the time complexity of PCA is $O(m^2 n+n^3)$ with m as data length and n as the dimension of the dataset, and such time complexity is not applicable for large datasets. For example, the dimension of the large-scale dataset is reduced by PCA, but all data have to be stored in the memory, which may result in memory overflow. In addition, the runtime of PCA grows exponentially with the continued accretion of the size of datasets. For example, reducing the dimension of a 2.6 GB dataset requires only 28 s by PCA, while the runtime will grow to 360 s when processing a 5.2 GB dataset.

In view of the defects of previous algorithms, such as the slow speed of processing large-scale datasets or even invalidation in handling those datasets, a straightforward way is to sample some data for dimension reduction, such as simple random sampling and stratified sampling. To achieve better performance suitable for processing huge data, we introduce a new method of dimensionality reduction algorithm based on sampling. The length of the original dataset is reduced by simple random sampling or stratified sampling, which reduces the computational burden on storage pressure and simplifies the next process. Dimension reduction is accomplished by PCA. In other words, PCA based on sampling can process small samples, and the sampling algorithms chosen in our new algorithms ensure that the sampling error is small enough by choosing an appropriate sampling rate, which means that our new algorithms can not only process large-scale datasets but also ensure a high speed and small error level compared with the original PCA.

The main contributions of this paper are as follows.

1. We propose PCA based on simple random sampling. Although sample error is inevitable, the whole dataset can be well represented by the small sample by finding an appropriate sample rate with its sample error at a sufficiently

small level. Furthermore, simple random sampling can be computed in time $O(r * N)$ with r as the sample rate and N as the total samples. This algorithm not only confirms good results but also speeds up the dimension reduction.

2. We propose PCA based on stratified sampling. We choose the most representative tuples by stratified sampling, which maximally covers the types of the whole dataset. When combining PCA and stratified sampling, we reduce the dimension on the most representative sample and thereby retain the principal information of the original dataset. Meanwhile, the sample can be small enough when we control the sample rate of each category, and then the speed of dimension reduction can be improved.

3. To demonstrate the efficiency and relatively low error level of the proposed algorithms, we conduct extensive experiments on real data. The experimental results show that those two algorithms can accelerate dimension reduction within the allowable range of error.

The rest of the paper is structured as follows. Background is discussed in Sect. 2. Section 3 describes our approach in detail. We discuss the pros and cons in Sect. 3.2. Our experimental results are presented in Sect. 4. Section 5 concludes the paper.

2   Background

In this paper, we propose two PCA approaches based on sampling. Therefore, the sample rate is the main factor affecting the error of PCA based on simple random sampling. In this section, we analyse the relationship between the sample rate and sample size to determine the interval of the best sample rate of simple random sampling, thereby optimizing the error of PCA based on a relatively small sample size.

Let N be the total number of samples, n be the sample size, f is the sample ratio, and S^2 is the population variance. Then, the variance of the sample mean $\overline{y}$ is [3]

$$V(\overline{y}) = \frac{1-y}{n}S^2 \tag{1}$$

When the population is known, S^2 is obvious. Hence the sample error only depends on n and f. Then we solve the functional expression about f as follows.

$$g(f) = (1-f)\frac{1}{n} = (1-f)\frac{1}{fN} \tag{2}$$

Obviously, according to (2), the value of $g(f)$ should be smaller to make the sample error small. When the population is determined, the sample size N is also determined. Thus, $g(f)$ can also be simplified as follows.

$$f(x) = \frac{1-x}{x} \tag{3}$$

where x means the sample rate, and $x \in [0, 1]$. Function $f(x)$ replaces $V(\overline{y})$.

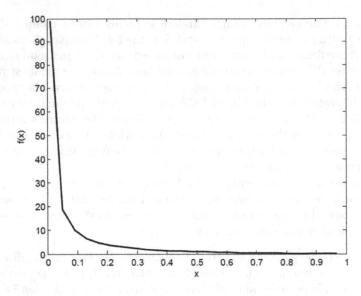

Fig. 1. f(x)

Then, we determine the interval of the best minimum sample rate via function $f(x)$, as Fig. 1 shows.

As observed from Fig. 1, when the sample rate is in [30%, 40%], the function curve achieves stability. When the sample rate is larger than 30%, the function curve approaches zero with the sample rate. Therefore, the best sample rate of a simple random sample is [30%, 40%] if the records need to have a high sampling accuracy.

3 Principal Component Analysis Based on Sampling

PCA is the simplest and most popular dimension reduction method [10]. It has been the most classical dimensionality reduction method. However, singular value decomposition or eigen decomposition is the main process in PCA and is very time-consuming [6]. When the matrix becomes large, matrix transformation will occupy too much time or even cannot be computed by general computers. PCA, as a general rule, can be computed in time $O(m^2 n + n^3)$ and has to take too much time to compute the covariance matrix and decompose it. Due to the high time complexity, when the original dataset occupies 1 TB or more, dimension reduction with PCA will take several days.

To solve this problem, we propose two sampling-based PCAs to perform PCA. We combine the sampling algorithm and PCA such that the number of tuples can be reduced by the sampling algorithm to accelerate the dimension reduction.

3.1   PCA Based on Simple Random Sampling

In this subsection, we propose PCA based on simple random sampling. Although sample error is inevitable, the whole dataset can be well represented by a small sample by finding an appropriate sample rate. Furthermore, simple random sampling can be computed in time $O(r * N)$ with r as sample rate and N as the number of total tuples. This algorithm not only confirms good results but also speeds up the dimension reduction. In addition, the dataset is divided into several logical partitions before sampling. When sampling the records, there are only part of partitions, but not all, to be sampled. In this way, not all records are scanned when sampling, which increases the speed of sampling.

The pseudo code of PCA based on simple random sampling is shown in Algorithm 1. First, we partition the dataset into b partitions and randomly sample b' partitions(Line 1-2). Then we randomly sample r tuples from each partition (Lines 3-5). After all partitions are sampled, all samples will become the input of PCA, and the dimension reduction results are returned as the answer (Lines 6-8).

Algorithm 1. PCA based on simple random sampling

Require: Data:$X_1, X_2, ..., X_n$,
    b: the number of data partitions,
    b': the number of sample data partitions,
    r: the number of sampled subsets from each partition
Ensure: M: low dimensional dataset.
1: Partition by $key(column\_list)$ partitions b
2: Randomly sample a partition set $I' = \{i_1, i_2, ..., i_{b'}\}$
3: **for** $j \leftarrow 1$ to b' **do**
4:    Randomly sample a set $N'_j = \{l_1, l_2, ..., l_r\}$ without replacement
5: **end for**
6: combine each sampled subsets: $N \leftarrow \{N'_1, N'_2, ..., N'_{b'}\}$
7: $M \leftarrow PCA(N)$
8:
9: **return** M

To show that PCA based on simple random sampling is faster than the original PCA, we analyse its time complexity. In PCA based on simple random sampling, the sampled dataset is much smaller than the original dataset. We suppose the sampled data have m' records, $m' << m$. Hence, the size of the simple random sampling is $O(m')$, and PCA based on simple random sampling can be computed in time $O(m' + m'^2 n + n^3)$ with m records and n dimension. Since $m' << m$, PCA based on simple random sampling has better performance in terms of runtime. Moreover, we analyse the error of PCA based on simple random sampling. The error of PCA based on simple random sampling, compared with the original PCA, is mainly influenced by sampling. The error of simple random sampling is represented by the variance of the sample mean, as shown in Theorem 1.

Theorem 1. With simple random sampling, the variance of $\hat{Y} = N\bar{y}$, as an estimation of the population total Y, is [3]

$$V(\hat{Y}) = E(\hat{Y} - Y)^2 = \frac{N^2 S^2}{n} \frac{(N-n)}{n} = \frac{N^2 S^2}{n}(1-f) \tag{4}$$

Then, according to Eq. 4, we obtain an accurate error of PCA based on simple random sampling. It shows that such estimation is unbiased. The error of PCA based on simple random sampling is controlled in an acceptable scope when the sample rate is appropriate. However, there are still some problems. The records are chosen randomly. Therefore, those records, which are selected by simple random sampling, may not represent the original dataset well. Thus, sampling approaches other than random sampling are in demand. We will introduce stratified sampling next.

3.2   PCA Based on Stratified Sampling

To obtain a better result of PCA based on sampling, we analyse the variance of the mean with different samplings, which can represent the sampling accuracy. Our goal is to find a more accurate sampling algorithm whose time complexity is not higher than $O(n)$. First, we obtain the variance of the sample mean from a simple random sampling (as shown in Theorem 2). Then, we analyse the sampling accuracy of stratified sampling since it can choose the records according to the records' category, as shown in Theorems 3 and 4.

Theorem 2. The variance of the sample mean $\overline{y_{ran}}$ from a simple random sampling is [3]

$$V_{ran} = E(\bar{y} - \overline{Y})^2 = \frac{S^2}{n} \frac{N-n}{n} = \frac{1-f}{n} S^2 \tag{5}$$

where $f = n/N$ is the sampling fraction.

Theorem 3. For stratified random sampling, the variance of the mean $\overline{y_{prop}}$ is [3]

$$V_{prop} = \sum W_h^2 \frac{1-f}{n_h} S_h^2 \tag{6}$$

Theorem 4. With proportional allocation, the variance of the mean $\overline{y'_{prop}}$ is [3]

$$V'_{prop} = \frac{1-f}{n} \sum W_h S_h^2 \tag{7}$$

Theorem 4 can also be represented as [3]:

$$V'_{prop} = \frac{1-f}{n} S^2 - \frac{1-f}{n} \sum W_h (\overline{Y_h} - \overline{Y})^2 \tag{8}$$

According to Eq. 8, if and only if $\overline{Y_h} = \overline{Y}$, namely, the mean of each layer is equal, stratified random sampling has the same effect as simple random sampling.

However, if $\overline{Y_h} \neq \overline{Y}$ and $V'_{prop} < V_{ran}$, stratified random sampling is better than simple random sampling in terms of accuracy.

According to the discussions above, we propose a PCA approach based on stratified sampling. We choose the most representative tuples via stratified sampling, which maximally covers the types of the whole dataset. When combining PCA and stratified sampling, we reduce the dimension on the most representative sample, thereby retaining the principal information of the original dataset. Meanwhile, the sample is minimum as much as possible for the subjective factor controlled. As a result, the speed of dimension reduction is improved. Furthermore, we improve the stratified sampling. It obtains samples in one pass [9] and stops the sampling when those requisite records are sampled enough, which makes this algorithm more suitable for large data. The whole algorithm is shown in Algorithm 2.

First, we partition the dataset into b partitions, and randomly sample b' partitions(Lines 1-2). Before stratified sampling, we need to decide the number of sampled category r and how many tuples s are sampled from each category on the basis of the original data. Then, we scan each sampled partition. In each scanned partition, we need a list to record the sampled classes. If the class of scanned tuples is not in the recording list, this tuple is put in the sampled dataset N(Lines 5-7); if the class of scanned tuples is already in the recording list, we need to judge whether the number of tuples in this class is over s. If so, this tuple will be ignored and not put in N; otherwise, this tuple will be put in N(Lines 8-12). Meanwhile, the length of the recording list cannot be over r. If it is up to r, scan the next partition. Finally, the input of PCA is the sampled dataset $\Gamma = \{N_1, N_2, \ldots, N_{b'}\}$, and we obtain the low-dimensional dataset.

To prove that the PCA based on stratification is more efficient than the original PCA, we first analyse its time complexity. In PCA based on stratified sampling, suppose we sample r records from the original dataset, $\triangle$ is the number of skimmed over records such that $r + \triangle << m$. The time complexity of stratified sampling is $O(r + \triangle)$. Therefore the time complexity of PCA based on stratified sampling is $O(r + \triangle + r^2 n + n^3)$. Because of greatly decreasing m, the complexity of PCA based on sampling accelerates the process of dimension reduction.

Then, we demonstrate the accuracy of the proposed approach. We analyse the sampling accuracy of stratified sampling as shown in Theorem 5.

Theorem 5. With stratified sampling, if the estimation of the population total Y is $\hat{Y}_{st} = N\overline{y_{st}}$, the variance of $\hat{Y}_{st}$ is [3]

$$V(\hat{Y}_{st}) = \sum N_h (N_h - n_h) \frac{S_h^2}{n_h} \qquad (9)$$

Then, According to Eq. 9, we obtain the error of PCA based on stratified sampling. This estimation is also unbiased. The error of PCA based on stratified sampling is controlled in an acceptable scope when the sample rate is appropriate.

4   Experiments

To verify the effectiveness and efficiency of the proposed algorithms, we conduct extensive experiments.

Algorithm 2. PCA based on stratified sampling

Require: Data:$X_1, X_2, ..., X_n$
 b: the number of data partitions
 b': the number of sampled data partitions
 $l_1, l2, ..., l_{b'}$: the length of each data partition
 r: the number of sampled categories from each partition
 R: a set of sampled category labels
 s: the number of sampled tuples from each category
 p: the tuple
Ensure: M: low-dimensional dataset
 Partition by $key(column\_list)$ partitions b
 Randomly sample a partition set $\Gamma = \{N_1, N_2, ..., N_{b'}\}$
 for $j \leftarrow 1$ to b' **do**
   while $i \leq l_j$ **do**
     if $p[i].label$ not in R **then**
       put $p[i]$ into the set N_i
       $R[p[i].label] \leftarrow 1$
     else
       if $R[p[i].label] \leq s$ **then**
         put $p[i]$ into the set N_i
       end if
     end if
     if $length(R) > r$ **then**
       break
     end if
   end while
 end for
 $M \leftarrow PCA(\Gamma)$

 return M

All of our experiments are performed on a PC with quad core, 64-bit, 2.6 GHz CPU and 32 GB memory. We use the Covertype dataset in [2] and adjust its categorical distribution, data size and length. We test our algorithms in three aspects: sample rate, data size and dimension. We compare the effectiveness of PCA based on sampling with the original PCA. The effectiveness is represented by the categorical accuracy of datasets after dimension reduction. Equation 10 shows the calculation method of categorical accuracy with n as the number of records classified correctly and N as the number of total records. Our classification algorithm is a multilayered perception network [5].

$$categorical - accuracy = \frac{n}{N} \tag{10}$$

4.1   Varying Sample Rates

In this section, we compare our algorithms with the original PCA in terms of efficiency and effectiveness on various sample rates. We choose three different distributions, namely, the categorical distribution, normal distribution, uniform distribution and Zipf distribution, to test our algorithms' performance. The function of zipf distribution is the "$1/f$" function. The length and dimension of the datasets are constant. The length of the datasets is 980,000. The dimension of the datasets is 55, and one of the dimensions is labels. The number of categories is 7. The categorical labels are 1,2,...,7. Table 1 shows the categorical distribution and the number of records in each category. The effectiveness of our algorithms is shown by their categorical accuracy after dimensional reduction. The runtime and categorical accuracy of the original PCA are shown in Table 2.

Table 1. Categorical distribution of datasets

distribution labels	1	2	3	4	5	6	7
normal distribution	5890	59518	237261	375184	236442	59805	5746
uniform distribution	136430	139711	139991	140097	139912	140289	143570
zipf distribution	377961	188981	125987	94490	75593	62994	53994

Table 2. The performance of original PCA

distribution performance	runtime(s)	categorical accuracy(%)
normal distribution	2.71	92.1
uniform distribution	2.75	85.3
zipf distribution	2.47	81.2

We test our algorithms with various sample rates. Figure 2(a) and (b) show the runtime and categorical accuracy of PCA based on simple random sampling, respectively. For the runtime, we observe that the speed of PCA based on simple random sampling is still faster than that of the original PCA even though the sample rate is 30%.The speed can be improved by at least 14.5% when the sample rate is 30% and the categorical distribution is a uniform distribution. Furthermore, from Fig. 2(b), the categorical accuracy of PCA based on simple random sampling is lower than that of the original PCA, but it then becomes a constant with increasing sample rate. The reason is that some categories are too small to be sampled by simple random sampling. However, those small categories have a limited effect on the error of PCA based on simple random sampling because their quantity is so small that it is not representative of the large-scale dataset.

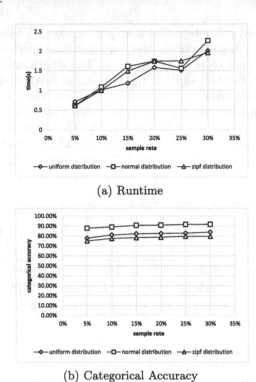

(a) Runtime

(b) Categorical Accuracy

Fig. 2. PCA based on simple random sampling

Figure 3(a) and (b) show the runtime and categorical accuracy of PCA based on stratified sampling, respectively. From these results, the PCA based on stratified sampling reduces the runtime by at least 28%, while the categorical accuracy is not obviously influenced. Another observation is that the performance of PCA based on stratified sampling is stabilized in these three categorical distributions. The reason is that stratified sampling can always screen out the representative tuples. Furthermore, stratified sampling is based on categories, which makes those small categories more likely to be sampled.

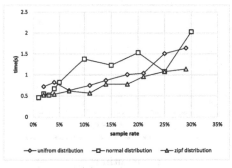

(a) Runtime

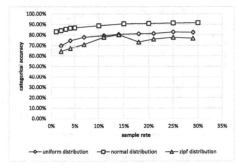

(b) Categorical Accuracy

Fig. 3. PCA based on stratified sampling

4.2   Varying Data Size

In this section, we compare our algorithms with original PCA in terms of efficiency and effectiveness on various data size. The effectiveness of our algorithms is shown by its categorical accuracy after dimensional reduction. The runtime represents its efficiency. We generate five datasets with different data sizes and the same dimension. The data sizes of those datasets are respectively 604 M, 725 M, 1.3 G, 2.6 G and 5.2 G. The categorical distribution is normal distribution. The sample rate is 15%. The experimental results are shown in Fig. 4. From Fig. 4(a), the runtime of PCA based on sampling is much less than original PCA, because the computation time is mainly influenced by the number of records and dimension, and the new dataset sampled by sampling algorithms is much smaller than original dataset after sampling. In addition, from Fig. 4(b), the categorical accuracy of new algorithms is better than original PCA. The reason is that those representative records are more likely to be chosen by simple random sampling although the dataset is small. And when the dataset is large enough, these two sampling algorithms can both choose the representative records and discard those categories in small numbers, which then leads to a better categorical accuracy.

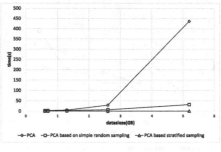

(a) Runtime

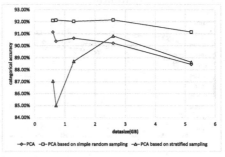

(b) Categorical Accuracy

Fig. 4. Varying data size

4.3 Varying Dimensions

In this section, we compare our algorithms with the original PCA in terms of efficiency and effectiveness on various dimensions. The effectiveness of our algorithms is shown by their categorical accuracy after dimensional reduction. The runtime represents its efficiency. We generate six datasets with various dimensions as follows. Given the different numbers of dimensions, we generate datasets with various dimensions, while they have the same number of tuples, and each feature is correlated with the labels. Each dataset has the same number of records. The experimental results are shown in Fig. 5.

Comparisons of runtime between our algorithms and the original PCA are shown in Fig. 5(a). The results show that the runtime of our algorithms is shorter compared to the original PCA. Thus, the PCA based on sampling is obviously faster than the original PCA.

Figure 5(b) shows the categorical accuracy. As Fig. 5(b) shows, the accuracy of our algorithms is much higher than that of the original PCA. In most cases, their accuracy can be 100%, although the original accuracy is not 100%. The reason is that the new dataset includes only the most representative records, and those small categories, the main factor producing the errors, are discarded by sampling algorithms.

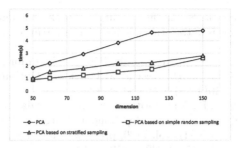

(a) Runtime

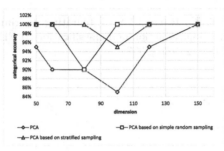

(b) Categorical Accuracy

Fig. 5. Varying dimension

5   Conclusion

In this paper, we study the dimension reduction method and combine it with sampling algorithms to accelerate its computation speed. To process large-scale datasets efficiently, we combine the dimension reduction algorithm and sampling algorithms and propose two methods, PCA based on random sampling and PCA based on stratified sampling. To ensure effectiveness and efficiency, we analyse the relationship between the sample rate and sampling accuracy with two sampling algorithms. Experimental results show that those two algorithms accelerate dimension reduction within the allowable range of error. Our future work includes reducing the randomness of sampling to make the results of dimension reduction more accurate and stable.

References

1. Abdi, H., Williams, L.J.: Principal component analysis. Wiley Interdiscip. Rev. Computat. Statist. **2**(4), 433–459 (2010)
2. Blackard, J.A., Dean, D.J., Anderson, C.: The forest covertype dataset (1998)
3. Cochran, W.G.: Sampling techniques, 3rd edition. DBLP (1977)
4. Deshpande, A.J.: Sampling-based algorithms for dimension reduction, Ph. D. thesis, Massachusetts Institute of Technology (2007)

5. Gardner, M.W., Dorling, S.: Artificial neural networks (the multilayer perceptron)-a review of applications in the atmospheric sciences. Atmos. Environ. **32**(14), 2627–2636 (1998)

6. Jolliffe, I.: Principal component analysis. Wiley Online Library (2002)

7. Kirkpatrick, S., Gelatt, C.D., Vecchi, M.P., et al.: Optimization by simulated annealing. Science **220**(4598), 671–680 (1983)

8. Łukasik, S., Kulczycki, P.: An algorithm for sample and data dimensionality reduction using fast simulated annealing. In: Tang, J., King, I., Chen, L., Wang, J. (eds.) ADMA 2011. LNCS (LNAI), vol. 7120, pp. 152–161. Springer, Heidelberg (2011). https://doi.org/10.1007/978-3-642-25853-4_12

9. Vitter, J.S.: Random sampling with a reservoir. ACM Trans. Math. Softw. (TOMS) **11**(1), 37–57 (1985)

10. Yu, J., Tian, Q., Rui, T., Huang, T.S.: Integrating discriminant and descriptive information for dimension reduction and classification. IEEE Trans. Circuits Syst. Video Technol. **17**(3), 372–377 (2007)

Complex Time Series Analysis Based on Conditional Random Fields

Yanjie Wei[1], Haifeng Guo[1], Donghua Yang[1($\boxtimes$)], Mengmeng Li[1], Bo Zheng[2], and Hongzhi Wang[1]

[1] Harbin Institute of Technology, Harbin, China
yang.dh@hit.edu.cn
[2] ConDB, Beijing, China

Abstract. A fundamental problem with complex time series analysis involves data prediction and repair. However, existing methods are not accurate enough for complex and multidimensional time series data. In this paper, we propose a novel approach, a complex time series prediction model, which is based on the conditional random field (CRF) and recurrent neural network (RNN). This model can be used as an upper-level predictor in the stacking process or be trained using deep learning methods. Our approach is more accurate than existing methods in some suitable scenarios, as shown in the experimental results.

Keywords: complex time series · missing data · conditional random field · Stacking · deep learning

1 Introduction

A time series is a sequence of data points measured in a fixed time interval. With the continuous development of computer networks, data storage and data mining, time series data are generated more frequently. However, time series data are usually accompanied by noise and data loss. Detecting and repairing time series with noise or missing values is a very challenging task, and complex time series such as multidimensional industrial time series data make the challenge even more difficult.

To repair the missing time series data, some time series repair algorithms have been proposed. In the existing time series repair algorithm, a common approach to detect anomalies and data repair is to use the established model to learn from the complete data and then use the predicted values to detect and fill. The autoregressive (AR) model is a common model, and another common method is constraint-based repair.

Although these methods have the advantages of simple assumptions and few parameters, they cannot describe the relationship between complex time series dimensions. Since these methods are designed for simple time series of a single dimension, they are not suitable for solving complex time series problems.

Supported by The National Key Research and Development Program of China (2020YFB1006104).

To avoid the drawbacks of existing approaches and detect time series data more accurately, we propose a complex time series prediction model based on the conditional random field (CRF) and recurrent neural network (RNN). Our algorithm generally follows the following processes:

1. A conditional random field is used as the upper classifier model, and the stacking training method is combined to enhance time series trend prediction.
2. A conditional random field is implemented by a neural network, so it can be combined with LSTM, GRU and other networks. Stacking, pretraining, fine-tuning and end-to-end training techniques are available to solve the problem of regression and repair of complex time series with neural networks.

The main contributions of this paper are as follows.

– We propose a novel approach that can improve the accuracy of data prediction.
– To demonstrate the accuracy of the proposed approaches, we conduct extensive experiments on real data. The experimental results show that the proposed algorithm outperforms existing approaches in terms of accuracy.

The rest of the paper is organized as follows: Background is discussed in Sect. 2. Section 3 introduces how to combine an image segmentation approach with a conditional random field for stacking time series prediction. In Sect. 4, we introduce the continuous value fully connected conditional random field implemented by the recurrent neural network and describe how to use the conditional random field for stacking-based integrated learning and further deep learning. Our experimental results are presented in Sect. 5. Section 6 concludes the paper. We discuss related work in the final section.

2   Background

In this paper, we propose a complex time series prediction model based on a conditional random field. In this section, we introduce the definition of time series, the preprocessing approach and the assumptions used in this paper.

We define the time series as follows:

Definition 1. *A time series S is described as a sequence of pairs, i.e., $S = [(t_1, t_2, ..., t_n)]$, in which t_i is the time stamp of s_i , satisfying $t_i < t_j$ if $i < j$. s_i is a d-dimension data point.*

A complex time series S has nonlinear and nonstationary features. We cannot guarantee that it is produced by a linear addition of several attributes or several previous time values, nor does it guarantee that its statistical properties will not change with time. In addition, the time series data studied in this paper are also characterized by multiple dimensions.

We define the conditional random field (CRF) as follows.

Definition 2. *In general, the formal form of a conditional random field is defined as follows: G = (V, E) is used to represent a graph. In this graph, each vertex represents a variable, and we use* $\mathbf{Y} = (Y_v)_{v \in V}$ *to represent the vertex. The upper variable, in turn, represents the observed value; then,* $(\mathbf{X}, \mathbf{Y})$ *is a conditional random field in the case where the random variable* Y_v *depends on* X, *where* $p(Y_v|X, Y_w, w \neq v) = p(Y_v|X, Y_w, w \sim v)$, $\mathbf{w} \sim \mathbf{v}$ *indicates that the two vertices are adjacent.*

For the convenience of our discussions, we make the following assumptions:

Assumption 1. *We have a series of predictors that predict the trend of each dimension of a complex time series.*

3 Stacking Integrated Learning Using Conditional Random Fields

3.1 Stacking

In [10], the conditional random field is used to solve the image segmentation problem. The nature of the image segmentation problem is the classification problem, so the conditional random field can be used to solve the trend problem of complex time series. Using the conditional random field as the upper classifier, we combine the training method of stacking to propose a complex time series prediction algorithm based on the conditional random field and stacking.

Definition 3. *Stacking means training a learning algorithm (upper-level predictor) that combines several other learning algorithms (base predictors).*

First, several selected learning algorithms are trained using the available data, and then the upper predictor uses the results of the base predictor as input.

With the appropriate upper-level predictor, stacking can be used to represent a range of integrated learning methods, including bagging [3] and boosting [2].

Model fusion through stacking can often achieve good results. Stacking has been successfully applied to supervised and unsupervised [8] learning tasks such as regression and classification [7] [4].

4 Trend Prediction of Complex Time Series Based on Conditional Random Field and Stacking

4.1 Assumptions and Main Ideas

As we assumed in Background, we have a series of predictors that can predict the trend of each dimension of a complex time series. The key issue that we focus on is how to combine the predictions of existing predictors with the correlation between the various dimensions of a complex time series to make predictions more accurate. For example, the meteorological centres in each city predict the rise or fall of the weather on the second day based on their respective data, and we need to use the relationship between the cities to correct this prediction.

Our algorithm roughly follows the following process:

1. Obtaining a part of the data as a validation set.
2. Using the validation set, the existing classifier and the results between different dimensions were stacked.
3. Validate the effect on the test data.

4.2   Algorithm

Using a fully connected conditional random field as an upper layer predictor, we propose a model fusion algorithm based on a conditional random field and stacking. The calculation process is shown in Algorithm 1.

Algorithm 1. Complex Time Series Trend Prediction Algorithm Based on Conditional Random Field and Stacking

---

Require: basic regressors' result P,kernel function $\left\{ \overrightarrow{\kappa^1} \ldots \kappa^M \right\}$.

Ensure: enhanced results Q.

1: **while** (not converged) **do**

2:    $\tilde{Q}_{i,k}^m \leftarrow \sum_{l \neq i} \kappa^m \left(f_i, f_j \right) Q_{l,k}$

3:    $\check{Q}_{i,k} \leftarrow \sum_m w^m \tilde{Q}_{i,k}^m$

4:    $\check{Q}_{i,k} \leftarrow \sum_{k'=1}^{K} \mu \left(k', k \right) \check{Q}_{i,k}$

5:    $\overline{Q}_{i,k} \leftarrow U_{i,k} - \check{Q}_{i,k}$

6:    $Q_{i,k} \leftarrow \frac{1}{Z_i} \overline{Q}_{i,k}$

7: **end while**

P is a two-dimensional vector, and $P_{i,k}$ represents the probability that the i-dimensional data predicted by the existing predictor belongs to class k. U is also a two-dimensional vector that represents the initial potential energy before the conditional random field is stabilized, obviously resulting from the initial predictor. After the third line, the algorithm enters the mean field iteration so that the conditional random field eventually becomes stable. The K^m function of the fourth line is a kernel function to describe the relationship between the two different dimensions. More kernel functions such as this can be created, where f_i represents the feature correlated with dimension i, and the predictions of the base predictors are included in stacking. The W^m in the fifth row is the weight corresponding to the kernel function, which is used to adjust the influence of the characterization relationship of the kernel function on the final result, which is a parameter to be learned. In line 6, the function μ is used to describe the relationship between classes. A simple and effective choice in practice is

$$\mu \left(k', k \right) = \begin{cases} 0, & k' \neq k \\ 1, & k' = k \end{cases} \tag{1}$$

The equation above means that only those intermediate results that are in the same class can have an impact on the final results. We used this option in experiments against weather data. Line 8 performs a softmax function on the result, which works as regularization.

The softmax function is shown below.

$$Z_i = \sum_k \exp\left(\overline{Q}_{i,k}\right) \tag{2}$$

4.3   Implementation and Experiment

First, we use CUDA to implement the conditional random field described in the previous section. Most of the calculations focus on the loop part starting from line 3, and this part can be calculated in parallel. By observing the operation of this implementation, it is found that similar to the literature [12], the third line begins to converge after 5 to 10 rounds of iteration, which makes it possible to implement iteration of conditional random fields using a structure similar to a recurrent neural network. To reduce the amount of calculation, we modify the fourth line of Algorithm 1.

$$\tilde{Q}_{i,k}^m \leftarrow \sum_{l \neq i} \kappa^m \left(\boldsymbol{f}_i, \boldsymbol{f}_j\right) Q_{l,k} N(i,j) \tag{3}$$

where

$$N(i,j) = \begin{cases} 0 & i \, is \, close \, to \, j \\ 1 & i \, is \, not \, close \, to \, j \end{cases} \tag{4}$$

The distance between the definition and the dimension that is closer and further is used as a hyperparameter. Generally, several dimensions closest to a certain dimension are relatively close dimensions, and others are further dimensions. This definition acts like a truncated Gaussian filter. We use a one-dimensional convolutional layer to implement row 5 and use the softmax layer commonly used in neural networks to implement row 8. When training, we use cross entropy as the loss function and Adam and other methods for training.

We captured the weather data from 2011 to 2017 from the weather post website (www.tianqihoubao.com), as shown in Fig. 1, and the data were scaled according to the time ratio of 3:1:1. Divided into training sequences, validation sequences and test sequences. We choose Adaboost, random forest, and gradient boosting decision tree to obtain the base classifier on the training set and predict the rise or fall of the city's weather on the next day based on the weather data of a city in the previous 10 d. Then, using our algorithm, we corrected the results of these classifiers.

According to the algorithm description in the previous section, we need to design the kernel function κ^m and the feature $\boldsymbol{f}_i$. The kernel function we designed is as follows:

$$\kappa^1 \left(\boldsymbol{f}_i \boldsymbol{f}_j\right) = \exp\left(-\frac{|\boldsymbol{p}_i - \boldsymbol{p}_j|^2}{2\theta_\alpha^2}\right) = \exp\left(-\frac{distancc_{i,j}^2}{2\theta_\alpha^2}\right) \tag{5}$$

$$\kappa^2\left(\boldsymbol{f}_i, \boldsymbol{f}_j\right) = \exp\left(-\frac{\left|\boldsymbol{p}_i - \boldsymbol{p}_j\right|^2}{2\theta_\beta^2} - \frac{\left|\boldsymbol{c}_i - \boldsymbol{c}_j\right|^2}{2\theta_\gamma^2}\right) \tag{6}$$

$\boldsymbol{p}_i$ and $\boldsymbol{c}_i$ constitute $\boldsymbol{f}_i$, where $\boldsymbol{p}_i$ represents the location of the city, and $\boldsymbol{c}_i$ represents the prediction of the temperature of the second day of the city by each base classifier, that is, stacking. Compared with the simple averaging of the prediction results of the primordial predictor. After stacking, the accuracy of the prediction is improved by more than 2%, considering that the selected base predictor has been based on a more capable model, such an increase is considerable.

5 Solving Regression Problems and Deep Learning with Conditional Random Fields

The algorithm introduced in the previous section is applicable to classification problems, and the problem of regression is the first thing to be solved based on the prediction of complex time series repair. This section will introduce several methods to solve this problem.

5.1 Transforming the Regression Problem into a Classification Problem

To solve the regression problem, we transform the regression problem into a classification problem by dividing the regression value into several value intervals. Then, we use the prediction value and accuracy of the base predictor to establish a truncated Gaussian distribution. The probability value of the distribution is taken as the probability value of each value interval.

Assume that this truncated Gaussian distribution is $Gaussian(a, b, u, \sigma)$, where a is the starting value of this distribution (the lower bound), b is the ending value of this distribution (the upper bound), and a and b are determined by our data range.

u is the mean of the truncated Gaussian distribution, determined by the result of the base predictor (or the average result), σ is the standard deviation representing the Gaussian distribution, and its value is inversely proportional to our confidence in the base predictor. When we consider the base, when the accuracy of the predictor is high, we set a lower standard deviation; conversely, when we think that the accuracy of the base predictor is low, we set a higher standard deviation. Finally, we calculate the probability of this distribution within the value interval as the final probability, as shown in Fig. 1.

5.2 Conditional Random Field for Regression Problems

Prediction Algorithm. Combining the approach in [6] with [5], we propose a conditional random field-based regression algorithm for regression problems.

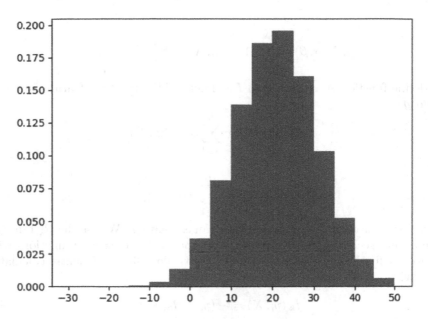

Fig. 1. Schematic diagram of interval probability obtained by truncating normal distribution.

Algorithm 2. Continuous condition random field iterative process

---

Require: basic regressors' results p, kernel matrix $\overrightarrow{K^1} \ldots \overrightarrow{K^m}$
Ensure: enhanced results Q.

$\quad q_i \leftarrow p_i$
$\quad u_i \leftarrow p_i$
$\quad$ **while** (not converged) **do**
$$q_i \leftarrow \frac{u_i + 2 \sum_{m=1}^{M} w^m \sum_{j \neq i} \kappa^m \left(\vec{f}_i, \vec{f}_j \right) q_i}{1 + 2 \sum_{m=1}^{M} w^m \sum_{i \neq j} \kappa^m \left(\vec{f}_i, \vec{f}_j \right)}$$
$\quad$ **end while**

Since the result predicted by the upper predictor is a numerical value, the corresponding input p output q is a one-dimensional vector, and the average field iteration is implemented by the loop starting from the row.

The following is a brief description of the derivation of the fourth line. We use a continuous conditional random field to characterize the conditional distribution $P(\mathbf{y}|\mathbf{X})$, where $\mathbf{y} \in R^N$:

$$P(y|X) = \frac{1}{Z(X, \alpha, \beta)} \exp(\phi(y, X, \alpha, \beta)) \tag{7}$$

where

$$Z(X, \alpha, \beta) = \int_y \exp(\phi(y, X, \alpha, \beta)) \tag{8}$$

and

$$\phi(y, X, \alpha, \beta) = \sum_{i=1}^{N} A\left(\alpha, y_i, X\right) + \sum_{i \neq j} I\left(\beta, y_i, y_j, X\right) \tag{9}$$

We define function A and function I as linear additions of the feature functions f and g.

$$A\left(\alpha, y_i, X\right) = \sum_{k=1}^{K} \alpha_k f_k\left(y_i, X\right) \tag{10}$$

and

$$I\left(\beta, y_i, y_j, X\right) = \sum_{l=1}^{L} \beta_l g_l\left(y_i, y_j, X\right) \tag{11}$$

Suppose we have a series of models that can estimate y_i. We use the R_k function to represent such a model. Reference [9] proposes that quadratic functions with A and I defined as y can make conditional random fields efficiently calculated and derived. Therefore, we suppose:

$$f_k\left(y_i, X\right) = -\left(y_i - R_k(X)\right)^2 \tag{12}$$

and

$$g_l\left(y_i, y_j, X\right) = -k_l\left(p_i, p_j\right)\left(y_i - y_j\right)^2 \tag{13}$$

To calculate the inverse matrix, an accurate continuous conditional random field requires $O\left(N^3\right)$ to be derived. Under field-average theory, a method is proposed to approximate the distribution $P(\mathbf{y}|X)$ by using KL divergence of $P(\mathbf{y}|\mathbf{X})$ and minimum distribution $Q(\mathbf{y}|X) = \prod_{i=1}^{N} Q_i\left(y_i|X\right)$.

$$\begin{array}{c} \log\left(Q_i\left(y_i|X\right)\right) = -\sum_{k=1}^{K} \alpha_k\left(y_i^2 - 2y_i R_k(X)\right) \\ -2\sum_{l=1}^{L} \beta_l \sum_{i=j} k_l\left(p_i, p_j\right)\left(y_i^2 - 2y_i E\left[y_j\right]\right) + const \end{array} \tag{14}$$

Therefore $\log\left(Q_i\left(y_i|X\right)\right)$ can be expressed as a normal distribution, and its mean is:

$$\mu_i = \frac{\sum_{k=1}^{K} \alpha_k R_k(X) + 2\sum_{l=1}^{L} \beta_l \sum_{i \neq j} k_l\left(p_i, p_j\right) \mu_j}{\sum_{k=1}^{K} \alpha_k + 2\sum_{l=1}^{L} \beta_l \sum_{i \neq j} k_l\left(p_i, p_j\right)} \tag{15}$$

We let $\alpha_k = \frac{1}{K}$ and then convert the corresponding symbol. Then, we obtain the fourth line in Algorithm 2. The reason why the value of α_k is fixed is to draw on the idea of a residual neural network [1] and use conditional random field learning residuals to make training easier.

Data Repair Algorithm. Simply improve the algorithm in the previous section to fix complex time series data, as shown in Algorithm 3. where t represents data with missing data, and we use $*$ for missing data. By determining whether the data at the corresponding location are missing, the algorithm determines whether the location oscillates during the calculation of the average field iteration.

Algorithm 3. Continuous condition random field iterative process

---

Require: basic regressors' results p, kernel matrix $\overrightarrow{K^1} \dots \overrightarrow{K^m}$, data with missing value t

Ensure: enhanced results q.

$$q_i \leftarrow \begin{cases} p_i, t_i = * \\ t_i, t_i \neq * \end{cases}$$

$u_i \leftarrow q_i$

while (not converged) **do**

$$q_i \leftarrow \begin{cases} \dfrac{u_i + 2\sum_{m=1}^{M} w^m \sum_{j \neq i} \kappa^m\left(\overrightarrow{f_i}, \overrightarrow{f_j}\right) q_i}{1 + 2\sum_{m=1}^{M} w^m \sum_{i \neq j} \kappa^m\left(f_i, f_j\right)}, t_i = * \\ t_i, t_i \neq * \end{cases}$$

end while

We find that if we consider the result of the kernel function as a fixed input and write it as a symmetric matrix:

$$K_{i,j}^m = \begin{cases} 0, & i = j \\ k^m\left(\boldsymbol{f}_i, \boldsymbol{f}_j\right), & i \neq j \end{cases} \tag{16}$$

Line 4 in Algorithm 2 can be implemented by matrix multiplication so that the gradient can propagate backwards.

Experiment on Weather Data. The same kernel function is used, and the conditional random field is used as the upper predictor to carry out stacking training. Adaboost, random forest and gradient lifting tree are used to train the base predictor. Based on the weather data of the previous 10 d, we predict the temperature of the next day and then use Algorithm 2 to enhance the prediction result; that is, the result of the base prediction is taken as input, and the mean square error is used as the loss on the verification set. The Adam algorithm is used to train the random field conditions described in Algorithm 2. Using the mean square error as a measure, the enhanced result on the test set has a 9.1% improvement over the simple average. On the basis of this experiment, we verify the validity of Algorithm 3 according to a certain probability that the temperature of a certain city is missing on a certain day. We use the mean square error of the data and the original data as the standard and find that after the enhancement of Algorithm 3, the result is a 10% improvement over the simple average of the predicted data. We record the results of the experiment in the table.

Table 1. ME and MAPE of different models

question type	Benchmark improvement
forecast	9.1%
repair	10.0%

Whether it is applied to the prediction problem or the repair problem, after the conditional random field fully captures the features of the time series dimension of the time to be predicted (repaired), the process of stacking this model makes the prediction and repair results more accurate.

Experiments on Industrial Data. Since the weather data contain some discrete attributes, it is difficult to directly use the neural network to process it. Therefore, predictors such as Adaboost, random forest and gradient lifting tree are selected as the base predictors. To explore whether our proposed algorithm is suitable for deep learning, we conducted experiments on industrial wind turbine data.

The wind unit data are numerical data, but the range of each dimension is very different, and we normalize it before use.

Similar to experiments with weather data, we divide industrial data into training sets, validation sets, and test sets in a 3:1:1 ratio. We use GRU and LSTM as the main network structure to construct the base classifier and use the data of the first 10 moments to predict the next moment. The network structure of the base classifier is shown in Fig. 2.

Since industrial data do not have a clear interdimensional relationship, such as geographical location and weather data, we use the Pearson correlation coefficient as the distance term of the kernel function to design a kernel function similar to the previous one.

On the training set, using the mean square error as the loss, we train the two networks of GRU and LSTM as RNN units. After convergence and the performance of different base predictors are similar, we use several methods to enhance the results by using Algorithm 2.

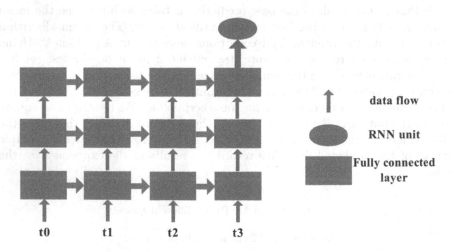

data flow

RNN unit

Fully connected layer

t0   t1   t2   t3

Fig. 2. Base predictor network structure.

- Stacking method, that is, using the results of the two models as the base predictor on the verification set as input, only the training condition random field.
- The pretraining-fine tuning method transmits the gradient of the conditional random field to the network as the base classifier, thereby realizing the fine tuning of the base classifier network.

Based on the results of the base predictor network, the experimental results are shown in the table.

Table 2. The experimental results

GRU+LSTM \ Stacking	GRU \ Fine-Tune	Lstm \ Fine-Tune
12%	4%	3%

Based on the results of the base predictor network, the experimental results are shown in the table.

The use of stacking for model fusion improves the prediction effect compared to the optimization of a single model. One possible reason is that the integration of the results of the two models alleviates the overfitting problem of the neural network to a certain extent, and the single model easily falls into this problem.

6   Conclusions

Using the cyclic neural network to characterize the time series features of complex time series and using the conditional random field to describe the features of complex time series dimensions, we describe the complex time series prediction model based on deep learning, which further solves the problem of complex time series prediction. Based on this, complex time series repair problems can be solved.

We have experimentally demonstrated that the conditional random field is applicable to complex time series data after the rational design of the kernel function in addition to the image data. In addition to the convolutional neural network in [11], it can also be used to enhance Adaboost, tree models such as random forests, and recurrent neural networks such as GRU and LSTM in stacking or pretraining-fine tuning. In addition to classification problems, continuous conditional random fields can also be used to solve regression problems, which are then used for data repair. Using weather data and industrial time series data, we verified the effectiveness of this method.

References

1. He, K., Zhang, X., Ren, S., Sun, J.: Deep residual learning for image recognition. In: Proceedings of the IEEE Conference on Computer Vision and Pattern Recognition, pp. 770–778 (2016)
2. Kégl, B.: The return of AdaBoost.MH: multiclass Hamming trees. arXiv preprint arXiv:1312.6086 (2013)
3. Liang, G., Zhang, C.: Empirical study of bagging predictors on medical data. In: Proceedings of the Ninth Australasian Data Mining Conference, vol. 121, pp. 31–40. Australian Computer Society, Inc. (2011)
4. Ozay, M., Vural, F.T.Y.: A new fuzzy stacked generalization technique and analysis of its performance. arXiv preprint arXiv:1204.0171 (2012)
5. Radosavljevic, V., Vucetic, S., Obradovic, Z.: Continuous conditional random fields for regression in remote sensing. In: ECAI, pp. 809–814 (2010)
6. Ristovski, K., Radosavljevic, V., Vucetic, S., Obradovic, Z.: Continuous conditional random fields for efficient regression in large fully connected graphs. In: Twenty-Seventh AAAI Conference on Artificial Intelligence (2013)
7. Sill, J., Takács, G., Mackey, L., Lin, D.: Feature-weighted linear stacking. arXiv preprint arXiv:0911.0460 (2009)
8. Smyth, P., Wolpert, D.: Linearly combining density estimators by stacking. Mach. Learn. 36(1–2), 59–83 (1999)
9. Xin, X., King, I., Deng, H., Lyu, M.R.: A social recommendation framework based on multiscale continuous conditional random fields. In: Proceedings of the 18th ACM Conference on Information and Knowledge Management, pp. 1247–1256. ACM (2009)
10. Zhang, Y., et al.: Sequential click prediction for sponsored search with recurrent neural networks. In: Twenty-Eighth AAAI Conference on Artificial Intelligence (2014)
11. Zhang, Y., et al.: Sequential click prediction for sponsored search with recurrent neural networks. In: Twenty-Eighth AAAI Conference on Artificial Intelligence (2014)
12. Zheng, S., et al.: Conditional random fields as recurrent neural networks. In: Proceedings of the IEEE International Conference on Computer Vision, pp. 1529–1537 (2015)

Feature Extraction of Time Series Data Based on CNN-CBAM

Jiaji Qin, Dapeng Lang[✉], and Chao Gao

Harbin Engineering University, Harbin 150001, China
langdapeng@hrbeu.edu.cn

Abstract. Methods for extracting features from time series data using deep learning have been widely studied, but they still suffer from problems of severe loss of feature information across different network layers and parameter redundancy. Therefore, a new time-series data feature extraction model (CNN-CBAM) that integrates convolutional neural networks (CNN) and convolutional attention mechanisms (CBAM) is proposed. First, the parameters of the CNN and BiGRU prediction models are optimized through uniform design methods. Next, the CNN is used to extract features from the time series data, outputting multiple feature maps. These feature maps are then subjected to feature re-extraction by the CBAM attention mechanism at both the spatial and channel levels. Finally, the feature maps are input into the BiGRU model for prediction. Experimental results show that after CNN-CBAM processing, the stability and accuracy of the BiGRU prediction model improved by 77.6% and 76.3%, respectively, outperforming other feature extraction methods. Meanwhile, the training time of the model has only increased by 7.1%, demonstrating excellent time efficiency.

Keywords: Uniform Design · CNN · CBAM · Time-series Data Feature Extraction

1 Introduction

Time series data refer to data obtained by sampling at a predetermined frequency based on the chronological order of events. With the development of big data and artificial intelligence, the processing and application of time series data have gained increasing attention and have been widely used in finance, healthcare, weather forecasting, and other fields, providing strong support for intelligent decision-making and prediction.

However, with the growing popularity of time series data in various industries, the problems posed by their massive data volume, high dimensionality, and noise interference have become more prominent. Therefore, feature extraction is necessary to improve data interpretability and accuracy. Through feature extraction, time series data can be transformed into a set of more concise and representative feature vectors, making it easier for data analysis and modelling. Additionally, feature extraction can remove noise and redundant information from the data, thereby improving model performance.

Time series data usually contain numerous features, and traditional feature extraction methods often require manual feature selection, which is subject to individual domain

Z. Yu et al. (Eds.): ICPCSEE 2023, CCIS 1879, pp. 233–245, 2023.
https://doi.org/10.1007/978-981-99-5968-6_17

expertise and experience. This approach can cause the feature extraction results to be influenced by human factors, leading to important information being overlooked or unnecessary noise being introduced. In contrast, deep learning models can automatically extract relevant features by learning the data's characteristics and patterns. They also capture nonlinear features in the time series data through multilevel nonlinear transformations, enhancing their ability to perform feature extraction. Thus, many researchers choose to use deep learning methods in studying feature extraction for time series data. The main work of this paper is as follows: a feature extraction model combining a convolutional neural network (CNN) and convolutional attention mechanism (CBAM) is proposed to improve the problem of loss in the process of transferring the original features of the time-series data between each network layer, and the effectiveness and feasibility of the model are verified by experimental comparison on the public dataset of the Investing Financial Database.

2 Relative Works

Currently, there are two main methods for extracting features from time series data: statistical-based and model-based approaches. The time-frequency domain feature method is a popular statistical-based approach that converts time series data into a two-dimensional time-frequency plane through time-frequency analysis, which extracts time-frequency features such as spectrograms, instantaneous frequency, and instantaneous amplitude for classification or prediction purposes. For example, Wan et al. [1] used a short-time Fourier transform (STFT) to extract time-frequency features from vibration signals collected by a pair of fibre Bragg grating (FBG) sensors during the normal operation time of trains and used four common unsupervised learning algorithms to monitor the health status of train wheels. Similarly, Zheng et al. [2] processed the raw vibration signals collected from an acceleration sensor with wavelet packet transform (WPT) to reduce noise and unrelated information, reconstructed the wavelet packets containing vibration information, computed 14 time-frequency domain features of the reconstructed vibration signal, and used them as a multifeature vector for torsional vibration detection.

However, the aforementioned statistical-based feature extraction methods are usually applied to linear time series data, and their feature extraction capabilities may be relatively weak for nonlinear time series data. Model-based methods can better extract features from nonlinear time series data. Yang et al. [3] extracted the interest region of aerial images using a selective search method and calculated the radial gradient of the interest area, which yielded rotation-invariant feature descriptors. They then filtered out noise from the raw data using a deep denoising autoencoder and extracted the deep features of the feature descriptors. Similarly, Lei et al. [4] used a basic deep learning network, an autoencoder, to extract features from the surface defect images of refractory bricks and improved the recognition effect of refractory brick surface defects.

In recent years, researchers have tended to use deep learning for feature extraction in time series data. Compared with the research method of autoencoders, convolutional neural networks (CNNs) are more effective in extracting local features due to their convolution layer characteristics [5–7], making them more suitable for feature extraction in time series data. Yin et al. [8] constructed a feature extraction structure based on a single CNN and multilight source fusion called SingleCNN-MergeRGB to extract comprehensive features of defects under different illumination conditions. Bhimavarapu et al. [9] utilized CNN to extract features from images of affected areas of plants and proposed an improved activation and optimization function (IAOF) to minimize losses, enhancing the accuracy of rice disease recognition and classification.

Although current methods have achieved good results, they only utilize CNNs to autonomously learn features from the data without fully utilizing the information within and between multiple feature maps [10]. Therefore, there is a problem of loss of feature information during transmission between network layers. Building on existing research, this paper introduces the CBAM mixed attention mechanism and proposes the CNN-CBAM model for feature extraction in time series data. This model analyses and highlights the intrinsic feature weights of time series data from both channel and spatial dimensions, suppressing nonimportant information expression and thereby improving the stability and accuracy of time series data prediction.

3　Model Construction

3.1　BiGRU Prediction Model

BiGRU is a commonly used method for time series data prediction [11], which benefits from its bidirectional structure that allows it to extract features of the time series data from both forward and backwards directions. This approach effectively addresses the vanishing and exploding gradient issues while also avoiding parameter redundancy. BiGRU consists of two unidirectional, reverse GRU networks. At each time step, input variables are simultaneously input into both reverse GRU networks, and the output of BiGRU is determined by these two GRUs working together (Fig. 1).

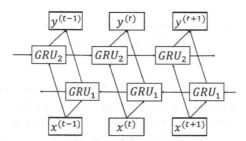

Fig. 1. BiGRU structure

The calculation formula of BiGRU is as follows:

$$\overrightarrow{h_t} = GRU\left(x_t, \overrightarrow{h_{t-1}}\right) \tag{1}$$

$$\overleftarrow{h_t} = GRU\left(x_t, \overleftarrow{h_{t-1}}\right) \tag{2}$$

$$y_t = \overrightarrow{h_t} \oplus \overleftarrow{h_t} \tag{3}$$

where y_t, $\overrightarrow{h_t}$, and $\overleftarrow{h_t}$ are the model output, forward output and reverse output, respectively. Because of its bidirectional structure, BiGRU has higher prediction accuracy than GRU.

3.2 CBAM Mixed Attention Mechanism

Convolutional neural networks can generate multiple feature maps, each containing distinct feature information. Spatial attention and channel attention mechanisms are two commonly used methods for re-extracting features from these maps. However, the spatial attention mechanism treats each feature map's information equally, disregarding any correlation between them. Similarly, the channel attention mechanism globally averages information within each feature map, neglecting how information weights vary within feature maps. To address this issue, this research has introduced the CBAM module for further re-extracting features from CNN-generated feature maps. The CBAM module structure (shown in Fig. 2) includes two components: a channel attention module (CAM) and a spatial attention module (SAM). These components extract data features from both the channel and spatial dimensions, preserving the relative positional information of the data and reducing the loss of information caused by flattening operations. This approach has improved model prediction performance.

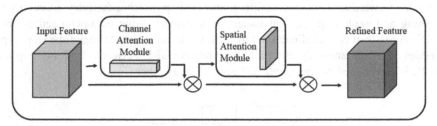

Fig. 2. CBAM structure

Channel Attention Module. As shown in Fig. 3, first, the input feature map is processed by global maximum pooling and global average pooling, and the obtained two feature maps are used as the inputs of the weight sharing multilayer perceptron. Then, their outputs are summed and normalized by the sigmoid function to obtain the input feature map MC required by the SAM module. The calculation formula is as follows:

$$M_c(F) = \sigma(MLP(AvgPool(F)) + MLP(MaxPool(F))) \tag{4}$$

In the formula, F refers to the feature map input to the CBAM module. Mc (F) represents the feature map processed by the CAM component, where σ represents the sigmoid operation, *AvgPool* denotes average pooling, and *MaxPool* denotes max pooling. The MLP layer refers to the multilayer perceptron.

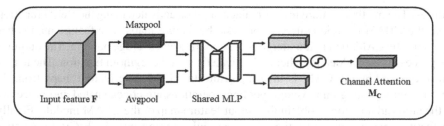

Fig. 3. CAM module structure

Spatial Attention Module. As shown in Fig. 4, first, the feature map output by the CAM module is taken as the input feature map of SAM, and the two feature maps obtained by maximum pooling and average pooling are spliced and integrated. Then, the dimension is reduced by the convolution operation, and the feature map MS is generated after normalization by the sigmoid function. Finally, after multiplying the characteristic map output by the CAM and SAM modules, the characteristic map processed by the CBAM module is obtained, and the calculation formula is as follows:

$$M_s(F) = \sigma\left(f^{k \times k}([AvgPool(F); MaxPool(F)])\right)$$
$$= \sigma\left(f^{k \times k}\left(\left[F_{avg}^s; F_{max}^s\right]\right)\right) \tag{5}$$

where F represents the input feature map, $Ms(F)$ represents the feature map processed by the SAM module, F_{avg}^s represents the feature map after global average pooling, F_{max}^s represents the feature map after global maximum pooling, σ represents the sigmoid operation, f represents the convolution operation, and $k \times k$ represents the size of convolution kernel.

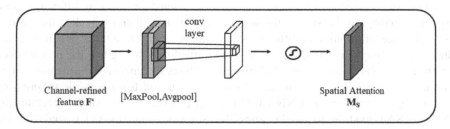

Fig. 4. SAM module structure

3.3 CNN-CBAM Model

The CNN-CBAM model proposed in this study consists of CNN and CBAM attention mechanisms, and the BiGRU model is used to predict time series data. First, the original time series data are input into a convolutional neural network to extract effective data features. The multiple feature maps obtained are then activated using the ReLU function. Next, the CBAM module is used to process the feature maps. The feature map is first input into the CAM, where it undergoes max pooling and average pooling. The output of this process is then added together and activated using the sigmoid function. The feature map processed by the CAM is then input into the SAM. The feature maps from the previous max pooling and average pooling operations are concatenated and processed by the sigmoid function to obtain the output feature map of the CBAM module. Finally, the feature map processed by the CBAM module is input into the BiGRU prediction model. The predicted results are then input into the fully connected layer to obtain the final output (Fig. 5).

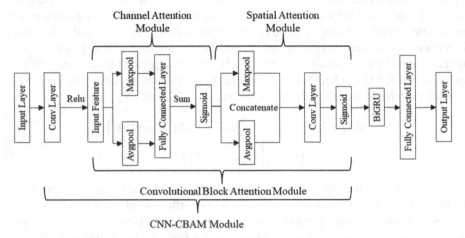

Fig. 5. CNN-BiGRU-CBAM model structure

The CAM is capable of learning the weight relationships between different feature maps. Depending on the size of the weights, it can strengthen or weaken the expression of feature information in different feature maps for different tasks. When there are many feature maps in a model, the CAM can effectively strengthen the performance of important feature maps. The SAM module analyses the importance of feature information within a single feature map and strengthens useful features while suppressing useless ones. Therefore, the CNN-CBAM model proposed in this study integrates the CAM and SAM modules to avoid treating all feature maps equally with spatial attention mechanisms, improving the inherent connectedness between feature maps and resolving the issue of channel attention's equal treatment of global information within a single feature map, which leads to ignored features. This results in better feature extraction and preservation of time series data, thus optimizing the prediction ability of the model.

4   Experimental Design and Result Analysis

The experiment of this study was completed on the Windows platform based on the PyTorch framework. To enhance the effectiveness of the experimental results and reduce the contingency, the average of 10 experimental results was taken as the final result of the experiment, and the ratio of the training set, verification set and test set in the experiment was divided into 8:1:1.

4.1 Dataset and Pretreatment

Stock price data in the financial field have distinct time series characteristics, so this study chooses the exchange rate of the US dollar against RMB as the research object. We select the stock price for 11 consecutive months from July 2019 to May 2020 to construct a dataset, which contains 26,277 sets of stock price data. Each set of data has a time interval of 5 min and contains 123 sets of data every day. The variables of each set of data are mainly the opening price, the lowest price, the closing price and the highest price. The above data are collected from the Investing Financial Information Database [12], as shown in Table 1 of the experimental section.

Table 1. Training set data

Date	Opening prices	Lowest price	Closing price	Highest price
2020-3-19 19:00	7.1099	7.1099	7.115	7.115
2020-3-19 19:04	7.12	7.116	7.116	7.12
2020-3-19 19:13	7.115	7.115	7.1176	7.1176

Data normalization is a preprocessing technique used to restrict sample data within a certain range, thereby eliminating adverse effects caused by distorted data and facilitating the use of a standard deviation between 0 and 1. Currently, there are two main data normalization methods: Z score normalization and max-min normalization. This study opted for the max-min normalization method because it facilitates the transformation of raw data into a scale easily interpretable by machine learning models.

$$x = (x' - x_{min})/(x_{max} - x_{min}) \tag{6}$$

where x_{max} and x_{min} are the maximum and minimum values in the selected dataset, x' is the original sample data, and x is the normalized data, and its value range is [0,1].

4.2 Optimization Parameters of the Uniform Design Method

The uniform design method is a commonly used parameter optimization technique in multiparameter modelling [13]. It enables the selection of representative solutions from all candidate options and reflects the main characteristics of the system. The network parameters of CNN and BiGRU have a significant impact on the predictive performance of the model. To avoid possible bias introduced by human factors, this study adopted the uniform design method to determine the parameters of the CNN network and the BiGRU prediction model.

The parameters to be optimized include the number of CNN convolutional kernels, the stride of the CNN, and the number of hidden layers in the BiGRU. The commonly used values for each parameter, as reported by Liu [14] and Liu [15], are as follows:

Number of CNN convolutional kernels (A): 16, 32, 64.

Number of hidden layers in BiGRU (B): 5, 10, 15.

CNN stride (C): 1, 2.

Since there are three factors in the experiment and the number of levels is not equal, a U_6 ($3^2 \times 2^1$) mixed-level orthogonal table was adopted to design the experiment, resulting in a total of six experimental schemes, as shown in Table 2.

Table 2. Orthogonal table U_6 ($3^2 \times 2^1$) results

Scheme number	A	B	C	MAE	RMSE
1	16	5	2	0.0236	0.0293
2	16	10	1	0.0286	0.0349
3	32	15	2	0.0182	0.0218
4	32	5	1	0.0153	0.0184
5	64	10	2	0.0321	0.0392
6	64	15	1	0.0266	0.0337

The evaluation metrics chosen for model training are the root mean square error (RMSE) and mean absolute error (MAE). According to the results in Table 2, scheme 4 has the best parameter selection. Based on the above analysis, the optimal values for the CNN network's convolutional kernel count and stride are 32 and 1, respectively. Additionally, the optimal number of hidden layers in BiGRU is 5.

4.3 Model Performance Evaluation Index

To overcome the limitations associated with a single evaluation index, this study uses two performance evaluation metrics, namely, the mean absolute error (MAE) and root mean square error (RMSE), to evaluate the stability and effectiveness of various models. The formulas for calculating these indices are as follows:

$$RMSE = \sqrt{\frac{1}{n} \sum_{i=1}^{n} (\hat{y}_i - y_i)^2} \tag{7}$$

$$MAE = \frac{1}{n} \sum_{i=1}^{n} |\hat{y}_i - y_i| \tag{8}$$

In the formula, $\hat{y}_i$ is the predicted stock price, y_i is the actual stock price, and n is the number of samples in the test set. The ranges for MAE and RMSE are from 0 to infinity, where a smaller MAE value indicates better model stability, while a smaller RMSE value indicates better predictive performance of the model.

4.4   Ablation Experiment

To validate the effects of each module in the model, ablation experiments were performed. Four models, namely, BiGRU, CNN-BiGRU, CBAM-BiGRU, and CNN-CBAM-BiGRU, were used to predict stock price data. When testing the four models, all other parameters except the model structures should be kept constant so as not to affect the performance of the models. The settings of the various parameters are shown in scheme 4 in Table 2.

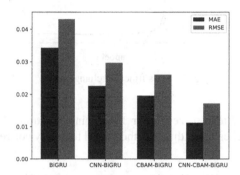

Fig. 6. Comparison of prediction performance after adding each module

The comparison of the prediction performance evaluation index values for the four models is shown in Fig. 6. It can be seen from the figure that the BiGRU model has the highest MAE and RMSE, indicating lower stability and accuracy of the model during prediction. However, after separately implementing CNN and CBAM modules to extract features from stock price data, both MAE and RMSE of the models showed a decrease. In particular, the CNN-CBAM feature extraction method proposed in this study showed a remarkable reduction in the MAE and RMSE of the BiGRU benchmark model by 77.6% and 76.3%, respectively, demonstrating superior stability and accuracy among the four models. This success can be attributed to the ability of the CNN-CBAM model to extract features of stock price data from spatial and channel perspectives, preserve the relative position information of the data, minimize the information loss due to flattening processing, and improve the predictive performance of the model.

4.5   Timeliness Analysis of the Model

The adaptability of a feature extraction method is affected by its up-to-dateness. If the model is poorly timed, it will take a long time to train, making it impossible to quickly make predictions and to verify and improve the model. This greatly reduces the practical value of the model. To evaluate the timeliness of the proposed CNN-CBAM model, a comparative experiment was conducted to compare its performance with other prediction models from the ablation experiments.

Figure 7 shows the change curve of the relationship between the loss function and the number of training iterations of the model in this study. It can be seen from the figure that the values of the loss function on both the validation set and the training set

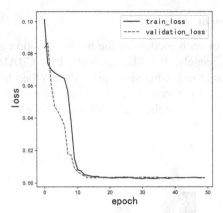

Fig. 7. Loss function change curve

gradually decrease and tend to zero after 10 training iterations. When the number of training iterations reaches more than 15, the model basically converges.

Table 3. Comparison of prediction performance after adding each module

Model	MAE	RMSE	training time of the model/s
BiGRU	0.0343	0.0431	358.25
CNN-BiGRU	0.0226	0.0297	366.22
CBAM-BiGRU	0.0196	0.0261	372.76
CNN-CBAM-BiGRU	0.0077	0.0102	383.98

Table 3 shows that by adding the CNN-CBAM module for temporal data feature extraction on top of the BiGRU baseline model, the training time of the model increased by 25.73 s compared to the BiGRU baseline model, which is a 7.1% increase in training time. Therefore, it can be considered that the CNN-CBAM method has excellent timeliness.

4.6 Comparison of Methods

To verify the superiority of the proposed method, the CNN-CBAM model proposed in this study is compared with other feature extraction models. The prediction results of each model are shown in Fig. 8, in which the black curve represents the real stock price data, the blue curve represents the predicted value processed by the CNN-CBAM model, and the pink curve represents the predicted value processed by other feature extraction models. The vertical axis represents the standardized stock price, and the horizontal axis represents the sample serial number.

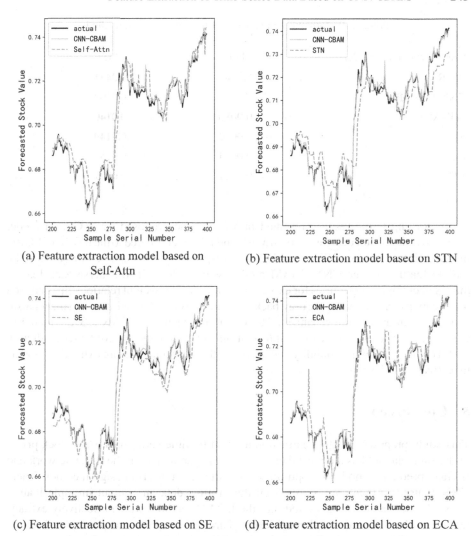

(a) Feature extraction model based on Self-Attn

(b) Feature extraction model based on STN

(c) Feature extraction model based on SE

(d) Feature extraction model based on ECA

Fig. 8. Comparison of prediction accuracy after processing by different feature extraction models

By comparing our proposed CNN-CBAM model with Self-Attn [16], STN [17], SE [18] and ECA [19], four existing feature extraction models, the latter four models were found to have the ability to extract features from stock prices and to fit price fluctuations to some degree. However, their fitting accuracy was found to be lower than that of our proposed CNN-CBAM model. This is because the model feeds the feature map output from the CNN into the CBAM module, uses the channel attention mechanism to filter out the feature maps that have a greater impact on the prediction performance from these feature maps, and then retains the valid feature information in each feature map through the spatial attention mechanism. The prediction model processed by the CNN-CBAM model exhibits better fitting accuracy.

Table 4. Performance comparison of feature extraction methods

Feature extraction model	MAE	RMSE
BiGRU	0.0343	0.0431
CNN-CBAM	0.0077	0.0102
Self-Attn	0.0086	0.0140
SE	0.0090	0.0149
ECA	0.0084	0.0153
STN	0.0121	0.0201

According to the results presented in Table 4, the proposed CNN-CBAM model achieved a significant decrease in both MAE and RMSE compared to the baseline BiGRU model, with decreases of 77.6% and 76.3%, respectively. This can be attributed to the unique features of the CNN-CBAM model, which allows it to analyse each feature map, select important features for expression, and consider weight relationships between feature maps, thus avoiding the drawbacks of using only the channel attention mechanism or the spatial attention mechanism. These features enable the CNN-CBAM model to extract more efficient and relevant information and ultimately achieve better prediction performance with higher stability and accuracy than any other feature extraction models investigated in this study.

5   Conclusion

This study proposes a feature extraction method for time-series data using stock prices in the financial field as a typical dataset. First, the parameters of the CNN network and BiGRU prediction model are optimized by horizontal uniform design methods. Then, CNN is used to identify features from the original data and output multiple feature maps. These feature maps are fed into the CBAM module, which effectively extracts features at both spatial and channel levels. Through comparison experiments with the Self-Attn, STN, SE, and ECA models, the CNN-CBAM model was found to produce the most accurate and stable results after feature extraction. In addition, experiments comparing the timeliness of the CNN-CBAM model and benchmark prediction models such as BiGRU highlight the excellent timeliness and effectiveness of the proposed CNN-CBAM model in extracting features from time series data. In future work, we will consider using the CNN-CBAM model for feature extraction and identification of atypical time-series data such as access traffic to explore the applicability of the model in the direction of data security and trustworthiness assessment.

References

1. Wan, T.H., Tsang, C.W., Hui, K., et al.: Anomaly detection of train wheels utilizing short-time Fourier transform and unsupervised learning algorithms. Eng. Appl. Artif. Intell. **122**, 106037 (2023)
2. Zheng, Q., Chen, G., Jiao, A.: Chatter detection in milling process based on the combination of wavelet packet transform and PSO-SVM. Int. J. Adv. Manuf. Technol. **120**(1–2), 1237–1251 (2022)
3. Yang, F., Ma, B., Wang, J., et al.: Target detection of UAV aerial image based on rotational invariant depth denoising automatic encoder. Xibei Gongye Daxue Xuebao/J. Northwestern Polytech. Univ. **38**(6), 1345–1351 (2020)
4. Lei, F., Wang, X., Liu, Y.J., et al.: Study on feature extraction and classification of surface defects of refractory tiles based on self-encoder. Manuf. Autom. **44**(12), 28–31+67 (2022)
5. Zha, W., Liu, Y., Wan, Y., et al.: Forecasting monthly gas field production based on the CNN-LSTM model. Energy 124889 (2022)
6. Kirisci, M., Cagcag, Y.O.: A new CNN-based model for financial time series: TAIEX and FTSE stocks forecasting. Neural Process. Lett. **54**(4), 3357–3374 (2022)
7. Wang, D., Gan, J., Mao, J., et al.: Forecasting power demand in China with a CNN-LSTM model including multimodal information. Energy **263**, 126012 (2023)
8. Yin, Z., Chen, M., Zhao, L., et al.: A novel automatic classification approach for microflaws on the large-aperture optics surface based on multilight source fusion and integrated deep learning architecture. J. Intell. Manuf. 1–16 (2022)
9. Bhimavarapu, U.: Prediction and classification of rice leaves using the improved PSO clustering and improved CNN. Multimedia Tools Appl. 1–14 (2023)
10. Agga, A., Abbou, A., Labbadi, M., et al.: CNN-LSTM: an efficient hybrid deep learning architecture for predicting short-term photovoltaic power production. Electr. Power Syst. Res. **208**, 107908 (2022)
11. Zhang, C., Wang, D., Wang, L., et al.: Temporal data-driven failure prognostics using BiGRU for optical networks. J. Opt. Commun. Network. **12**(8), 277–287 (2020)
12. Investing Homepage. https://cn.investing.com/economic-calendar. Accessed 23 Oct 2022
13. Cao, G., Wu, H., Wang, G., et al.: Selection of a suitable electrolyte for electrochemical grinding of high-speed steel roll material based on electrochemical techniques and uniform design machining experiments. Int. J. Adv. Manuf. Technol. **122**(7–8), 3129–3147 (2022)
14. Liu, Y.M., Zhao, Z.Y., Zhang, S., et al.: Identification of abnormal processes with spatial-temporal data using convolutional neural networks. Processes **8**(1), 73 (2020)
15. Liu, C., Hou, W.Y., Liu, D.Y.: Foreign exchange rates forecasting with convolutional neural network. Neural Process. Lett. **46**(3), 1095–1119 (2017)
16. Zhang, Y., Yao, L., Zhang, L., et al.: Fault diagnosis of natural gas pipeline leakage based on 1D-CNN and self-attention mechanism. In: 2022 IEEE 6th Advanced Information Technology, Electronic and Automation Control Conference (IAEAC), pp. 1282–1286. IEEE (2022)
17. He, X., Chen, Y.: Optimized input for CNN-based hyperspectral image classification using spatial transformer network. IEEE Geosci. Remote Sens. Lett. **16**(12), 1884–1888 (2019)
18. Zhang, X., Ding, G., Li, J., et al.: Deep learning empowered MAC protocol identification with squeeze-and-excitation networks. IEEE Trans. Cogn. Commun. Network. **8**(2), 683–693 (2021)
19. Yechuri, S., Vanambathina, S.: A nested U-net with efficient channel attention and D3Net for speech enhancement. Circuits Syst. Sig. Process. 1–21 (2023)

Optimization of a Network Topology Generation Algorithm Based on Spatial Information Network

Peng Yang[1](✉) ⓘ, Shijie Zhou[1], and Xiangyang Zhou[2]

[1] School of Information and Software Engineering,
University of Electronic Science and Technology of China, Chengdu 610054,
People's Republic of China
acm@uestc.edu.cn

[2] Huawei Technologies Co., Ltd., Chengdu 611730, People's Republic of China

Abstract. Spatial information network (SIN) is a network with high speed and periodicity of node operation. In recent days, China will build a complete asteroid monitoring and warning system and a near-Earth asteroid defense system. This requires launching more low-Earth orbit satellites. In order to adapt to the increase in the number of near-Earth satellites, the dynamic optimization of space information network topology between satellites will have research significance. Considering the visibility of satellite networking, the connectivity of satellite nodes, and the number of links connected to the whole network, with the goal of minimizing the end-to-end delay between satellite nodes in the network as the optimization goal, a network topology optimization model that meets multiple constraints is constructed, and the model is solved using greedy algorithm and simulated annealing algorithm. In the process of simulated annealing, the network flow algorithm is innovatively proposed for neighborhood solution. Experiments show that the simulated annealing hybrid neighborhood algorithm is significantly better than the simulated annealing random neighborhood algorithm.

Keywords: Spatial Information Network · Dynamic Optimization of Network Topology · Network Flow Algorithm · Simulated Annealing Algorithm

1 Introduction

Space Information Network (SIN) plays an important role in the fields where it is not convenient to lay the ground network, such as the sky and the sea. At the same time, it can also expand and improve the ground information network, provide coverage anytime and anywhere, and optimize and enhance the connectivity of the entire sky-ground system. Compared[1] with the traditional ground information network, SIN network has many advantages, such as small ground restrictions, flexible networking, wide coverage area, and high efficiency of available link communication. Through the space information

---

[1] Supported by Sichuan Science and Technology Program (2023YFG0155).

network designed by networking, its wireless communication signal can cover the entire earth surface. Therefore, designing a space satellite network to meet various needs has become a research hotspot of scholars at home and abroad.

There are two main types of algorithms for solving spatial information network topology optimization models. One is based on random algorithms, such as simulated annealing, genetic algorithm or ant colony algorithm, and the other is related algorithms with minimum spanning tree construction as the core.The random algorithm calculates the connectivity of the corresponding topology of the solution set and the optimization rate for the optimization target by randomly enumerating the solution set, and then iterates to make it converge at the optimal solution to obtain the final solution set topology. The algorithm based on the spanning tree is to ensure connectivity by establishing the spanning tree, and then change or optimize the shape of the spanning tree or spanning subgraph by strictly proving the correctness of the strategy, so as to obtain the topology matrix of the final result.

The purpose of these algorithms is to construct excellent spatial information network topology to improve its invulnerability. However, the optimization objectives in the optimization models proposed in relevant studies are different. For example, taking the average information transmission time delay and the maximum transmission time delay between satellite nodes in the inter-satellite link as the optimization objectives, the improved Multi-Objective Simulated Annealing (IMOSA) algorithm is adopted to solve the multi-objective model [4]. With the goal of improving the algebraic connectivity of the topological matrix as the optimization objective, the optimization model is solved by controlling the subtree merging and maximizing algebraic connectivity algorithm (SMMAC) [3]. There is also an optimization algorithm similar to the objective of this paper, which aims to reduce the average edge weight of spanning trees or spanning subgraphs, that is, the cost of establishing inter-satellite links.

Building an inter-satellite topology for a space information network can shorten the point-to-point delay between satellites, enhance the survivability and reduce the cost of communication links. For example, the precise algorithm and approximate algorithm based on the spatial information network are mentioned in the on board online task scheduling planning [12]. The greedy algorithm based on a single time slice is proposed to solve the inter-satellite link allocation problem [14]. The immune algorithm based on the spatial information network invulnerability model is proposed [11], and the simulated annealing algorithm that considers multiple time slices as states under multiple constraints is proposed [13]. A simulated annealing optimization algorithm [15] is proposed for the generation model of multi-objective and multi-state extension structure.

This article is organized as follows. Firstly, the topological structure and optimization of spatial information network are briefly introduced. Secondly, Sect. 2 introduces the concept of time slice and proposes a network topology optimization model. Then, Sect. 3 proposes an improved simulated annealing algorithm. After that, Sect. 4 conducts experiments on the dataset. Finally, Sect. 5 concludes and summarizes the whole paper.

2 Related Work

2.1 Time Slice Concept

In the spatial information network model, all satellite nodes keep running at a high speed along their respective orbits, resulting in constant changes in the distance and relative position between nodes. Therefore, the whole operation cycle of the satellite network can be divided into several time slices. Each time slice corresponds to the topology structure of satellite network in a period of time and is calculated separately. In the traditional time slice segmentation method, the length of a time slice is usually selected between 15 s and 120 s. The network topology in such a time period called a network topology snapshot is considered fixed and unique. There is also an intermediate state between the two adjacent time slices, which is the time when the links between some satellite nodes in the previous time slice is disconnected and new links is established. When the communication links between satellite nodes are changed, a new topology snapshot is formed, that is, the next time slice is entered.

In this way, it is necessary to detect whether the node link changes in the spatial information network in real time. Once the old links is broken and new links is established, it is necessary to split the time slice immediately. Since the topology of spatial information network changes rapidly and dynamically, this partition method will produce a large number of fragmental time slices, which will lead to the reduction of the efficiency of the algorithm, and also affect the computing performance of other links. Therefore, the equal long time interval division method will be used to divide the time period. That is, the time is divided into slices of equal length, and the data is sampled at the initial moment of each slice. The sampled topology is used as the network topology snapshot of the time slice. Due to the characteristics of spatial information network, the topological structure changes of two adjacent time slices are very subtle, so the sampled data can be approximately used to represent the topological snapshot of the current time slice. The topology snapshot obtained by this method is fast and concise. The division of equal-length interval slices is shown in Fig. 1:

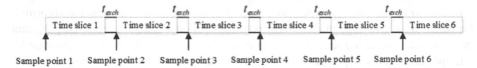

Fig. 1. Time slices of the same interval length

2.2 Optimization Model

Graph $G = (V, E)$ can be described as a space information network composed of N satellites and E communication links, where $V = \{v_i | i \in 1, 2, ..., N\}$ is the node set and $E = \{e_{ij} | i, j \in V, i \neq j\}$ is the edge set. In view of the topological characteristics, physical conditions and actual performance of satellites in the space information network, the survivability optimization model has the following constraints:

(1) Solution of spatial constraints: for the adjacency matrix A corresponding to the spatial information network, it represents the maximum possible link of satellite nodes at a certain time. However, due to the maximum constraint, each satellite cannot establish all communication links. Therefore, in the optimization model, it is necessary to generate the objective matrix X that meets the actual link conditions to replace A. Obviously, $X <= A$, X is the sub-matrix of A.

(2) Connectivity constraint: connectivity graph is the premise to ensure the normal information exchange of the whole network, the constraint condition that the normal networking of the spatial information network must meet, and the target network for model solution. The Laplace matrix of graph G is a symmetric positive semi-definite singular matrix, and its corresponding eigenvalue satisfies $\lambda_n \geq \cdots \geq \lambda_2 \geq \lambda_1 = 0$. If the second smallest eigenvalue of Laplace matrix $\lambda_2 > 0$ if and only if graph G is connected. The connectivity of the graph can be described by the eigenvalue. The larger the eigenvalue, the better the robustness of the graph [8].

(3) Maximum degree constraint: considering the actual performance of the satellite, it is not possible to establish communication links with any satellite node in the visible range at the same time. At present, the number of communication links that can be actually established for satellites in orbit is basically no more than 8, and some constellations in the early stage are basically no more than 4, that is, nodes in the network topology have the maximum degree limit.

Let the matrix V be the visible matrix, and $v_{i,j}$ represents the distance between the two nodes of i, j in the time slice. If two nodes are visible, $v_{i,j} = 1$, otherwise $v_{i,j} = \infty$. Let matrix C be the link selection matrix. If two nodes have established communication links, then $c_{i,j} = 1$, otherwise $c_{i,j} = 0$. On this basis, finding a sub-topology network to minimize the average transmission delay between nodes becomes an optimization problem. Let the function $d_{i,j}(V, C)$ be the shortest path length from node i to node j. The specific mathematical model is as follows:

$$\min \ \tau(A) = \frac{2}{n \times (n+1)} \sum_{i=1}^{n} \sum_{j=1}^{i} d_{i,j}(V, C)$$

$$s.t. \ \sum_{i=1}^{n} c_{j,i} \leq \max \text{degree}, \quad \forall j \in \{1, 2, \cdots, n\}$$

$$\sum_{i=1}^{n} \sum_{j=i+1}^{n} c_{i,j} \leq \max \text{edge}$$

$$c_{i,j} = 0, \quad \forall v_{i,j} = \infty$$

$$c_{i,j} \in \{0, 1\}, \quad \forall v_{i,j} \neq \infty \tag{1}$$

where τ is the average value of communication transmission delay of all satellite nodes in the whole network. Max degree represents the upper limit of the link that each satellite node in the network can establish. Max edge represents the upper limit of the total number of communication links in the entire network.

3 Simulated Annealing Algorithm

3.1 Algorithm Flow

The simulated annealing algorithm does not limit the form of the solution. It only needs to put forward the objective function and constraint conditions, and construct a method to find the neighborhood solution, which can effectively find the global optimal solution at a lower time complexity. Therefore, simulated annealing algorithm provides more efficient solutions to many optimization problems. The flow chart of simulated annealing algorithm is shown in Fig. 2.

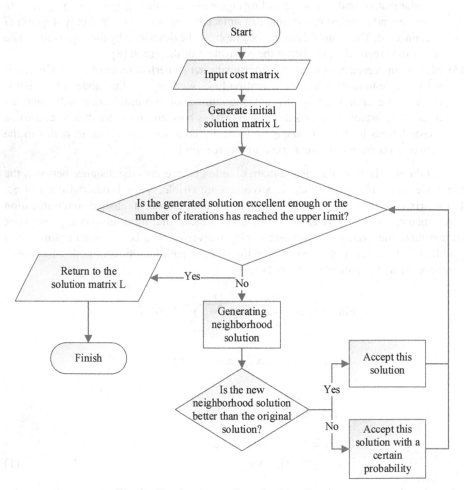

Fig. 2. Simulated annealing algorithm flowchart

3.2 Neighborhood Solution

The selection of neighborhood solution needs to be carried out under the condition that the space-space information network is fully connected. When the network is disconnected, the distance between satellite nodes in two different connected blocks will be set to infinity. At this time, the average time delay between nodes will also become infinite. This ensures that this solution will not be selected as a new solution by simulated annealing algorithm. Therefore, in the process of neighborhood solution, it is not necessary to consider how to maintain the connectivity of the space-space information network, but only to limit the number of satellite node links.

One way is to use the spanning tree random traversal method to solve the neighborhood solution [17]. The other way is to use the dual-link random switching method, which is more concise. [16].

For a connected graph G, its edge connectivity is the size of its minimum cut. Generally, the greater the connectivity of a graph, the more stable the network is and the stronger its survivability is, because the size of the minimum cut can represent the number of unrelated paths from one node to another node. The more the number of incoherent paths, the stronger their ability to maintain communication and be repaired after being attacked.

In this paper, the objective function of optimization is the average minimum delay between nodes. With the enhancement of network survivability, it is beneficial to shorten the average minimum delay between nodes to a certain extent. Because the stronger the survivability of a network, the more paths between two nodes, and its topological geometry will tend to a hypersphere, and a hypersphere can minimize the delay between points. Therefore, this paper proposes a neighborhood solution method based on network flow, which can ensure that the edge connectivity of the solution is not reduced each time the neighborhood solution is found. The specific operation process is shown in Fig. 3.

4   Experimental Simulation and Result Analysis

This paper selects Iridium constellation with 66 satellites and Globalstar constellation with 48 satellites to generate visual matrix data in each time slice through STK. In the simulation experiment, a time slice is divided every 60 s for sampling, and then the data of each time slice is preprocessed. The larger the amount of data, the less chance the experimental results will be, and the more accurate the results will be. However, if too many time slices are selected, the operation will be slow. Therefore, 100 time slices from 00:00 on January 1, 2021 are selected as the experimental data. The spatial information network models mentioned in this paper meet the following requirements:

$$\text{max edge} = 1.8 \times n, \quad \text{max degree} = 4 \qquad (2)$$

It can be seen from Table 1 that the spanning tree random traversal algorithm cannot obtain satisfactory results no matter which constellation it is in. This is because the randomness of this algorithm is not high enough and there is no definite optimization direction.

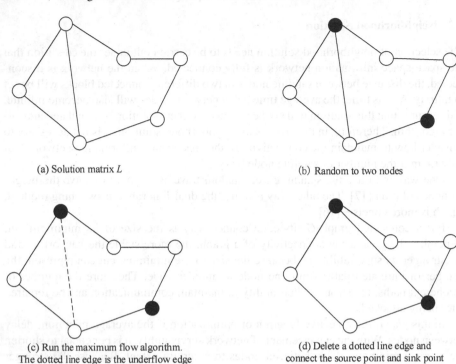

(a) Solution matrix L

(b) Random to two nodes

(c) Run the maximum flow algorithm.
The dotted line edge is the underflow edge

(d) Delete a dotted line edge and
connect the source point and sink point

Fig. 3. Network flow neighborhood solution algorithm

Table 1. Comparison of operation efficiency (sum of 100 time slices)

Constellation Neighborhood solving algorithm	Iridium constellation		Globalstar constellation	
	Iteration steps	Run time	Iteration steps	Run time
Spanning tree random traversal	171664	31.166s	258971	21.970s
Dual link random switching	32183	6.679s	52877	4.286s
Network maximum flow	9092	1.972s	156944	11.300s

As shown in Fig. 4 and Fig. 5, the running time changes of the three neighborhood algorithms in 100 time slices show that the dual-link random switching method and the maximum flow selection method have their own advantages and disadvantages in the two constellations. The characteristics of strong symmetry of global constellation and few effective states make the dual-link random switching method with stronger randomicity have advantages. The characteristics of Iridium constellation, such as rapid change, multiple satellites and multiple states, are more conducive to the maximum flow selection method to play its ability to maintain connectivity. With the increasing number of LEO satellites in the future, the network topology after networking will become

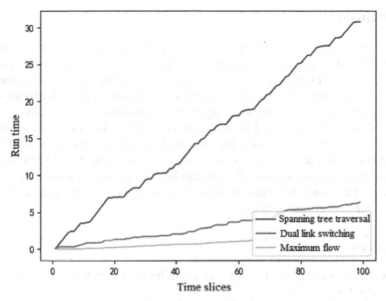

Fig. 4. Running time of neighborhood solution algorithm (Iridium)

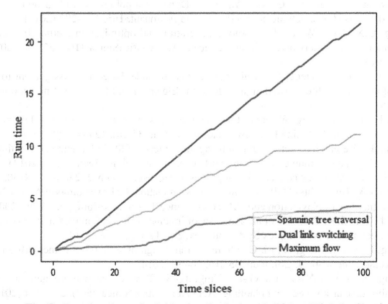

Fig. 5. Running time of neighborhood solution algorithm (Globalstar)

increasingly complex, and the maximum flow selection method will have more and more advantages over the dual-link random switching method.

5   Conclusion

This paper analyzes and models the complex spatial information network, focusing on how to build and optimize the topological structure of the spatial information network to minimize the average value of the satellite node communication transmission delay. The network maximum flow selection method is proposed as the neighborhood solution algorithm of simulated annealing algorithm. Experiments show that the maximum flow selection method proposed in this paper as a neighborhood solution algorithm is better than the spanning tree random traversal method, and has advantages and disadvantages with the dual-link random selection method. When the number of LEO satellites is increasing and the network topology is constantly changing, the solution of the optimization model of the space and space information network in this paper can be effectively applied to the networking strategy.

References

1. Qi, X., Ma, J., Wu, D., et al.: A survey of routing techniques for satellite networks. J. Commun. Inf. Netw. 1(4), 66–85 (2016)
2. Sui, T., Mo, Y., Marelli, D., et al.: The vulnerability of cyber-physical system under stealthy attacks. IEEE Trans. Autom. Control 66(2), 637–650 (2020)
3. Xianfeng, L., Xiaoqian, C., Lei, Y., et al.: Dynamic topology control in optical satellite networks based on algebraic connectivity. Acta Astronaut. 165, 287–297 (2019)
4. Xing, G.X., Qi, Y.W., et al.: Topology generation and optimization method in multi-state space information network. Acta aeronautica et AstronauticaSinica 41(4): 323546 (2020). (in Chinese)
5. Zhe, L., Guo, W., Deng, C., et al.: Perfect match model-based link assignment to design topology for satellite constellation system. In: Electrical and Computer Engineering. IEEE (2015)
6. Sun, H., Hao, X., Feng, W., et al.: Inter-satellite links topology scenario based on minimum PDOP criterion. J. Beijing Univ. Aeronaut. Astronaut. 37(10), 1245–1249 (2011)
7. Wang, D.: Research on networking of navigation inter-satellite link oriented the optimization of ranging and communication. National University of Defense Technology (2014)
8. Fiedler, M.: Algebraic connectivity of graphs. Czechoslov. Math. J. 23(98), 298–305 (1973)
9. Rajan, J.A.: Highlights of GPS ii-r autonomous navigation. In: Proceedings of Annual Meeting of the Institute of Navigation and CIGTF Guidance Test Symposium, pp. 24–26 (2002)
10. Wu, J., Tan, S.Y., Tan, Y.J., et al.: Analysis of invulnerability in complex networks based on natural connectivity. Complex Syst. Complex. Sci. 11(1), 77–86 (2014)
11. Dong, F., Lv, J., Gong, X., Li, C.: Optimization design of structure invulnerability in space information network. J. Commun. 35(10), 50–58 (2014)
12. Xiang, S., Chen, Y.-G., Li, G.-L., Xing, L.-N.: Review on satellite autonomous and collaborative task scheduling planning. Acta Automatica Sinica 45(2), 252–264 (2019)
13. Yan, H., Zhang, Q., Sun, Y.: Link assignment problem of navigation satellite networks with limited number of inter-satellite links. Acta Aeronautica et AstronauticaSinica 36(7), 2329–2339 (2015)
14. Shi, L.: A link assignment algorithm applicable to crosslink ranging and data exchange for satellite navigation system. J. Astronaut. 32(9), 1971–1977 (2011)
15. Dong, M., Lin, B., Liu, Y., Zhou, L.: Topology dynamic optimization for inter-satellite laser links of navigation satellite based on multi-objective simulated annealing method. Chin. J. Lasers 45(07), 217–228 (2018)

16. Pan, C., Xing, G., Qi, Y., Yang, L.: Topological generation and optimization method in multi-state space information network. Acta Aeronautica et Astronautica Sinica **41**(4), 323546 (2020)
17. Yang, P., Zhuo, M., Tian, Z., Liu, L., Hu, Q.: Optimization of space information network topology based on spanning tree algorithm. Commun. Comput. Inf. Sci. **1587**, 668–679 (2022)

Data Visualization

MBTIviz: A Visualization System for Research on Psycho-Demographics and Personality

Yutong Yang[1], Xiaoju Dong[2(✉)], Xuefei Tian[2], Yanling Zhang[2], and Meng Zhou[2]

[1] SJTU Paris Elite Institute of Technology, Shanghai Jiao Tong University, Shanghai 200240, China

[2] Department of Computer Science, Shanghai Jiao Tong University, Shanghai 200240, China
xjdong@sjtu.edu.cn

Abstract. The increasing interest in exploring the correlation between personality traits and real-life individual characteristics has been driven by the growing popularity of the Myers–Briggs Type Indicator (MBTI) on social media platforms. To investigate this correlation, we conduct an analysis on a Myers–Briggs Type Indicator (MBTI)-demographic dataset and present MBTIviz, a visualization system that enables researchers to conduct a comprehensive and accessible analysis of the correlation between personality and demographic variables such as occupation and nationality. While humanities and computer disciplines provide valuable insights into the behavior of small groups and data analysis, analysing demographic data with personality information poses challenges due to the complexity of big data. Additionally, the correlation analysis table commonly used in the humanities does not offer an intuitive representation when examining the relationship between variables. To address these issues, our system provides an integrated view of statistical data that presents all demographic information in a single visual format and a more informative and visually appealing approach to presenting correlation data, facilitating further exploration of the linkages between personality traits and real-life individual characteristics. It also includes machine learning predictive views that help nonexpert users understand their personality traits and provide career predictions based on demographic data. In this paper, we utilize the MBTIviz system to analyse the MBTI-demographic dataset, calculating age, gender, and occupation percentages for each MBTI and studying the correlation between MBTI, occupation, and nationality.

Keywords: Data Visualization · Psychodemographic Data · Myers–Briggs Type Indicators · Correlation Analysis

1 Introduction

The growing popularity of the Myers–Briggs Type Indicator (MBTI) on social media platforms has prompted increased interest in exploring the correlation between personality traits and real-life individual characteristics. To investigate this association, we conducted an analysis on the MBTI-demographic dataset obtained from the openpsychometrics project [1]. The dataset contains demographic information, such as age, gender, nationality, and occupation, along with individual MBTI types.

Z. Yu et al. (Eds.): ICPCSEE 2023, CCIS 1879, pp. 259–276, 2023.
https://doi.org/10.1007/978-981-99-5968-6_19

A comprehensive visualization and analysis of large-scale MBTI surveys has not yet been conducted. Nonetheless, integrating demographic data with personality information often results in big data, characterized by its high volume and great diversity. Analysing this type of data presents two primary issues: 1. Statistical data tables contain a large amount of redundant information, making them difficult to display due to their size. Therefore, researchers often use visual charts to present and analyse the data. However, visual representations of data, such as statistical charts including line graphs and bar charts, often have a relatively simple format and can only represent a limited number of dimensions. 2. When analysing the correlation between variables, the correlation analysis table may not provide an intuitive representation [2]. Therefore, a more informative and visually appealing approach is required to facilitate the comprehension of the relationship between occupation/personality and geography/personality. [3].

To address this issue, we have employed digital humanities [4, 5] methods to integrate psychological data with a visualization system that enables the visualization and analysis of data in a more effective manner. Specifically, we have developed a psycho-demographic visualization system that has two main contributions:

1. A comprehensive demographic data view is implemented through an integrated view of statistical data that presents all demographic information in a single visual format.
2. To present and analyse correlation data, suitable visualization techniques, including a correlation parallel coordinate system view and a correlation heatmap, are employed in lieu of conventional data tables.

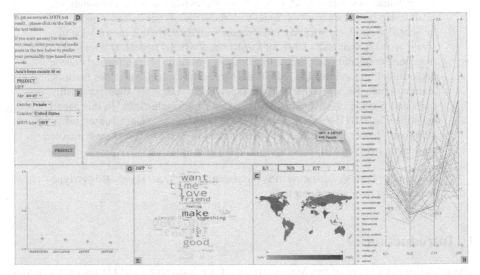

Fig. 1. MBTIviz: A psycho-demographics data analysis system.

By using the MBTIviz system (Fig. 1), we analyse the MBTI-demographic dataset, present demographic information for all samples in an integrated way and conduct correlation analyses for personality with career choice and nationality.

The remainder of this paper is organized as follows: Sect. 2 briefly reviews the related works on the topic of personality and demographic analysis; Sect. 3 introduces the data being utilized and the humanities data processing methods; Sect. 4 presents the system design and how it works; Sect. 5 conducts a correlation analysis of personality and demographics by using the system; Sect. 6 gives the feedback of three users; Sect. 7 is the conclusion and future work plan.

2 Related Work

2.1 Myers–Briggs Type Indicator (MBTI)

"Personality" is defined by the American Psychological Association as "individual differences in patterns of thinking, feeling and behaving." [6] As such, measuring personality is an important subject and research tool in this field, and the Myers–Briggs Type Indicator (MBTI) is one of the most widely used personality measures in the world due to its simplicity and ease of understanding. The MBTI divides personality into four dimensions: attitudes (I, introversion/E, extraversion), perceiving functions (S, sensing/N, intuition), judging functions (T, thinking/F, feeling), and lifestyle preferences (J, judging/P, perception). [7–9].

To be more precise, attitudes refer to the general orientation of an individual's energy, whether it is directed inwardly (Introversion) or outwardly (Extraversion). Perceiving functions, however, determine how people gather information about the world around them. Sensing types tend to focus on the present and rely on their real senses, while intuitive types are more interested in the future and rely on their intuition and imagination. Thinking types rely on logic and analysis, while feeling types rely on empathy and subjective values. Finally, lifestyle preferences refer to how individuals interact with the outside world. Judgment types tend to be organized and decisive, while perception types tend to be more open-ended and adaptable. [8].

2.2 Psychodemographic Research Background

The correlation between personality test [10] and demographic characteristics (age, gender, nationality, occupation, etc.) [11] has a broad research base. In the context of personality analysis and prediction of social network users, the Text Analysis and Knowledge Engineering Lab at the University of Zagreb proposed a dataset with both personality and demographic labels, PANDORA. The authors used PANDORA to exploit the correlations between the traits of the different models and their manifestations in text and perform a confirmatory and exploratory analysis between propensity for philosophy and certain psycho-demographic variables [12]. Psychologist Adrian Fountain et al. examined the associations between sociodemographic variables, the Big Five personality traits, and the extent of political interest as well as voting behavior [13]. "Personality Type Preferences of Asian Managers: A Cross-Country Analysis Using the MBTI Instrument" analysed the personality characteristics of managers from four countries, India, the Philippines, Malaysia and Indonesia, and concluded that there is a relationship between the influence of geographical and cultural background on personality [14].

In this paper, a visual analysis is selected to present and analyse the correlations between personality and occupation, personality and geography, personality and gender, and personality and age using different views to provide a comprehensive study of the relationship between personality trait variables and demographic data.

2.3 Data Visualization in Demography

The visualization of demographic data has a long history, with various methods and techniques developed over time to present information in a clear and accessible manner. As an example, Abel et al. (2014) proposed a circular migration plot that employs a circular display to exhibit the approximations of directional flows among 123 countries that experienced a migration volume of at least 100,000 individuals in two or more of the four temporal intervals [15]. In the work of Riffe et al. (2021), seven notable papers on demographic data visualization were collected as a "special collection", which address both the expansion of conventional demographic data representations and the illustrative applications of less common visualizations [16–22]. The authors also proposed ten practical guidelines for creating effective data visualizations in scientific manuscripts and presentations, including using a standard chart type, aiming for clarity and simplicity, and visual encoding that is deliberate and transparent. Figures should be appealing and relevant to the manuscript's purpose, while avoiding unnecessary charts [23].

3 Data Preprocessing

3.1 Data Presentation

In this paper, we use the dataset from the Open-Source Psychometrics Project [1]. The original dataset consisted of 70575 samples, but due to the focus on occupation choice, the samples from students were removed. We also merged similar occupations and deleted incomplete sample data. As a result, 16,868 complete samples remained for analysis.

The dataset contains five fields: nationality, age, gender, occupation, and MBTI personality type. Among them, age is divided into three stages (13–19 years old; 20–27 years old; more than 27 years old). Gender is treated as a binary variable. The four dimensions of the MBTI are regarded as four binary variables (extraversion/introversion; intuition/sensing; feeling/thinking; judgment/perception).

This dataset has three limitations:

1. There is a biased sample due to the overrepresentation of people who preferred intuition in the population interested in the MBTI on social media. As a result, the personality sample size cannot reflect the actual proportions of different personality types in real life. Therefore, the MBTI proportion is not reflected in the visualizations, and all types are coded into rectangles of equal length.
2. The overrepresentation of young people among the population on social media makes the age distribution of each personality type not representative. Therefore, this study only focuses on the relative age distribution of personality types for horizontal comparison.

3. The questionnaire was only distributed to internet users in certain countries, resulting in missing data for some countries. These missing data are represented as white in the heatmap visualization.

3.2 Humanistic Data Analysis Methods

In this study, a logistic regression analysis model [24] was employed to process the correlation data. The four binary dimensions of the MBTI were used as independent variables, and the dependent variables were occupation and nationality.

Logistic regression analysis models use the odds ratio (OR) to indicate the degree of correlation and measure the influence of an independent variable on the dependent variable. The formula for calculating the odds ratio is as follows, where π represents the probability of success:

$$Odds = \frac{\pi}{1 - \pi} \tag{1}$$

In the logistic regression analysis model, let X_p indicate the independent variable and p represent the number of independent variables. For $X_1 = X_1^* + 1$, with the other independent variables as $X_2, \ldots, X_p$, the logarithm of the odds score is calculated as follows, where β represents the correlation coefficient of the independent variables:

$$\text{Ln}\left(Odds_{X_1^* + 1}\right) = \ln\left(\frac{\pi}{1 - \pi}\right)\Big|_{X_1^* + 1} = \beta_0 + \beta_1(X_1^* + 1) + \beta_2 X_2 + \cdots + \beta_p X_p \tag{2}$$

For $X_1 = X_1^*$:

$$\ln\left(Odds_{X_1^*}\right) = \ln(\frac{\pi}{1 - \pi})\Big|_{X_1^*} = \beta_0 + \beta_1(X_1^*) + \beta_2 X_2 + \cdots + \beta_p X_p \tag{3}$$

Therefore, the odds ratio of $X_1 = X_1^* + 1$ to $X_1 = X_1^*$ is:

$$\ln\left(OR_{X_1}\right) = \ln(\frac{Odds_{X_1^* + 1}}{Odds_{X_1^*}}) = Odds_{X_1^* + 1} - Odds_{X_1^*} = \beta_1 \tag{4}$$

$$OR_{X_1} = \exp(\beta_1) \tag{5}$$

The correlation between an occupation and the two variables in a dimension can be determined by calculating the odds ratio. For example, the correlation between the E/I dimension and a given occupation is determined by the probability of being extraverted (π) to the probability of being introverted ($1 - \pi$). When the OR value is less than one (i.e., β is negative), the dependent variable is negatively correlated with the independent variable. When the OR value is greater than one (i.e., β is positive), the dependent variable is positively correlated with the independent variable.

In addition, the significance analysis (p value) indicates whether the correlation between the dependent variable and the independent variable is significant or not. "**" indicates $0.01 < \text{p value} < 0.05$, "***" indicates a p value < 0.01.

4 System Overview

Figure 1 shows the front-end interface of the system. The system is divided into three parts. Comprehensive demographic data view; humanity analysis part, which contains the correlation parallel coordinate system view and the correlation heatmap, for visualizing and representing the correlation analysis data; interactive part, including MBTI prediction, occupational prediction, MBTI word cloud map and scatter plot of occupational prediction results.

4.1 Demographic Data Visualization Method

Adhering to Edward Tufte's principles of information graphics design, we have adopted high standards of "data density" and a high "data-ink ratio" while designing our graphics, with a focus on efficiently representing data while avoiding unnecessary elements [23]. As shown in Fig. 1A, from bottom to top, the demographic information provided in the view is the total number of populations for each occupation, the total number of populations for each MBTI personality type, the gender ratio and the age ratio for each MBTI personality type.

Among them, the baseline graph for the population ratio of MBTI personality type to occupation is a Sankey graph; for the gender ratio is a bar graph; and for the age ratio is a line graph.

1. MBTI-Occupation statistics are presented with a Sankey diagram. In Fig. 1A, the 16 rectangles of equal length at the top of the Sankey diagram represent the 16 personality types of the MBTI; the rectangles of unequal length at the bottom represent different occupations, and the lengths of the rectangles reflect the number of people in that occupation. The lines between the rectangles reflect the number of people in a certain occupation for a certain personality type.

    When the cursor is moved to a rectangle, the personality type and the number of people are displayed, and the related lines are highlighted; when the cursor is moved to a line, the number of people working in a certain occupation in that personality type is displayed.

2. MBTI-Gender statistics are presented with bar charts. To show the sex ratio, this view renders the rectangles (bars) in the Sankey diagram using different colors so that they represent the percentage of different genders in one personality type. The lighter color in the upper part of the bar represents the percentage of males and the darker color in the lower part represents the percentage of females. When the cursor is moved to the bar graph, the view shows the corresponding number of gender statistics. At the same time, the bar chart can intuitively show the percentage of personalities in each MBTI personality type, such as the relatively higher percentage of females in the "ISFJ" type.

3. MBTI-Age statistics are presented with line charts. To show the age ratio, this view uses line graphs corresponding to the personality type rectangles separately to facilitate comparison of the age ratios in each type. The ages are divided into three groups. From left to right, they are ages from 13 to 19 (group 1), 20 to 27 (group 2) and over 27 (group 3). For example, in Fig. 1A, we can visually observe that within the

"ESTP" group, the percentage of individuals aged between 13–19 years is approximately 60%, which is higher than the corresponding percentage in the "ISFJ" group, which is approximately 40%.

4.2 Correlation Visualization Methods

Research shows that using values alone to indicate statistical significance is not advisable and should be avoided. When displaying data using a single significance level, the use of absolute values requires the most time and effort. The use of visual techniques such as bolding, intensity, and color tones leads to faster and more accurate identification of significant values. [3].

Our approach to visualizing the correlation between personality and demographic variables uses a heatmap and parallel coordinate system plots. These plots make it easier to identify patterns and relationships between variables, as opposed to traditional correlation coefficient tables that can be difficult to interpret. The heatmap adds color-coded visualizations to highlight the strength of the correlations. This approach makes complex data more accessible and user-friendly, providing valuable insights into the relationship between personality and demographic variables.

1. In Fig. 1B, we see the correlation parallel coordinate system view optimized to present correlation data based on the parallel coordinate system. The view presents data on the correlation between occupation and the four dimensions of MBTI. The dark line indicates correlations between Analyst and the four dimensions, with a ratio of Analyst's extravert value to introvert value (E/I) of 0.881 relative to the baseline data. The analyst is more introverted and therefore more correlated with the I value. The other three dimensions are similar.

   The representation of significance in this view is the same as the traditional table representation. "**" indicates significant correlation and "***" indicates highly significant correlation. Data processing involves filtering and replacement work performed by front-end data analysis. For example, if the correlation p value between analyst and F/T is <0.05, "***" is displayed next to the F/T data in the black tip box of analyst.

2. The correlation heatmap is depicted in Fig. 1C, where nationality is a crucial demographic variable. For geolocation data, a more intuitive visualization method in the form of a heatmap is utilized. This view illustrates the correlations of each region with the four dimensions of the MBTI. The four options located on the top of the view enable the selection of which dimension's correlation data are displayed. In this case, the second button, N/S value (the ratio of N value (intuition) to S value (sense)), is selected, and the view presents the correlations of each region with the N/S value rendered in light to dark colors based on the values from small to large.

4.3 Machine Learning Prediction View

We introduce two functional views: the MBTI prediction view and the career prediction view.

1. The MBTI prediction view contains a link to the 16 personalities site and a text input window, as shown in Fig. 1D. The link provides users with an accurate test, while

the input window provides them with a relatively less accurate but more convenient NLP-based test [25]. The model utilizes the spaCy library to create a blank language model and adds a text classifier to the pipeline with four labels: INTROVERTED, INTUITIVE, THINKING, and JUDGEMENTAL. It is trained on the MBTI dataset, which includes a large number of people's MBTI types and content written by them [26]. The probabilities of four variables are computed by the model. A string representing an MBTI type is constructed by checking the probabilities of each dimension and appending the corresponding character to the string.

When users enter content that they have posted on social media and click the PREDICT button, the system will display the predicted MBTI personality type and the word cloud corresponding to the personality type. In Fig. 1E, we choose a post by singer Billie Eilish on her Ins, "Asia's been cuuute, see you tonight;)" as an example, and the model predicts the result as "ISFP".

2. The career prediction window, as shown in Fig. 1F, provides users with four information choices, including age, gender, nationality, and MBTI. The model used is a decision tree classifier that is trained on the MBTI-demographic dataset to make predictions about one's occupation. The dataset is split into a feature matrix X and a target vector y. The feature matrix X consists of the "country", "age group", "gender", and "MBTI" columns, while the target vector y represents the "occupation" column. The decision tree classifier is fitted to the training set, allowing it to learn patterns and relationships between the features and target variable. Subsequently, the model is employed to make predictions on the test set.

After selecting their information and clicking the PREDICT button, the system returns the top four occupations with the highest predicted probabilities and displays the results in the scatter plot shown in Fig. 1G. For example, when a user selects the profile of a 20- to 27-year-old female from the United States with the MBTI personality type ISFP, the prediction result shows that the potential suitable occupations for her are marketing, educator, artist and barista.

5 Case Study: Correlation Analysis of MBTI and Demographics

5.1 Visual Analytics of Statistical Data

Observing demographic data visualizations (Fig. 1A), we obtained the following statistical information:

1. Gender distribution. The highest proportion of female participants with MBTI types ISFJ and ESFJ was observed, while the highest proportion of male participants with MBTI types INTP and INTJ was observed. Additionally, according to the statistical data, females are more likely to lean towards feeling (F) than males in the feeling/thinking dimension. During the preanalysis of the validity of the test dataset, we also conducted a preliminary analysis of gender and the four MBTI dimensions. The results showed that females are twice as likely to lean towards feeling compared to males, and the correlation is significant. This analysis result is consistent with the statistical results and existing research results [27], validating the effectiveness of the dataset.

2. Age distribution. We can observe that the sample has a higher proportion in age group 1. Age group 1 with MBTI types ESFP, ISTP, ISFP, ESTP, and ENTP has the highest proportion. Age group 2 with MBTI types ISFJ, INFJ, and ENFJ has the highest proportion, while age group 3 with MBTI types INTJ and ISTJ has the highest proportion. The age line chart reflects the more mature personality types among MBTI types. According to the analysis, in the judging (J)/perceiving (P) dimension, people with higher age tend to have a higher likelihood of judging than those with lower age.

3. Occupation distribution. Both INFP and ENFP individuals are most represented in arts and marketing. INFJ individuals are found in the field of education and arts, and the highest proportion of individuals in the ENFJ type are most commonly found in education. The INTP type is most represented in engineering, and for ENTP individuals, it is education and engineering, followed by acting. ISFP individuals are primarily in arts while ESFP individuals are mostly in education. The highest proportion of INTJ individuals are in engineering and education, and ENTJ individuals are primarily found in law and engineering. ISFJ individuals are mostly in education, and ESFJ individuals are mostly in education and nursing. ISTP individuals are most commonly found in engineering, while ESTP individuals are most represented in education and marketing. ISTJ individuals are primarily found in engineering and education, ESTJ individuals are primarily found in engineering and law.

5.2 Correlation Analysis: MBTI and Career

We selected a portion of occupations (those whose names begin with A) to display in the correlation table (Table 1). As mentioned before, the numbers can be difficult to observe, so we used the correlation views proposed in this paper for further analysis of the correlations.

Table 1. A sample correlation table: correlation between occupations and MBTI dimensions

Occupation	E/I	N/S	F/T	J/P
ACCOUNTANT	0.755	0.610***	0.833	2.791***
ACTOR_ACTRESS	2.994***	1.236	2.706***	1.234**
ADMINISTRATOR	1.207	0.851	1.039	1.656***
ANALYST	0.881	0.823	0.613***	2.564***
ARCHITECT	0.761	1.727	0.473***	1.344
ARTIST	0.803***	1.804***	1.413***	0.848***
ASSISTANT	0.838	0.723	1.714***	1.549***

Associations of MBTI preferences with career choice are shown in Fig. 1B. To enable centralized observation of the relationships between occupations and the eight variables within four dimensions, we present the top six occupations with the highest values for each variable in a tree diagram (Fig. 2). These occupations are arranged in order of their values.

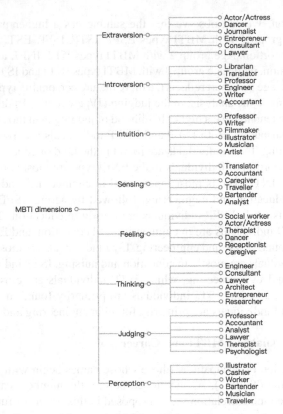

Fig. 2. Tree diagram of the top six highly correlated occupations for each variable

The E/I dimension, which stands for extraversion/introversion, is shown by the first axis from the left (Fig. 3). The odds of selecting careers that involve social interaction and communication, such as performing and consulting are greater for people who preferred extraversion. For example, actor (, actress) has an E/I value of 2.994, where extroverts are 2.994 times more likely to work as actors than introverts. This is because extroverts tend to be outgoing and confident, which makes them naturally charismatic and able to communicate effectively with others. These traits are particularly useful in fields that require a high degree of interpersonal interaction and relationship building. Extroverts also tend to excel in leadership positions due to their natural ability to inspire and motivate others. Their outgoing and confident nature allows them to take charge and guide their team towards success. [28] For those with the opposite preference, i.e., introversion, jobs that require less social interaction and can be done personally are more attractive. For example, librarian is the most typical career choice for introverts compared to extroverts, with an E/I value of 0.41, where the likelihood of an extrovert pursuing a career as a librarian is 0.41 times that of an introvert.

The N/S dimension, which stands for intuition/sensing, is represented by the second axis from the left in the MBTI assessment (Fig. 4). Individuals who score high on the Intuition coefficient are more suited for careers in the arts. For instance, the likelihood of

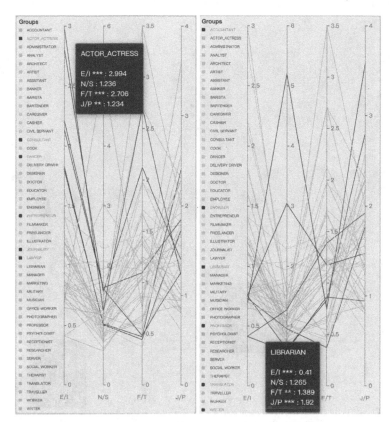

Fig. 3. High Extraversion and Introversion occupations

a person with an intuition preference becoming an artist is 1.804 times that of a person with a sensing preference. This is because intuitive individuals tend to be creative, imaginative, and innovative, making them well suited for careers that require originality and innovation. In contrast, individuals who score high on the sensing coefficient tend to be more practical, detail-oriented, and focused on the present. Therefore, they may be better suited for careers that require attention to detail and practical problem-solving skills, such as analysts or accountants (with an N/S score of 0.61). Moreover, professors are significantly represented in the N/S dimension due to the abstract nature of intuition and the perceived preference for possibility, which make research an attractive career path [29]. They can also see the potential of new ideas and concepts, making them well suited for research and development roles.

The feeling/thinking dimension is represented by the third axis from the left (Fig. 5). Individuals who have a feeling preference tend to be more in tune with their emotions and are often gifted with empathy. This makes them well suited for careers that involve human interaction, such as caregiving, social work (with an F/T score of 3.49), or other people-focused roles. They tend to prioritize the emotional needs of others and are skilled at providing support and developing relationships. On the other hand, individuals who have

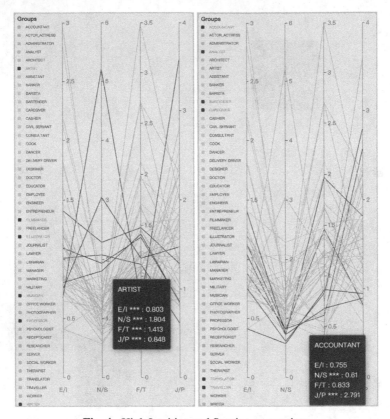

Fig. 4. High Intuition and Sensing occupations

a preference for thinking tend to be more objective and rational in their decision-making. They are typically analytical and enjoy problem solving, making them well-suited for careers that require critical thinking and attention to detail. Jobs that involve complex problem solving, such as engineering (with an F/T score of 0.372), scientific research, or technology, are often attractive to individuals with a thinking preference.

The judging/perception dimension is shown by the fourth axis from the left (Fig. 6). Individuals with a judging preference tend to be structured and programmatic in their decision-making, being more accustomed to planning and organization [30]. Their natural strengths make them well suited for managerial or administrative positions, including those in business, finance, or analysis. For example, those who prefer judging are 2.564 times more likely to become analysts than those with a perception preference. Conversely, individuals with a preference for perception tend to be adaptable and flexible in their decision-making. They enjoy exploring new possibilities and are more likely to be drawn to careers that offer freedom and opportunities for creativity, such as arts. Additionally, they may be drawn to careers that allow them to work independently or in a less structured environment, such as freelance work or travel (traveller, with a J/P score 0.892).

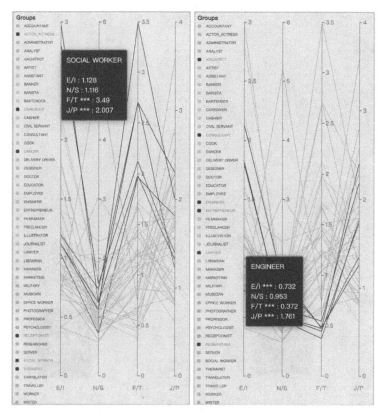

Fig. 5. High Feeling and Thinking occupations

5.3 Correlation Analysis: MBTI and Nationality

The reference value for the correlation analysis is set to the country with the largest number of people in the dataset (United States), i.e., with the United States' value as 1, and the dimension preference value for each country is their ratio compared to the United States. We present eight heatmaps showing the distribution of regions with high scores for the eight MBTI preferences in Fig. 7. From left to right, top to bottom, they are regional distributions with the preference of Extroversion, Introversion, Intuition, Sensing, Feeling, Thinking, Judging, Perception.

Compared to the general American population, Turkey has the largest extraversion preference, with a ratio of 1.652 between extraversion and introversion. Other countries, such as Ecuador and Panama, also exhibit higher extraversion preferences. Analysed by regional distribution, South America, Central America, Eastern Europe, the Middle East, and the Balkans show a higher preference for extraversion.

Latvians are significantly overrepresented for Introversion. Introversion is indeed a stereotype of Latvians, even Latvia itself seems to embrace and appreciate the introverted characteristics of its population [31–33]. Other countries, such as Paraguay, Slovakia,

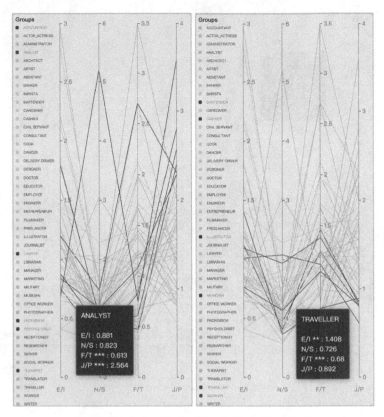

Fig. 6. High Judging and Perception occupations

the Philippines and Finland, also have relatively high introversion preferences. Observing the screening area (green area) in Fig. 7, it is clear that Central Europe, Northern Europe, North America, and Eastern Asia have higher introversion preference coefficients. Among them, Asia has a smaller data sample size, but for the countries with data support (areas with colors), a larger proportion of Asian countries (Philippines, Japan, Singapore) have higher introversion preference coefficients.

In the intuitive dimension, the analysis reveals that the intuition coefficients of the ancient civilizations are generally higher, including Greece, Egypt and India. This observation aligns with the prevailing understanding of these civilizations, which are often associated with abstract thinking, creativity, and imagination. Lacking the scientific knowledge and tools that we have today, in the absence of empirical evidence, subjective judgments are often the only means of making decisions and solving problems. The ancient Greeks, for example, believed that creativity was a gift from the gods and that intuition was an important tool for discovering hidden truths. This concept still has an impact in modern times, as evidenced by current research in the field [34].

Figure 7 shows that South Central America, the Middle East, and Northern Europe have higher coefficients of sensing preference. Interestingly, the analysis also reveals that in one region, more affluent countries are more likely to have high coefficients of sensing

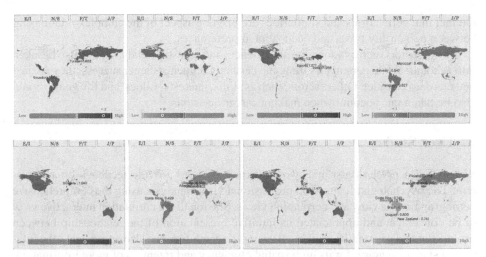

Fig. 7. Distribution of countries with high preferences for Extroversion, Introversion, Intuition, Sensing, Feeling, Thinking, Judging, Perception (from left to right, top to bottom)

preference. This could be attributed to the fact that individuals in these countries may be more concerned with stability, security, and practicality, which are all traits associated with sensing preference.

There is no significant regional distribution of countries with higher values of F/T preference and J/P preference. By simply observing the regional distribution, it is learned that Southwest Europe, East Asia, North America, and Oceania are the regions with higher feeling preference coefficients, while Eastern Europe, Middle East, Balkans and Southern America have higher thinking preference coefficients. Among them, Slovakia has the highest feeling preference, which is 1.240 times more than thinking, and Ukraine (Eastern Europe) has the highest thinking preference with a value of 0.388 for the F/T dimension. The analysis indicates that the Middle East, Balkans, North America, and Asia have higher judging preference coefficients, while Central and Eastern Europe and South America have higher preference for perception.

5.4 Discussion

The analysis of the correlation between MBTI and occupation, as well as MBTI and nationality, provides valuable insights into the differences and commonalities of MBTI variables in relation to career choice and geographical distribution. This information can be effectively utilized to make informed decisions about career paths and aid in the selection of appropriate professions.

Further observation of the comprehensive demographic data view reveals that all MBTI personality types have a diversity of career choices and that all occupations correspond to multiple MBTI personality types. This suggests that personality types can be used as a valid tool for career selection but not as a stereotypical categorization criterion. It is important to acknowledge the existence of diversity in career choices and the fact that correlations and similarities are just one aspect of the complex relationship between personality types and occupational preferences.

Other factors such as education, experience, and industry trends, should also be taken into account. While personality traits can certainly influence career success, they are just one factor to consider. Other factors, such as skills, interests, values, and life goals should also be taken into account when making career decisions.

6   User Studies

We invited a psychologist and two nonexpert users to provide feedback on our system. The psychologist focused on professional visual analysis using the comprehensive demographic data view and correlation views. She found the correlation interactive view to be convenient and appreciated its intuitive presentation of the relationship between profession and personality, making it easier to search for. She also appreciated the integrated statistical graph for its high spatial efficiency and retention of more information. According to her, this design reasonably solved the problem of information redundancy in statistical data and the lack of intuitive correlation charts, making analysis more convenient with a favorable information encoding method. She also raised some suggestions. First, she expressed concern that while the system interface has sufficient information, it may be complex for nonexpert users. She also pointed out some issues with the dataset, such as the lack of data (missing samples of some countries in the dataset), which meant that the analysis in the MBTI and Nationality part could not make sufficient judgments. To improve this, she suggested that we include more complete sample data in the system for further analysis in our future work. Additionally, she mentioned that we could consider adding a dynamic input channel to connect to a database for continuous iteration and precollection of information to form an iterative dynamic model optimization system.

The nonexpert users focused on the MBTI prediction and career prediction windows. They found the personality prediction window useful for users who are not familiar with MBTI, as it allowed them to quickly grasp relevant information. The career prediction window provided better career selection references for users who are already familiar with MBTI. At the same time, they pointed out some shortcomings of the system, such as the low accuracy of natural language personality prediction. To address this, we included a link to the official MBTI website in the window.

7   Conclusion

In this paper, we introduce MBTIviz, a personality visualization system for psycho-demographics research. The system has two main contributions: a comprehensive demographic data view and correlation views that can help present and analyse correlation data in a more accessible manner than conventional data tables. With the help of this system, users can effectively perform research on the relationship between personality traits and demographic variables.

We have demonstrated the effectiveness of the system through a case study of the correlation analysis of MBTI and demographics. The analysis of MBTI variables in relation to career choice provides valuable insights for us to understand the complex relationship between personality traits and occupational preferences. By utilizing appropriate visualization techniques and data processing models, complex data can be made more accessible and user-friendly, providing researchers and users with a deeper understanding of the relationship between variables.

Limitations of the dataset, which include a lack of samples from several countries and age-biased samples that may not be representative of the whole population, have the potential to reduce the effectiveness of the analysis. In the future, efforts will be made to expand the dataset to encompass more diverse populations and demographics, improve machine learning predictive views, and explore potential applications in areas such as education and career counselling.

Acknowledgements. The paper is supported by the National Nature Science Foundation of China (Grant No. 61100053) and a research grant from Intel Asia-Pacific Research and Development Co., Ltd. We would like to express our gratitude to Tingfei Zhu at the Shanghai Jiao Tong University Psychological Counselling Center for her expert contribution to our user studies, as well as to Tianxiang Ye and Jixuan Wang for their participation as nonexpert users. Our sincere thanks also go to the anonymous reviewers for their insightful feedback and constructive suggestions that helped improve the quality of this paper.

References

1. Open Source Psychometrics Project. https://openpsychometrics.org/. Accessed 29 Mar 2023
2. Wyeld, T., Nakayama, M.: The structural equation model diagram as a visualization tool. In: 2019 23rd International Conference in Information Visualization – Part II, pp. 78–81 (2019). https://doi.org/10.1109/IV-2.2019.00024
3. Patil, V.H., Franken, F.H.: Visualization of statistically significant correlation coefficients from a correlation matrix: a call for a change in practice. J. Market. Anal. **9**, 286–297 (2021). https://doi.org/10.1057/s41270-021-00120-z
4. Drucker, J.: Visualization and Interpretation: Humanistic Approaches to Display. MIT Press, Cambridge (2020)
5. Callaway, E., Turner, J., Stone, H., Halstrom, A.: The push and pull of digital humanities: topic modelling the "what is digital humanities?" Genre. Digit. Human. Q. **14** (2020)
6. Personality. https://www.apa.org/topics/personality. Accessed 28 Mar 2023
7. Myers, I.B.: The Myers–Briggs type indicator: Manual (1962)
8. Myers, I.B., Myers, P.B.: Gifts differing: Understanding personality type. Nicholas Brealey (2010)
9. MBTI I Myers–Briggs Type Indicator. https://www.psychometrics.com/assessments/Myers-Briggs-type-indicator/. Accessed 29 03 2023
10. Cattell, R.B.: Personality and Mood by Questionnaire. Jossey-Bass, Oxford (1973)
11. Lambert, T.W., Goldacre, M.J., Parkhouse, J.: Doctors who qualified in the UK between 1974 and 1993: age, gender, nationality, marital status and family formation. Med. Educ. **32**, 533–537 (1998). https://doi.org/10.1046/j.1365-2923.1998.00244.x
12. Gjurković, M., Karan, M., Vukojević, I., Bošnjak, M., Šnajder, J.: PANDORA talks: personality and demographics on reddit. arXiv:2004.04460 [cs]. (2021)
13. Furnham, A., Cheng, H.: Personality traits and socio-demographic variables as predictors of political interest and voting behavior in a British Cohort. J. Individ. Differ. **40**, 118–125 (2019). https://doi.org/10.1027/1614-0001/a000283

14. Boonghee, Y., Neelankavil, J.P., de Guzman, G.M., Lim, R.A.: Personality type preferences of asian managers: a cross-country analysis using the MBTI instrument. Int. J. Glob. Manag. Stud. **5**, 1–23 (2013)
15. Abel, G.J., Sander, N.: Quantifying global international migration flows. Science **343**, 1520–1522 (2014). https://doi.org/10.1126/science.1248676
16. Acosta, E., van Raalte, A.A.: APC curvature plots: displaying nonlinear age-period-cohort patterns on Lexis plots. Demogr. Res. **41**, 1205–1234 (2019)
17. Cimentada, J., Klüsener, S., Riffe, T.: Exploring the demographic history of populations with enhanced Lexis surfaces. Demogr. Res. **42**, 149–164 (2020)
18. Kashnitsky, I., Aburto, J.M.: Geofaceting: aligning small multiples for regions in a spatially meaningful way. Demogr. Res. **41**, 477–490 (2019)
19. Nowok, B.: A visual tool to explore the composition of international migration flows in the EU countries, 1998–2015. Demogr. Res. **42**, 763–776 (2020)
20. Pattaro, S., Vanderbloemen, L., Minton, J.: Visualizing fertility trends for 45 countries using composite lattice plots. Demogr. Res. **42**, 689–712 (2020)
21. Riffe, T., Aburto, J.M.: Lexis fields. Demogr. Res. **42**, 713–726 (2020)
22. Schöley, J.: The centered ternary balance scheme: a technique to visualize surfaces of unbalanced three-part compositions. Demogr. Res. **44**, 443–458 (2021)
23. Riffe, T., Sander, N., Klüsener, S.: Editorial to the special issue on demographic data visualization: getting the point across – reaching the potential of demographic data visualization. Demogr. Res. **44**, 865–878 (2021)
24. Wright, R.E.: Logistic regression. In: Reading and Understanding Multivariate Statistics, pp. 217–244. American Psychological Association, Washington, DC, US (1995)
25. Liang, A.M.: MBTI text classifier (2023). https://github.com/Neoanarika/MBTI
26. (MBTI) Myers–Briggs Personality Type Dataset. https://www.kaggle.com/datasets/datasnaek/mbti-type. Accessed 29 Mar 2023
27. Psychometrics Canada: Myers–Briggs Type Indicator (MBTI) instrument in French and English Canada (2008). https://www.psychometrics.com/wp-content/uploads/2015/02/mbti-in-canada.pdf
28. Do, M.H., Minbashian, A.: A meta-analytic examination of the effects of the agentic and affiliative aspects of extraversion on leadership outcomes. Leadersh. Q. **25**, 1040–1053 (2014). https://doi.org/10.1016/j.leaqua.2014.04.004
29. Goetz, M.L., et al.: An examination of Myers-briggs type indicator personality, gender, and career interests of Ontario veterinary college students. J. Vet. Med. Educ. **47**, 430–444 (2020). https://doi.org/10.3138/jvme.0418-044r
30. Briggs-Myers, I.: Introduction to Type, 6th edn. CPP Inc., Palo Alto (1998)
31. "Latvia to be introduced in the London book fair as a nation of introverts," Latvian Literature (2018). http://www.latvianliterature.lv/en/news/latvia-to-be-introduced-in-the-london-book-fair-as-a-nation-of-introverts
32. Ro, C.: Latvia: Europe's nation of introverts. https://www.bbc.com/travel/article/20180611-latvia-europes-nation-of-introverts. Accessed 29 Mar 2023
33. Auers D.: Who Are We? Latvia's International Image Today, and Tomorrow. The Centenary of Latvia's Foreign Affairs: Scenarios for the Future. 96–113 (2018). https://liia.lv/en/publications/the-centenary-of-latvias-foreign-affairs-scenarios-for-the-future-760?get_file=2#page=96
34. Chun, C.: Scientific approaches, the origin of civilization and the argument about the history of the xia dynasty. J. Guangxi Teach. Educ. Univ. (Philos. Soc. Sci. Ed.). **56**, 128–140 (2020). https://doi.org/10.16088/j.issn.1001-6597.2020.03.011

Data-Driven Security

Distributed Implementation of SM4 Block Cipher Algorithm Based on SPDZ Secure Multi-party Computation Protocol

Xiaowen Ma[1], Maoning Wang[1,2(✉)], and Zhong Kang[1]

[1] School of Information, Central University of Finance and Economics,
Beijing 100081, China
[2] Engineering Research Center of State Financial Security, Ministry of Education,
Central University of Finance and Economics, Beijing 102206, China
1385413929@139.com

Abstract. SM4 is a block cipher algorithm among Chinese commercial cryptographic algorithms, which is advanced in terms of efficiency and theoretical security and has become national and international standards successively. However, existing literature shows that SM4 was not designed with an emphasis on key storage, which means that in today's world where a single trusted hardware device with the built-in key faces challenges such as vulnerability, high cost, and unreliability, the usability of SM4 may be limited. Therefore, this paper proposes an implementation scheme for SM4 based on secure multi-party computation (MPC) technology. The scheme involves dispensing the key among multiple users' devices in a distributed manner, and when using the SM4 algorithm for encryption, multiple users perform joint computation without opening the full key. Specifically, this paper employs the MP-SPDZ framework, which satisfies security requirements in the presence of a dishonest majority of active adversaries. In view of the fact that this framework can only perform basic linear operations such as addition and multiplication, this paper focuses on the algebraic analysis of Sbox, which is the only non-linear component in SM4, and reconstructs it using the bit decomposition method. Furthermore, this paper demonstrates the conversion between the SM4-Sbox field $GF(2^8)$ and the SPDZ parameter field $GF(2^{40})$ through the isomorphic mapping, making it possible to perform joint calculations throughout the entire SM4 algorithm. Complexity analysis shows that this scheme has advantages in terms of data storage and communication volume, reaching a level of usability.

Keywords: SM4 algorithm · SPDZ protocol · MPC

1 Introduction

SM4 [23], as the first officially published commercial symmetric block cipher algorithm in China, was announced by the State Cryptography Administration

Supported by the National Natural Science Foundation of China under Grant No. 61907042 and Beijing Natural Science Foundation under Grant No.4194090.

(sca.gov.cn) in 2006 and has greatly advanced cryptographic algorithms' commercialization in China. It is designed based on a generalized unbalanced Feistel network, and both of its block size and key size are 128 bits. The algorithm utilizes a 32-round nonlinear iterative structure, with each iteration consisting of a transformation function F that includes non-linear word transformation τ, linear word transformation L and other operations. From the design principle standpoint, its security has been thoroughly demonstrated by various theoretical methods [5, 19, 20, 25, 28].

However, from the practical standpoint, the security of cryptosystems often depends on the security of their keys. In particular, SM4, as a block cipher, faces the challenges of secure key storage and management. Currently, for block ciphers, key storage mostly requires hardware security modules [1] or relies on a higher-level trusted environment [24, 26], such as using a USB hardware device with built-in keys. However, such solutions are subject to the risk of key leakage due to loss, and at the same time, they also face the economic pressure caused by the high price of hardware security modules, especially for mobile terminals and other applications where it is difficult to connect external USB devices. Further, some of the existing devices have been proven to have security vulnerabilities that have been exploited by attackers to obtain or export keys through elevated privileges, making it difficult to meet the key management requirements [16]. Therefore, finding a practical, affordable and secure key storage and usage method is the most urgent task for the current SM4 key management. To solve the above problem, this paper proposes an SM4 computation scheme based on secure multi-party computation(MPC) technology, which is oriented to the distributed key storage scenarios, i.e., the keys are separately stored in the devices of multiple users, and two or more parties jointly compute without opening the complete key when performing encryptions via SM4 algorithm. This measure helps to extend the security boundary beyond a single user/device.

The concept of secure multi-party computation(MPC) was proposed by Yao et al. [27] in 1982 and improved in the follow-up work [4, 21], which has now been implemented in program libraries and frameworks [3, 13, 22] reaching an engineering level of efficiency. Secure multi-party computation means that in the absence of a trusted third party, all participants can collaboratively compute an agreed function and ensure that each participant only gets its own computation result and cannot infer the input or output of any others through the data interaction during the computation. That is to say, each participant has absolute control over the data in its possession. It covers various types of protocols based on technologies such as secret sharing, oblivious transfer and garbled circuits. It can support different security models such as active security, passive security and covert security, as well as different types of adversarial assumptions, including majority-honest and majority-dishonest.

Since the main point of MPC is to "distribute" trust among participants of the protocol, one of the significant applications of MPC is to protect long term secret keys. This allows participants in a computation to manage their secrets without requiring hardware security modules or without relying off-the-shelf secure environments. In this setting, the secret key is distributed among participants by splitting it into shares such that only a qualified subset of participants can encrypt or decrypt data by running the MPC protocol without ever opening the key. The current MPC implementation of block ciphers represented by AES has attracted the attention of academic researchers [7, 14] and they have achieved partial optimization in efficiency [10], which is gradually approaching the practical application requirements.

One of the most effective MPC protocols at present is the SPDZ protocol [9]. SPDZ protocol is based on homomorphic encryption, secret sharing and other techniques to achieve secure multi-party computation. It mainly consists of two phases, in which the offline phase is based on homomorphic encryption technique to generate the parameters needed in the online phase, and the online phase is based on secret sharing to complete various types of computations. Compared with the previous BGW protocol [2], the efficiency of its online phase has been largely improved. In terms of security performance, SPDZ protocol satisfies the static adaptive adversary attack model under the UC security framework, i.e., joint computation results remain valid even if up to $n - 1$ of the n participants violate the protocol or collude intentionally. SPDZ protocol uses Message Authentication Code(MAC) [7, 9] to verify whether each participant is honestly participating in the computation. Although SPDZ protocol does not detect which participant is an adaptive participant, it suspends the protocol as soon as it detects a violation to prevent data leakage. Overall, compared with GMW protocol [12], BGW protocol [2] and other protocols, SPDZ protocol is considered more mature.

Therefore, in this paper, MP- SPDZ framework [13] (which is usually used for benchmarking) is utilized to reconstruct SM4. Since MPC can only perform linear operations such as addition and multiplication, the reconstruction of SM4 poses a technical challenge due to the non-linear nature of Sbox. The implementation method of Sbox in the standard SM4 algorithm is SM4-LT [23], i.e., it is implemented by using table lookup. However, in the case of MPC, the SM4-LT method needs to calculate a large table in advance for the operation of Sbox, resulting in a huge circuit size and very expensive storage requirements. Therefore, this paper adopts the SM4-BD method, i.e., the Sbox is implemented by using bit decomposition. In order to implement the MPC operations of SM4, this paper focuses on the single Sbox implementation of SM4-BD in the MP-SPDZ framework, analyzes the Sbox from an algebraic perspective, and disassembles the Sbox operations into algebraic expressions with only additions and multiplications.

2 Preliminaries

2.1 Sbox in SM4

SM4 consists of 32 rounds of nonlinear iterative transformations, and the nonlinear word transformation τ in its round transformation function F is composed of four Sbox operations in parallel. Sbox is a nonlinear transformation that acts independently on the state byte, and its specific operation process is shown as follows [18].

(1) Perform the following affine transformation on the input $w \in GF(2^8) = \frac{GF(2)[X]}{(X^8+X^7+X^6+X^5+X^4+X^2+1)}$ to obtain the result $v \in GF(2^8)$, where the elemental components of w in $GF(2^8)$ are $(w_7, w_6, w_5, w_4, w_3, w_2, w_1, w_0)$.

$$v = L_a \times w + \text{const} = \begin{bmatrix} 1\,1\,0\,1\,0\,0\,1\,1 \\ 1\,1\,1\,0\,1\,0\,0\,1 \\ 1\,1\,1\,1\,0\,1\,0\,0 \\ 0\,1\,1\,1\,1\,0\,1\,0 \\ 0\,0\,1\,1\,1\,1\,0\,1 \\ 1\,0\,0\,1\,1\,1\,1\,0 \\ 0\,1\,0\,0\,1\,1\,1\,1 \\ 1\,0\,1\,0\,0\,1\,1\,1 \end{bmatrix} \begin{bmatrix} w_7 \\ w_6 \\ w_5 \\ w_4 \\ w_3 \\ w_2 \\ w_1 \\ w_0 \end{bmatrix} + \begin{bmatrix} 1 \\ 1 \\ 0 \\ 1 \\ 0 \\ 0 \\ 1 \\ 1 \end{bmatrix} \qquad (1)$$

(2) Find the multiplication inverse in $GF(2^8)$. i.e., for the input $v \in GF(2^8)$, find $u \in GF(2^8)$ satisfying the following equation.

$$v * u = 1 \mod (X^8 + X^7 + X^6 + X^5 + X^4 + X^2 + 1) \qquad (2)$$

then

$$u = v^{-1} = \begin{cases} v^{254}, & \text{when} \quad v \neq 0, \\ 0, & \text{when} \quad v = 0. \end{cases} \qquad (3)$$

(3) Let the element components of u in $GF(2^8)$ be $(u_7, u_6, u_5, u_4, u_3, u_2, u_1, u_0)$ and perform the following affine transformation on u.

$$z = L_a \times u + \text{const} \qquad (4)$$

The matrix L_a and vector const used for the affine transformation in this step are the same as in step 1.

2.2 The SPDZ Protocol

SPDZ protocol is one of the protocol frameworks for MPC, first proposed in 2012 [9] and then optimized successively as SPDZ2 [8], MASCOT [14] and SPDZ2k [6,11]. It is based on homomorphic encryption, linear secret sharing and other techniques [9,15] to enable secure sharing and computation of secret values in the dishonest majority setting. The detection mechanism of the SPDZ protocol for adaptive adversary is achieved through an information-theoretic Message Authentication Code(MAC). Specifically, all parties generate corresponding MACs of the secret shared values and commit to them, and then, each participant verifies the MACs before opening the result after joint calculation and stops calculating if inconsistency is found. Therefore, SPDZ protocol guarantees that if the protocol terminates then the output received by the honest parties is correct, except with negligible probability. In this paper (similar to the previous standardized engineering framework work [13]), this probability is set to 2^{-40}.

We denote the secret shared value s as $[s]$. It is understood that for n parties $P_i, i = 1 \cdots n$, $s = \sum_i [s]_i$ where $[s]_i$ are random shares of s. SPDZ protocol uses authenticated secret sharing, i.e., each secret s is attached with a MAC, computed as $[\alpha s]$, where α is a global MAC key. In this setting, each P_i owns a share of the MAC key $[\alpha]_i$. At this point the secret s containing authentication information is denoted as $< s >= ([s], [\alpha s])$.

SPDZ protocol is divided into two phases: offline phase and online phase. The offline phase is used to generate MAC key α, beaver multiplication triple (a, b, c) and other information to assist subsequent calculation and verification, and does not involve input data and logic in the online phase. The online phase mainly carries out logical calculation of input data and MAC verification of results.

Offline Phase. In the offline phase, all parties compute the authenticated shares of the beaver multiplication triples and their MACs in such a way that no party learns about (a, b, c) and α. This is achieved in the pre-computation phase using homomorphic encryption(HE) and zero-knowledge(ZK) schemes. To this end, each P_i obtains $(< a >_i, < b >_i, < c >_i)$, where $< a >_i= ([a]_i, [\alpha a]_i)$ and similarly $< b >_i= ([b]_i, [\alpha b]_i)$ and $< c >_i= ([c]_i, [\alpha c]_i)$. The beaver multiplication triple (a, b, c) needs to satisfy $c = a * b$, where a, b are random values in $GF(2^{40})$. In other words, the beaver multiplication triples between all parties need to satisfy the following condition, for all parties $P_i, i = 1 \cdots n$, each P_i has a triple share $(< a >_i, < b >_i, < c >_i)$ which satisfies $(\sum_i < a >_i) * (\sum_i < b >_i) = (\sum_i < c >_i)$.

Online Phase. In the online phase, calculations and MAC verification of intermediate values and results are performed. Its basic operations include addition, multiplication, etc., which involve homomorphic cryptographic properties.

(1) Addition. The process of addition is more straightforward and can be calculated directly via the expression of secret shared values, whose formula is as follows.

$$< s_1 + s_2 > \leftarrow < s_1 > + < s_2 >$$

That is, in SPDZ protocol, given additive shares $[s_1], [s_2]$ of s_1, s_2 and $\eta \in GF(2^{40})$, each P_i can locally compute the additive share of sum and scalar multiplication as follows.

$$[s_1] + [s_2] = [s_1 + s_2], \quad [\eta s] = \eta[s].$$

(2) Multiplication. First, in order to multiply two secret value $[s_1]$ and $[s_2]$ with beaver multiplication triple $([a], [b], [c])$, each P_i computes $[\epsilon]_i \leftarrow [s_1]_i - [a]_i$ and $[\rho]_i \leftarrow [s_2]_i - [b]_i$ locally. All parties come together to open ϵ and ρ by broadcasting $[\epsilon]_i$ and $[\rho]_i$. Finally, each P_i computes locally

$$[s_1 s_2]_i \leftarrow [c]_i + \epsilon[b]_i + \rho[a]_i + \epsilon\rho.$$

Then, to verify the MAC of the multiplication, during the multiplication protocol above, each P_i also gets

$$[\alpha s_1 s_2]_i \leftarrow [\alpha c]_i + \epsilon[\alpha b]_i + \rho[\alpha a]_i + \epsilon\rho.$$

(3) MAC verification. To improve efficiency, SPDZ protocol is based on aggregating intermediate computation results into a polynomial to verify MAC centrally only once after opening multiple secret shared values. Specifically, when opening the computation result $< s >$, each P_i sets the batch intermediate computation results $[t_j]_i$, $[\alpha t_j]_i$ as polynomial coefficients, committing the value of this polynomial at the random point e and sends it to others. Finally, each P_i verifies whether α (generated in the offline phase and its secret shared values are held by all parties) can be reconstructed. If it can, the MAC verification is passed, otherwise it means that at least one party violates the protocol.

The Algorithm 1 shows the online phase of SPDZ protocol.

3   Method

3.1   SM4-Sbox in SPDZ

The Fig. 1 provides a brief overview of the SPDZ implementation for the SM4-Sbox. Prior to the start of the protocol, according to the structure of Sbox in SM4, the input state $w \in GF(2^8)$ has been divided into n splits $w_{(0)}, w_{(1)}, \cdots, w_{(n)} \in GF(2^8)$ where $w = \Sigma_i w_{(i)}$ and each P_i holds $w_{(i)}$. According to the commonly used parameters' security level, in MP-SPDZ framework calculations are performed in $GF(2^{40})$. So each P_i is required to utilize the mapping matrix T (mentioned in Sect. 3.3) on $w_{(i)} \in GF(2^8)$ to obtain $< s_w >_i \in GF(2^{40})$ locally at the beginning of the protocol execution and the secret shared value $< s_w >_i$ in $GF(2^{40})$ satisfies $< s_w >= \Sigma_{i=1}^n < s_w >_i$. Then the secret value $< s_w >$ is input into the algorithm 1(mentioned in Sect. 3.4) to participate in the execution of the SPDZ implementation of the SM4-Sbox. And the output of algorithm 1 is the opened value s_z of $< s_z >\in GF(2^{40})$, where each P_i holds the share $< s_z >_i$ and z, as the isomorphic element of s_z in $GF(2^8)$, satisfies $z = \text{Sbox}(w)$. In other words, $< s_z >_i$ is required to be mapped to $z_{(i)} \in GF(2^8)$ using the isomorphism mapping matrix T^{-1} (mentioned in Sect. 3.3) for the continued preservation in each P_i for the next round of the iterative transformation.

Based on the description above, this paragraph presents a top-down description of the SPDZ implementation for the whole SM4 algorithm. SM4 employs a 32-round generalized unbalanced Feistel network with both the block size and key size being 128 bits, and data operations are carried out in units of bytes(8 bits) and words(32 bits). When initiating the SM4 computation, the 128-bit initial status $\mathcal{W}$ is first assigned by the plaintext data. Considering the secure multi-party computation setting, the status $\mathcal{W}$ is divided into n splits using

Initialize: Each P_i invokes the preprocessing in the offline phase to get the global MAC key $[\alpha]$, a sufficient number of beaver multiplication triples of the form $(<a>,<b>,<c>)$, as well as random values $[m]$, $[e]$. (P_i holds shares $[\alpha]_i$, triples' shares like $(<a>_i,<b>_i,<c>_i)$, $[m]_i$ and $[e]_i$)

Input: $<s>$, and each P_i holds the secret shared value $<s>_i$ such that
$$<s>=\sum_i<s>_i.$$
Add: The parties locally follow $<s_1+s_2>\leftarrow<s_1>+<s_2>$ for the addition of secret shared values $<s_1>$ and $<s_2>$.

Multiply:

1. To calculate the multiplication of the secret shared values $<s_1>$ and $<s_2>$, we need to use the beaver multiplication triple $(<a>,<b>,<c>)$, but before multiplication, we need to check the beaver multiplication triple to see if it satisfies $<c>=<a>*<b>$. Therefore, an additional beaver multiplication triple $(<a'>,<b'>,<c'>)$ is needed for the test of $(<a>,<b>,<c>)$, which is done as follows:
   (1)Open a random value $[m]$.
   (2)Partially open $\theta \leftarrow m<a>-<a'>$ to get θ and $\sigma \leftarrow<b>-<b'>$ to get σ.
   (3)Evaluate $m<c>-<c'>-\theta<b'>-\sigma<a>-\theta\sigma$, and partially open the result.
   (4)Check the result of the previous calculation step. If the result is non-zero the players abort, otherwise go on with $(<a>,<b>,<c>)$.
2. The parties partially open $\epsilon \leftarrow<s_1>-<a>$ to get ϵ and $\rho \leftarrow<s_2>-<b>$ to get ρ.
3. Calculate the multiplication $<s_3>$ of the secret shared values $<s_1>$ and $<s_2>$ by $<s_3>\leftarrow<c>+\epsilon<b>+\rho<a>+\epsilon\rho$

Output: The output result to be obtained by each P_i is s, and the result of its calculation by the above operations is $<s>$, so it is necessary to open $<s>$, the specific steps of which are as follows.

1. Let $t_1,\cdots,t_K$ be all values publicly to open so far, where $<t_j>=(([t_j]_1,\cdots,[t_j]_n),([\alpha t_j]_1,\cdots,[\alpha t_j]_n))$. Now, the random value $[e]$ is opened, and each P_i sets $e_i=e^i, i=1,...,K$. All parties compute $t \leftarrow \sum_j e_j t_j$ to verify MAC of t.
2. Each P_i commits to $\gamma_i \leftarrow \sum_j e_j[\alpha t_j]_i$. For the output value $<s>$, P_i also commits that its secret shared share $[s]_i$ and its share in the corresponding MAC $[\alpha s]_i$ are true.
3. Open global MAC key $[\alpha]$.
4. Each P_i asks the commit protocol to open γ_i, and all parties verify that $\alpha t = \sum_i \gamma_i$. If this does not hold, the protocol aborts. Otherwise the parties conclude that the intermediate values are correctly computed.
5. To get the output s, the commitments to $[s]_i$ and $[\alpha s]_i$ are opened. Now, s is defined as $s := \sum_i[s]_i$ and each party verifies that $\alpha s = \sum_i[\alpha s]_i$, if so, s is the output.

Algorithm 1: The pseudocode for the online phase of SPDZ protocol

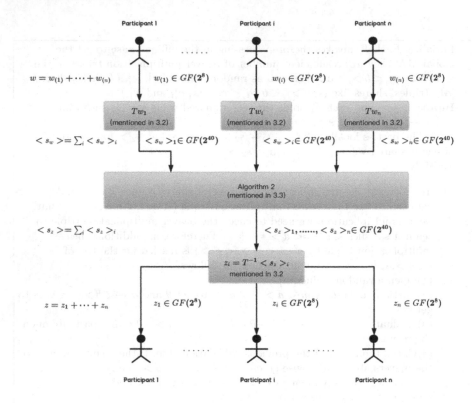

Fig. 1. SM4-Sbox in SPDZ

the following method(where each P_i holds $\mathcal{W}_{(i)}$ respectively and the symbol $\oplus$ denotes the operation XOR).

$$\mathcal{W}_{(i)} = \begin{cases} \text{128-bit random number,} & \text{when} \quad i = 1, 2, \cdots, n-1 \\ \mathcal{W} \oplus \bigoplus_{j=1}^{n-1} \mathcal{W}_{(j)}, & \text{when} \quad i = n \end{cases} \qquad (5)$$

After that, the state value $\mathcal{W}$ are entered into the round transformation function F, noting $\mathcal{W} = W_1 \| W_2 \| W_3 \| W_4$ where the symbol $\|$ denotes the cascade and $W_j \in GF(2^{32}), j = 1, 2, 3, 4$. At this point, corresponding to $\mathcal{W}_{(i)}$, each P_i holds $W_{j,(i)}$. In the transformation function of the rth round, first, the rth round key K_r will be XORed with the computation result W of $W_2 \oplus W_3 \oplus W_4$. In the case of secure multi-party computation, each P_i locally XOR its share of the round key $K_{r,(i)}$ held by itself (the generation method is explained in the next paragraph) with the share $W_{(i)}$ of the intermediate state word W for which the operation needs to be performed, at which point there is:

$$W \oplus K_r = (W_{(1)} \oplus K_{r,(1)}) \oplus \cdots \oplus (W_{(n)} \oplus K_{r,(n)})$$

Without affecting understanding, we still write down the result of this step as W. Next, this word W will be transformed into a new intermediate word by the action of the nonlinear transformation τ and the linear transformation L. In the τ transformation, the word W is split into a cascade of four bytes, where each byte, denoted as w, will be input into the Sbox respectively to obtain the output z (mentioned in the Fig. 1 and its interpretation is in the last paragraph). After this step, each P_i holds $z_{(i)}$ such that $\Sigma_i z_{(i)} = z$ and z can be obtained after the opening of SPDZ. After the four bytes are substituted by the distributed computation Sbox, they will be cascaded, after which the L transformation continues. The state word entering the L transformation is still denoted as W, at this time, each P_i holds $W_{(i)}$, and it is necessary to compute the following equation.

$$L(W) = W \oplus (W <<< 2) \oplus (W <<< 10) \oplus (W <<< 18) \oplus (W <<< 24)$$

The linearity of L enables that each P_i executes the corresponding cyclic shift operation and the XOR operation on its share $W_{(i)}$ is equivalent to that of the state word W. Finally, let $W = W_2||W_3|||W_4||W_1 \oplus L(W)$ be the computation result of this round of the round transformation function, in which, still, no interaction or joint computation between each P_i is involved, and hence all operations can be performed by P_i locally. In summary, after 32 rounds of computation, the 128-bit state value $\mathcal{W}$ is ciphertext, and each P_i holds $\mathcal{W}_i$ satisfying $\mathcal{W} = \bigoplus_{i=1}^{n} \mathcal{W}_{(i)}$.

In the aforementioned explanation, the round key K_r involved in different rounds of the SM4 algorithm are derived from the key $\mathcal{K}$ using a key expansion process. In the distributed scenario, firstly, using a similar method to the equation(5), $\mathcal{K}$ is divided into n splits and each P_i holds $\mathcal{K}_{(i)}$ respectively. Secondly, the SM4 key expansion process still involves only the generalized unbalanced Feistel network, the XOR operation, τ transformation and L transformations, and so on, in a similar manner to its encryption process. Therefore, each P_i is able to generate $K_{r,(i)}$ satisfying $K_r = \bigoplus_{i=1}^{n} K_{r,(i)}$ after similar steps including interactions and local computations.

3.2 Algebraic Expression for Sbox

Sbox, as the only nonlinear layer in SM4, implements the obfuscation function of the cryptographic design guidelines and is the most critical part of SM4 security. Considering the computation process of the affine transformation L_a, this paper uses the bit decomposition method [17], which is an algebraic method that uses the elements themselves in $GF(2^8)$ as a set of basis to represent the element components, to solve the algebraic expressions of Sbox.

In SM4, the reduction modulus for $GF(2^8)$ is defined as $g(X) = X^8 + X^7 + X^6 + X^5 + X^4 + X^2 + 1$, and let $\beta \in GF(2^8)$ be the root of $g(X)$, then there is a set of basis in $GF(2^8)$ represented by $B = (1, \beta, \beta^2, \beta^3, \beta^4, \beta^5, \beta^6, \beta^7)$. At this point, the element x is expressed in the base B as

$$x = (x_0, x_1, x_2, x_3, x_4, x_5, x_6, x_7)_B = \sum_{i=0}^{7} x_i \beta^i, \tag{6}$$

where $x_i \in GF(2), i = 0, 1, 2, ..., 7$.

$\forall f, h \in GF(2^8)$, $(f+h)^{2^j} = f^{2^j} + h^{2^j}$, $j \in \mathbb{N}$. Therefore, we can get eight equations from the equation (6), namely $x^{2^j} = \sum_{i=0}^{7} x_i \beta^{i2^j}$, $j = 0, ..., 7$. The matrix representation of the above eight equations is shown as follows.

$$
\begin{bmatrix}
1 & \beta & \beta^2 & \beta^3 & \beta^4 & \beta^5 & \beta^6 & \beta^7 \\
1 & \beta^2 & \beta^4 & \beta^6 & \beta^8 & \beta^{10} & \beta^{12} & \beta^{14} \\
1 & \beta^4 & \beta^8 & \beta^{12} & \beta^{16} & \beta^{20} & \beta^{24} & \beta^{28} \\
1 & \beta^8 & \beta^{16} & \beta^{24} & \beta^{32} & \beta^{40} & \beta^{48} & \beta^{56} \\
1 & \beta^{16} & \beta^{32} & \beta^{48} & \beta^{64} & \beta^{80} & \beta^{96} & \beta^{112} \\
1 & \beta^{32} & \beta^{64} & \beta^{96} & \beta^{128} & \beta^{160} & \beta^{192} & \beta^{224} \\
1 & \beta^{64} & \beta^{128} & \beta^{192} & \beta^{256} & \beta^{320} & \beta^{384} & \beta^{448} \\
1 & \beta^{128} & \beta^{256} & \beta^{384} & \beta^{512} & \beta^{640} & \beta^{768} & \beta^{896}
\end{bmatrix}
\begin{bmatrix}
x_0 \\ x_1 \\ x_2 \\ x_3 \\ x_4 \\ x_5 \\ x_6 \\ x_7
\end{bmatrix}
=
\begin{bmatrix}
x \\ x^2 \\ x^4 \\ x^8 \\ x^{16} \\ x^{32} \\ x^{64} \\ x^{128}
\end{bmatrix}
\tag{7}
$$

Let A denote the coefficient matrix of the components of x in equation (6), the determinant of A is

$$
\prod_{0 \le j < i \le 7} (\beta^{2^i} - \beta^{2^j}) \ne 0
\tag{8}
$$

Therefore, there must be a matrix A^{-1} satisfying

$$
[x_0, x_1, x_2, x_3, x_4, x_5, x_6, x_7]^T = A^{-1}[x, x^2, x^4, x^8, x^{16}, x^{32}, x^{64}, x^{128}]^T
$$

That is, the element component $x_i \in GF(2)$ can be expressed in terms of the element $x \in GF(2^8)$ itself.

Since the element β^{i2^j} of the matrix A is in $GF(2^8)$ and $g(\beta) = \beta^8 + \beta^7 + \beta^6 + \beta^5 + \beta^4 + \beta^2 + 1 = 0$, the matrix A can be expressed in the following form.

$$
A =
\begin{bmatrix}
1 & \beta & \cdots & \beta^7 \\
1 & \beta^2 & \cdots & \beta^3 + \beta \\
1 & \beta^4 & \cdots & \beta^6 + \beta^2 \\
\vdots & \vdots & & \vdots \\
\vdots & \vdots & & \vdots \\
1 & \beta^7 + \beta^6 + \beta^4 + \beta + 1 & \cdots & \beta^7 + \beta^5 + \beta^4 + \beta^3 + \beta
\end{bmatrix}
\mod g(\beta)
\tag{9}
$$

After converting the elements in the matrix A to the byte representations in $GF(2^8)$, the corresponding matrix derived from the equation (9) is as follows.

$$
A =
\begin{bmatrix}
01 & 02 & 04 & 08 & 10 & 20 & 40 & 80 \\
01 & 04 & 10 & 40 & f5 & 3e & f8 & a \\
01 & 10 & f5 & f8 & 28 & 9f & 79 & 44 \\
01 & f5 & 28 & 79 & 7e & aa & 72 & e8 \\
01 & 28 & 7e & 72 & 67 & 70 & 37 & 8c \\
01 & 7e & 67 & 37 & d3 & 33 & de & 5a \\
01 & 67 & d3 & de & 02 & ce & 53 & 49 \\
01 & d3 & 02 & 53 & 04 & a6 & 08 & b9
\end{bmatrix}
\tag{10}
$$

By performing primary transformations on the matrix A, we get the matrix A^{-1}, which satisfies the following equation.

$$
\begin{bmatrix} x_0 \\ x_1 \\ x_2 \\ x_3 \\ x_4 \\ x_5 \\ x_6 \\ x_7 \end{bmatrix} = \begin{bmatrix} 07 & e7 & d9 & 46 & ec & 9c & af & 61 \\ 99 & be & 95 & ee & 98 & bf & 94 & ef \\ 71 & 32 & cf & a7 & 21 & 3f & 9e & ab \\ c2 & f6 & 2d & 6f & 93 & fa & 7d & 62 \\ a6 & 20 & 3e & 9f & aa & 70 & 33 & ce \\ 94 & ef & 99 & be & 95 & ee & 98 & bf \\ 8d & 5b & 48 & 68 & 81 & b & 45 & e9 \\ 7b & 76 & 27 & 2b & 7b & 76 & 27 & 2b \end{bmatrix} \begin{bmatrix} x \\ x^2 \\ x^4 \\ x^8 \\ x^{16} \\ x^{32} \\ x^{64} \\ x^{128} \end{bmatrix} \quad \mathrm{mod} \quad g(\beta) \quad (11)
$$

From the equation (11), it follows that the element component x_i of any element in $GF(2^8)$ can be represented by the element x itself. Let $y = L_a \times x$, where L_a is defined in the same way as in the equation (1), bringing the above algebraic formula for the elemental component x into $y = L_a \times x$ gets the following result.

$$
y = \begin{bmatrix} y_0 \\ y_1 \\ y_2 \\ y_3 \\ y_4 \\ y_5 \\ y_6 \\ y_7 \end{bmatrix} = \begin{bmatrix} c0 & f2 & 3d & 9a & bb & 84 & 1a & b1 \\ 60 & c6 & e6 & d8 & 47 & ed & 9d & ae \\ f7 & 2c & 6e & 92 & fb & 7c & 63 & c3 \\ 46 & ec & 9c & af & 61 & c7 & e7 & d9 \\ e4 & dc & 57 & 18 & b5 & d0 & 07 & 15 \\ b5 & d0 & 07 & 15 & e4 & dc & 57 & 18 \\ 67 & d3 & 02 & 04 & 10 & f5 & 28 & 7e \\ e & 54 & 1d & a4 & 24 & 2e & 6a & 82 \end{bmatrix} \begin{bmatrix} x \\ x^2 \\ x^4 \\ x^8 \\ x^{16} \\ x^{32} \\ x^{64} \\ x^{128} \end{bmatrix} \quad (12)
$$

Calculating the components of y to one byte by the formula $y = \sum_{i=0}^{7} y_i \beta^i$ yields the following equation.

$$
y =' 77' x^{128} +' e6' x^{64} +' c8' x^{32} +' f4' x^{16} +' e9' x^8 +' 91' x^4 +' 06' x^2 +' 18' x \quad (13)
$$

Further, considering the full expression covering the constant vector const, there is

$$
L_a \times x + \mathrm{const} =
$$
$$
'77' x^{128} +' e6' x^{64} +' c8' x^{32} +' f4' x^{16} +' e9' x^8 +' 91' x^4 +' 06' x^2 +' 18' x +' 53' \quad (14)
$$

Therefore, to obtain the algebraic expression for Sbox of SM4, according to equation (4), it is necessary to compute $z = L_a \times u + \mathrm{const}$. i.e., x of equation (14) is replaced by u, and according to equation (3), there is $u = v^{-1} = v^{254}$, thus there is

$$
\begin{aligned}
z =' & 77'(v^{254})^{128} +' e6'(v^{254})^{64} +' c8'(v^{254})^{32} +' f4'(v^{254})^{16} \\
& +' e9'(v^{254})^8 +' 91'(v^{254})^4 +' 06'(v^{254})^2 +' 18'(v^{254}) +' 53'
\end{aligned} \quad (15)
$$

According to equation (1), where $v = L_a \times w + \text{const}$, so v can be calculated by the equation (14) similarly.

$$v =' 77'w^{128} +' e6'w^{64} +' c8'w^{32} +' f4'w^{16} +' e9'w^8 +' 91'w^4 +' 06'w^2 +' 18'w +' 53' \tag{16}$$

According to the equation (14) and (15), the algebraic expression for Sbox of SM4 is as follows.

$$z =' a7'w^{254} +' 06'w^{253} +' 71'w^{252} + \cdots\cdots +' 46'w^2 +' 8e'w +' d6' \tag{17}$$

See the Table 2 in the Appendix A for the detailed form.

3.3 Isomorphic Mapping of $GF(2^8)$ and $GF(2^{40})$

MP-SPDZ framework enables us to implement functions in binary finite field (such as $GF(2^{40})$) as well as prime finite field(such as Z_p). Standard SM4 arithmetic is defined with Galois field $GF(2^8)$ with a reduction modulus $g(X) = X^8 + X^7 + X^6 + X^5 + X^4 + X^2 + 1$. In order to satisfy the secure parameter requirements, MP-SPDZ needs to be computed in the binary finite field $GF(2^{40})$ and hence we need to compute in advance the isomorphic mapping between $GF(2^8)$ elements and $GF(2^{40})$ elements. According to the general method [9] in the existing SPDZ engineering framework, the reduction modulus to define $GF(2^{40})$ is $Q(Y) = Y^{40} + Y^{20} + Y^{15} + Y^{10} + 1$.

In this section, the elements in a finite field, i.e., polynomials, are expressed in vector form. For example, the element in $GF(2^8)$, i.e., the polynomial $X^7 + X^6 + X^5 + X^2 + 1$, is expressed as the vector $[1, 1, 1, 0, 0, 0, 1, 0, 1]^T$. Suppose β is the root of the reduction modulus $g(X) = X^8 + X^7 + X^6 + X^5 + X^4 + X^2 + 1$ in $GF(2^8)$ and $c_7 X^7 + c_6 X^6 + c_5 X^5 + c_4 X^4 + c_3 X^3 + c_2 X^2 + c_1 X + c_0$ is a polynomial in the above finite field, then the polynomial is equivalent to $c_7 \beta^7 + c_6 \beta^6 + c_5 \beta^5 + c_4 \beta^4 + c_3 \beta^3 + c_2 \beta^2 + c_1 \beta + c_0$, and its matrix representation is $[\beta^7, \beta^6, \beta^5, \beta^4, \beta^3, \beta^2, \beta, 1] \times [c_7, c_6, c_5, c_4, c_3, c_2, c_1, c_0]^T$, set vector $c = [c_7, c_6, c_5, c_4, c_3, c_2, c_1, c_0]^T$. Similarly, suppose λ is a root of the reduction modulus $Q(Y) = Y^{40} + Y^{20} + Y^{15} + Y^{10} + 1$ in $GF(2^{40})$ and $d_{39} Y^{39} + d_{38} Y^{38} + \cdots + d_1 Y + d_0$ is a polynomial in the above finite field, then the polynomial can be equivalent to $d_{39} \lambda^{39} + d_{38} \lambda^{38} + \cdots + d_1 \lambda + d_0$, and its matrix representation is $[\lambda^{39}, \lambda^{38}, \cdots, \lambda, 1] \times [d_{39}, d_{38}, \cdots, d_1, d_0]^T$, set vector $d = [d_{39}, d_{38}, \cdots, d_1, d_0]^T$. In order to realize the isomorphic mapping between $GF(2^8)$ and $GF(2^{40})$, it is necessary to find a mapping matrix T to satisfy $Tc = d$ for all vectors c, d.

The calculation of the mapping matrix can be done by finding the mapping relation $\beta \leftarrow \lambda'$, i.e. finding an element $\lambda' \in GF(2^{40})$ with the reduction modulus $Q(Y) = Y^{40} + Y^{20} + Y^{15} + Y^{10} + 1$ such that it satisfies $g(\lambda') = 0$. Due to the linearity of the isomorphic relation, there is $\beta^i \leftarrow \lambda'^i$, so its isomorphic mapping relation can be expressed as follows.

$$c_7 \beta^7 + c_6 \beta^6 + c_5 \beta^5 + c_4 \beta^4 + c_3 \beta^3 + c_2 \beta^2 + c_1 \beta^1 + c_0 \beta^0$$
$$\rightarrow c_7 \lambda'^7 + c_6 \lambda'^6 + c_5 \lambda'^5 + c_4 \lambda'^4 + c_3 \lambda'^3 + c_2 \lambda'^2 + c_1 \lambda'^1 + c_0 \lambda'^0$$

Let the matrix corresponding to β be A and the matrix corresponding to λ' be B, then the mapping matrix T needs to satisfy $TAc = Bc$, that is, $T = BA^{-1}$. For the convenience of calculation, $\beta = \overline{X} \bmod g(X)$ is generally chosen. So the matrix A corresponding to β is I and then the mapping matrix $T = B$. The isomorphic mapping relation is $\beta \leftarrow \lambda^{25} + \lambda^{15} + \lambda^5 + 1$ by the composite field isomorphic mapping and then the equivalence relation is as follows.

$$\beta^0 \cong \lambda'^0 \cong (\lambda^{25} + \lambda^{15} + \lambda^5 + 1)^0 = 1$$

$$\beta^1 \cong \lambda'^1 \cong (\lambda^{25} + \lambda^{15} + \lambda^5 + 1)^1 = \lambda^{25} + \lambda^{15} + \lambda^5 + 1,\, or\,(0x2008021)_{16}$$

$$\beta^2 \cong \lambda'^2 \cong (\lambda^{25} + \lambda^{15} + \lambda^5 + 1)^2 = \lambda^{25} + \lambda^{20} + 1,\, or\,(0x2100001)_{16}$$

$$\beta^3 \cong \lambda'^3 \cong (\lambda^{25} + \lambda^{15} + \lambda^5 + 1)^3 = \lambda^{35} + \lambda^{25} + \lambda^{15},\, or\,(0x802008000)_{16}$$

$$\beta^4 \cong \lambda'^4 \cong (\lambda^{25} + \lambda^{15} + \lambda^5 + 1)^4 = \lambda^{30} + \lambda^{25} + \lambda^{15},\, or\,(0x42008000)_{16}$$

$$\beta^5 \cong \lambda'^5 \cong (\lambda^{25} + \lambda^{15} + \lambda^5 + 1)^5 = \lambda^{30} + \lambda^{20} + \lambda^{15} + \lambda^{10} + \lambda^5,\, or\,(0x40108420)_{16}$$

$$\beta^6 \cong \lambda'^6 \cong (\lambda^{25} + \lambda^{15} + \lambda^5 + 1)^6 = \lambda^{25} + \lambda^{10} + \lambda^5 + 1,\, or\,(0x2000421)_{16}$$

$$\beta^7 \cong \lambda'^7 \cong (\lambda^{25} + \lambda^{15} + \lambda^5 + 1)^7 = \lambda^{35} + \lambda^{30} + \lambda^{20} + \lambda^{15},\, or\,(0x840108000)_{16}$$

In summary, the isomorphic mapping matrix T from $GF(2^8)$ to $GF(2^{40})$ can be derived as follows.

$$
T = \begin{bmatrix}
0\,0\,0\,0\,0\,0\,0\,0\,0\,0\,0\,0\,0\,0\,0\,0\,0\,0\,0\,0\,0\,0\,0\,0\,0\,0\,0\,0\,0\,0\,0\,0\,0\,0\,0\,0\,0\,0\,0\,1 \\
0\,0\,0\,0\,0\,0\,0\,0\,0\,0\,0\,0\,0\,0\,0\,1\,0\,0\,0\,0\,0\,0\,0\,0\,0\,1\,0\,0\,0\,0\,0\,0\,0\,0\,1\,0\,0\,0\,0\,1 \\
0\,0\,0\,0\,0\,0\,0\,0\,0\,0\,0\,0\,0\,0\,1\,0\,0\,0\,0\,1\,0\,0\,0\,0\,0\,0\,0\,0\,0\,0\,0\,0\,0\,0\,0\,0\,0\,0\,0\,1 \\
0\,0\,0\,0\,1\,0\,0\,0\,0\,0\,0\,0\,0\,1\,0\,0\,0\,0\,1\,0\,0\,0\,0\,0\,0\,0\,0\,0\,0\,0\,0\,0\,0\,0\,0\,0\,0\,0\,0\,0 \\
0\,0\,0\,0\,0\,0\,0\,0\,0\,1\,0\,0\,0\,0\,1\,0\,0\,0\,0\,1\,0\,0\,0\,0\,0\,0\,0\,0\,0\,0\,0\,0\,0\,0\,0\,0\,0\,0\,0\,0 \\
0\,0\,0\,0\,0\,0\,0\,0\,0\,1\,0\,0\,0\,0\,0\,0\,0\,0\,1\,0\,0\,0\,0\,1\,0\,0\,0\,0\,1\,0\,0\,0\,0\,1\,0\,0\,0\,0\,0\,0 \\
0\,0\,0\,0\,0\,0\,0\,0\,0\,0\,0\,1\,0\,0\,0\,0\,0\,0\,0\,0\,0\,0\,0\,0\,0\,1\,0\,0\,0\,0\,1\,0\,0\,0\,0\,1\,0\,0\,0\,1 \\
0\,0\,0\,0\,1\,0\,0\,0\,0\,1\,0\,0\,0\,0\,0\,0\,0\,0\,0\,1\,0\,0\,0\,0\,1\,0\,0\,0\,0\,0\,0\,0\,0\,0\,0\,0\,0\,0\,0\,0
\end{bmatrix}^{T}
$$

The pseudo-inverse T^{-1} of the matrix T can be derived by calculating as follows.

$$
T^{-1} = \begin{bmatrix}
0\,0\,0\,0\,1\,0\,0\,0\,0\,1\,0\,0\,0\,0\,0\,0\,0\,0\,1\,0\,0\,0\,0\,1\,0\,0\,0\,0\,1\,0\,0\,0\,0\,0\,0\,0\,0\,0\,0\,1 \\
0\,0\,0\,0\,0\,0\,0\,0\,0\,0\,0\,0\,0\,0\,0\,0\,0\,0\,0\,0\,0\,0\,0\,0\,0\,0\,0\,1\,0\,0\,0\,0\,1\,0\,0\,0\,0\,0 \\
0\,0\,0\,0\,1\,0\,0\,0\,0\,1\,0\,0\,0\,0\,1\,0\,0\,0\,0\,0\,0\,0\,0\,0\,0\,0\,0\,0\,0\,0\,0\,0\,1\,0\,0\,0\,0\,0 \\
0\,0\,0\,0\,0\,0\,0\,0\,1\,0\,0\,0\,0\,0\,0\,0\,0\,0\,0\,0\,1\,0\,0\,0\,0\,1\,0\,0\,0\,0\,1\,0\,0\,0\,0\,0 \\
0\,0\,0\,0\,1\,0\,0\,0\,0\,0\,0\,0\,0\,1\,0\,0\,0\,0\,1\,0\,0\,0\,0\,0\,0\,0\,0\,0\,0\,0\,0\,1\,0\,0\,0\,0\,0 \\
0\,0\,0\,0\,0\,0\,0\,0\,0\,0\,0\,0\,0\,0\,1\,0\,0\,0\,0\,1\,0\,0\,0\,0\,1\,0\,0\,0\,0\,1\,0\,0\,0\,0\,0\,0\,0 \\
0\,0\,0\,0\,0\,0\,0\,0\,0\,0\,0\,0\,0\,0\,1\,0\,0\,0\,0\,1\,0\,0\,0\,0\,1\,0\,0\,0\,0\,0\,0\,0\,0\,0\,0\,0 \\
0\,0\,0\,0\,1\,0\,0\,0\,0\,1\,0\,0\,0\,0\,0\,0\,0\,0\,0\,0\,0\,0\,1\,0\,0\,0\,0\,1\,0\,0\,0\,0\,1\,0\,0\,0\,0\,0
\end{bmatrix}
$$

3.4 Complexity of SPDZ Modules for SM4

Algorithm 2 describes the full pseudocode for Sbox computation on the input $< s_w >$ using SM4-BD proposed in Sect. 3.2. In step 1, a linear transformation composed of the isomorphic mapping matrix T and the matrix A in the equation (10) is applied to the secret value $< s_w >$ to generate $\{< s_w^{128} >, \cdots, < s_w^2 >\}$ in preparation for the subsequent steps. In step 2, the results obtained in step

Input: The input secret state is $< s_w >\in GF(2^{40})$.
Output: The output secret state is $< s_z >\in GF(2^{40})$.
1 Calculate $\{< s_w^{128} >, < s_w^{64} >, < s_w^{32} >, < s_w^{16} >, < s_w^8 >, < s_w^4 >, < s_w^2 >\}$ using $< s_w >$ through the isomorphic mapping matrix T and the matrix A in the equation (10).
2 Substitute the results in step 1 into the affine transformation expression $< s_v >=' 77' < s_w^{128} > +... +' 18' < s_w > +'53'$ to get the result $< s_v >\in GF(2^{40})$.
3 Calculate $\{< s_v^{128} >, ..., < s_v^2 >\}$ using $< s_v >$ through the isomorphic mapping matrix T and the matrix A in the equation (10).
4 Calculate $< s_u >=< s_v^{254} >=< s_v^2 > * < s_v^4 > * < s_v^8 > * < s_v^{16} > * < s_v^{32} > * < s_v^{64} > * < s_v^{128} >$.
5 Calculate $\{< s_u^{128} >, ..., < s_u^2 >\}$ using $< s_u >$ through the isomorphic mapping matrix T and the matrix A in the equation (10).
6 Substitute the results in step 5 into the affine transformation expression $< s_z >=' 77' < s_u^{128} > +... +' 18' < s_u > +'53'$ to get the finial result $< s_z >\in GF(2^{40})$.

Algorithm 2: The pseudocode for Sbox of SPDZ protocol

1 are substituted into equation (14) to get the affine transformation result $< s_v >\in GF(2^{40})$, via additions and scalar multiplications. In step 3, the same linear transformation as in step 1 is used to the secret value $< s_v >$ to compute $\{< s_v^{128} >, \cdots, < s_v^2 >\}$. Step 4 computes $< s_u >=< s_v^{254} >\in GF(2^{40})$ using six multiplications. In step 5, the same linear transformation as in step 1 is used again to the secret value $< s_u >$ to compute $\{< s_u^{128} >, \cdots, < s_u^2 >\}$. Step 6 is similar to step 2, which substitutes the results obtained in step 5 into equation (14) to get the finial result $< s_z >\in GF(2^{40})$.

The overall complexity of the scheme is analyzed as follows:

Offline Phase. SPDZ protocol needs to generate 4 random values and 6 triplets for one Sbox. This requires 512 random values and 768 triplets for the full SM4.

Online Phase.

1. Storage:
   (a) Multiplication: The multiplication operation (in step 4) is implemented using the beaver multiplication triple, so it needs to store a triplet data of size 30 bytes for each multiplication operation. Each Sbox has 6 multiplication operations and hence needs to store 6*30 bytes.
   (b) Powers' calculation: Each power needs 40 bits storage, therefore the powers (in step 1, step3 and step 5) have $8 * 2 * 40$ bits $= 80$ bytes (as each value requires a 40-bit MAC additionally).
   So for a single Sbox, the algorithm stores 180+80*3=420 bytes. For the full SM4 algorithm with 32 rounds and 4 S-boxes per round, it stores 52.5 KB.

2. Round Trip:
   (a) Open: In theory, opening one secret $GF(2^{40})$ element requires a round trip communication of 10 bytes.
   (b) Multiplication: In the multiplication operation, it is necessary to partially open two elements obtained by the secret shared value subtracted by one component of beaver multiplication triple when performing multiplication operations, i.e., 2 round trips are required for each multiplication operation, consuming 20 bytes.
   In summary, for a single Sbox, 12 round trips are required. For the full SM4, 1537 round trips are required, as the batch verification mechanism of the SPDZ protocol opens only extra once at the last step when all operations are completed.
3. Communication: The round trips for 6 multiplications require 20*6 bytes. In total, each Sbox requires 120 bytes of communication, for the full SM4, the amount of communication is 15 KB (Table 1).

Table 1. Complexity of the SPDZ module for a single Sbox/the full SM4

	storage(KB)	round trip	communication(KB)
Single Sbox	0.41	12	0.117
Full SM4	52.5	1537	15

4   Conclusion

This paper introduces a distributed key storage scheme for SM4 algorithm based on secure multi-party computation technology to address its key management problem. The scheme utilizes the MP-SPDZ framework to reconstruct the SM4 computation process, and the complexity analysis results indicate that the scheme can guarantee the security in computation while ensuring the efficiency in performance. From the technical standpoint, this paper focuses on describing the technique for reconstructing the nonlinear structure Sbox in the SM4 algorithm. By using bit decomposition technique, specially the algebraic representation expressing element components, the Sbox is decomposed into a combinatorial algorithmic circuit that only includes addition and multiplication operations. Meanwhile, the calculation method of the isomorphic transformation between the SM4-Sbox byte field elements and the SPDZ parameter field elements is also given. From the application standpoint, this scheme extends and explores the application scenarios of secure multi-party computation. Additionally, it provides a feasible scheme for the key management and security implementation problems of block ciphers and other similar cryptographic algorithms.

Acknowledgements. This work is supported by the National Natural Science Foundation of China under Grant No. 61907042 and Beijing Natural Science Foundation under Grant No.4194090.

A   Appendix

Table 2. Algebraic Expression of Sbox

a7	w^{254}	f2	w^{209}	2d	w^{164}	6f	w^{118}	bb	w^{73}	31	w^{28}
6	w^{253}	70	w^{208}	82	w^{163}	39	w^{117}	c5	w^{72}	e3	w^{27}
71	w^{252}	60	w^{207}	b3	w^{162}	db	w^{116}	5a	w^{71}	ae	w^{25}
f7	w^{251}	2e	w^{206}	a8	w^{161}	1e	w^{115}	b9	w^{70}	6b	w^{24}
43	w^{250}	7a	w^{205}	65	w^{160}	fc	w^{114}	96	w^{69}	1c	w^{23}
7b	w^{249}	f3	w^{204}	fe	w^{159}	e4	w^{113}	8d	w^{68}	99	w^{22}
6c	w^{248}	19	w^{203}	b2	w^{158}	50	w^{112}	a	w^{67}	29	w^{21}
a5	w^{247}	c5	w^{202}	a7	w^{157}	b7	w^{111}	de	w^{66}	c0	w^{20}
15	w^{246}	e7	w^{201}	64	w^{156}	8b	w^{110}	de	w^{65}	5d	w^{19}
b6	w^{245}	de	w^{200}	88	w^{155}	1a	w^{109}	78	w^{64}	c7	w^{18}
32	w^{244}	2c	w^{199}	c3	w^{154}	1a	w^{108}	2b	w^{63}	6f	w^{17}
81	w^{243}	22	w^{198}	83	w^{153}	57	w^{107}	c	w^{62}	18	w^{16}
8b	w^{242}	f3	w^{197}	2f	w^{152}	9e	w^{106}	7f	w^{61}	96	w^{15}
32	w^{241}	1d	w^{196}	aa	w^{151}	c4	w^{105}	58	w^{60}	48	w^{14}
7d	w^{240}	68	w^{195}	b2	w^{150}	75	w^{104}	17	w^{59}	4d	w^{13}
c1	w^{239}	2b	w^{194}	f3	w^{149}	37	w^{103}	ba	w^{58}	44	w^{12}
46	w^{238}	a6	w^{193}	95	w^{148}	26	w^{102}	bc	w^{57}	80	w^{11}
7	w^{237}	73	w^{192}	d0	w^{147}	7f	w^{101}	51	w^{56}	de	w^{10}
7b	w^{236}	70	w^{191}	91	w^{146}	7d	w^{100}	c5	w^{55}	9d	w^{9}
5d	w^{235}	99	w^{190}	d5	w^{145}	c2	w^{99}	63	w^{54}	99	w^{8}
dd	w^{234}	72	w^{189}	d1	w^{144}	6e	w^{98}	d2	w^{53}	e4	w^{7}
a4	w^{233}	8f	w^{188}	6b	w^{143}	42	w^{97}	a9	w^{52}	26	w^{6}
b6	w^{232}	89	w^{187}	36	w^{142}	44	w^{96}	5c	w^{51}	d6	w^{5}
7e	w^{231}	f	w^{186}	d6	w^{140}	5d	w^{95}	e	w^{50}	e1	w^{4}
7d	w^{230}	2c	w^{185}	58	w^{139}	db	w^{94}	55	w^{49}	3e	w^{3}
6	w^{229}	7e	w^{184}	59	w^{138}	8b	w^{93}	2d	w^{48}	46	w^{2}
3e	w^{228}	6	w^{183}	13	w^{137}	ff	w^{92}	4f	w^{47}	8e	w
e	w^{227}	a6	w^{182}	54	w^{136}	8e	w^{91}	6a	w^{46}	d6	1
16	w^{226}	6b	w^{181}	be	w^{135}	51	w^{90}	ce	w^{45}		
3d	w^{225}	4f	w^{180}	12	w^{134}	71	w^{89}	fa	w^{44}		
8	w^{224}	c1	w^{179}	a5	w^{133}	4e	w^{88}	5d	w^{43}		
a5	w^{223}	27	w^{178}	c2	w^{132}	16	w^{87}	1d	w^{42}		
2f	w^{222}	c8	w^{177}	5b	w^{131}	5f	w^{86}	21	w^{41}		
76	w^{221}	8c	w^{176}	3b	w^{130}	f6	w^{85}	19	w^{40}		

(continued)

Table 2. (*continued*)

4b	w^{220}	67	w^{175}	81	w^{129}	4	w^{84}	2a	w^{39}
84	w^{219}	61	w^{174}	9f	w^{128}	9c	w^{83}	ea	w^{38}
3f	w^{218}	a9	w^{173}	72	w^{127}	a8	w^{82}	37	w^{37}
3c	w^{217}	ce	w^{172}	6e	w^{126}	a1	w^{81}	9e	w^{36}
86	w^{216}	d3	w^{171}	2e	w^{125}	c3	w^{80}	a6	w^{35}
f1	w^{215}	18	w^{170}	30	w^{124}	97	w^{79}	e3	w^{34}
e4	w^{214}	c	w^{169}	9f	w^{123}	b0	w^{78}	b9	w^{33}
a	w^{213}	52	w^{168}	d0	w^{122}	9c	w^{77}	96	w^{32}
b5	w^{212}	66	w^{167}	94	w^{121}	44	w^{76}	f2	w^{31}
b6	w^{211}	36	w^{166}	53	w^{120}	bc	w^{75}	6	w^{30}
c	w^{210}	91	w^{165}	36	w^{119}	41	w^{74}	c0	w^{29}

References

1. Aaraj, N., Raghunathan, A., Jha, N.K.: Analysis and design of a hardware/software trusted platform module for embedded systems. ACM Trans. Embedded Comput. Syst. (TECS) **8**(1), 1–31 (2009)
2. Ben-Or, M., Goldwasser, S., Wigderson, A.: Completeness theorems for non-cryptographic fault-tolerant distributed computation. In: Providing Sound Foundations for Cryptography: On the Work of Shafi Goldwasser and Silvio Micali, pp. 351–371 (2019)
3. Bogdanov, D., Laur, S., Willemson, J.: Sharemind: a framework for fast privacy-preserving computations. In: Jajodia, S., Lopez, J. (eds.) ESORICS 2008. LNCS, vol. 5283, pp. 192–206. Springer, Heidelberg (2008). https://doi.org/10.1007/978-3-540-88313-5_13
4. Chaum, D., Crépeau, C., Damgard, I.: Multiparty unconditionally secure protocols. In: Proceedings of the Twentieth Annual ACM Symposium on Theory of Computing, pp. 11–19 (1988)
5. Chen, J.: A note on the impossible differential attacks on block cipher SM4. In: 2016 12th International Conference on Computational Intelligence and Security (CIS), pp. 551–554. IEEE (2016)
6. Cramer, R., Damgård, I., Escudero, D., Scholl, P., Xing, C.: SPD$\mathbb{Z}_{2^k}$: efficient MPC mod 2^k for dishonest majority. In: Shacham, H., Boldyreva, A. (eds.) CRYPTO 2018. LNCS, vol. 10992, pp. 769–798. Springer, Cham (2018). https://doi.org/10.1007/978-3-319-96881-0_26
7. Damgård, I., Keller, M., Larraia, E., Miles, C., Smart, N.P.: Implementing AES via an actively/covertly secure dishonest-majority MPC protocol. In: Visconti, I., De Prisco, R. (eds.) SCN 2012. LNCS, vol. 7485, pp. 241–263. Springer, Heidelberg (2012). https://doi.org/10.1007/978-3-642-32928-9_14
8. Damgård, I., Keller, M., Larraia, E., Pastro, V., Scholl, P., Smart, N.P.: Practical covertly secure MPC for dishonest majority – or: breaking the SPDZ limits. In: Crampton, J., Jajodia, S., Mayes, K. (eds.) ESORICS 2013. LNCS, vol. 8134, pp. 1–18. Springer, Heidelberg (2013). https://doi.org/10.1007/978-3-642-40203-6_1

9. Damgård, I., Pastro, V., Smart, N., Zakarias, S.: Multiparty computation from somewhat homomorphic encryption. In: Safavi-Naini, R., Canetti, R. (eds.) CRYPTO 2012. LNCS, vol. 7417, pp. 643–662. Springer, Heidelberg (2012). https://doi.org/10.1007/978-3-642-32009-5_38

10. Durak, F.B., Guajardo, J.: Improving the efficiency of AES protocols in multi-party computation. In: Borisov, N., Diaz, C. (eds.) FC 2021. LNCS, vol. 12674, pp. 229–248. Springer, Heidelberg (2021). https://doi.org/10.1007/978-3-662-64322-8_11

11. Escudero, D., Xing, C., Yuan, C.: More efficient dishonest majority secure computation over Z_{2^k} via Galois rings. In: Advances in Cryptology-CRYPTO 2022: 42nd Annual International Cryptology Conference, CRYPTO 2022, Santa Barbara, CA, USA, August 15–18, 2022, Proceedings, Part I, pp. 383–412. Springer (2022). https://doi.org/10.1007/978-3-031-15802-5_14

12. Goldreich, O., Micali, S., Wigderson, A.: How to play any mental game, or a completeness theorem for protocols with honest majority. In: Providing Sound Foundations for Cryptography: On the Work of Shafi Goldwasser and Silvio Micali, pp. 307–328 (2019)

13. Keller, M.: Mp-spdz: A versatile framework for multi-party computation. In: Proceedings of the 2020 ACM SIGSAC Conference on Computer and Communications Security, pp. 1575–1590 (2020)

14. Keller, M., Orsini, E., Rotaru, D., Scholl, P., Soria-Vazquez, E., Vivek, S.: Faster secure multi-party computation of AES and DES using lookup tables. In: Gollmann, D., Miyaji, A., Kikuchi, H. (eds.) ACNS 2017. LNCS, vol. 10355, pp. 229–249. Springer, Cham (2017). https://doi.org/10.1007/978-3-319-61204-1_12

15. Keller, M., Pastro, V., Rotaru, D.: Overdrive: making SPDZ great again. In: Nielsen, J.B., Rijmen, V. (eds.) EUROCRYPT 2018. LNCS, vol. 10822, pp. 158–189. Springer, Cham (2018). https://doi.org/10.1007/978-3-319-78372-7_6

16. Kocher, P., et al.: Spectre attacks: exploiting speculative execution. Commun. ACM **63**(7), 93–101 (2020)

17. Lidl, R., Niederreiter, H.: Finite fields. No. 20, Cambridge University Press (1997)

18. Liu, F., et al.: Analysis of the SMS4 block cipher. In: Pieprzyk, J., Ghodosi, H., Dawson, E. (eds.) ACISP 2007. LNCS, vol. 4586, pp. 158–170. Springer, Heidelberg (2007). https://doi.org/10.1007/978-3-540-73458-1_13

19. Liu, Y., Liang, H., Wang, W., Wang, M.: New linear cryptanalysis of Chinese commercial block cipher standard SM4. Security and Communication Networks 2017 (2017)

20. Miao, X., Guo, C., Wang, M., Wang, W.: How fast can SM4 be in software? In: Deng, Y., Yung, M. (eds.) Information Security and Cryptology, pp. 3–22. Springer Nature Switzerland, Cham (2023). https://doi.org/10.1007/978-3-031-26553-2_1

21. Micali, S., Goldreich, O., Wigderson, A.: How to play any mental game. In: Proceedings of the Nineteenth ACM Symposium on Theory of Computing, STOC. pp. 218–229. ACM New York, NY, USA (1987)

22. Songhori, E.M., Hussain, S.U., Sadeghi, A.R., Schneider, T., Koushanfar, F.: TinyGarble: highly compressed and scalable sequential garbled circuits. In: 2015 IEEE Symposium on Security and Privacy, pp. 411–428. IEEE (2015)

23. of State Commercial Cipher Administration, O.: SMS4cipher for WLAN products (2006)

24. Tischer, M., et al.: Users really do plug in USB drives they find. In: 2016 IEEE Symposium on Security and Privacy (SP), pp. 306–319. IEEE (2016)

25. Wang, R., Guo, H., Lu, J., Liu, J.: Cryptanalysis of a white-box sm4 implementation based on collision attack. IET Inf. Secur. **16**(1), 18–27 (2022)

26. Wilkins, R., Richardson, B.: UEFI secure boot in modern computer security solutions. In: UEFI forum, pp. 1–10 (2013)
27. Yao, A.C.: Protocols for secure computations. In: 23rd Annual Symposium on Foundations of Computer Science (SFCS 1982), pp. 160–164. IEEE (1982)
28. Zhang, J., Wu, W., Zheng, Y.: Security of SM4 against (Related-Key) differential cryptanalysis. In: Bao, F., Chen, L., Deng, R.H., Wang, G. (eds.) ISPEC 2016. LNCS, vol. 10060, pp. 65–78. Springer, Cham (2016). https://doi.org/10.1007/978-3-319-49151-6_5

DP-ASSGD: Differential Privacy Protection Based on Stochastic Gradient Descent Optimization

Qiang Gao, Han Sun, and Zhifang Wang[✉]

Department of Electronic Engineering, Heilongjiang University, Harbin 150080, China
wangzhifang@hlju.edu.cn

Abstract. Recently, differential privacy algorithms based on deep learning have become increasingly mature. Previous studies provide privacy mostly by adding differential privacy noise to the gradient, but it will reduce the accuracy, and it is difficult to balance privacy and accuracy. In this paper, the DP-ASSGD algorithm is proposed to counterpoise privacy and accuracy. The convergence speed is improved, the number of optimized iterations is decreased, and the privacy loss is significantly reduced. On the other hand, by using the postprocessing immunity characteristics of the differential privacy model, the Laplace smoothing mechanism is added to make the training process more stable and the generalization ability stronger. The experiment uses the MNIST dataset, with the same privacy budget, and compared with the existing differential privacy algorithms, the accuracy is improved by 1.8% on average. When achieving the same accuracy, the DP-ASSGD algorithm consumes less privacy budget.

Keywords: Differential privacy protection · Learning rate adaptation · Laplace smoothing

1 Introduction

With the advent of the big data era, the privacy leakage problem of deep learning is becoming increasingly serious. To solve this problem, researchers have proposed a series of privacy protection methods, such as differential privacy protection, homomorphic encryption, and secure multiparty computation [1]. Among them, differential privacy protection technology [2] provides strict mathematical theoretical support and has become a practical standard for privacy protection. It is of great significance to investigate privacy leakage in deep learning.

In 2016, Abadi et al. proposed the differential privacy stochastic gradient descent (DPSGD) algorithm. Differential privacy is first applied to protect the parameters of neural networks, and a moments accountant is proposed to calculate the privacy cost. It has become the mainstream algorithm in differential privacy protection [3]. DPSGD introduces the Gaussian noise mechanism on the gradient of the neural network, trains the original dataset in batches, and clips the normal form of the gradient [4], which

© The Author(s), under exclusive license to Springer Nature Singapore Pte Ltd. 2023
Z. Yu et al. (Eds.): ICPCSEE 2023, CCIS 1879, pp. 298–308, 2023.
https://doi.org/10.1007/978-981-99-5968-6_21

greatly protects the privacy of the training data. Due to the remarkable effectiveness of DPSGD in privacy protection, a series of algorithms based on DPSGD optimization have been proposed. For example, periodic decay allocation, linear and exponential decay allocation [5] and other methods of dynamically allocating the privacy budget are proposed, which realizes the more accurate allocation of the privacy budget and improves the accuracy of the model. In addition, there are some improvements for the amount of noise added. The hierarchical correlation propagation protocol was used to establish a relationship between the amount of noise added and the output correlation of the model, and adaptive noise addition was realized [6]. These algorithms based on DPSGD optimization effectively solve the privacy leakage of the model training data.

At present, although the optimization algorithm based on DPSGD performs well in privacy disclosure, there are still shortcomings. Differential privacy technology provides different degrees of privacy protection for privacy units, which is generally realized by controlling the amount of noise added [7]. However, blindly introducing noise into the model will affect the accuracy. The more noise that is introduced, the lower the accuracy of the model. And vice versa. Since DPSGD adds noise to the training gradient [8] of the neural network, the accuracy is lower than that of the nondifferential privacy algorithm. Predecessors proposed the DP-LSSGD [9] algorithm to smooth the added noise to improve the convergence of the differential privacy model [2]; the subsequent DP-CD [10] algorithm, because of the larger step size, makes use of the imbalance of gradient coordinates for optimization. Therefore, how to balance the privacy and accuracy of the model has become a research hotspot. Inspired by the large step size of DP-CD, this paper proposes DP-ASSGD to accelerate model convergence, which achieves a balance between privacy and accuracy. Compared with the traditional differential privacy algorithm, the learning rate adaptive feature of DP-ASSGD allows the learning rate to be updated according to the actual situation of the model. On the one hand, it can accelerate the convergence speed of the model and reduce the privacy loss. On the other hand, the training process of the model is more flexible, alleviating the difficulty of parameter adjustment in the differential privacy model [11]. In addition, DP-ASSGD's Laplace smoothing mechanism uses the postprocessing immune feature of differential privacy to Denoise the added Gaussian noise, which greatly ensures the stability of the model training process and obtains stronger generalization ability. The main contributions of this paper can be summarized as follows:

1) The learning rate is updated adaptively, the convergence speed of the model is improved, the number of iterations to optimize the model is reduced, and the loss of privacy is effectively decreased.

2) With the addition of the Laplace smoothing mechanism, the training process of the model is more stable and has better generalization ability.

This paper first introduces the overall flow of the DP-ASSGD algorithm in Sect. 2.1 and then describes the main functions and specific algorithm flow of Learing Rate Adaptation and Laplacian Smoothing Noise in Sects. 2.2 and 2.3, respectively. In Sect. 3.2, the experimental results of DP-ASSGD and existing differential privacy algorithms are compared. Finally, the experimental conclusions are drawn in Sect. 4.

2 Proposed Method

2.1 Algorithm Process

The algorithm flow of DP-ASSGD is shown in Fig. 1. It consists of five parts. Among them, the purpose of gradient clipping is to alleviate gradient explosion and gradient disappearance in the process of updating the gradient. However, clipping is generally performed after all the gradients are calculated, which can improve the problem of poor convergence. Common gradient clipping norms are L_0, L_1 and L_2. The norms L_0 and L_1 are mainly used to realize the sparsity of parameters, and the wide application of L_1 is wider than that of L_0 because of its better optimization solution characteristics. The L_2 norm can limit the range of the solution space, prevent overfitting, and improve the generalization ability of the model. Therefore, in DP-ASSGD, the L_2 norm is used to clip the gradient vector. In DP-ASSGD, each gradient vector is pruned by the L_2 norm. When the gradient is less than the clipping threshold, that is, $\|g\|_2 \leq C$, the gradient is retained; when $\|g\|_2 > C$, the gradient will be replaced by the clipping threshold C. By setting the clipping threshold to constrain the gradient vector, the influence of each sample on the model parameters is controlled, and the sensitivity of the algorithm is controlled. DP-ASSGD has the characteristic of an adaptive learning rate, which can reduce the iterations of model optimization and decrease the loss of privacy. Noise addition adds random noise to the gradient value. It ensures that the attacker cannot determine whether certain data are in the training dataset, according to the final output of the model. In this paper, Gaussian noise $N\left(0, C^2\sigma^2 I\right)$ [12] is added, and the noise distribution follows the Gaussian distribution with mean 0 and standard deviation $C \cdot \sigma$. I is the number of units whose dimension is related to the number of samples and the number of gradients. The size of the noise is determined by the gradient clipping threshold C and noise parameter σ. The Laplace noise smoothing mechanism [13] in DP-ASSGD is a denoising technique for postprocessing Gaussian noise in gradients, which makes the training process of the model more stable.

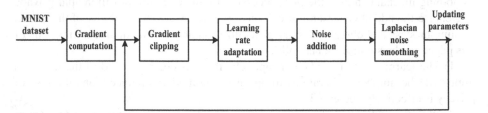

Fig. 1. Flowchart of the DP-ASSGD algorithm

2.2 Learning Rate Adaptation

DP-ASSGD is based on the gradient descent algorithm. In the process of gradient descent, the choice of learning rate is the main factor that affects the convergence and iteration number of the algorithm. If the learning rate is set too high, the weight update range will be too large, and the model is difficult to converge; when the learning rate is set too small, the parameter update speed is slow, resulting in the model not finding the optimal value quickly, the number of iterations increases, and the privacy burden increases. Therefore, when the learning rate cannot be updated adaptively according to the actual situation, it will consume more privacy loss, which is harmful to balance privacy and accuracy.

The main motivation for learning rate adaptation comes from the numerical solution of ordinary differential equations: the Euler method [14]. Let g be a differentiable function, η_t be the step size, θ be the initialization parameter, and the gradient descent in this case is

$$\theta_{t+1} \leftarrow \theta_t - \eta_t \nabla g_t \tag{1}$$

It is a first-order method for finding the minimum of a function g, which can be viewed as an Euler method [14] with order length η_t. It is suitable for the system of ordinary differential equations $ODEs\theta\prime(t) = -\nabla g(\theta)$, $\theta(0) = \theta_0$, also known as the gradient flow corresponding to g.

To estimate the error in (1), the following extrapolation is made. An Euler step of size η is applied to the gradient flow:

$$\theta_1 \leftarrow \theta_0 - \eta \nabla g(\theta_0) \tag{2}$$

Use the improved Euler formula to calculate, according to its calculation formula:

$$\bar{y}_{n+1} = y_n + hf(x_n, y_n), \quad y_{n+1} = y_n + \frac{h}{2}\left(f(x_n, y_n) + f(x_{n+1}, \bar{y}_{n+1})\right)$$

The parameters θ are the result of two steps with a step size of $\frac{\eta}{2}$: $\theta_{1/2} \leftarrow \theta_0 - \frac{\eta}{2}\nabla g(\theta_0)$, $\widehat{\theta_1} \leftarrow \theta_{1/2} - \frac{\eta}{2}\nabla g(\theta_{1/2})$.

Using Taylor's expansion, $\theta(\eta) - \theta_1 = \frac{\eta^2}{2}J_g(\theta_0)\nabla g(\theta_0) + O(\eta^3)$ is obtained, and $2\left(\widehat{\theta_1} - \theta_1\right)$ gives the local error $O(\eta^3)$ generated in (1). If iterated t times, the error estimate is as follows:

$$err_t \leftarrow \left\| err\left(\theta_{t+1}, \widehat{\theta_{t+1}}\right) \right\|_2 \tag{3}$$

If a local error of size tol is needed, the simple mechanism for updating the step size is as follows:

$$\eta_{t+1} \leftarrow \min\left(\max\left(\frac{tol}{err_t}, \alpha_{\min}\right), \alpha_{\max}\right) \cdot \eta_t \tag{4}$$

Applying η_t to DPSGD, the following can be obtained: $\theta_T = \theta_0 - \sum_{t=0}^{T-1} \eta_t g(\theta_t) + N\left(0, \sum_{t=0}^{T-1} \eta_t^2 C^2\sigma^2 I\right)$

where $g(\theta_t) = \sum_{i \in L} \nabla_{\theta_t} L(\theta_t, x_i)$.

It holds that $\left\| \theta_{t+1} - \widehat{\theta_{t+1}} \right\|_2 = \frac{\eta_t}{2}\left\| g(\theta_t) - g\left(\widehat{\theta_t}\right) + N\left(0, \sum_{t=0}^{T-1} \eta_t^2 2C^2\sigma^2 I\right)\right\|$,

assuming $\sqrt{d} \gg |L|$ ($\theta \in R^d$), approximately obtaining $\left\| \theta_{t+1} - \widehat{\theta_{t+1}} \right\| \approx \eta_t \sigma C \sqrt{\frac{d}{2}}$. If this estimate is set to tol, $\eta_t^2 = \frac{2tol^2}{\sigma^2 C^2 d}$ is obtained, substituted into the above formula and each element is obtained as $\frac{\sqrt{2T}\,tol}{\sqrt{d}}$. For example, if the noise is required to be $O(1)$, then $\frac{2T}{d} = O(1)$, and the appropriate value of tol is 1.

where $\alpha_{\min} < 1$ and $\alpha_{\max} > 1$. After testing, it is found that $\alpha_{\min} = 0.9, \alpha_{\max} = 1.1$ consumes the least privacy budget. Therefore, $\alpha_{\min} = 0.9$, and $\alpha_{\max} = 1.1$ is chosen in this paper.

The pseudocode form of applying the adaptive learning rate to the differential privacy mechanism is depicted in Algorithm 1. In DP-ASSGD, the 2-norm (4) of the function is used as the error estimate instead of (3):

$$err\left(\theta, \hat{\theta}\right)_i = \frac{\left|\theta_i - \hat{\theta}_i\right|}{\max(1, |\theta_i|)} \tag{5}$$

Because (4) performs better numerically, in the case of differential privacy learning, the algorithm is stable without X, so this paper omits it.

Algorithm 1: learing rate adaptation

Input: Sample $\{x_1, \ldots, x_N\}$; Batch size L_1 , L_2 ; The gradient pruning norm C ; Gradient g ; Learning rate η ; Initializing parameters θ ; Number of iterations T ; Loss function $L(\theta) = \frac{1}{N}\sum_i L(\theta, x_i)$

For $t \in [T]$ do

Gradient calculation: $g_1 \leftarrow \sum_{i \in L_1} \nabla_{\theta_t} L(\theta_t, x_i) + N(0, C^2\sigma^2 I)$

Update the parameters with η_t : $\theta_{t+1} \leftarrow \theta_t - \eta_t g_1$

$\frac{\eta_t}{2}$ Updating parameters : $\theta_{t+1/2} \leftarrow \theta_t - \frac{\eta_t}{2} g_1$

Gradient computation : $g_2 \leftarrow \sum_{i \in L_2} \nabla_{\theta_{t+1/2}} L(\theta_{t+1/2}, x_i) + N(0, C^2\sigma^2 I)$

Update the parameters with $\frac{\eta_t}{2}$: $\theta_{t+1} \leftarrow \theta_{t+1/2} - \frac{\eta_t}{2} g_2$

Error estimation: $err_t \leftarrow \left\| err\left(\theta_{t+1}, \theta_{t+1}\right) \right\|_2$

Update the learning rate: $\eta_{t+1} \leftarrow \min\left(\max\left(\frac{tol}{err_t}, \alpha_{\min} \right), \alpha_{\max} \right) \cdot \eta_t$

Output: η_{t+1}

2.3 Laplacian Smoothing Noise

Laplace smoothing (LS) is a denoising technique for postprocessing the random gradient of injected Gaussian noise. In DP-ASSGD, the error of stochastic gradient descent can be dynamically reduced, which makes the training process of the model more stable and has better generalization ability.

The loss function of DP-ASSGD is given by:

$$L(\theta) = \frac{1}{N} \sum_i L(\theta, x_i) \tag{6}$$

The goal of this paper is to find a tool that minimizes the loss function for the LS of (5), solved by the following formula:

$$\tilde{g}_t \leftarrow \frac{1}{L} \sum_i \left(g_t(x_i) + N\left(0, C^2\sigma^2 I\right) \right)$$

$$\theta_{t+1} = \theta_t - \eta_{t+1} A_\sigma^{-1} \tilde{g}_t \tag{7}$$

Let $A_\sigma = I - \sigma L$ and σ be constants greater than 0, where I and L are the identity and discretization of a one-dimensional Laplacian matrix with periodic boundary conditions, respectively. Therefore,

$$A_\sigma = \begin{bmatrix} 1+2\sigma & -\sigma & 0 & \cdots & 0 & -\sigma \\ -\sigma & 1+2\sigma & -\sigma & \cdots & 0 & 0 \\ 0 & -\sigma & 1+2\sigma & \cdots & 0 & 0 \\ \cdots & \cdots & \cdots & \cdots & \cdots & \cdots \\ -\sigma & 0 & 0 & \cdots & -\sigma & 1+2\sigma \end{bmatrix} \tag{8}$$

A_σ is positive definite under the condition $1+4\sigma$ and is independent of the dimension of A_σ. Let $u = A_\sigma^{-1} v$ be $v = A_\sigma u$, where A_σ is the convolution matrix, so $v = A_\sigma u = u - \sigma d * u, d = [-2, 1, 0, \cdots, 0, 1]^T, *$ is the convolution operator. Using fast Fourier transform (FFT) [15], we obtain:

$$A_\sigma^{-1} v = u = ifft(fft(v)/(1 - \sigma \cdot fft(d))) \tag{9}$$

Matrix A_σ^{-1} can remove the noise in v. Assuming that v is the original signal, $A_\sigma^{-1} v$ can be considered an approximate dissipation step of the original noise signal, thereby removing the noise in v. The pseudocode that applies the Laplace smoothing mechanism to differential privacy is shown in algorithm 2.

Algorithm 2: laplacian smoothing noise

Input: Privacy budget (ε,σ); Learning rate η_{t+1}; Number of iterations T

For $t \in [T]$ do

Noise addition: $g_t \leftarrow \frac{1}{L}\sum_i \left(g_t(x_i) + N\left(0, C^2\sigma^2 I\right)\right)$

Laplacian noise smoothing: $\theta_{t+1} = \theta_t - \eta_{t+1}A_\sigma^{-1}g_t$

Output: θ_t; (ε,σ)

2.4 Privacy Preserving Properties of the Method

Differential privacy is defined as follows: the condition of differential privacy is that if the random algorithm $M : D \rightarrow R$ satisfies $(\varepsilon, \delta) - DP$, if and only if the adjacent datasets d and $d\prime$, all possible output subsets $S \in R$ of algorithm M satisfy inequalities:

$$Pr[M(d) \in S] \leq e^\varepsilon Pr[M(d\prime) \in S] + \delta \tag{10}$$

Differential privacy has the postprocessing immunity property. For privacy algorithm M, let $M:\mathbb{N}^{|x|} \rightarrow R$, a randomized algorithm satisfies (ε, δ) differential privacy, and let $f: R \rightarrow R\prime$ be an arbitrary random mapping; then, $f \circ M :\mathbb{N}^{|x|} \rightarrow R\prime$ is (ε, δ) differential private.

According to the differential privacy definition, the following corollary can be obtained: Given the function q: $S^n \rightarrow R$, the Gaussian mechanism M $= q(s) + N\left(\mu, \sigma^2 C^2 I\right)$ satisfies RDP differential privacy. At this time, if we apply the mechanism to the sample subset sampled uniformly without replacement, the mechanism still satisfies RDP differential privacy. Among them, RDP differential privacy is the classical differential privacy algorithm when α tends to infinity, which further broadens the definition of differential privacy.

According to the proposed algorithm and the postprocessing immune properties of differential privacy, let $q = |L|/N, \sigma \geq 1$ and $C > 0, \alpha_M(\lambda)$ is the moments accountant, and $\widetilde{M}$ denotes the mechanism applying the proposed algorithm, which can be obtained as follows:

$$\alpha_{\widetilde{M}}(\lambda) \leq 2\alpha_M(\lambda) \tag{11}$$

Using the same parameter values, under the same privacy protection,
The algorithm in this paper runs half the number as the DPSGD algorithm.
After k iterations at given points θ^{k+1} and $\widehat{\theta}^{k+1}$, the gradient of the parameters is
$G_k = \frac{1}{m}\nabla_\theta \sum_i^m L(f(x_i, \theta), y_i), \widetilde{G}_k = \frac{1}{m}\nabla_\theta \sum_i^m L(f(x_i, \widetilde{\theta}), y_i)$. The updates of DPSGD

and DPLSSGD are expressed as follows:

$$\theta^{k+1} = \theta^k - \eta_k \left(G_k + N\left(\mu, \sigma^2 C^2 I\right)\right)$$
$$\widetilde{\theta}^{k+1} = \widetilde{\theta}^k - \eta_k A_\sigma^{-1}\left(\widetilde{G}_k + N\left(\mu, \sigma^2 C^2 I\right)\right)$$

(12)

$N\left(\mu, \sigma^2 C^2 I\right)$ is the added noise. According to the postprocessing immunity property and corollary, DPSGD satisfies privacy, and the algorithm in this paper also satisfies privacy.

3 Experiments

3.1 Dataset and Experimental Environment Implementation Details

In the field of differential privacy protection, the MNIST dataset is often used to verify the performance of the algorithm. To ensure the impartiality of the experiment, the same dataset is chosen in this paper. There are four files in the MNIST dataset: the training dataset, the training dataset label, the test dataset, and the test dataset label. The MNIST dataset contains 60000 training data points and 10000 test data points, and each image is a $0 \sim 9$ handwritten digit image with $28 \times 28 \times 1$ pixels.

DP-ASSGD is implemented on Python3.6, Pytorch1.5.0 and Nvidia 2080Ti Gpus. During the experiment, the training times are set to 2000, the training batch samples L is 100, the gradient pruning threshold C is 1, the noise parameter σ is 4, the initial learning rate η is 0.01, and the privacy parameter δ is $1e - 5$.

3.2 Experimental Results

The evaluation index of this experiment mainly uses the privacy budget parameter ε and the prediction accuracy of the model.

Privacy budget ε represents a parameter used to control the difference between two adjacent datasets. If the value of privacy budget ε is smaller, the similarity between two adjacent datasets is higher, and it is difficult for an attacker to distinguish which dataset a record exists in. Therefore, the smaller the privacy budget ε is, the stronger the privacy protection provided. In traditional database protection, when $\varepsilon \in (0, 1)$ thinks that privacy strength is effective, but in the field of deep learning, $\varepsilon \in (0, 10)$ thinks that it is acceptable. This paper mainly studies the accuracy of the model under high privacy protection, so the privacy budget ε is selected as 0.30, 0.40, 0.50, 0.60, and 0.70. To prevent the situation in which the privacy budget is too small, the model is not trained. Alternatively, the privacy budget is too large, and the privacy of the model cannot be guaranteed. Table 1 compares the experimental results of DP-ASSGD with those of DP-SGD, DP-LSSGD and DP-CD on the MNIST dataset. First, the utility of the algorithm is analysed. Compared with DP-SGD, the accuracy of DP-ASSGD is improved by 8.6% on average. Compared with DP-LSSGD, the average increase is 5.6%. Compared with DP-CD, the average improvement is 1.8%. Second, when the privacy budget ε is 0.30 and 0.40, the accuracy improved by DP-ASSGD is higher than the average accuracy improved, indicating that DP-ASSGD better balances the privacy and accuracy of the model. Because DP-ASSGD has the learning rate adaptive feature, the learning rate is updated according to the actual situation of the model, and the optimal value of accuracy can be found with a smaller privacy budget. Then, this paper selects the accuracy of the four algorithms within 300 iterations and draws Fig. 2. It can be intuitively seen that the accuracy of DP-ASSGD has a significant improvement compared with the other three algorithms.

Table 1. Accuracy of the algorithm under different privacy budgets

Privacy budget ε	0.30	0.40	0.50	0.60	0.70
DP-SGD	71.39 ± 0.30	75.62 ± 0.11	79.33 ± 0.12	81.42 ± 0.13	82.63 ± 0.10
DP-LSSGD	74.29 ± 0.49	78.44 ± 0.11	82.16 ± 0.10	84.70 ± 0.09	85.98 ± 0.12
DP-CD	81.73 ± 0.17	84.67 ± 0.11	85.88 ± 0.11	86.99 ± 0.04	87.51 ± 0.10
DP-ASSGD	**86.26 ± 0.07**	**86.36 ± 0.10**	**86.59 ± 0.09**	**86.81 ± 0.10**	**87.62 ± 0.05**

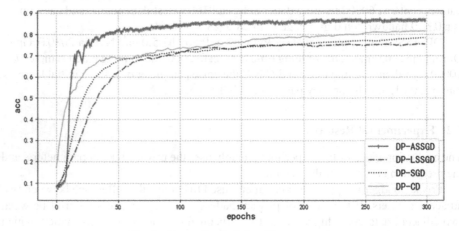

Fig. 2. Comparison of algorithm accuracy

Next, the privacy of the algorithms is compared. Figure 3 compares the privacy budget consumption of the algorithm within 300 iterations. In the field of differential privacy, researchers strive to find the optimal parameters of the model with the smallest privacy budget. The less the privacy budget is consumed, the more private the model is. Among them, the privacy budget consumed by DP-SGD is significantly higher than that of the other algorithms. As the iterations increase, the privacy budget consumed by DP-CD is also gradually higher than DP-ASSGD. Because of the Laplace noise smoothing mechanism in DP-ASSGD, the training process of the model is more stable, and the extra privacy budget is avoided.

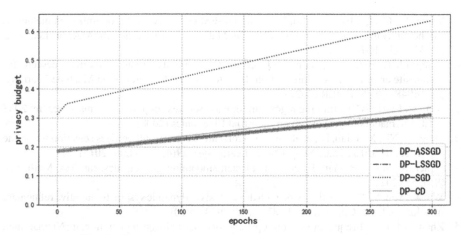

Fig. 3. Comparison of algorithm privacy budget consumption

4 Conclusion

In this paper, the DP-ASSGD algorithm is proposed to balance the privacy and accuracy of the differential privacy model. The DP-ASSGD algorithm has the characteristics of learning rate adaptation and Laplacian smoothing denoising, which effectively improves the accuracy of the model, reduces the consumption of the privacy budget, and makes the model have better generalization ability. On the MNIST dataset, compared with the existing differential privacy algorithms, the DP-ASSGD algorithm improves the accuracy by 1.8% on average. At the same accuracy, the DP-ASSGD algorithm consumes less privacy budget.

References

1. Li, L., Qin, Q., Hua, L., et al.: Data fusion algorithm of privacy protection based on QoS and multilayers hierarchically. Int. J. Distrib. Sens. Netw. **9**(12), 926038 (2013)
2. Dwork, C.: Differential privacy. In: Bugliesi, M., Preneel, B., Sassone, V., Wegener, I. (eds.) ICALP 2006. LNCS, vol. 4052, pp. 1–12. Springer, Heidelberg (2006). https://doi.org/10.1007/11787006_1
3. Phan, N.H., Wu, X., Hu, H., et al.: Adaptive laplace mechanism: Differential privacy preservation in deep learning. In: 2017 IEEE International Conference on Data Mining (ICDM), pp. 385–394. IEEE (2017)
4. Asi, H., Duchi, J., Fallah, A., et al.: Private adaptive gradient methods for convex optimization (2021). https://doi.org/10.48550/arXiv.2106.13756
5. Yu, L., Liu, L., Pu, C., et al.: Differentially private model publishing for deep learning. In: 2019 IEEE Symposium on Security and Privacy (SP), pp. 332–349. IEEE (2019)
6. Chen, Y., Gu, H.H., Perl, Y., et al.: Structural group-based auditing of missing hierarchical relationships in UMLS. J. Biomed. Inform. **42**(3), 452–467 (2009)
7. Amian, M.: Improving the algorithm of deep learning with differential privacy. arXiv preprint arXiv:2107.05457 (2021)
8. Andrew, G., Thakkar, O., McMahan, B., et al.: Differentially private learning with adaptive clipping. Adv. Neural. Inf. Process. Syst. **34**, 17455–17466 (2021)

9. Wang, B., Gu, Q., Boedihardjo, M., et al.: DP-LSSGD: a stochastic optimization method to lift the utility in privacy-preserving ERM. In: Mathematical and Scientific Machine Learning, pp. 328–351. PMLR (2020)

10. Mangold, P., Bellet, A., Salmon, J., et al.: Differentially private coordinate descent for composite empirical risk minimization. In: International Conference on Machine Learning, pp. 14948–14978. PMLR (2022)

11. Xia, X., Li, K.: A fast training algorithm for least-squares support vector machines. In: 2008 IEEE International Symposium on Industrial Electronics (2008)

12. Ding, S., Nie, X., Qiao, H., et al.: A fast algorithm of convex hull vertices selection for online classification. IEEE Trans Neural Netw. Learn. Syst. **29**(4), 792–806 (2017)

13. Osher, S., Wang, B., Yin, P., et al.: Laplacian smoothing gradient descent. Res. Math. Sci. **9**(3), 55 (2022)

14. Wu, X., You, X.: Extended version with the analysis of dynamic system for iterative refinement of solution. Int. J. Comput. Math. **87**(4), 920–934 (2010)

15. Zhang, M., et al.: Frequency diffeomorphisms for efficient image registration. In: Niethammer, M., Styner, M., Aylward, S., Zhu, H., Oguz, I., Yap, P.-T., Shen, D. (eds.) Information Processing in Medical Imaging, pp. 559–570. Springer International Publishing, Cham (2017). https://doi.org/10.1007/978-3-319-59050-9_44

Study on Tourism Workers' Intercultural Communication Competence

Dan Xian[✉]

Sanya Aviation and Tourism College, Hainan 572000, China
25297641@qq.com

Abstract. Under the background of the development of Hainan tourism and consumption destinations to increase the reception level of international tourism in Hainan, an empirical study is carried out to investigate the intercultural communication competence of tourism workers in Hainan from cognitive, operational and affective dimensions. In this study, the cognitive dimension is reflected by the knowledge of domestic and foreign culture; the operational dimension focuses on verbal and nonverbal communication abilities; and the affective dimension is represented by emotional level, which mainly focuses on the participants' ability to understand and appreciate another culture as well as their willingness to interact with people from other cultures. The results indicate that tourism workers' knowledge of domestic culture is the best, followed by their emotional level, which are all above average. Next are nonverbal communication ability and knowledge of foreign culture, both of which are below average. Their verbal communication ability is the lowest among the five factors. The intercultural communication competence of Hainan tourism workers is not satisfactory. The analytical results also show that each factor among the three dimensions is positively correlated with each other.

Keywords: Intercultural Communication Competence · Tourism Workers · Hainan · Empirical Study

1 Introduction

Since 2018, China has rolled out a plan for building its southern island province of Hainan into a pilot free trade zone [1]. China's Hainan Province will have achieved the goal of becoming an international tourism and consumption destination by 2025 according to its development plan for the 14th Five-Year period (2021–25) [2]. By 2025, tourists' consumption potential will be further unleashed with the expansion of the high-end tourism market. In addition, Hainan is the only province in the Chinese mainland that offers one-way visa-free entry to 59 countries, including major tourist sources such as the United States, Russia, Britain and France. These measures will enable more people overseas to travel, stay and work in Hainan [3], and Hainan will become a world-renowned paradise for tourists. However, with the continuous opening-up of the inbound tourism market in the future, challenges have been posed for Hainan's international

Z. Yu et al. (Eds.): ICPCSEE 2023, CCIS 1879, pp. 309–328, 2023.
https://doi.org/10.1007/978-981-99-5968-6_22

tourism reception and the intercultural communication competence of tourism workers who are playing a very important role in providing considerate service for international tourists.

Intercultural communication refers to communication between people of different cultural backgrounds [4]. Since the concept of communication competence was put forward by D.H. Hymes in 1972 [5], scholars at home and abroad have performed a great deal of research, in which scholars mainly focus on studying its connotation and application. Many scholars, such as Byram, Lustig and Koester (1999) [6], try to define and distinguish intercultural communication competence from intercultural competence. Intercultural communication competence is usually defined as the knowledge, motivation, and skills to interact effectively and appropriately with members of a host culture [7]. Chen and Starosta (1996) emphasize that intercultural communication competence is the ability of communicators to communicate appropriately to ensure their multiple identities in a given context [8]. Byram (1997) believes that intercultural competence consists of knowledge, attitudes, skills, and critical cultural awareness, in addition to linguistic, sociolinguistic and discursive skills [9]. Kim (1991) combines knowledge of sociological psychology, applied linguistics and sociology and recognizes intercultural communication competence in the cognitive dimension, affective dimension, and operational dimension. The cognitive dimension refers to discerning meaning, the affective dimension refers to emotions involved with a willingness to accommodate different cultural ways, and the operational dimension is mainly about behavioural flexibility and resourcefulness in intercultural interaction [10]. Despite different definitions, scholars generally believe that intercultural communication competence can be analysed from the cognitive, affective and operational dimensions of participants during intercultural communication [11].

Chinese scholar Daokuan He introduced intercultural communication as a course to China in 1983. In the early stage of the development of intercultural communication in China, domestic studies emphasized language teaching, seldom involving psychology [12]. In recent years, Chinese scholar Yanfang Liu (2014) conducted an empirical study on the intercultural communication competence of tourism workers in Hubei [13]. Jing Zhang (2021) developed a scale for college students in western regions in China [14]. Qian Li and Xuan Zheng (2021) analysed the effect of cross-culture simulation games in cultivating college students' intercultural communication competence and provide suggestions on the effectiveness of the games in promoting the students' intercultural communication competence [15]. Jing Yuan (2022) developed an intercultural communication competence scale for non-English majors to improve students' intercultural communication competence through the college English teaching process [16].

This paper mainly focuses on the intercultural communication between international tourists and tourism workers in Hainan and studies the current situation of the intercultural communication competence of Hainan tourism workers.

2 Research Design

2.1 Variables

Based on Kim's theory, Chinese scholar Yanfang Liu (2014) defined the intercultural communication competence of tourism workers as an inherent ability that enables tourism workers to appropriately and effectively communicate with tourists of foreign culture in intercultural communication. It mainly consists of verbal and nonverbal communication abilities, which are the operational dimension of intercultural communication competence. Verbal communication ability refers to listening, speaking, vocabulary, grammar, etc.; nonverbal communication ability, which includes every way of communication except verbal communication, such as body language, use of strategies, etc.; and the cognitive dimension of intercultural communication competence is cultural awareness, which is the mastery and comprehension of cultural connotation and environment, such as values, social rules, way of life, etc.; emotional level, which mainly refers to sympathy, is considered the affective dimension of intercultural communication competence [13].

2.2 Questionnaire Design and Data Statistical Method

Mainly based on the current situation of the intercultural communication competence of tourism workers in Hainan, a sampling questionnaire survey is conducted on tourism workers in Hainan Province, China, and SPSS 26 is used to analyse the statistics.

The questionnaire is divided into two parts. The first part is an investigation of the demographic information of the questionnaire participants, including gender, age, workplace, posts, years of service, educational background, intercultural communication competence training and foreign languages they speak at work. The second part includes 21 items that investigate the current situation of intercultural communication competence of the participants from cognitive, operational and affective dimensions. In the cognitive dimension, the study mainly focuses on the cultural awareness of tourism workers, which mainly includes knowledge of their own culture and foreign culture. In the operational dimension, it mainly focuses on the verbal and nonverbal communication abilities of tourism workers. In the affective dimension, it mainly concentrates on their ability to understand and appreciate another culture, as well as their willingness to interact with people from other cultures. A Likert scale is adopted in this part, and the format of each item is as follows: "never", "seldom", "sometimes", "usually", and "always", with each option scoring 1 point, 2 points, 3 points, 4 points, and 5 points, respectively, and with 3 points being the critical value. A high score represents a high level of intercultural communication competence.

2.3 Research Participants

The research is designed to study the intercultural communication competence (hereafter referred to as ICC in tables) of tourism workers in Hainan. They are mainly from different sectors of tourism industries in Hainan from travel agencies, tourist attractions, hotels, yacht services and other tourism-related service sectors. A total of 335 tourism workers participated in the survey, and 309 questionnaires were valid. Among the 309 participants, 162 were females, accounting for 52.43%. The participants come from different positions from front-line posts to management with different educational backgrounds (Table 1).

Table 1. Analysis of demographic characteristics

Variables	Attribute	Frequency	Percentage
Gender	Male	147	47.57
	Female	162	52.43
Age	19–24	48	15.53
	25–35	145	46.93
	36–45	100	32.36
	46–55	15	4.85
	Above 55	1	0.32
Workplace	Tourist attractions	163	52.75
	Hotel	94	30.42
	Travel agency	24	7.77
	Yacht company	9	2.91
	Others	19	6.15
Post	Front line	172	55.66
	Management	77	24.92
	Administration	16	5.18
	Marketing & Advertising	29	9.39
	Others	15	4.85
Years of service	5 or below 5	115	37.22
	6–10	103	33.33
	11–20	81	26.21
	Above 21	10	3.24
Education level	Junior school or below	47	15.21
	High school/Vocational secondary school	130	42.07
	Senior college	73	23.62
	Undergraduate	36	11.65
	Postgraduate	23	7.44
ICC training	Yes	106	34.30
	No	203	65.70

(*continued*)

Table 1. (*continued*)

Variables	Attribute	Frequency	Percentage
Foreign languages	Never speak	69	22.33
	English	185	59.87
	Russian	11	3.56
	Korean	3	0.97
	Two or more foreign languages	41	13.27
Total		309	100.00

3 Empirical Research Procedure

3.1 Reliability and Validity Test

Reliability Test. Cronbach's α coefficient is used here as an index to test the reliability. SPSS 26 is used to analyse the statistics of the survey, and Table 2 reports the reliability statistics. As shown in Table 2, the Cronbach's α coefficient of the whole questionnaire is 0.953, which is greater than 0.6, indicating that the reliability of the questionnaire is very good.

Table 2. Reliability statistics

	Cronbach's Alpha	Case number	Number of items
ICC	0.953	309	21

Validity Test. Construct validity is used here as an evaluation index to test the validity of the questionnaire. As shown in Table 3, the KMO value is 0.939; the chi-square value of Bartlett's test is 6099.197; the degree of freedom is 210; and its significance level is approximately 0, suggesting that this sample has good validity.

Table 3. KMO and Bartlett's tests

Sampling sufficient KMO measures		0.939
Bartlett's test	Approximate chi-square	6099.197
	df	210
	Sig	0.000

3.2 Descriptive Statistics

Descriptive statistics are used to analyse the current situation of tourism workers' intercultural communication competence. The mean value and standard deviation can be seen in Table 4, which can be used to evaluate tourism workers' intercultural communication competence. A higher mean value represents better competence. As shown in Table 4, the mean value of the tourism workers' intercultural communication competence is 3.055, which shows that their intercultural communication competence is not very good. The mean values of domestic culture and emotional level are above average, with mean values of 3.759 and 3.112, respectively. Their knowledge of foreign culture, verbal communication ability and nonverbal communication ability are below average, with mean values of 2.724, 2.508, and 2.806, respectively.

Table 4. Descriptive analysis

	Case number	Minimum value	Maximum value	Mean value	Standard deviation
ICC	309	1.000	5.000	3.055	0.879
Domestic culture	309	1.000	5.000	3.759	0.920
Foreign culture	309	1.000	5.000	2.724	1.162
Verbal communication ability	309	1.000	5.000	2.508	1.177
Nonverbal communication ability	309	1.000	5.000	2.806	1.091
Emotional level	309	1.000	5.000	3.112	0.942

3.3 Tourism Workers Test

Tourism workers' tests are conducted to test whether there is a significant difference among participants of different genders, ages, workplaces, positions, years of service, intercultural communication competence training, education levels and spoken foreign languages.

Gender. As shown in Table 5, there is no significant difference between male participants and female participants in knowledge of domestic culture ($t = 0.289$, $p > 0.05$) and foreign culture ($t = 0.444$, $p > 0.05$), verbal communication ability ($t = 0.513$, $p > 0.05$), nonverbal communication ability ($t = -0.048$, $p > 0.05$) and emotional level ($t = -1.151$, $p > 0.05$).

Table 5. Test on gender

	Gender	Case number	Mean value	Standard deviation	t	df	p
Domestic culture	Male	147	3.78	1.03	0.289	275.838	0.773
	Female	162	3.74	0.81			
	Total	309	3.76	0.92			
Foreign culture	Male	147	2.76	1.20	0.444	307.000	0.658
	Female	162	2.70	1.13			
	Total	309	2.72	1.16			
Verbal communication ability	Male	147	2.54	1.24	0.513	307.000	0.608
	Female	162	2.48	1.12			
	Total	309	2.51	1.18			
Nonverbal communication ability	Male	147	2.80	1.15	-0.048	307.000	0.962
	Female	162	2.81	1.03			
	Total	309	2.81	1.09			
Emotional level	Male	147	3.35	1.08	-1.151	307.000	0.250
	Female	162	3.48	1.02			
	Total	309	3.42	1.05			

Age. As shown in Table 6, there is a significant difference in knowledge of domestic culture ($p < 0.05$). Basically, the mean value of domestic culture is higher as participants get older except for two age groups: 25–34 and above 55. The mean value of the 25–34 age group is lower than that of the 19–24 age group; there is only one participant who is in the above 55 age group, and the statistics of the age group may not be valid. No significant difference exists in the knowledge of foreign culture, verbal communication ability, nonverbal communication ability and emotional level for participants from different age groups.

Table 6. Test on age

	Age	Case number	Mean value	Standard deviation	F	p
Domestic culture	19–24	48	3.65	0.97	2.498	0.043*
	25–35	145	3.63	0.95		
	36–45	100	3.96	0.84		
	46–55	15	4.00	0.69		
	Above 55	1	4.33	null		
	Total	309	3.76	0.92		

(*continued*)

Table 6. (*continued*)

	Age	Case number	Mean value	Standard deviation	F	p
Foreign culture	19–24	48	2.96	1.14	1.395	0.236
	25–35	145	2.58	1.17		
	36–45	100	2.82	1.13		
	46–55	15	2.80	1.29		
	Above 55	1	1.80	null		
	Total	309	2.72	1.16		
Verbal communication ability	19–24	48	2.87	1.24	2.306	0.058
	25–35	145	2.50	1.19		
	36–45	100	2.45	1.11		
	46–55	15	1.96	1.06		
	Above 55	1	1.33	null		
	Total	309	2.51	1.18		
Nonverbal communication ability	19–24	48	3.10	1.03	1.085	0.364
	25–35	145	2.75	1.13		
	36–45	100	2.74	1.05		
	46–55	15	2.80	1.13		
	Above 55	1	3.00	null		
	Total	309	2.81	1.09		
Emotional level	19–24	48	3.65	0.96	1.042	0.386
	25–35	145	3.32	1.11		
	36–45	100	3.43	1.02		
	46–55	15	3.59	0.94		
	Above 55	1	3.00	null		
	Total	309	3.42	1.05		

Workplace. As shown in Table 7, a significant difference exists in foreign culture ($p < 0.05$). The knowledge of foreign culture of participants from travel agencies is significantly higher than that of the rest of the participants, with a mean value of 3.42, successively followed by participants from other tourism-related service sectors, tourist attractions, hotels and yacht services, with mean values of 2.93, 2.69, 2.58 and 2.56, respectively. There is a significant difference in verbal communication ability ($p < 0.05$). Participants from other tourism-related service sectors have the best verbal communication ability, with a mean value of 3.14, followed by participants from travel agencies, tourist attractions, hotels and yacht services, whose mean values are all below average. A significant difference can also be seen in nonverbal communication ability ($p < 0.05$).

Participants from other tourism-related service sectors have the best ability, with a mean value of 3.28, followed by participants from travel agencies, with a mean value of 3.22. The mean values of participants from tourist attractions, hotels and yacht services are below average, at 2.79, 2.72, and 1.97, respectively. There is no significant difference in tourism workers' knowledge of domestic culture and emotional level.

Table 7. Test on workplace

	Workplace	Case number	Mean value	Standard deviation	F	p
Domestic culture	Tourist attractions	163	3.71	0.92	0.996	0.410
	Hotel	94	3.73	1.00		
	Travel agency	24	4.08	0.61		
	Yacht services	9	3.93	0.98		
	Others	19	3.84	0.80		
	Total	309	3.76	0.92		
Foreign culture	Tourist attractions	163	2.69	1.12	2.793	0.026*
	Hotel	94	2.58	1.19		
	Travel agency	24	3.42	1.02		
	Yacht services	9	2.56	1.52		
	Others	19	2.93	1.11		
	Total	309	2.72	1.16		
Verbal communication ability	Tourist attractions	163	2.55	1.16	3.118	0.016*
	Hotel	94	2.35	1.17		
	Travel agency	24	2.63	1.14		
	Yacht services	9	1.67	0.93		
	Others	19	3.14	1.28		
	Total	309	2.51	1.18		
Nonverbal communication ability	Tourist attractions	163	2.79	1.09	3.317	0.011*
	Hotel	94	2.72	1.06		

(*continued*)

Table 7. (*continued*)

	Workplace	Case number	Mean value	Standard deviation	F	p
	Travel agency	24	3.22	1.06		
	Yacht services	9	1.97	0.94		
	Others	19	3.28	1.15		
	Total	309	2.81	1.09		
Emotional level	Tourist attractions	163	3.40	1.03	0.911	0.458
	Hotel	94	3.38	1.05		
	Travel agency	24	3.78	0.91		
	Yacht services	9	3.13	1.72		
	Others	19	3.41	1.08		
	Total	309	3.42	1.05		

Post. As shown in Table 8, there are significant differences in intercultural communication competence ($p < 0.05$), domestic culture ($p < 0.05$), foreign culture ($p < 0.05$), verbal communication ability ($p < 0.05$), nonverbal communication ability ($p < 0.05$) and emotional level ($p < 0.05$) for tourism workers in different posts. Tourism workers at management posts have better domestic and foreign cultural knowledge, verbal and nonverbal communication ability and emotional level than workers from other posts, followed by front line tourism workers who have a higher mean value in every variable except in domestic culture. The intercultural communication competence of participants in different posts is listed in descending order as follows: management, front line, marketing and advertising, other tourism-related service sectors and administration.

Table 8. Test on post

	Post	Case number	Mean value	Standard deviation	F	p
ICC	Front line	172	2.92	0.85	14.240	0.000**
	Management	77	3.64	0.74		
	Administration	16	2.62	0.67		
	Marketing & Advertising	29	2.74	0.89		

(*continued*)

Table 8. (*continued*)

	Post	Case number	Mean value	Standard deviation	F	p
	Others	15	2.70	0.66		
	Total	309	3.06	0.88		
Domestic culture	Front line	172	3.76	0.94	4.709	0.001**
	Management	77	3.98	0.75		
	Administration	16	3.08	1.12		
	Marketing & Advertising	29	3.41	0.99		
	Others	15	3.96	0.64		
	Total	309	3.76	0.92		
Foreign culture	Front line	172	2.55	1.11	14.039	0.000**
	Management	77	3.49	0.99		
	Administration	16	2.24	0.87		
	Marketing & Advertising	29	2.32	1.21		
	Others	15	2.08	0.95		
	Total	309	2.72	1.16		
Verbal communication ability	Front line	172	2.29	1.12	12.987	0.000**
	Management	77	3.28	1.09		
	Administration	16	1.98	0.98		
	Marketing & Advertising	29	2.20	1.01		
	Others	15	2.18	1.10		
	Total	309	2.51	1.18		
Nonverbal communication ability	Front line	172	2.63	1.07	13.446	0.000**
	Management	77	3.51	0.94		
	Administration	16	2.31	0.91		
	Marketing & Advertising	29	2.59	0.91		
	Others	15	2.18	1.01		
	Total	309	2.81	1.09		
Emotional level	Front line	172	3.30	1.06	5.160	0.000**
	Management	77	3.87	0.87		

(*continued*)

Table 8. (*continued*)

Post	Case number	Mean value	Standard deviation	F	p
Administration	16	3.24	1.05		
Marketing & Advertising	29	3.13	1.16		
Others	15	3.20	1.00		
Total	309	3.42	1.05		

Years of Service. As shown in Table 9, there is a significant difference in domestic culture for participants with different years of service. Participants who have worked for 11–20 years have the best knowledge of domestic culture ($p < 0.05$). For the rest of the participants, those who stay in their posts longer have a better knowledge of domestic culture. No significant difference exists in foreign culture ($p > 0.05$), verbal communication ability ($p > 0.05$), nonverbal communication ability ($p > 0.05$) or emotional level ($p > 0.05$).

Table 9. Test on years of service

	Years of service	Case number	Mean value	Standard deviation	F	p
Domestic culture	5 or below 5	115	3.56	0.98	4.991	0.002**
	6–10	103	3.73	0.93		
	11–20	81	4.05	0.73		
	Above 21	10	4.00	0.83		
	Total	309	3.76	0.92		
Foreign culture	5 or below 5	115	2.72	1.17	0.360	0.782
	6–10	103	2.67	1.16		
	11–20	81	2.82	1.17		
	Above 21	10	2.50	1.10		
	Total	309	2.72	1.16		
Verbal communication ability	5 or below 5	115	2.50	1.14	0.817	0.485
	6–10	103	2.50	1.23		
	11–20	81	2.58	1.18		

(*continued*)

Table 9. (*continued*)

	Years of service	Case number	Mean value	Standard deviation	F	p
	Above 21	10	1.97	1.09		
	Total	309	2.51	1.18		
Nonverbal communication ability	5 or below 5	115	2.82	1.08	0.189	0.904
	6–10	103	2.81	1.12		
	11–20	81	2.82	1.10		
	Above 21	10	2.55	0.99		
	Total	309	2.81	1.09		
Emotional level	5 or below 5	115	3.43	1.05	0.200	0.896
	6–10	103	3.40	1.05		
	11–20	81	3.39	1.08		
	Above 21	10	3.65	0.84		
	Total	309	3.42	1.05		

Educational Background. As shown in Table 10, there are significant differences in foreign culture ($p < 0.05$), verbal communication ability ($p < 0.05$), nonverbal communication ability ($p < 0.05$) and emotional level ($p < 0.05$). On verbal and nonverbal communication ability, the statistics show that participants' ability on the two abilities is higher when they are at a higher education level. In terms of the knowledge of foreign culture, participants' knowledge increases when they obtain a higher education level, except for those graduating from senior college who have higher scores than those with a bachelor's degree. The mean values of the emotional level of participants from different educational backgrounds are all above average, and those graduating from senior college are higher than the rest, with a mean value of 3.72, followed by those with a master's degree and a bachelor's degree. The emotional level of participants who only finished high school or vocational secondary school is the lowest, with a mean value of 3.21. There is no significant difference in domestic culture ($p > 0.05$).

Table 10. Test on educational background

	Educational level	Case number	Mean value	Standard deviation	F	p
Domestic culture	Below high school	47	3.65	0.89	1.604	0.173
	High school/Vocational secondary school	130	3.67	1.01		

(*continued*)

Table 10. (*continued*)

	Educational level	Case number	Mean value	Standard deviation	F	p
	Senior college	73	3.92	0.86		
	Undergraduate	36	3.73	0.80		
	Postgraduate	23	4.04	0.71		
	Total	309	3.76	0.92		
Foreign culture	Below high school	47	2.22	1.21	12.061	0.000**
	High school/Vocational secondary school	130	2.42	1.14		
	Senior college	73	3.17	1.06		
	Undergraduate	36	3.07	0.94		
	Postgraduate	23	3.52	0.77		
	Total	309	2.72	1.16		
Verbal communication ability	Below high school	47	2.00	1.06	13.630	0.000**
	High school/Vocational secondary school	130	2.20	1.11		
	Senior college	73	2.81	1.17		
	Undergraduate	36	2.99	1.00		
	Postgraduate	23	3.57	0.91		
	Total	309	2.51	1.18		
Nonverbal communication ability	Below high school	47	2.38	1.08	10.756	0.000**
	High school/Vocational secondary school	130	2.56	1.07		
	Senior college	73	3.07	1.05		
	Undergraduate	36	3.14	0.86		
	Postgraduate	23	3.74	0.76		
	Total	309	2.81	1.09		
Emotional level	Below high school	47	3.29	0.99	3.608	0.007**
	High school/Vocational secondary school	130	3.21	1.15		
	Senior college	73	3.72	0.97		
	Undergraduate	36	3.54	0.79		
	Postgraduate	23	3.70	0.88		
	Total	309	3.42	1.05		

ICC Training. As shown in Table 11, there are significant differences in intercultural communication competence ($p < 0.05$), foreign culture ($p < 0.05$), verbal communication ability ($p < 0.05$), nonverbal communication ability and emotional level ($p <$

0.05). Obviously, participants who have received intercultural communication competence training have better intercultural communication competence than those who have not. No significant difference exists in domestic culture (p > 0.05).

Table 11. Test on intercultural communication competence training

	ICC training	Case number	Mean value	Standard deviation	t	df	p
ICC	Yes	106	3.37	0.81	4.779	307.000	0.000**
	No	203	2.89	0.87			
	Total	309	3.06	0.88			
Domestic culture	Yes	106	3.84	0.92	1.152	307.000	0.250
	No	203	3.72	0.92			
	Total	309	3.76	0.92			
Foreign Culture	Yes	106	3.19	1.06	5.294	307.000	0.000**
	No	203	2.48	1.14			
	Total	309	2.72	1.16			
Verbal communication ability	Yes	106	2.90	1.08	4.342	307.000	0.000**
	No	203	2.30	1.18			
	Total	309	2.51	1.18			
Nonverbal communication ability	Yes	106	3.19	0.95	4.600	307.000	0.000**
	No	203	2.61	1.11			
	Total	309	2.81	1.09			
Emotional level	Yes	106	3.66	0.99	2.927	307.000	0.004**
	No	203	3.29	1.06			
	Total	309	3.42	1.05			

Foreign Languages Spoken. According to Table 12, there is a significant difference in foreign culture (p < 0.05), verbal communication ability (p < 0.05), nonverbal communication ability (p < 0.05) and emotional level (p < 0.05) for participants who speak different foreign languages at work. It shows that participants who speak Russian have a relatively higher mean value in every variable than the rest of the participants, followed by participants who speak two or more languages, English, and Korean. Those who do not speak any foreign languages significantly have the lowest mean value. No significant difference exists in domestic culture (p > 0.05).

Table 12. Test on spoken foreign languages

	Foreign languages	Case number	Mean value	Standard deviation	F	p
Domestic culture	Never use	69	3.63	0.95	1.280	0.278
	English	185	3.81	0.86		
	Russian	11	3.48	1.16		
	Korean	3	3.11	1.84		
	Two or more foreign languages	41	3.88	0.98		
	Total	309	3.76	0.92		
Foreign culture	Never use	69	1.91	1.02	14.710	0.000**
	English	185	2.87	1.09		
	Russian	11	3.45	1.13		
	Korean	3	2.53	1.86		
	Two or more foreign languages	41	3.26	1.01		
	Total	309	2.72	1.16		
Verbal communication ability	Never use	69	1.67	0.92	14.579	0.000**
	English	185	2.72	1.12		
	Russian	11	3.24	1.17		
	Korean	3	1.67	0.67		
	Two or more foreign languages	41	2.82	1.16		
	Total	309	2.51	1.18		
Nonverbal communication ability	Never use	69	2.06	1.01	14.140	0.000**
	English	185	2.99	0.99		
	Russian	11	3.73	1.03		
	Korean	3	2.08	0.95		
	Two or more foreign languages	41	3.04	1.07		
	Total	309	2.81	1.09		
Emotional level	Never use	69	2.95	1.20	6.191	0.000**
	English	185	3.48	0.95		
	Russian	11	3.85	1.01		
	Korean	3	3.00	2.00		

(*continued*)

Table 12. (*continued*)

Foreign languages	Case number	Mean value	Standard deviation	F	p
Two or more foreign languages	41	3.83	0.90		
Total	309	3.42	1.05		

3.4 Correlation Analysis

The Pearson correlation coefficient is used to analyse the correlation between each two dimensions by analysing the mean value of the items in each dimension. The value range of the correlation coefficient is between -1 and 1, denoted by r. If $|r|=1$, complete correlation exists between the variables. If $0.7 \leq |r| < 1$, a high correlation exists between the variables. If $0.4 \leq |r| < 0.7$, a moderate correlation exists between variables. If $0.1 \leq |r| < 0.4$, a low correlation exists between variables. If $|r| < 0.1$, no correlation exists between the variables. The conclusions mentioned above can be true only when the significance value $p < 0.05$.

As shown in Table 13, a positive correlation exists between each pair of variables of intercultural communication competence. High positive correlations exist between the following groups of variables: foreign culture and nonverbal communication ability (r = 0.785, p < 0.05), verbal communication ability and nonverbal communication ability (r = 0.779, p < 0.05). Moderate positive correlations exist between the following groups of variables: domestic culture and foreign culture (r = 0.480, p < 0.05), domestic culture and emotional level (r = 0.477, p < 0.05), foreign culture and verbal communication ability (r = 0.616, p < 0.05), and foreign culture and emotional level (r = 0.599, p < 0.05). Low positive correlations exist between the following groups of variables: domestic culture and verbal communication ability (r = 0.221, p < 0.05), domestic culture and nonverbal communication ability (r = 0.371, p < 0.05), and verbal communication ability and emotional level (r = 0.382, p < 0.05).

Table 13. Correlation analysis between each variable

		Domestic culture	Foreign culture	Verbal communication	Nonverbal communication	Emotional level
Domestic culture	Pearson Correlation	1	0.480**	0.221**	0.371**	0.477**
	Sig. (2-tailed)		0.000	0.000	0.000	0.000
	Number of cases	309	309	309	309	309
Foreign culture	Pearson Correlation	0.480**	1	0.616**	0.785**	0.599**

(*continued*)

Table 13. (*continued*)

		Domestic culture	Foreign culture	Verbal communication	Nonverbal communication	Emotional level
	Sig. (2-tailed)	0.000		0.000	0.000	0.000
	Number of cases	309	309	309	309	309
Verbal communication	Pearson Correlation	0.221**	0.616**	1	0.779**	0.382**
	Sig. (2-tailed)	0.000	0.000		0.000	0.000
	Number of cases	309	309	309	309	309
Nonverbal communication	Pearson Correlation	0.371**	0.785**	0.779**	1	0.599**
	Sig. (2-tailed)	0.000	0.000	0.000		0.000
	Number of cases	309	309	309	309	309
Emotional level	Pearson Correlation	0.477**	0.599**	0.382**	0.599**	1
	Sig. (2-tailed)	0.000	0.000	0.000	0.000	
	Number of cases	309	309	309	309	309

4 Some Suggestions to Improve Tourism Industry Employees' Intercultural Communication Competence

The development of Hainan international tourism and consumption destinations requires more tourism workers with better intercultural communication competence. To cultivate excellent talent to better meet the needs of developing international tourism and consumption destinations, some suggestions are given based on the results of this research.

4.1 Encourage Employers and Local Tourism Authorities to Provide Systematic and Targeted ICC Training

The overall educational level of tourism workers is low, and most of them are high school graduates or do not even receive a high school education. It is also found in the survey that more than half of the tourism workers do not participate in any intercultural communication competence training; therefore, it is of great help if the employers of tourism industries and the local tourism authorities could provide systematic intercultural communication competence training. Given the condition that the mean value in foreign culture, verbal and nonverbal communication ability is far below average, with the lowest mean value in verbal communication ability, verbal and nonverbal communication training and intercultural communication strategies in real situations should be provided in small classes for trainees to get enough practice. Training on Chinese and Western cultural differences and intercultural communication etiquette are also needed. Moreover, different training can be carried out according to different posts and different educational backgrounds to make intercultural communication competence training more targeted and effective. Long-term targeted systematic training plays a positive role in improving the overall level of intercultural communication competence of tourism workers.

4.2   Measures from Employers and Local Tourism Authorities

Education plays a necessary part in the intercultural communication competence of tourism workers. Incentive policies can be used to mobilize the enthusiasm of tourism workers to actively raise their level of education so that they can better meet the needs of providing high-quality services. Employers could award employees who study further to raise their education level and take into consideration assessing employees' intercultural communication competence. In addition, local tourism authorities can regularly hold intercultural communication meetings for workers from tourism industries to communicate with each other and improve their intercultural communication competence.

5   Conclusion

Through this empirical study, it can be concluded that tourism workers' intercultural communication competence is not as good, especially in their knowledge of foreign culture and verbal and nonverbal communication ability. Significant differences exist in different posts, education levels, intercultural communication competence training and spoken foreign language on intercultural communication competence. Tourism workers who are in management posts have a relatively higher intercultural communication competence, followed by front line posts, marketing & advertising, other tourism related service sectors and administration. The intercultural communication competence of tourism workers is significantly higher when their educational level is higher, except for the fact that tourism workers graduating from senior college have a relatively higher competence than those with a bachelor's degree. The intercultural communication competence of tourism workers who attend intercultural communication competence training is higher than that of those who do not receive the training. The intercultural communication competence of tourism workers who speak Russian is better than those who speak other languages, followed by those who can speak two or more foreign languages and then those who speak English or Korean at work. Those who do not speak any foreign languages have the lowest intercultural communication competence. There is no significant difference in gender, age, workplaces and years of service on intercultural communication competence. By Pearson correlation analysis, it comes to the conclusion that positive correlation exists between each two variables of the three dimensions of intercultural communication competence.

Cultivating and improving tourism workers' intercultural communication competence is a very long journey considering their current education level and intercultural communication competence training that they might lack. Both employers and tourist authorities need to put more resources into helping tourism workers raise their levels of intercultural communication competence and better meet the needs of the intercultural communication that is required in their job to improve the level of international tourism reception in Hainan Province.

References

1. Chinadaily Homepage. http://www.chinadaily.com.cn/a/201810/16/WS5bc58ef6a310eff3 03282aaa.html. Accessed 16 Oct 2022
2. Chinadaily Homepage. https://cn.chinadaily.com.cn/a/202111/18/WS61961615a3107be4 979f8dad.html. Accessed 11 Nov 2022
3. Liming, W.: Free trade port policies promote the development of Hainan tourism and consumption destination. China Territory Today 6, 14–18(2020)
4. Dan, L.: A study of intercultural communicative competence model and development. Foreign Lang. Res. 6, 127–131 (2015)
5. Ying, Y., Enping, Z.: Frame construction of intercultural communication competence in foreign language teaching. Foreign Lang. World 4(121), 13–21 (2007)
6. Lustig, M.W., Koester, J.: Intercultural Competence: Interpersonal Communication across Cultures, 3rd edn. Addison Westerly Longman Inc., New York (1999)
7. Barker, G.G.: Cross-cultural perspectives on intercultural communication competence. J. Intercult. Commun. Res. 45(1), 13–30 (2015)
8. Yongchen, G.: The development of a conceptual framework for assessing Chinese college students' intercultural communication competence. Foreign Lang. World 4, 80–88 (2014)
9. Byram, M.: Teaching and Assessing Intercultural Communicative Competence. Multilingual Matters, New York (1997)
10. Arasaratnam, L.A., Doerfel, M.L.: Intercultural communication competence: identifying key components from multicultural perspectives. Int. J. Intercult. Relat. 29(2), 137–163 (2005)
11. Munezane, Y.: A new model of intercultural communicative competence: bridging language classrooms and intercultural communicative contexts. Stud. High. Educ. 46(8), 1664–1681 (2019)
12. Dan, R., Dan, X.: Study on vocational college students' communicative competence of intercultural communication. In: Jianchao, Z., Pinle, Q., Weipeng, J., Xinhua, S., Zeguang, L. (eds.) CONFERENCE 2021, ICPCSEE, vol. 1452, pp. 443–455. Springer, Singapore (2021)
13. Yanfang, L.: The empirical study on intercultural communication competence of tourism workers in Hubei. J. Yangtze Univ. 37(12), 75–78 (2014)
14. Jing, Z.: Research on the construction of communicative competence scale for the college students in western China. Foreign Lang. Test. 4, 50–59 (2021)
15. Qian, L., Xuan, Z.: Applying cross-cultural simulation games in college English classrooms to promote intercultural communication competence. Intercultural Stud. Forum 5, 41–51 (2021)
16. Jing, Y.: The development of intercultural communication competence scale for non-English majors, pp. 23–25. Heilongjiang University (2022)

A Novel Federated Learning with Bidirectional Adaptive Differential Privacy

Yang Li[1,2(✉)], Jin Xu[1,2], Jianming Zhu[1,2], and Youwei Wang[1,2]

[1] School of Information, Central University of Finance and Economics, Beijing 100081, China
liyang@cufe.edu.cn
[2] Engineering Research Center of State Financial Security, Ministry of Education, Central University of Finance and Economics, Beijing 102206, China

Abstract. With the explosive growth of personal data in the era of big data, federated learning has broader application prospects, in order to solve the problem of data island and preserve user data privacy, a federated learning model based on differential privacy (DP) is proposed. Participants share the parameters after adding noise to the central server for parameter aggregation by training local data. However, there are two problems in this model: on the one hand, the data information in the process of broadcasting parameters by the central server is still compromised, with the risk of user privacy leakage; on the other hand, adding too much noise to parameters will reduce the quality of parameter aggregation and affect the model accuracy of federated learning finally. Therefore, a novel federated learning approach with bidirectional adaptive differential privacy (FedBADP) is proposed, it can adaptively add noise to the gradients transmitted by participants and central server, and protects data security without affecting model accuracy. In addition, considering the performance limitations of the participants' hardware devices, this model samples their gradients to reduce communication overhead, and uses RMSprop to accelerate the convergence of the model on the participants and central server to improve the ac-curacy of the model. Experiments show that our novel model can not only obtain better results in accuracy, but also enhance user privacy preserving while reducing communication overhead.

Keywords: bidirectional adaptive noise · RMSprop · gradient sampling · differential privacy · federated learning

1 Introduction

In recent years, machine learning related technologies have developed rapidly, and the demand for data volume, communication overhead and data processing capabilities has increased significantly. For example, some ecommerce platforms collect user data to train recommendation algorithms in order to provide customers with a better experience. But some users refuse to share their data, they think that action will leak their privacy, so machine learning is difficult to achieve higher model accuracy without enough data to train the model.

Based on the above considerations, federated learning is an effective solution to solve the problem of data privacy protection. The main idea is to send the machine learning model to several clients for training. After each round of training, the model parameters are uploaded to the central server. Then, the central server chooses an appropriate algorithm to aggregate the parameters, which are broadcast to each client to continue training until the model converges. Compared with traditional machine learning, federated learning has a wide range of application prospects in the fields of smart cities [1], electronic medical [2], edge computing, etc. [3, 4]. In summary, federated learning can achieve the accuracy of traditional machine learning through several iterations while protecting user data better.

2 Related Work

In the training process of federal learning, the server and clients need to exchange parameters frequently, which require a lot of communication. In order to reduce communication overhead, scholars have proposed a variety of solutions. McMahan [5, 6] proposed FedAvg, which is a large number of local clients that use the stochastic gradient descent (SGD) method to calculate the gradient repeatedly, and send the sampled data to the server to reduce communication overhead. However, this method has a slow convergence speed for non-independent and identically distributed (Non-IID) data, and it is difficult to achieve high accuracy. FedProx was proposed by Li [7], who makes it possible to perform a variable number of SGD according to the number of clients, which shortens the convergence time and ensures accuracy when the data are Non-IID.

Although federated learning solves the problem of data islands under the premise of privacy preserving, the research of Melis [8] shows that attackers of federated learning can recover user sensitive information from gradient inference attacks. Therefore, Abadi [9] proposed to apply differential privacy to the upload gradient process to enhance privacy preserving capabilities in federated learning, and made mathematical derivations to give a stricter privacy budget which is moment accountant. However, the scheme uses the most primitive gradient descent scheme (SGD) and it converges slowly. Wei [10, 11] considered that both the clients and the server were dishonest. Therefore, they added noise to the federated learning in clients and server. In addition, this scheme adjusts the privacy budget according to the number of clients. When the convergence trend slows down, the privacy preserving capability is increased by reducing the number of communication rounds, but the final model accuracy can still be improved. In Reference [12], the author improved the convergence speed and reduced the privacy budget by adaptive gradient descent and adaptive gradient layering noise, but this will inevitably lead to the reduction of model accuracy. Through the above combination, our main contributions are as follows:

First, this paper uses Gaussian noise (ε, δ) to protect all data on the clients and server.

Second, we sample the client gradient and then upload it to the server, and this approach can better reduce the communication overhead while achieving a high data protection capability.

Third, this paper uses the RMSprop optimizer for adaptively adjusting the learning rate to accelerate the training speed on the client side, and also uses RMSprop on the server side to stabilize the training results.

Finally, by setting a noise range, adaptive noise is added at each dimension of parameters in the client and server to improve the ability of protecting privacy data. Then, through the comparison of different simulation experiments, the model accuracy, noise and other targets are compared with other algorithms to show the effectiveness of this program.

3 Preliminaries

In this section, to illustrate our FedBADP algorithm, we present definitions and formulas of DP, Gaussian noise based on federated learning and composition theorem in DP.

Definition 1: Given a query function $q = D \to R^d$, where D is the input dataset and R^d is d-dimensional real vector returned for the query function. For any two adjacent datasets D_1, D_2 that differ by only one data record, the global sensitivity of the query function q is as follows:

$$\Delta q = \max_{D_1, D_2} q(D_1) - q(D_2) \tag{1}$$

Definition 2: A randomized algorithm M with domain $\mathbb{N}^{|x|}$ is (ε, δ)- differential privacy if for all $S \subseteq Range(M)$ and for all $x, y \in \mathbb{N}^{|x|}$ such that $x - y_1 \leq 1$:

$$\Pr[M(x) \in S] \leq \exp(\varepsilon) \Pr[M(y) \in S] + \delta \tag{2}$$

where the probability space is over the coin flips of the algorithm M. If $\delta = 0$, then the random algorithm is ε-differential privacy, and $\varepsilon \in (0, 1)$.

Traditional differential privacy federated learning [14] uses SGD to train the model of the client, which clips the gradient of each sample first, and then adds noise to the clipped gradient. Finally, clients upload the gradient to the central server for parameter aggregation. The gradient with noise will be used for the next round of model optimization. The gradient clipping formula is as follows:

$$\overline{w_t^i} = w_t^i / \max\left(1, \frac{w_{t2}^i}{C}\right) \tag{3}$$

where w_t^i is the client i in the $t - th$ round of communication, w_{t2}^i is the l_2 norm of the model parameter, and C is the clipping range, usually taking the median of the unclipped gradient.

By adding differential privacy noise to the gradient of federated learning, user information is protected from adversary attacks. Laplace noise and Gaussian noise are commonly used in noise adding methods. The formula is as follows:

$$M(d) \triangleq f(d) + N\left(0, C^2\sigma^2\right) \tag{4}$$

where $N\left(0, C^2\sigma^2\right)$ is the Gaussian distribution or the Laplacian distribution, the mean is 0, the variance is $C^2\sigma^2$, C is gradient norm bound, and σ is the noise level.

4 Method Design

4.1 Our Approach FedBADP

In this section, we propose a novel federated learning with bidirectional adaptive differential privacy, which uses local noise and central noise to enhance privacy preserving. At the same time, in order to reduce the communication overhead and improve the accuracy of the model, we use partial data to train model and upload parameter to the server. Besides, we use the RMSprop optimizer to optimize the model in client, and also uses the RMSprop optimizer in the central server to stabilize the model parameters under the influence of Gaussian mechanism (ε, δ)-differential privacy.

This paper adds noise to the randomly selected clients by setting different sampling rates and the training samples are randomly sampled with sampling rate $q = L/N$, where N is the total number of local samples and L is the number of samples participating in the training. Through data sampling, the calculation overhead of the client can be reduced and it can reach a better model accuracy.

Using the RMSprop optimizer on the local client can accelerate the convergence speed, and it can adaptively adjust the learning rate to reach the global optimal solution. In each communication round, the gradient is updated once in a batch, and the previous batch data are added to the gradient optimization of the current batch. Adding the same optimizer to the central server can stabilize the parameters uploaded by the clients and reach a better accuracy. The formulas of the RMSprop optimizer are as follows:

$$E[g^2] = \beta E[g^2]_{t-1} + (1 - \beta)g_t^2 \tag{5}$$

$$g_{t+1} = g_t - \frac{\eta}{\sqrt{E[g^2]_t + \tau}} g_t \tag{6}$$

where $E[g^2]_t$ is the mean of the sum of the gradients' squares in the current round, g_t is the gradient of the current round, and τ takes 1×10^{-8} to prevent the denominator from being 0. We sum the previous and current round gradients by weight, and the weight is usually 0.9. η is the learning rate, by $\frac{g_t}{\sqrt{E[g^2]_t + \tau}}$ adaptively adjusting the learning rate to achieve rapid convergence, and the adaptive noise formulas are as follows:

$$\sigma_{l_i} = \frac{g_t}{\sqrt{E[g^2]_t + \tau}} \sigma \tag{7}$$

4.2 Design of DP Constraint Mechanism

According to *Definition 1*, considering two adjacent datasets, D_i and $D_{i'}$, this two datasets are only one data difference, then we can obtain t − *th* client's sensitivity is:

$$\Delta s^{D_i} = \frac{2C_L}{|D_i|} \tag{8}$$

Based on the above analysis, to achieve the minimum global sensitivity, we set all clients to have the same sample size m. The standard deviation of the added Gaussian

Table 1. Algorithm of Bidirectional Adaptive Differential Privacy Federated Learning

Algorithm 1 Bidirectional Adaptive DP-Federated Learning
1:Initialize: randomly set w_t, total noise σ
2: Client:
3: For each round of iterations do:
4:     Random sampling of c models participating from all clients
5:     Central server broadcasts initial model parameters to c servers
6:     Each client c Using differential privacy federated learning training model:
7:     Every client using sample rate q
8:     Train the model to get the model gradient $g_{t+1} = g_t - \dfrac{\eta}{\sqrt{E[g^2]_t} + \tau} g_t$
9:     Taking Gradient Median to Update Parameters w_t $g_{t+1} = median(g_{t+1})$
10:    Calculate $C_L = median(\|w_1\|_2, \cdots, \|w_t\|_2)$
11:    For i in all dimensions:
12:       clip gradient $\overline{w_t^i} = w_t^i / max\left(1, \dfrac{w_{t,2}^i}{C}\right)$
13:       add noise $\overline{\overline{w_t^i}} = \overline{w_t^i} + N\left(0, C_L^2 \sigma_{t_i}^2\right)$
14:    End For
15:    $\sigma_L^2 = \sum_{i=1,\cdots,k} \sigma_{ti}^2$
16: End For
17: Central server:
18: Receive all client uploaded parameters w_i
19: Calculate central server's parameters $w_t = \sum\limits_{i=1}^{N} \dfrac{n}{m} w_t^i$
20: Using RMSprop stabilize parameters
21: Calculate $\sigma_C^2 = \sigma_A^2 - \sigma_L^2$
22: Add noise $w_{t+1} = w_t + N\left(0, C_C^2 \sigma_C^2\right)$
23: Send parameters to clients

noise is calculated by global sensitivity Δs^{D_i}. Furthermore, the noise added to the server is:

$$\sigma_C = \sqrt{(\frac{2cT}{b\varepsilon m})^2 - \sigma_L^2} \qquad (9)$$

We can obtain (8) from the DP definition:

$$\left| \ln \frac{Pr[M(D\prime) = S]}{Pr[M(D) = S]} \right| \leq \varepsilon \qquad (10)$$

Then, according to [9], μ_0 represents the Gaussian distribution $N(0, \sigma_L^2)$, μ_1 represents $qN(\Delta s, \sigma_L^2) + (1 - q)N(0, \sigma_L^2)$, which is a finite mixture distribution, and q is clients' sampling rate, The formulas of μ_0 and μ_1 are (16) and (17).

$$\mu_0(z) = \frac{1}{\sqrt{2\pi}} \exp(-\frac{z^2}{2\sigma_A^2}) \tag{11}$$

$$\mu_1(z) = \frac{q}{\sqrt{2\pi}} \exp(-\frac{(z + \Delta s)^2}{2\sigma_A^2}) + \frac{1 - q}{\sqrt{2\pi}} \exp(-\frac{z^2}{2\sigma_A^2}) \tag{12}$$

When we obtain the above formulas, we can substitute formulas (16) and (17) into (15) and add T rounds to get the following formula:

$$\left| \sum_{i=1}^{T} \ln \frac{q\exp(-\frac{(z+\Delta s)^2}{2\sigma_A^2}) + (1 - q)\exp(-\frac{z^2}{2\sigma_A^2})}{\exp(-\frac{z^2}{2\sigma_A^2})} \right|$$

$$= \left| \sum_{i=1}^{T} \ln(1 - q + q\exp(-\frac{2z\Delta s + \Delta s^2}{2\sigma_A^2})) \right| \tag{13}$$

$$= \left| \prod_{i=1}^{T} \ln(1 - q + q\exp(-\frac{2z\Delta s + \Delta s^2}{2\sigma_A^2})) \right|$$

This quantity is bounded by ε, then

$$T \ln(1 - q + q\exp(-\frac{2z\Delta s + \Delta s^2}{2\sigma_A^2})) \geq -\varepsilon \tag{14}$$

Considering the independence of adding noises, we know

$$z \leq -\frac{\sigma_A^2}{\Delta s_D} \ln(\frac{\exp(-\frac{\varepsilon}{T})}{q} - \frac{1}{q} + 1) - \frac{\Delta s_D}{2} \tag{15}$$

We can obtain the result

$$b = -\frac{T}{\varepsilon} \ln(\frac{\exp(-\frac{\varepsilon}{T}) - 1}{q} + 1) \tag{16}$$

$$-\frac{b\varepsilon}{T} = \ln(\frac{\exp(-\frac{\varepsilon}{T}) - 1}{q} + 1) \tag{17}$$

Note that ε and T should satisfy

$$\left\{ \begin{array}{l} \varepsilon < -T \ln(1 - q) \\ T > \dfrac{-\varepsilon}{\ln(1 - q)} \end{array} \right\} \tag{18}$$

Then,

$$z \leq \frac{\sigma_A^2 b\varepsilon}{T \Delta s_D} - \frac{\Delta s_D}{2} \tag{19}$$

We can obtain the formula for δ based on the tail bound [14]

$$\Pr[z > \beta] \leq \delta \tag{20}$$

$$\Pr[z > \beta] \leq \frac{\sigma_A}{\beta\sqrt{2\pi}} \exp(-\frac{\beta^2}{2\sigma_A^2}) \tag{21}$$

Combine (25) and (26)

$$\frac{\sigma_A}{\beta\sqrt{2\pi}} \exp(-\frac{\beta^2}{2\sigma_A^2}) < \delta \tag{22}$$

Then

$$\ln \frac{\sigma_A}{\sqrt{2\pi}} \frac{1}{\beta} \exp(-\frac{\beta^2}{2\sigma_A^2}) < \ln \delta \tag{23}$$

$$\ln \frac{\beta}{\sigma_A} + \ln \sqrt{2\pi} + \frac{\beta^2}{2\sigma_A^2} > \ln \frac{1}{\delta} \tag{24}$$

$$\ln(\frac{\beta}{\sigma_A}) + \frac{\beta^2}{2\sigma_A^2} > \ln(\sqrt{\frac{2}{\pi}}\frac{1}{\delta}) \tag{25}$$

If $\frac{b\varepsilon}{T} \in (0, 1)$ and $\sigma_A = \frac{c\Delta_{SD}T}{b\varepsilon}$, the result for c is

$$c^2 \geq 2\ln(\frac{1.25}{\delta}) \tag{26}$$

Additionally, ε and T should satisfy

$$\left\{\begin{array}{l} \varepsilon < -T\ln(1 - q + \frac{q}{e}) \\[2mm] T > \dfrac{-\varepsilon}{\ln(1 - q + \frac{q}{e})} \end{array}\right\} \tag{27}$$

If $\frac{b\varepsilon}{T} > 1$, we can adjust the value of c. The standard deviation of requiring noises is given as

$$\sigma_A \geq \frac{2cT}{b\varepsilon m} \tag{28}$$

Hence, if Gaussian noises are added at the client sides, we can obtain the additive noise scale in the server as

$$\sigma_C = \sqrt{(\frac{2cT}{b\varepsilon m})^2 - \sigma_L} \tag{29}$$

5   Simulation Results and Discussion

In this part, we design a series of simulations to verify the performance of the algorithm proposed in this paper with the Non-IID MNIST handwritten dataset, which includes 60,000 training images and 10,000 test images. Firstly, we construct the data as Non-IID, the training dataset is divided into several shards. Each shard is the same label and each client contains two shards for training. A client can only learn the results of two labels, but cannot learn the training samples of other data. It can better simulate the federal learning dataset in real scenes.

Then, the model is tested by a normal dataset, in which every image's format is 28×28 grayscale image. Before training, each pixel of image in the training dataset and the test dataset is subtracted from the mean of the sample, and then divided by a sample variance to accelerate the convergence speed of the model.

DP-SGD [9], CRD [10], CDP-FL [13] and DPOPT [14] are used to compare with our algorithm. The experimental results are shown in Table 2 by comparing the communication rounds (CR), client sampling rate (CSR), sample sampling rate (SR), noise level and accuracy (Acc). In our algorithm, the training model of federated learning uses convolutional neural network to improve the result.

By setting the different client sampling rates, sample sampling rates and noise, we compare with the classical algorithm and the recent DP algorithm. DP-SGD [9] first uses PCA to reduce the data's dimension, which makes the image into 60 dimensions, then clips all samples. Besides, it selects 1% of the local samples to train the model, which uses the SGD optimizer and a specific clipping threshold. However, it is difficult to clip the gradient to a reasonable range, and the model needs to greatly increase the number of communication rounds on the Non-IID dataset to achieve higher accuracy. CRD [10] sets an adaptive communication round T, which is adjusted by the variation of the loss function and it can reduce communication overhead. CDP-FL [13] proposes that the central server is honest, and partial clients are malicious, so the algorithm only adds noise before broadcast model parameters to resist malicious attacks. DPOPT [14] proposes a new optimized noise-adding mechanism to improve the overall learning accuracy while meeting the differential privacy (Table 1).

Table 2. Experimental results contrasting

Algorithm	CR	CSR	SR	Acc	DP
DP-SGD[9]	700	1	0.01	97%	(8,1e-5)-DP
FedBADP	81	0.22	0.1	96.85%	(8,1e-288)-DP
CRD[10]	30	0.6	1	90%	(8,1e-5)-DP
FedBADP	32	0.5	0.1	90.91%	(8,1e-288)-DP
DPOPT[14]	100	1	1	96%	(5.48,1e-4)-DP
FedBADP	60	1	1	96.29%	(5.48,1e-135)-DP
CDP-FL[13]	54	0.22	1	92%	(8,1e-5)-DP
FedBADP	42	0.22	0.1	94.4%	(8,1e-288)-DP

6 Conclusion

In this paper, we propose a novel federated learning algorithm that can achieve bidirectional adaptive DP. This algorithm samples client and local data, and uses the RMSprop optimizer and adaptive noise for all devices. We designed different experiments to compare the privacy budget, accuracy, communication rounds and sampling rate on the Non-IID MNIST dataset. The experimental results show that the FedBADP proposed in this paper has better results than other algorithms, and can greatly reduce communication overhead by sampling.

Acknowledgement. This research is funded by the 2022 Central University of Finance and Economics Education and Teaching Reform Fund (No. 2022ZXJG35), Emerging Interdisciplinary Project of CUFE, the National Natural Science Foundation of China (No. 61906220) and Ministry of Education of Humanities and Social Science project (No. 19YJCZH178).

References

1. Jiang, J.F., Kantarci, B.S., Oktug, S.F.T.: Federated learning in smart city sensing: challenges and opportunities. Sensors **20**(21), 6230 (2020)
2. Xu, J.F., Glicksberg, B.S.S., Su, C.T.: Federated learning for healthcare informatics. J. Healthcare Inform Res. **5**, 1–19 (2021)
3. Chen, J.F., Sun, C.S., Zhou, X.T.: Local privacy protection for power data prediction model based on federated learning and homomorphic encryption. Inf. Secur. Res. **9**(03), 228–234 (2023)
4. Su, Y.F., Liu, W.S.: Secure protection method for federated learning model based on secure shuffling and differential privacy. Inf. Secur. Res. **8**(03), 270–276 (2022)
5. Mcmahan, H.B.F., Moore, E.S., Ramage, D.T.: Communication-efficient learning of deep networks from decentralized data. In: Artificial Intelligence and Statistics. PMLR, pp. 1273–1282 (2017)
6. Konen, J.F., Mcmahan, H.B.S., Yu, F.X.T.: Federated Learning: Strategies for Improving Communication Efficiency. arXiv preprint, arXiv:1610.05492 (2016)
7. Li, T.F., Sahu, A.K.S., Zaheer, M.T.: Federated Optimization in Heterogeneous Networks. arXiv preprint, arXiv:1812.06127 (2018)
8. Melis, L.F., Song, C.S., Cristofaro, E.D.T.: Inference Attacks Against Collaborative Learning. CoRR abs, 1805.04049 (2018)
9. Abadi, M.F., Chu, A.S.: Deep learning with differential privacy. In: Proceedings of the 2016 ACM SIGSAC Conference on Computer and Communications Security, pp. 308–318 (2016)
10. Wei, K.F., Li, J.S., Ding, M.T.: Federated learning with differential privacy: algorithms and performance analysis. IEEE Trans. Inf. Forensics Secur. **PP**(99), 1–1 (2020)
11. Wei, K.F., Li, J.S., Ding, M.T.: User-level privacy-preserving federated learning: analysis and performance optimization. IEEE Trans. Mob. Comput. **21**(9), 3388–3401 (2022)
12. Xu, Z.F., Shi, S.S., Liu, A.X.T.: An adaptive and fast convergent approach to differentially private deep learning. In: Proceedings of the IEEE INFOCOM, pp. 1867–1876 (2020)
13. Geyer, R.C.F., Klein, T.S., Nabi, M.T.: Differentially Private Federated Learning: A Client Level Perspective. arXiv preprint, arXiv:1712.07557 (2017)
14. Xiang, L.F., Yang, J.S., Li, B.T.: Differentially-private deep learning from an optimization perspective. In: Proceedings of the IEEE Conference on Computer Communications, pp. 559–5672015 (2019)

Chaos-Based Construction of LWEs in Lattice-Based Cryptosystems

Nina Cai[1,2], Wuqiang Shen[1,3], Fan Yang[1,2], Hao Cheng[1,2], Huiyi Tang[1,2], Yihua Feng[4], Jun Song[1,2], and Shanxiang Lyu[1,2(✉)]

[1] Joint Laboratory On Cyberspace Security, China Southern Power Grid, Guangzhou, China
lsx07@jnu.edu.cn
[2] College of Cyber Security, Jinan University, Guangzhou 510632, China
[3] Guangdong Power Grid Company Limited, Guangzhou, China
[4] School of Computer Science and Technology, Guangdong University of Technology, Guangzhou 510006, China

Abstract. In postquantum cryptography, the problem of learning with errors (LWE) has been widely used to create secure encryption algorithms. Nevertheless, the transmission of a large-dimensional public key matrix brings heavy overhead to communication systems. Addressing this problem, we propose a simpler scheme to generate the public key matrix with elements admitting uniform distributions. From the perspective of chaos, we employ logistic mapping and modulo lattice operations to generate uniform random numbers that feature good randomness. The public key with a large number of elements can be described by only a few parameters, which significantly reduces the transmission cost. On the basis of uniformly distributed random numbers, one can also construct random numbers admitting discrete Gaussian distributions.

Keywords: postquantum cryptography · LWE · chaos

1 Introduction

The learning with error (LWE) problem was first proposed by Oded Regev [13], which refers to finding an unknown solution from a sequence of "approximate" random linear equations. There is a great deal of research based on this principle due to its great potential for cryptographic applications and feasibility in constructing classical public key cryptosystems [1, 14, 15].

Many LWE-based encryption schemes have been developed in the past two decades; nevertheless, the transmission of large public-key matrices imposes a heavy burden on communication systems. Therefore, how to reduce a large amount of public information to a few parameters has become an open issue worth studying. The "seeding" method is an efficient and simple way to address this problem, where only a few seeds are needed to generate a large number of random numbers. Inspired by this method, this paper focuses on the seeding-based generation of random numbers.

Z. Yu et al. (Eds.): ICPCSEE 2023, CCIS 1879, pp. 338–349, 2023.
https://doi.org/10.1007/978-981-99-5968-6_24

Approaches to generating random numbers can be divided into physical and mathematical methods. Random number generators based on physical methods cannot generate exactly the same random numbers again, cannot be verified, and the physical components are poorly stable and require frequent maintenance. Therefore, mathematical methods are currently widely used for random number generation. This means that the random number series is computed according to a certain calculation method and can be reproduced exactly. It is not truly random numbers and becomes pseudorandom numbers.

Presently, the most widely used random number generators [6] include linear congruential generators [3, 7], nonlinear congruential generators [2] and random number generators based on chaotic systems [16]. Most of them are constructed for certain specific occasions and are not well suited to the construction of random parameter matrices for LWE cryptosystems. In this regard, Zhang et al. [4] proposed a chaotic system-based scheme to generate a random matrix for LWE encryption, where the random sequence generated by a one-dimensional logistic map was further adjusted to a pseudo-random sequence admitting a uniform distribution. Parallel to their scheme, we propose a more concise way of generating random matrices. The contributions of our work are summarized as follows:

- Starting from a logistic mapping of chaotic systems, by employing the flat-host assumption [11, 17], we present a method for generating uniform random numbers to build a random matrix. It is simpler, faster, and has great true randomness, including uniformity and independence. The method can flexibly generate random number series of multiple dimensions according to the dimensionality of the desired random numbers. This makes it possible to have a much wider range of applications.
- We employ several parameters as the key to generate series admitting Gaussian distribution, which appears for the first time in generating Gaussian distribution to the best of the authors' knowledge and can be studied as the backdoor of the systems in the follow-up work.

2 Preliminaries

2.1 Logistic Map

A chaotic system can be generated by a nonlinear iterative process, which has the characteristics of bifurcation, aperiodicity, and sensitivity to initial conditions called the "butterfly effect". This means that a small change can make the final output of the system unpredictable. This property gives it the potential to be a random number generator [12]. The one-dimensional logistic map is one of the most common chaos systems and is widely used in the encryption field since it has low complexity [5, 10], which can be expressed by the following iterative process:

$$z_{n+1}' = rz_n'(1 - z_n') \tag{1}$$

where r denotes the bifurcation parameter. The logistic map takes 2 parameters as input, r and z_0'. The system exhibits a chaos state for $r \in (3.57, 4]$ and $z_0' \in (0, 1]$.

When $r = 4$, almost all initial values cause chaos in the logistic series, and the probability density function is defined as:

$$\rho(y) = \begin{cases} \frac{\pi^{-1}}{[y(1-y)]^{1/2}}, & 0 < y < 1 \\ 0, & \text{others} \end{cases} \tag{2}$$

Then, an initial set of random numbers can be generated from this simple setting.

2.2 Modulo Lattice

An n-dimensional lattice Λ is a discrete point subset in $\mathbb{R}^n$ that can be defined as:

$$\Lambda = \left\{ \sum_{i=1}^{n} x_i \mathbf{b}_i | x_i \in \mathbb{Z} \right\} \tag{3}$$

where the basis of the lattice Λ consists of n linearly independent vectors $\mathbf{b}_1, \mathbf{b}_2, \ldots, \mathbf{b}_n$.

Some computationally hard problems have been defined on lattices. One associated with our work is known as the closest vector problem (CVP): given a random vector u, the closest vector problem refers to finding its closest lattice vector v from the point set Λ [8]. That is, for any vector $\mathbf{x}'$, the CVP can be expressed as follows:

$$Q_\Lambda(\mathbf{x}') = \arg\min_{\mathbf{u} \in \Lambda} ||\mathbf{x}' - \mathbf{u}|| \tag{4}$$

where $|| \cdot ||$ denotes the Euclidean norm. Based on the CVP quantizer, we can further define the modulo-Λ reduction of vector $\mathbf{x}'$ as:

$$\mathbf{x} = \mathbf{x}' \bmod \Lambda = \mathbf{x}' - Q_\Lambda(\mathbf{x}') \tag{5}$$

i.e., calculate the error distance of the vector from its nearest lattice point.

2.3 Flat-Host Assumption

Luis P'erez-Freire [11] provided a complete elaboration of the flat-host assumption based on previous work, which was originally a hypothesis proposed for the evaluation of the performance of quantum-based data hiding methods in analysis.

The flat-host assumption shows that the probability density function of the host signal can be considered to obey a uniform distribution in each quantization cell and has a sufficiently large host variance so that all the centroids occur with equal probability. That is, assuming that the host signal variance is sufficiently large, the resulting sequence will be uniformly distributed in the Voronoi region after performing the modulo lattice operation.

2.4 LWE and the Encryption Scheme

With $\mathbf{A} \in \mathbb{Z}_q^{m*n}$ from a uniform distribution and $\mathbf{e} \in \mathbb{Z}_q^m$ from a discrete Gaussian distribution χ, for any $\mathbf{s} \in \mathbb{Z}_q^n$, we define:

$$\mathbf{b} = \mathbf{As} + \mathbf{e} \in \mathbb{Z}_q^m \qquad (6)$$

The search version LWE refers to finding s for given A and b. The LWE problem can be used as a basis for constructing cryptographic systems. The complete process can be described as follows:

$-KeyGen(\lambda) \rightarrow (pk, sk)$:

generate $\mathbf{s} \overset{\$}{\leftarrow} \mathbb{Z}_q^n$, $\mathbf{A} \overset{\$}{\leftarrow} \mathbb{Z}_q^{m*n}$, $\mathbf{e} \overset{\$}{\leftarrow} \chi \in \mathbb{Z}_q^m$
compute $\mathbf{b} = \mathbf{As} + \mathbf{e} \in \mathbb{Z}_q^m$
return $pk = (\mathbf{A}, \mathbf{b})$, $sk = \mathbf{s}$

$-Enc(pk, m \in \{0, 1\}) \rightarrow$ cipher $c = (c_0, c_1)$:

generate $\mathbf{r} \overset{\$}{\leftarrow} \mathbb{Z}_2^m$
compute $c_0 = \mathbf{r}^T \mathbf{A}$, $c_1 = \mathbf{r}^T \mathbf{b} + \lfloor \frac{q}{2} \rfloor m$
return $c = (c_0, c_1)$

$-Dec(sk, c = (c_0, c_1)) \rightarrow m\prime$:

compute $v = |c_1 - c_0 \mathbf{s}|$

$$m' = \begin{cases} 0, & v < \frac{q}{4} \\ 1, & \text{others} \end{cases}$$

return $m\prime$

In the above process, the matrix $\mathbf{A}$ is a large random matrix, which imposes heavy computing and communication overhead on the cryptosystem. In response, we propose a method of generating uniform sequences with a logistic chaotic system based on the flat-host assumption that the matrix $\mathbf{A}$ is obtained by sampling from it.

3 Construction

3.1 Generating Uniform Chaotic Sequences

Equation (2) does not obey a uniform distribution, and the resulting sequence is not sufficiently random. Therefore, converting the nonuniformly distributed random variables to uniformly distributed random variables by the module lattice and flat-host assumption is necessary.

By setting an initial value $z_0\prime$ for the one-dimensional logistic map in Sect. 2.1, the mapping sequence $\mathbf{z}\prime = (z_1', z_2', \ldots, z_n')$ can be generated iteratively. Then, the sequence $z\prime$ is multiplied by a factor μ (as shown in Fig. 1).

$$z_i = \mu * z_i', i = 1, 2, \ldots, n \tag{7}$$

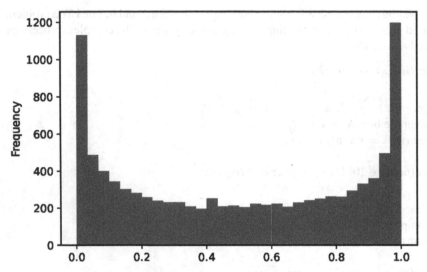

Fig. 1. Statistical histograms of the logistic sequence.

The sequence $\mu * z'$ is partitioned into a number of different vectors by setting the sampling intervals d (the dimension of the chosen fundamental cell).

$$\mathbf{x_j} = \left(z_{(j-1)*d+1}, z_{(j-1)*d+2}, \ldots, z_{(j-1)*d+d}\right), j = 1, 2, \ldots, n \tag{8}$$

and then a modulo lattice operation is performed.

$$\mathbf{y_i} = \mathbf{x_i} mod \, \Lambda = \mathbf{x_i} - Q_\Lambda(\mathbf{x_i}), i = 1, 2, \ldots, n \tag{9}$$

According to the flat-host assumption, the sequence $\mathbf{y} = (Y_1, Y_2, \ldots, Y_n)$ after the modulo lattice operation will obey a uniform distribution in the Voronoi region. For simplicity, we perform a modulo one-dimensional integer lattice $\mathbb{Z}$ on the series so that it ranges in the interval [0,1] (as shown in Fig. 2).

3.2 Generating Gaussian Distribution from Uniform Sequences.

The random array $\mathbf{y} = (Y_1, Y_2, \ldots.)$ generated in Sect. 3.1, which obeys a uniform distribution, is first expanded into one dimension and then grouped under the given positive integers len_χ. In this case, for any $y_i \in Y$ such that y_i satisfies $0 < y_i < 1$, the transformation is performed as follows.

$$r_i = \begin{cases} 1, \ y_i > 0.5 \\ 0, \ y_i \le 0.5 \end{cases} \tag{10}$$

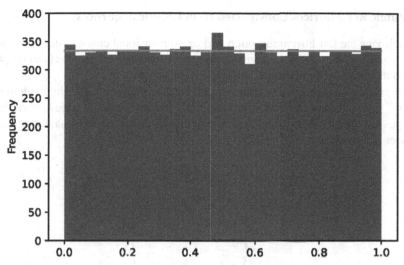

Fig. 2. The statistical histograms of the improved logistic sequence with the generator matrix [1].

$$\mathbf{r_j} = \left(r_{j*len_\chi+1}, r_{j*len_\chi+2}, \ldots\ldots, r_{j*len_\chi+len_\chi} \right), j = 1, 2, 3, \ldots\ldots \tag{11}$$

For a given positive integer p, the domain of the discrete Gaussian distribution χ is defined by $\{-p, -p+1, \ldots, -1, 0, 1, \ldots, p-1, p\}$. The table of cumulative distribution functions for the distribution χ is given as T_χ. For len_χ, the table satisfies the following cases:

$$\frac{T_\chi(0)}{2^{len_\chi}} = \frac{\chi(0)}{2} \tag{12}$$

And

$$\frac{T_\chi(z)}{2^{len_\chi}} = \frac{\chi(0)}{2} + \sum_{i=1}^{z} \chi(i) \quad 1 \le z \le p \tag{13}$$

Therefore, $T_\chi(p) = 2^{len_\chi-1}$. Note that table T_χ must satisfy this condition as closely as possible to bring the sampling closer to the distribution χ.

An example of the constructed discrete Gaussian distribution is shown in Fig. 3.

Algorithm 1: Frodo.Sample

Input: A (random) bit string $\mathbf{R} = (r_0, r_1, \ldots, r_{len_\chi-1}) \in \{0,1\}^{len_\chi}$ The table
$\quad\quad \mathbf{T}_\chi = (T_\chi(0), T_\chi(1), \ldots, T_\chi(s),)$
Output: A sample $e \in Z$

1   $t \leftarrow \sum_{i=1}^{len_\chi-1} r_i * 2^{i-1}$;
2   $e \leftarrow 0$;
3   **for** $z \leftarrow 0$ **to** s **do**
4   **if** $t > T_\chi(z)$ **then**
5       $\lfloor \; e \leftarrow e + 1$;

6   $e \leftarrow (-1)^{r_0} e$;
7   **return** e;

3.3 Public Key Matrices Constructed from Chaotic Sequences

By employing the flat-host assumption and one-dimensional chaos system to construct public key matrix $\mathbf{A} \in \mathbb{Z}_p^{m*n}$, we consider that our scheme not only enjoys low complexity but also the elements of matrix A satisfy uniform distribution.

The construction process is described in detail as follows. First, the random series $\mathbf{y} = (y_1, y_2, \ldots)$ is generated and unfolded to a one dimensional series. Then, for the given positive integers i and h, the public key matrix $\mathbf{A}$ is formed by sampling through the series as follows.

$$a_k = \lfloor (q-1) * y_{i+kh} \rfloor mod q, k = 1, 2, 3, \ldots. \tag{14}$$

$$\mathbf{A} = \begin{pmatrix} a_0 & \cdots & a_{m*(n-1)} \\ a_1 & \cdots & a_{m*(n-1)+1} \\ \vdots & \ddots & \vdots \\ a_m & \cdots & a_{m*n-1} \end{pmatrix} \tag{15}$$

By means of the above, the matrix $\mathbf{A}$ can be transmitted with just these few parameters:

$$\mathbf{A} \leftarrow W(z_{0'}, \mu, d = 1, q, i, h, m, n) \tag{16}$$

In this way, a sequence of random numbers with good randomness and uniform distribution can be generated quickly. The matrix generated on this basis also has fine randomness. The storage of large matrices is transformed into the storage of eight parameters, which effectively reduces the communication overhead. By doing so, the efficiency of the transmitted information is increased.

3.4 Public-Key Encryption Scheme Based on LWE and Chaos

Based on the LWE problem, the uniformly distributed sequence and the Gaussian distributed sequence generated by the chaos-based system mentioned above, we can implement a cryptographic system. Its main implementation is depicted as follows:

$-KeyGen(\lambda) \rightarrow (pk, sk)$

The generation of (pk, sk) can be represented in detail by the following steps:

i) An initial value $W = \left(z_0', \mu, 1, q, i, h, m, n \right)$ is set.

ii) The logistic map is used to produce a chaotic sequence $\mathbf{z}' = \left(z'_1, z'_2, \ldots. \right)$, and a positive integer μ is chosen and computed $\mathbf{x} = \mu * \mathbf{z}'$. The modulo lattice ope-ration is performed to produce a sequence $\mathbf{y} = (Y_1, Y_2, \ldots)$ that obeys a uniform distribution.

iii) Based on the given positive integers i, h, a random matrix $\mathbf{A}$ is formed by sampling from the sequence $\mathbf{y}$, which is from the lattice of a dimension.

iv) The Gaussian distribution $\mathbf{e} \in \mathbb{Z}_q^m$ is generated in Sect. 3.2.

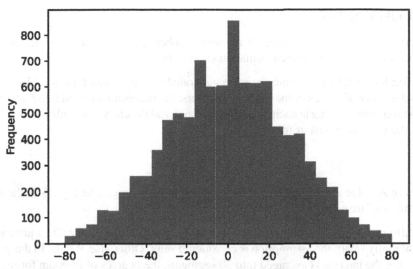

Fig. 3. The statistical histograms of the Gaussian sequence from the improved logistic sequence of Fig. 2

v) Randomly select a positive vector $s \in \mathbb{Z}_q^n$, and calculate $b = As + e$. The public key $pk = (W, b)$, and private key $sk = s$.

$-Enc(pk, m \in \{0, 1\}) \rightarrow c$

This cryptosystem independently encrypts and decrypts one bit of data. The ciphertext is calculated in the following way.

i) Choose the plaintext message $m \in \{0, 1\}$, use the public key pk, recover the matrix A from the parameter W.

ii) pick a random $r \in \mathbb{Z}_2^m$, and encrypt it by computing $c_0 = r^T A$, $c_1 = r^T b + \left|\frac{q}{2}\right| m$. The final output is the encrypted ciphertext $c = (c_0, c_1)$.

$-Dec(sk, c) \rightarrow m$

Decrypt the received ciphertext c with the private key sk to obtain the plaintext m as follows: Calculate $v = |c_1 - c_0 s| = \left|r^T e + \left|\frac{q}{2}\right| m\right|$. m equals 0 when v is close to 0. Otherwise, m equals 1.

4   Test

In this section, we use statistical analysis to analyse the true randomness of pseudo-random arrays generated in this way, including parametric tests, uniformity tests and independence tests.

Our results showed that the random sequence generated achieves good uniform randomness when a good combination of the initial value z_0' of the logistic mapping and the multiplier μ is chosen. For example (abbreviated as E1), when $n = 9300$, $z_0' \in [0.2, 0.9]$ (we test when \$step = 0.0005\$), we chose the μ in the interval $[700, 800]$ to implement the algorithm.

4.1 Uniformity Test

To verify whether the sequence of random numbers generated in Sect. 3.1 obeys a uniform distribution, we use chi-square tests as follows.

1. Given length l, l random numbers are generated by the approach in Sect. 3.1. Based on the range of values of the random numbers, the values are divided into f intervals.
2. The frequency falling in each interval is counted, and the chi-square value is calculated by the chi-square test as follows.

$$X^2 = \sum \frac{(A-T)^2}{T} \tag{17}$$

where A is the actual number of frequencies in each range and $T = \frac{l}{f}$ is the ideal number of frequencies.

Building on the assumption of obeying a uniform distribution, if the Chi-square value is sufficiently small, the assumption is considered valid. Otherwise, it fails. In this paper, supposing the interval is averaged into 30 segments, the degrees of freedom for the chi-square test are 29. In $E1$, taking $\alpha = 0.1$, the proportion that could pass the cardinality test was 83.81%. Namely, for any $z_0' \in [0.2, 0.9]$, there is a high probability that the corresponding z_0' with good uniform randomness can be found in the small range.

Our results showed that the random sequence generated achieves good uniform randomness when a good combination of the initial value z_0' of the logistic mapping and the multiplier μ is chosen.

4.2 Parametric Test

The parametric test for uniform random numbers is to test whether the mean, variance or matrix of each order, etc., of the generated random series is significantly different from the ideal value of a uniform distribution.

The mean, the variance and the mean square value of the random variables $R \sim U(0, 1)$ should theoretically be $E(R) = \frac{1}{2}$, $Var(R) = \frac{1}{12}$, $E(R^2) = \frac{1}{3}$.

In the above example of generating a sequence of random numbers, we calculate the three values of the random number generated in E1. The result is shown in Fig. 4, and we find that the three values remain very slightly different from the ideal value. We can assume that the sequence of random numbers generated in the above way passes the parameter test.

4.3 Independence Test

We use the correlation coefficient to test the independence of a sequence of random numbers. If two random variables are independent, their correlation coefficients $\rho = 0$. For a set of random numbers to be tested $r_1, r_2, \ldots, r_n$, we assume that the correlation coefficient $\rho = 0$ and calculate the autocorrelation coefficient of order j of the sample according to the following equation.

$$\rho = \frac{\frac{1}{n-j}\sum_{i=1}^{n-j}(r_i-\bar{r})}{\frac{1}{n}\sum_{i=1}^{n}(r_i-\bar{r})^2} \tag{18}$$

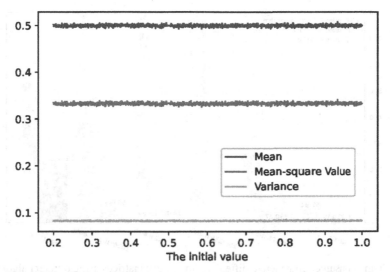

Fig. 4. Distribution of statistical parameters for a sequence of random numbers

If $n - j$ is large enough and $\rho = 0$ holds, $u_j = \rho(j)\sqrt{(n-j)}$ asymptotically obeys the distribution $N(0, 1)$. In practical terms, the test is generally $m \in [10, 20]$. This paper calculates the statistic u_i for the data selected in Sect. 4.2, and the outcome shows that the statistic is asymptotically distributed $N(0, 1)$, meaning that the random number series passes the independence test.

5 Comparison of Schemes

This section compares the time consumption of our scheme and two other algorithms in generating the public key matrix to demonstrate the efficiency of our scheme. To visualize performance comparisons, a histogram is shown in Fig. 5. It is obvious that the AES 128 encryption algorithm used in FrodoKEM [9] is computationally complex and takes much more time than the other two algorithms. As the author mentions in the literature, a small amount of matrix A reuse can be performed to reduce the computational burden. Compared to AES128, the proposed algorithm only involves simple arithmetic operations during public key generation, resulting in a significant improvement in computational efficiency.

Our scheme also has a slight time improvement over Zhang's scheme [4] for the same computational power (choosing the $i = 5, h = 3$ to sample to form a matrix), and the present scheme is more flexible in that the dimensionality of the moduli lattice can be adjusted as needed, thus producing random sequences of different dimensions. However, it should be noted that good uniformity requires a proper combination of $z_{0'}$ ($z_{0'} \neq 0.5$) and μ, which can be determined from a large number of experiments.

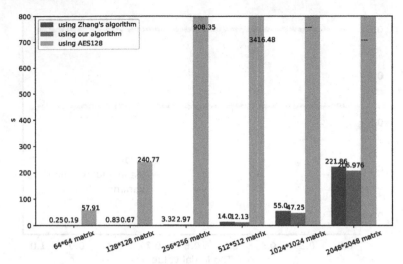

Fig. 5. Time consumed to generate different dimensional matrices using different algorithms.

6 Conclusion

In this paper, we have proposed a scheme for the generation of uniform random matrices for the LWE encryption scheme in accordance with the flat-host assumption. The scheme is simple to generate and, with the appropriate parameter selection, produces a sequence with very good true randomness that passes the chi-square test, the parametric test and the independence test.

The scheme is fast in generating one-dimensional random sequences, and the dimensionality of the moduli lattice can be adjusted as required to generate multidimensional sequences that are uniformly distributed in different dimensions. We can generate Gaussian distributions with keys as well based on the resulting uniformly distributed random series. Additionally, when constructing the encryption scheme, we use seeding to reduce the burden of matrix storage and transmission.

Acknowledgement. This work was supported in part by the Open Research Fund of Joint Laboratory on Cyberspace Security, China Southern Power Grid (Grant No. CSS2022KF03), and the Science and Technology Planning Project of Guangzhou, China (Grant No. 202201010388).

References

1. Becker, D.: LWE-based encryption schemes and their applications in privacy-friendly data aggregation, Ph.D. thesis, Technische Universität Hamburg (2018)
2. Eichenauer, J., Lehn, J.: A nonlinear congruential pseudo random number generator. Statistische Hefte **27**(1), 315–326 (1986)
3. H, L.D.: Mathematical methods in large-scale computing units. Annu. Comput. Lab. Harvard Univ. **26**, 141–146 (1951), https://cir.nii.ac.jp/crid/1573387451173414912

4. Kaiwei, Z., Ma, A., Shanxiang, L., Jiabo, W., Shuting, L.: Efficient Construction of Public-Key Matrices in Lattice-Based Cryptography: Chaos Strikes Again. In: Deng, R., Bao, F., Wang, G., Shen, J., Ryan, M., Meng, W., Wang, D. (eds.) ISPEC 2021. LNCS, vol. 13107, pp. 1–10. Springer, Cham (2021). https://doi.org/10.1007/978-3-030-93206-0_1
5. Kocarev, L., Jakimoski, G.: Logistic map as a block encryption algorithm. Physics Letters A **289**(4), 199–206 (2001). https://doi.org/10.1016/S0375-9601(01)00609-0, https://www.sci encedirect.com/science/article/pii/S0375960101006090
6. L'Ecuyer, P.: History of uniform random number generation. In: 2017 Winter Simulation Conference (WSC), pp. 202–230. IEEE (2017)
7. L'Ecuyer, P.: Tables of linear congruential generators of different sizes and good lattice structure. Math. Comput. **68**(225), 249–260 (1999)
8. Micciancio, D.: The hardness of the closest vector problem with preprocessing. IEEE Trans. Inf. Theory **47**(3), 1212–1215 (2001). https://doi.org/10.1109/18.915688
9. Naehrig,M., Alkim, E.J.e.a.: Frodokem. Tech. rep., National institution of Standards and Technology (2017)
10. Pareek, N.K., Patidar, V., Sud, K.K.: Image encryption using chaotic logistic map. Image Vis. Comput. **24**(9), 926–934 (2006)
11. Pérez-Freire, L., Pérez-González, F.: Spread-spectrum vs. quantization-based data hiding: misconceptions and implications. In: Security, Steganography, and Watermarking of Multimedia Contents VII, vol. 5681, pp. 341–352. SPIE (2005)
12. Phatak, S., Rao, S.S.: Logistic map: a possible random-number generator. Phys. Rev. E **51**(4), 3670 (1995)
13. Regev, O.: On lattices, learning with errors, random linear codes, and cryptography. J. ACM (JACM) **56**(6), 1–40 (2009)
14. Regev, O.: The learning with errors problem. Invited survey in CCC **7**(30), 11 (2010)
15. Roy, S.S., Karmakar, A., Verbauwhede, I.: Ring-LWE: applications to cryptography and their efficient realization. In: Carlet, C., Hasan, M., Saraswat, V. (eds.) Security, Privacy, and Applied Cryptography Engineering. SPACE 2016. Lecture Notes in Computer Science, vol 10076. Springer, Cham (2016). https://doi.org/10.1007/978-3-319-49445-6_18
16. Wang, L., Cheng, H.: Pseudorandom number generator based on logistic chaotic system. Entropy **21**(10), 960 (2019)
17. Wang, Y.G., Zhu, G., Li, J., Conti, M., Huang, J.: Defeating lattice-based data hiding code by decoding security hole. IEEE Trans. Circuits Syst. Video Technol. **31**(1), 76–87 (2021). https://doi.org/10.1109/TCSVT.2020.2971590

Security Compressed Sensing Image Encryption Algorithm Based on Elliptic Curve

Anan Jin[1], Xiang Li[1,2,3(✉)], and Qingzhi Xiong[1]

[1] School of Information Engineering, East China University of Technology, Nanchang 330013, Jiangxi, China
tom_lx@126.com

[2] Jiangxi Engineering Technology Research Center of Nuclear Geoscience Data Science and System, East China University of Technology, Nanchang 330013, Jiangxi, China

[3] Jiangxi Key Laboratory of Cybersecurity Intelligent Perception, East China University of Technology, Nanchang 330013, Jiangxi, China

Abstract. Aiming to address the problems of high costs associated with the storage and transmission of environmental monitoring images, as well as potential security risks, this paper proposes a security compressed sensing image encryption algorithm based on elliptic curve cryptography. The algorithm introduces elliptic curve encryption technology within a compressed sensing framework, using elliptic curve cryptography to encrypt the matrix during the compression perception acquisition process. This enables secure acquisition of environmental monitoring encrypted images. Therefore, this paper presents a security compressed sensing framework to address security gaps in the compression perception reconstruction process and improve the security of environmental monitoring images. Experimental results show that the proposed algorithm effectively encrypts and reconstructs environmental monitoring images, with high security and resistance to cracking.

Keywords: Compressed Sensing · Elliptic Curve · Image Reconstruction · Image Encryption · Environmental Monitoring

1 Introduction

With the rapid development of networks, environmental monitoring is no longer limited to traditional monitoring methods, such as manual monitoring [1], robots [2], and wireless sensor networks [3]. Therefore, network cameras have been very popularly used for environmental monitoring in recent years [4]. Infrared and visible light images are very common in environmental monitoring. With the growth of time, due to network cameras working 24 h a day, the stored pictures are very large, which is a huge burden for storage and transmission [5]. Furthermore, the quality of the images generated by the network cameras is poor, which is mainly caused by the environmental monitoring that has certain limitations, i.e., darkness or air pollution. Therefore, improving image quality is a research hotspot of many scholars at present [6]. In addition, some images

Z. Yu et al. (Eds.): ICPCSEE 2023, CCIS 1879, pp. 350–360, 2023.
https://doi.org/10.1007/978-981-99-5968-6_25

of environmental monitoring need to be encrypted and stored to prevent criminals from stealing due to security reasons. However, if there are a large number of images that are encrypted and stored one by one, it will cost a lot of money to encrypt storage, which is very unrealistic [7].

Image encryption technology is relatively mature at present and is mainly divided into two types, information hiding techniques and cryptography. Information hiding techniques also contain watermarking and steganography. However, these methods encrypt the image after the image acquisition is finished, which results in storing many images that occupy storage resources. For example, [8] proposed a new watermarking algorithm and image encryption scheme that improved the complex properties of the chaotic logistic system, i.e., nonperiodic motion and nonconvergence. Arora et al. proposed an enhanced symmetric hybrid digital cryptosystem algorithm that used the jigsaw transform to disturb the watermark image [9]. In addition, [10] proposed a new steganography technology for efficient image encryption that used Choquet's fuzzy integral sequences and hybrid DNA encoding. [11] used Sudoku scrambling and the queen traversal mode to locate the pixels on the image to propose image encryption with steganography for medical images in content confidentiality. Jinyuan Liu et al. proposed a 4D chaotic mapping and steganography algorithm for image encryption that mainly used two rounds of encryption and one embedded hash operation [12]. Furthermore, to support high-level security in medical images, Akkasaligar and Biradar proposed DNA cryptography and double hyperchaotic mapping techniques [13]. Yujia Liu et al. aimed to address the complex key secure transmission and distribution in optical transformation images by proposing an optical image encryption algorithm that used public-key cryptography and hyperchaos [14]. However, these image encryption methods encrypt the image itself, which contains considerable redundant information and has a large size, making it unfavorable for storage and transportation. Therefore, this paper focuses on the problem of much redundant information in previous image encryption work.

However, compressed sensing (CS) technology [15-17] can compress the image during the acquisition process, which greatly reduces the redundant information of the image and reduces the size of the image, which brings hope for the storage and transportation of images. Furthermore, elliptic curve encryption technology is widely used in blockchain technology, which makes elliptic curve encryption technology very prominent in the field of encryption and is currently favoured by scholars.

In summary, this paper proposes a secure compressed sensing image encryption algorithm based on elliptic curve encryption technology, which uses compressed sensing technology to sample environmental monitoring images and encrypt the observation matrix. The contributions of this work are as follows: 1) It reduces the number of observations and increases the sampling rate when using CS technology to sample the image. 2) It reduces the storage and transmission costs because the sampled image is of smaller size. 3) It improves the security of environmental monitoring images, preventing criminals from obtaining and cracking them.

The rest of this paper is organized as follows: Sect. 2 describes the background knowledge related to elliptic curve cryptography. The details of how to design the image encryption algorithm are presented in Sect. 3. Section 4 provides a detailed experiment of the proposed algorithm. Section 5 discusses the proposed algorithm from the analysis of feasibility and security. Finally, the paper concludes in Sect. 6.

2 Background and Methods

2.1 Elliptic Curve Cryptography

Elliptic curve cryptography (ECC) is developed based on elliptic curve mathematics. It is a public-key encryption algorithm [18], also known as asymmetric encryption, which uses the discrete logarithm problem to achieve encryption. ECC can use a smaller key, which provides a higher level of security and faster computing speed compared to other encryption algorithms. Currently, ECC is the most secure encryption method for a given key length [19]. ECC is also used as public-key encryption technology in blockchain.

Most of the encryption algorithms are based on the discrete logarithm problem, such as the well-known Rivest Shamir Adleman (RSA) encryption algorithm [20], which uses the characteristic that it is very easy to multiply two prime numbers, but it is extremely difficult to decompose their composite numbers. However, the ECC encryption algorithm is encrypted in the finite field $\mathbf{F}_p$, and its formula is defined as:

$$Q = kG \tag{1}$$

In formula (1), k is a large number, and Q and G are two points on the elliptic curve. If the large numbers k and G points are known, it is very easy to solve the Q point. However, if the Q point and G points are known, it becomes extremely difficult to solve the large number k. In other words, it is almost impossible, which is the traditional discrete logarithm problem [21]. The ECC encryption algorithm uses this feature to realize encryption. In the ECC encryption algorithm, the Q point is used as the public key, the large number k is used as the private key, and the G point is used as the base point of the elliptic curve.

The steps of elliptic curve encryption are as follows:

1) User A selects an elliptic curve $E_p(a, b)$ under the finite field $\mathbf{F}_p$ and chooses a base point G on it.
2) User A chooses a large number k as the encrypted private key and generates the corresponding public key Q, namely, $Q = kG$.
3) User A sends $E_p(a, b)$, Q, and G to User B together.
4) When user B has received the information sent by user A, user B will encode the plaintext that needs to be encrypted with $E_p(a, b)$ to obtain a point M on the elliptic curve. Furthermore, user B generates a random number r.
5) User B calculates point $C_1 = rG$ and point $C_2 = M + rQ$ and encrypts it with the public key Q to generate ciphertext C, which is a point pair, namely,$C = \{C_1, C_2\} = \{rG, M + rQ\}$.
6) User B sends the ciphertext C to User A.
7) User A uses the private key k to decrypt (or decode), which is calculated as $C_2 - kC_1$. The decrypted result is the point M. The calculation process is as follows: $C_2 - kC_1 = M + rQ - k(rG) = M + r(kG) - k(rG) = M$.

8) User A decodes points M to obtain plaintext.

In the above process, even if an attacker obtains the elliptic curve $E_p(a, b)$, the public key Q, the base point G, and the ciphertext point $C = \{C_1, C_2\}$, they cannot obtain the point M. In other words, the plaintext cannot be obtained by the attacker, because point M cannot be obtained without the private key k. The attacker cannot calculate the private key k using the public key Q and the base point G, or calculate the random number r using the ciphertext point $C = \{C_1, C_2\}$ and the base point G. Therefore, using ECC to encrypt data can ensure its security during transmission.

The ECC encryption algorithm can provide a higher level of security, faster processing speed, smaller consumption of network bandwidth, and require less storage space than the RSA algorithm [22]. Therefore, it is gradually replacing the RSA algorithm as the mainstream encryption algorithm [23]. This can be seen from the use of ECC in blockchain encryption technology as well [24].

3 Secure Compressed Sensing Image Encryption Algorithm Based on Elliptic Curve

It is worth mentioning that CS theory lacks security in image processing. This paper combines the processing process of CS, where the measurement matrix is encrypted during construction, and is used as the "key". Therefore, a security compressed sensing based on the elliptic curve image encryption algorithm is proposed, named SCSEC. In the encryption of images, this algorithm uses the Monte Carlo method to construct the chaotic matrix and then encrypts the chaotic factor λ with ECC to construct a secure compressed sensing chaotic measurement matrix Φ. This matrix is then used to perform a linear measurement to obtain the signal measurement value. In the decrypted images, the user decrypts using ECC to obtain the chaotic measurement matrix Φ, and finally, uses the reconstruction algorithm to reconstruct the image. The flow of this image encryption algorithm is shown in Fig. 1.

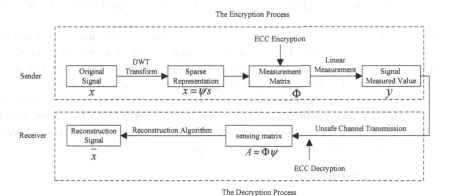

Fig. 1. Flowchart of the security compressed sensing image encryption algorithm based on the elliptic curve

The security compressed sensing chaotic measurement matrix constructed in this paper is based on our previous work [25]. The details of the process for constructing the matrix are not relevant to this work. Therefore, we treat the constructed matrix as a black box that generates a measurement matrix that meets the requirement for the measurement matrix in CS theory. λ denotes the multiplier in [25], but in this paper, we use it as a chaotic factor to encrypt the measurement matrix and obtain the secure compressed sensing chaotic measurement matrix Φ.

The framework flowchart for this algorithm is shown in Fig. 1 above. The SCSEC image encryption algorithm implemented in this paper follows the steps below:

1) Receiver: Select a suitable elliptic curve $E_p(a, b)$ under the finite field $\mathbf{F}_p$, and then select the base point G on it. In addition, the receiver selects the encrypted private key k to generate the public key Q that is named $Q = kG$. Finally, the receiver sends the elliptic curve $E_p(a, b)$, the public key Q, and the base point G to the sender.
2) Sender: Upon receiving the information, the sender selects a chaotic factor λ that is plaintext. The sender uses $E_p(a, b)$ to encode plaintext to obtain the point M. Furthermore, the sender generates a random number r. Last, the sender uses the public key Q and random number r to generate the ciphertext C.
3) The sender uses the discrete wavelet transform (DWT) to sparsely transform the environmental monitoring image x on a sparse basis ψ to obtain the sparse representation of the image, namely, $x = \psi s$. Thus, the sender obtains the sparse coefficient matrix s.
4) The sender uses the chaotic factor λ selected earlier to generate random numbers using the mixed congruence method. Next, the Box-Muller transformation is performed on these random numbers. Then the sender uses the Monte Carlo sampling method to generate the chaotic sequence and takes the first $M \times N$ term of the chaotic sequence to construct the chaotic matrix by column. Finally, the sender normalizes it to obtain the secure compressed sensing chaotic measurement matrix Φ.
5) The sender performs a linear measurement with the above constructed measurement matrix Φ to obtain the measured value of the signal $y = \Phi \psi s = As$, where $A = \Phi \psi$ is the sensing matrix.
6) The sender sends the signal measurement value y and the ciphertext C back to the receiver.
7) Receiver: the receiver uses the private key k to decrypt the ciphertext C to obtain the point M. Then, the point M is decoded to obtain the plaintext, which is the chaos factor λ. Last, the receiver restores the chaotic measurement matrix Φ using the same method in step (2), which can obtain the sensing matrix that is $A = \Phi \psi$.
8) The receiver uses the obtained sensing matrix $A = \Phi \psi$ and reconstruction algorithm to obtain the sparse coefficient matrix s. Then, the inverse DWT transformation is performed on the sparse coefficient matrix to obtain the reconstructed signal $\bar{x}$, thereby obtaining the environmental monitoring image.

4 Experimental Results

The images used in this experiment are infrared and visible light images that are commonly used in environmental monitoring. The experimental simulation environment is a Windows 10 64-bit operating system, with an IntelR CoreTM i5-6300HQ CPU, running at a main frequency of 2.30 GHz, and with a memory size of 8 GB. The programming software used is MATLAB2016a.

The results of applying the SCSEC encryption algorithm to environmental monitoring images are presented in Fig. 2. Figures 2(a), (b), and (c) depict the original environmental monitoring images, while Figs. 2(d), (e), and (f) show the decrypted (reconstructed) images obtained using the proposed SCSEC algorithm. Figure 2(g), (h), and (i) demonstrate the reconstruction failure when attempting to reconstruct the environmental monitoring images by randomly guessing the chaos factor λ to construct the chaotic matrix, without knowing the actual value of the chaos factor λ.

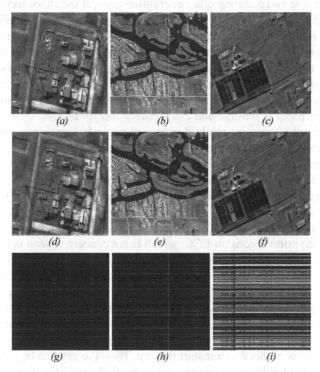

Fig. 2. Original image VS Reconstructed image VS Reconstructed failed image

The experimental results in Fig. 2 demonstrate that the reconstructed image has better quality when the SCSEC algorithm proposed in this paper is used for encryption. The decrypted images are closer to the original images and appear more realistic. Subjectively, the texture details of the reconstructed environmental monitoring images are clearly visible, making them closer to the original images. However, without knowing the chaotic factor λ randomly guessing a chaotic factor λ to construct a chaotic measurement matrix, reconstructing the result shows that it is a black image with white stripes that cannot see any information in the original image. In other words, it meets the requirements of image encryption [26]. Therefore, the SCSEC encryption algorithm proposed in this paper achieves a significant improvement in image encryption.

5 Discussion

5.1 Feasibility Analysis

The SCSEC image encryption algorithm proposed in this paper is different from other encryption algorithms, which encrypt a number rather than the image itself. The encryption method used in this paper is to "encrypt" the measurement matrix Φ in the process of signal acquisition, using the secp256k1 elliptic curve in Bitcoin. This mainly encrypts the chaotic factor λ instead of the image itself, resulting in almost zero cost during the acquisition process. Encryption is completed during the image acquisition process, saving encryption time and reducing other encryption costs. If the chaos factor λ cannot be cracked, which can be guaranteed by the ECC encryption algorithm, the image cannot be accurately reconstructed. Therefore, the SCSEC encryption algorithm achieves the purpose of encrypting environmental monitoring images. To reconstruct the signal, you need to obtain the measurement matrix Φ used for signal acquisition and use the reconstruction algorithm. However, it is almost impossible to reconstruct the signal when the measurement matrix Φ is unknown. Therefore, the SCSEC encryption algorithm proposed in this paper is feasible and can securely encrypt images.

The traditional encryption method uses symmetric key encryption to encrypt the image [27]. Although it has achieved certain results, it also has certain shortcomings, such as the problem of key distribution. When the ciphertext is transmited, it is also necessary to transmit the key, which raises security concerns. If the key is intercepted, the encryption of the image can be compromised. In addition, symmetric key encryption encrypts the image itself [28], which encrypts the image and is also an image in essence. Thus, it requires a large bandwidth during transmission. However, this paper uses the asymmetric encryption algorithm ECC, which is not concerned with key transmission. In addition, this paper encrypts the chaotic factors λ in the measurement matrix rather than the image itself. The chaotic factor λ is essentially a number. The file size of an encrypted number is obviously not on the same order of magnitude as the file size of the encrypted image itself. Therefore, the bandwidth requirements for the transmission of ciphertext in this paper are greatly reduced, which is conducive to the transmission of ciphertext. Furthermore, the ECC used in this paper does not have the key distribution problem, so there is no need to transmit the key. Thus, the key can be guaranteed to be secure, and the bandwidth requirements are further reduced. Therefore, the encryption algorithm proposed in this paper is feasible and an efficient transmission method, that can reduce the bandwidth and speed up the transmission of ciphertext.

5.2 Security Analysis

In information theory [29], the statistical independence between plaintext and ciphertext indicates the strength of security. Therefore, we can measure the security of a cryptographic system from this perspective. According to the definition of security in Shannon's theory [30], if a cryptographic system is a perfectly secure encryption system, it can guarantee that no information about the plaintext X can be obtained from the ciphertext Y, namely $P(X|Y) = P(X)$. In other words, the mutual information of plaintext X and ciphertext Y is 0, that is, $I(X; Y) = 0$. Therefore, as long as $I(X; Y) = 0$, it can be said that the system is secure.

As we all know, compressed sensing is defined as $y = Ax$. Assuming that x is a random vector with any probability distribution, $I(x; y)$ is used to represent the mutual information between x and y, and is defined $\varepsilon_x = \|x\|_2^2$. Therefore, we can obtain the following corollary:

Corollary: If the sensing matrix A satisfies the RIP requirement, then $I(x; y) = I(\varepsilon_x; y) = 0$.

Proof: The sensing matrix A satisfies the Gaussian distribution [25]. Thus, under the given probability distribution function $P(y|x)$, which is a multivariate Gaussian distribution with mean $\mu(y|x)$, the covariance matrix is $C(y|x)$, so we have the following result:

$$\mu(y|x) = E[y|x] = E[A]x = 0 \tag{2}$$

where $E[]$ represents expectation. Therefore, Eq. (2) can be rewritten as:

$$y = (I \otimes x^T)vec(A^T) \tag{3}$$

where $\otimes$ represents the Kronecker product and $vec(A^T)$ represents the column vectorization of the matrix by column superposition, so we can obtain:

$$\begin{aligned} E[y \cdot y^T |x] &= (I \otimes x^T)E[vec(A^T)vec(A^T)^T](I \otimes x) \\ &= \sigma_A^2(I \otimes x^T)(I \otimes x) \\ &= \sigma_A^2 x^T x I_m = \sigma_A^2 \varepsilon_x I_m \end{aligned} \tag{4}$$

In formula (4), m denotes the size of the measurement matrix, I_m represents the unit matrix with a size of $m \times m$, and σ_A^2 is the variance of the sensing matrix A.

It can be seen from the above results that ε_x independently obeys the $P(y|x)$ probability distribution, which can be obtained from [31]:

$$\begin{aligned} I(x; y) &= I(x, \varepsilon_x; y) \\ &= I(\varepsilon_x; y) + I(x; y|\varepsilon_x) = 0 \end{aligned} \tag{5}$$

Because $P(y|x) = P(y|\varepsilon_x)$, $I(x; y|\varepsilon_x) = 0$. Therefore, formula (10) can be transformed into $I(x; y) = I(\varepsilon_x; y) = 0$.

This paper aims to encrypt the measurement matrix Φ, in which the key is to use the secp256k1 elliptic curve in Bitcoin to encrypt the chaotic factor λ in the measurement matrix, rather than encrypting the image itself. Encrypting the chaotic factor λ is equivalent to encrypting the measurement matrix Φ. When we use the measurement matrix Φ to linearly measure the environmental monitoring image, we can obtain the measurement value $y = \Phi\psi s$, which indicates that the linear measurement value has a nonlinear mapping relationship with the measurement matrix Φ. Therefore, the reconstruction of the original signal itself becomes an NP-hard problem [32]. To facilitate the solution, the problem is converted from solving the l_0 minimum norm to solving the l_1 minimum norm problem. Under the condition that the measurement matrix cannot be known, the number of unknowns becomes even larger, which makes the number of unknowns much greater than the number of equations, making it almost impossible to solve. Therefore, without knowing the measurement matrix Φ, it is impossible to reconstruct the signal [33]. As long as the measurement matrix Φ cannot be deciphered, the chaotic factor λ cannot be deciphered, ensuring the safety of the measurement matrix. Therefore, guaranteeing the safety of the chaotic factor λ is the key to the security of the algorithm.

If you want to crack the chaotic factor λ, you need to crack the asymmetric encryption method (ECC encryption technology). However, it is almost impossible to crack the elliptic curve. The best time complexity of the current algorithm for cracking the ECC is $O(\sqrt{p})$, where p is the largest prime factor of the order n of the base point G. When p is sufficiently large, such as $p > 2^{160}$, it is impossible to crack the ECC based on the computing power of the current computer. As we all know, the prime192v curve is the "smallest" elliptic curve among the elliptic curves. To date, only 109-bit long elliptic curves were cracked in 1998 [34]. The most recent crack was in 2004 by a team of 2600 people. Consequently, to estimate how long it will take us to crack the prime192v1 curve with the same environment in 2004, we can make a rough calculation according to formula (6):

$$17 months \times \frac{\sqrt{2^{192}}}{\sqrt{2^{109}}} \approx 5.3 \times 10^{13} \qquad (6)$$

According to formula (6), it would take 5.3×10^{13} months to crack the "smallest" elliptic curve, which is equivalent to approximately 4.41×10^{12} years (441 billion years). However, the elliptic curve used in Bitcoin is secp256k1, in which the order n of the elliptic curve is 256. Thus, the time complexity of cracking the curve with the most efficient algorithm is $O(\sqrt{2^{256}}) = O(2^{128})$, and the time it takes to crack the curve is much more than 4410 years. Therefore, it is almost impossible to crack the elliptic curve in Bitcoin, which means it is impossible to crack the chaotic factor λ. In other words, the measurement matrix is safe. Consequently, the algorithm proposed in this paper can be guaranteed in terms of security performance.

6   Conclusions

In conclusion, this paper presents a novel solution to the challenges of environmental monitoring image lack of security. The proposed algorithm enhances the security of acquired images, effectively mitigates the security deficiencies in the compression perception reconstruction process, and demonstrates the suitability of elliptic curve cryptography for secure environmental monitoring image acquisition. Ultimately, this paper's contributions have the potential to advance the state of the art in environmental monitoring and offer new possibilities for secure and efficient storage and transmission of environmental monitoring data.

Acknowledgments. This paper was supported by the Jiangxi Province Network Space Security Intelligent Perception Key Laboratory Open Fund (No.: JKLCIP202205).

References

1. Li, J., et al.: A review of remote sensing for environmental monitoring in China. Remote Sens. **12**(7), 1130 (2020)
2. Dunbabin, M., Marques, L.: Robots for environmental monitoring: significant advancements and applications. IEEE Robot. Autom. Mag. **19**(1), 24–39 (2012)
3. Hakala, I., Tikkakoski, M., Kivelä, I.: Wireless sensor network in environmental monitoring-case foxhouse. In: 2008 Second International Conference on Sensor Technologies and Applications (sensorcomm 2008). IEEE (2008)
4. Xu, L.D., He, W., Li, S.: Internet of things in industries: a survey. IEEE Trans. Ind. Inform. **10**(4), 2233–2243 (2014)
5. Dong, B., et al.: A novel approach to improving the efficiency of storing and accessing small files on hadoop: a case study by powerpoint files. In: 2010 IEEE International Conference on Services Computing. IEEE (2010)
6. Zhou, X., Zhang, H., Wang, C.: A robust image watermarking technique based on DWT, APDCBT, and SVD. Symmetry **10**(3), 77 (2018)
7. Zhang, Y., et al.: HF-TPE: high-fidelity thumbnail-preserving encryption. IEEE Trans. Circuits Syst. Video Technol. **32**(3), 947–961 (2021)
8. Shah, T., Jamal, S.S.: An improved chaotic cryptosystem for image encryption and digital watermarking. Wireless Personal Commun. **110**(3): 1429–1442 (2020)
9. Arora, M., Khurana, M.: Secure image encryption technique based on jigsaw transform and chaotic scrambling using digital image watermarking. Opt. Quant. Electron. **52**(2), 1–30 (2020)
10. El-Khamy, S.E., Korany, N.O., Mohamed, A.G.: A new fuzzy-DNA image encryption and steganography technique. IEEE Access **8**, 148935–148951 (2020)
11. Bala Krishnan, R., et al.: An approach for attaining content confidentiality on medical images through image encryption with steganography. Wireless Personal Communications: 1–17 (2021)
12. Liu, J., et al.: A sensitive image encryption algorithm based on a higher-dimensional chaotic map and steganography. Int. J. Bifurcation Chaos **32**(01), 2250004 (2022)
13. Akkasaligar, P.T., Biradar, S.: Selective medical image encryption using DNA cryptography. Inf. Secur. J. Global Perspective **29**(2), 91–101 (2020)
14. Liu, Y., et al.: Optical image encryption algorithm based on hyper-chaos and public-key cryptography. Optics Laser Technol. **127**, 106171 (2020)
15. Donoho, D.L.: Compressed sensing. IEEE Trans. Inf. Theory **52**(4), 1289–1306 (2006)
16. Candès, E.J., Romberg, J., Tao, T.: Robust uncertainty principles: exact signal reconstruction from highly incomplete frequency information. IEEE Trans. Inf. Theory **52**(2), 489–509 (2006)
17. Baraniuk, R.G.: Compressive sensing [lecture notes]. IEEE Signal Process. Magaz. **24**(4), 118–121 (2007)
18. Koblitz, N., Menezes, A., Vanstone, S.: The state of elliptic curve cryptography. Des. Codes Crypt. **19**(2), 173–193 (2000)
19. Amara, M., Siad, A.: Elliptic curve cryptography and its applications. In: International workshop on systems, signal processing and their applications, WOSSPA. IEEE (2011)
20. Zhou, X., Tang, X.: Research and implementation of RSA algorithm for encryption and decryption. In: Proceedings of 2011 6th International Forum on Strategic Technology, vol. 2. IEEE (2011)
21. Corrigan-Gibbs, H., Kogan: The discrete-logarithm problem with preprocessing. In: Annual International Conference on the Theory and Applications of Cryptographic Techniques. Springer, Cham (2018). Doi: https://doi.org/10.1007/978-3-319-78375-8_14

22. Mahto, D., Yadav, D.K.: RSA and ECC: a comparative analysis. Int. J. Appl. Eng. Res. **12**(19), 9053–9061 (2017)
23. Yadav, A.K.: Significance of elliptic curve cryptography in blockchain IoT with comparative analysis of RSA algorithm. In: 2021 International Conference on Computing, Communication, and Intelligent Systems (ICCCIS). IEEE (2021)
24. Nakamoto, S.: Bitcoin: A peer-to-peer electronic cash system. Decentralized Bus. Rev., 21260 (2008)
25. Zhang, L., Zhang, J., Jin, A. Optimization and reconstruction for EPMA image compressed sensing based on chaotic measurement matrix. In: 2022 14th International Conference on Machine Learning and Computing (ICMLC), 02, pp. 474–482 (2022)
26. Kamali, S.H., et al.: A new modified version of advanced encryption standard based algorithm for image encryption. In: 2010 International Conference on Electronics and Information Engineering, vol. ol. 1. IEEE (2010)
27. Mohammad, O.F., et al.: A survey and analysis of the image encryption methods. Int. J. Appl. Eng. Res. **12**(23), 13265–13280 (2017)
28. Mandal, M.K., et al.: Symmetric key image encryption using chaotic Rossler system." Security and Communication Networks 7(11): 2145–2152 (2014)
29. Yockey, H.P.: Information theory, evolution and the origin of life. Inf. Sci. **141**(3–4), 219–225 (2002)
30. Shannon, C.E.: Communication theory of secrecy systems. Bell Syst. Techn. J. **28**(4), 656–715 (1949)
31. Thomas, M.T.C.A.J., Thomas Joy, A.: Elements of information theory. Wiley-Interscience (2006)
32. Zhang, Y., et al.: Signal reconstruction of compressed sensing based on alternating direction method of multipliers. Circuits Syst. Signal Process. **39**(1), 307–323 (2020)
33. Li, R.: Fingerprint-related chaotic image encryption scheme based on blockchain framework. Multimed. Tools Appl. **80**(20), 30583–30603 (2020). https://doi.org/10.1007/s11042-020-08802-z
34. Certicom, E. C. C.: Challenge and The Elliptic Curve Cryptosystem, pp. 409–418 (1998)

Infrastructure for Data Science

Two-Dimensional Code Transmission System Based on Side Channel Feedback

Han Sun, Qiang Gao, and Zhifang Wang[✉]

Department of Electronic Engineering, Heilongjiang University, Harbin 150080, China
`wangzhifang@hlju.edu.cn`

Abstract. Two-dimensional code technology is widely used in all walks of life due to its high reliability, large information capacity and relative security of information. As the carrier of data transmission, two-dimensional code has become a new form of near field communication, and mainly used for the flow of documents. The rapid spread of smart devices with cameras has further broadened the use of two-dimensional code. The combination of smartphones and two-dimensional code eliminates the dependence on special physical isolation devices and reduces transmission costs. In this paper, a two-dimensional code transmission system based on side channel is designed, which transmits confidential information through the visual main channel of two-dimensional code and retransmits feedback information through the acoustic side channel. This paper verifies the impact of QR code capacity on data integrity and improves file transfer integrity.

Keywords: QR code · Physical isolation · Information security · Smartphone · Acoustic side channel

1 Introduction

With the rise of cloud computing, the Internet of Things, mega data, 5G and other emerging technologies, the boundary of network information security is weakening, and data security and information security are facing great challenges. Data loss, illegal interception of information, and system intrusion occur frequently, and cyber threats continue to evolve and increase the concern of governments, enterprise, and individuals. To ensure the security of confidential information, governments and businesses often adopt authentication, encryption, and isolation techniques to prevent unauthorized access. Among them, the concept of physical isolation was initially proposed and adopted by the military of the United States, Israel and other countries. It restrains various security problems derived from the connection between classified networks and public networks to a certain extent.

A one-dimensional barcode is composed of a "bar" and "empty" arrangement to store information. It was widely used in commodity settlement, library management, warehousing and transportation and other fields in the 1970s. Two-dimensional code (2D code) was first studied in the United States and Japan in the 1980s. It is encoded horizontally and vertically to expand storage capacity. Compared with one-dimensional code, it can store not only letters and numbers, but also pictures, audio and other information. In addition, it increases the strong error correction function. There are more than 20 kinds of 2D code formats, such as Data Matrix, PDF417 and QR Code. The QR code has 4-level error correction capability and supports Chinese character coding. It has become the most popular 2D code at present. This paper is also based on QR codes.

In 2010, PixNet proposed a new form of near-field communication based on screen camera communication with 2D Code as the carrier, which encodes a large amount of data into a 2D Code stream [1]. The research aim of this new communication mode is usually the flow of documents and information. The 2D code encoding process adds the function of encryption to the data. In the transmission process, there are factors such as the capturing angle and ambient light, which further improve the communication security. For this form of communication with 2D Code stream as the carrier, there is no physical connection between the two systems. Each network is an independent channel, which realizes strong physical isolation. There is no influence on each other, and there is no data interaction. It has important practical significance for the transmission of confidential information scenarios.

Some researchers have studied the throughput of the 2D Code transmission system by adding color [2, 3], optimizing frame synchronization to improve the frame rate [4], and increasing the capacity of the black-and-white Code based on the pattern [5]. At present, reliability is also the focus of researchers. There are usually two methods to improve reliability, one is to use adaptive coding, such as RDCode [6] and ERSCC [7] and the other is to use retransmission feedback. The retransmission requests acoustic, optical and other channels [7, 8]. Because of the easy-to-use and relatively safe characteristic of 2D code, more research and applications have been made on the transmission of information in classified scenes. In 2018, automatic data transmission between classified information systems was realized [9]. Reference [7] designed a visual encryption algorithm to disguise the image transmitted between the screen and the camera. When the camera is set at the specified viewing position, the image is displayed. Only the camouflaged image can be seen from other locations.·

In this paper, a the two-dimensional code transmission system based on the acoustic side channel is designed. The visual channel is used for confidential information transmission, and the acoustic side channel is used for retransmission feedback. The reliability of the system is improved through the acoustic side channel, and security monitoring is studied to further improve the security of the system.

2 Systematic Design

2.1 System Overview

Figure 1 shows the overall framework of the system. It includes the QR code main channel and the acoustic side channel. In the QR code transmission channel, the sender preprocesses the data to be sent. It completes data segmentation, adds marks and packages data. The processed data is encoded into a QR code sequence, and it is played in order on the mobile phone screen. The receiver decodes the video and a QR code sequence by recording through the mobile phone camera. Finally, the data packets were split. The receiver synthesizes the original file to complete the data reception. In the sound side channel, the receiver encodes the sequence number of the missing data packet into an acoustic wave, and plays it through the loudspeaker. It can perform real-time retransmission feedback. The transmitter records audio through the microphone, and decodes the sound wave. In the end, it splits the data packet content and retransmits the missing images in real time.

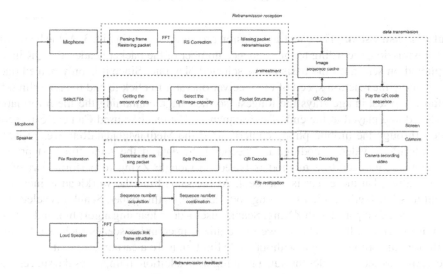

Fig. 1. System Architecture

2.2 Main Channel Transmission Process

Due to the limited capacity of a single QR code image, it is impossible to store a large amount of data in a single image, such as long files, pictures, and videos. For the purpose of data transmission, the data are segmented marked and encoded by the QR encoder, and transmitted in the form of an image sequence.

Data Preprocessing. First, the sent data were preprocessed, and the sender obtained the file size m and the selected QR code image capacity n. If $m\%n = 0$, the data were divided into m/n segments. If $m\%n > 0$, split the data into $(m/n) + 1$ segments. To improve the reliability of data transmission, the packet design of the segmented content was carried out, and the header was added as the packet sequence number, which was used as the packet identification and retransmission request content in the acoustic side channel. When the sender and the receiver communicate, the recognition of the QR code can only extract the data packet about the content of the file, and file attributes such as file name, file type, and the number of packets cannot be obtained. It cannot guarantee the integrity of the file information. Therefore, the data packet is set up to transmit the file attributes, which provides the basis for receiving the file. The filename and file type are used to synthesize the final file, and the number of packets is used to judge the missing packets. The file information packet structure is "fileName:type:codenums" with ":" as the subsection basis. The file content packet structure is shown in Fig. 2.

Packet Sequence Number	Segmented Content
6byte	(n-6)byte

Fig. 2. File content Packet structure

Data Transmission. The data transmission process is shown in Fig. 3. At the beginning of transmission, according to the sequence number of the packet header, the QR image is played on screen on the basis of the selected play rate. Since it is only played once, we observe the reaction speed of the receiver operator to click to end the recording. By default, the last image stays the longest, so the image containing the file information is placed and played at the end of the sequence. The receiver used CameraX to record video through the mobile phone camera, used the MediaMetadataRetriever provided by android for video decoding, and decoded the data at the same time. At present, the mainstream decoding method is based on Google Zxing to perform part of the optimization, but the effect is not ideal. At the same time the workload is increased. Scankit is HuaWei's unified scanning code service. In this paper, Scankit is selected as the encoder. Compared with ZXing, Scankit uses a deep learning algorithm, and extracts spatial invariance. It can also be well recognized in complex scenes such as reflection, pollution, distortion and blur without angle limitation. To a certain extent, it extends the use scenarios such as under the sun and at night. In addition, it improves the success rate of 2D code recognition.

Fig. 3. File content Packet structure

Data Restoration. The data recovery process is shown in Fig. 4. The file content packets were parsed into the collection, and the identified packet sequence number was obtained. The receiver parses the file information packets to obtain the number of packets, file name and file type. The number of data packets and the obtained data packet sequence number were used to determine the missing packet sequence number and wait for retransmission. If the data packet sequence number is complete, the data restoration is completed locally according to the file name and file type.

2.3 Acoustic Side Channel Transmission Process

In 1945, the United States Navy developed the first underwater acoustic communication system. It is mainly for submarine communication. Acoustic communication technology

Fig. 4. Data merging

is a new short-range communication technology, that uses acoustic waves as carriers to transmit data in the air medium. This method of communication is less dependent on the infrastructure, and only uses the smart device's speaker and microphone. The transmission of acoustic waves in air is slower than that in water, and the transmission rate of acoustic communication is slower for a large amount of data. Without considering the cloud server, it is more suitable for the transmission of a small amount of data. In this paper, acoustic communication is used as the side channel to transmit the missing packet sequence number.

Retransmit Feedback. The retransmission feedback module receives the missing data packet sequence number and concatenates it. "Rtrs000100040026" is taken as an example. It is part of the retransmission frame content. The data content is divided into segments, with each 4byte as a data segment, and the last less than 4byte also as a data segment. The system receiver splits the data into 9 byte frames for acoustic generation. Acoustic communication has been used to transmit feedback information in this paper. It will increase the complexity of the system to solve the error code problem by retransmissing. The way to restore the error data is to encode the data content to the greatest extent. The transmission frame structure is shown in Fig. 5, where the packet header is used to identify the beginning of a packet and record the data length, 4 byte in the middle of the packet is used to store the data content, and 4 byte at the end is used to store RS error correction bits. Audio modulation is implemented using FFT, 16 bit audio sampled at 16 kHz, and 10 transmission channels below 8 kHz are selected for transmission.

Header	Data Length	Segmented Content	RS Error Correction
1byte		4byte	4byte

Fig. 5. Acoustic link frame structure

Acoustic communication quality is affected by many factors such as environmental noise, multipath interference, and frequency selection of the speaker-microphone link. Frequency attenuation may be caused by distance, occlusion, and whether the microphone and speaker are relative to each other. To ensure the integrity of the data as much as possible, one of the channels is selected to transmit the parity information of the other nine channels.

Retransmit Receive. The receiver parses each frame data and restores data packets. Frame data are divided into 9 segments of 1 byte each as shown in Fig. 6. When there is an error in byte data, RS can correct the error. The receiver performs windowed FFT on the received signal section by section and analyses the frequency of the sound wave at which time, so as to obtain the transmission content.

Fig. 6. Diagram of frame parsing

Discussion. Security is the primary requirement for systems that transmit classified information. In the process of information transmission, information signals are released inadvertently by computers and smart devices, such as electromagnetic radiation, sound, and electricity [10]. Intruders use this information to decipher and steal information. With an increasing understanding of side-channel attacks, how to prevent them can be studied. This paper designs a QR code transmission system based on side channel. And the acoustic side channel is used for retransmission feedback. At the same time, it will also be further studied to carry out security monitoring through the side channel. By detecting the intrusion signal, the acoustic channel alarms in time to improve the security of the transmission system.

3 Experiments

3.1 Test Environment

Preparation was carried out before the test. Both the sender and the receiver use mobile devices. The device used for the transmitter is VIVO Y93 and Android version 8.1.0. The device used for the receiver is OPPO PEQM00 and the Android version is Android 12.

3.2 Data Integrity Test

The test was carried out with the same device in the indoor daytime light and indoor nighttime light environments. The receiver parses the video at 30 fps and 20 fps. The sender plays at rates of 10 fps, 15 fps, 20 fps, 25 fps and 30 fps. The capacity of a single QR code increases from 100 to 1800 every 100 bytes. Testing the number of missing packets for each transmission of the same file, translated into the received integrity rate is shown in Fig. 7 and Fig. 8.

With the increase in the capacity of a single QR code, the transmission integrity is exhibits a downwards trend. As shown in Fig. 7, the capacity of a single QR code is relatively stable before 1100 bytes, and the completeness can basically remain above 80%. The number of missing data packets accounts for a small proportion, and retransmission will not take too long. Beyond 1100 bytes, data completeness drops dramatically. Exposure, shooting jitter, the jelly effect and other factors can also have an impact. As

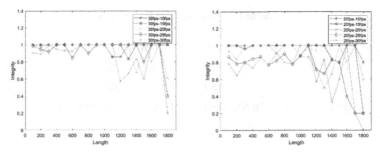

Fig. 7. Trend graph of completeness of received files in indoor daytime

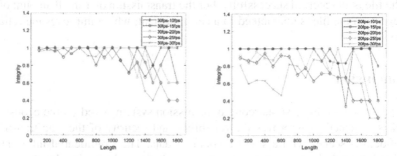

Fig. 8. Trend chart of receiving file completeness with lights on indoors at night

shown in Figs. 7 and 8, when the critical point is within 1100 bytes and the parsing rate is greater than or equal to the playback rate, the system performs better and more stably in both dark and overlight environments. Analysing videos with different playback rates at 30 fps and 20 fps, the slower the playback rate of the sender is, the more times the same QR code image will be captured repeatedly. It can offset the influence of other factors to a certain extent and improve the integrity of file transmission.

3.3 Transmission Success Rate Test

According to the above experimental results, the receiver parses the video at 30 fps. The sender plays at 15 fps, 20 fps and 25 fps. The capacity of a single QR code increases from 100 to 1100 every 100 bytes. We test the success rate of each transfer of the same file. If the transmission fails, we test the success rate of retransmission feedback and the success rate of the sender retransmission once. The success of retransmission feedback includes successful playback by the receiver and successful recognition by the sender. The tester holds two devices, the distance is kept at approximately 7-8cm, and the volume of the receiving system is fixed at 80%. The QR code playback rate is fixed at 2 fps when retransmitting. The test results are shown in Table 1.

As shown in Table 1, when the capacity of a single QR code is less than 1100 byte, the performance is relatively stable. When the parsing rate is greater than or equal to 2 times the play rate, the success rate is basically above 80%. When the parsing rate is less than 2 times the play rate, the success rate is almost more than half. The retransmission

Table 1. Success rate of reception

Frame Rate	100 (Byte)	200	300	400	500	600	700	800	900	1000	1100
15 fps	1	1	1	0.83	0.83	0.83	0.83	0.83	0.5	0.83	0.67
20 fps	1	0.5	1	0.83	1	1	0.67	0.67	0.67	0.83	0.5
25 fps	0.33	0.5	0.67	0.67	0.83	0.67	0.67	0.5	0.5	0.67	0.33

test is performed in the case of missing packets in Table 1. Under the above conditions, acoustic channel encoding and decoding are successful. After the sender retransmits once, the file is all received successfully. For the transmission of a small amount of data, the integrity of the file is guaranteed to a certain extent, which improves the reliability of the system.

4 Conclusion

In this paper, a two-dimensional code transmission system based on the acoustic side channel is designed. It improves the reliability and security of the data transmission for confidential scenes with a small amount of data. The Confidential information is transmitted through the main transmission channel of two-dimensional code in specific scenarios. The real-time retransmission information feeds back through the acoustic side channel. This article improves the integrity of transferred files without overcoding. This paper verifies the influence of QR code capacity on data integrity, and gives the QR code capacity range of the adaptive system. The transmission success rate is higher when the capacity of a single QR code is less than 1100 bytes. The acoustic feedback mechanism performs stably under fixed conditions, which lays the foundation for subsequent side-channel monitoring work and improving system safety.

References

1. Perli, S.D., Ahmed, N., Katabi, D.: 2010. PixNet: interference-free wireless links using LCD-camera pairs. In: Proceedings of the Sixteenth Annual International Conference on Mobile Computing and Networking (MobiCom 2010), pp. 137–148. Association for Computing Machinery, New York
2. Langlotz, T., Bimber, O.: Unsynchronized 4D barcodes: coding and decoding time-multiplexed 2D colorcodes. In: Proceedings of the 3rd international conference on Advances in visual computing - Volume Part I (ISVC'07), pp. 363–374. Springer, Heidelberg (2007). http://doi.org/10.1007/978-3-540-76858-6_36
3. Hao, T., Zhou, R., Xing, G.: COBRA: color barcode streaming for smartphone systems. In: Proceedings of the 10th International Conference on Mobile Systems, Applications, and Services (MobiSys '12), pp. 85–98. Association for Computing Machinery, New York (2012)
4. Hu, W., Gu, H., Pu, Q.: LightSync: unsynchronized visual communication over screen-camera links. In: Proceedings of the 19th Annual International Conference on Mobile Computing & Networking (MobiCom '13), pp. 15–26. Association for Computing Machinery, New York (2013)

5. Zhan, T., Li, W., Chen, X., Lu, S.: Capturing the shifting shapes: enabling efficient screen-camera communication with a pattern-based dynamic barcode. Proc. ACM Interact. Mob. Wearable Ubiquitous Technol. 2, 1, Article 52 (March 2018), 25 pages (2018)

6. Wang, A., Ma, S., Hu, C., Huai, J., Peng, C., Shen, G.: Enhancing reliability to boost the throughput over screen-camera links. In: Proceedings of the 20th Annual International Conference on Mobile Computing and Networking (MobiCom 2014), pp. 41–52. Association for Computing Machinery, New York (2014)

7. Zhang, O., Qian, Z., Mao, Y., Srinivasan, K., Shroff, N.B.: ERSCC: enable efficient and reliable screen-camera communication. In: Proceedings of the Twentieth ACM International Symposium on Mobile Ad Hoc Networking and Computing (Mobihoc 2019), pp. 281–290. Association for Computing Machinery, New York

8. Zhou, M., Wang, Q., Lei, T., Wang, Z., Ren, K.: Enabling online robust barcode-based visible light communication with realtime feedback. IEEE Trans. Wireless Commun. 17(12), 8063–8076 (2018). https://doi.org/10.1109/TWC.2018.2873731

9. Yang, H., Zhang, Y., Jiang, Y., Ling, H.: Design of confidential network data transmission system based on two-dimensional code. Comput. Technol. Automation 39(03) (2020)

10. Wang, Y., Guo, H., Yan, Q.: Network and Distributed Systems Security (NDSS) Symposium 202224-28 April 2022, San Diego, CA, USAISBN 1-891562-74-6

An Updatable and Revocable Decentralized Identity Management Scheme Based on Blockchain

Zhiping Wang, Meijiao Duan$^{(\boxtimes)}$, and Maoning Wang

School of Information, Central University of Finance and Economics, Beijing , China
duanmeijiao@cufe.edu.cn

Abstract. We study the decentralized identity management mechanism based on blockchain. Finally, we propose an updatable and revocable decentralized identity management scheme DIURS. In the scheme, we construct the DID management tree, which is a dynamic chameleon authentication tree essentially by using the chameleon hash function. We design algorithms in detail from four stages: system initialization, identity creation, identity update and revocation, and identity verification. We make the DID documents on the blockchain editable successfully and realize the update and revocation of DIDs. Then, we observe that DIURS can meet the structural stability and irreversibility requirements. The time of identity search and update is milliseconds. The length of the identity authentication path is short. There is no need to save the historical version of DID documents. These results indicate that DIURS is not only safe and reliable but also performs well and achieves functional optimization.

Keywords: Blockchain · decentralized identity · Chameleon hash function · DID management tree

1 Introduction

With the gradual maturity of Internet technology, the concept of digital identity has received more attention. The problem of user digital identity management has been widely considered by many researchers.

The development of digital identity management technology can be divided into four phases: the centralized identity management phase, which is controlled by third-party organizations [1]; the federated identity management phase, which connects different identity providers across multiple organizations [2]; the user-centric identity management phase, which can facilitate user control and reduce information leakage; and the self-sovereign identity management phase, which emphasizes that identity is fully controllable by the user, such as decentralized identity [3, 4].

It can be seen that the management mechanism of digital identity is optimized constantly and has achieved certain results. However, there is a large number of historical identity records leading to redundancy. Users' personal data may be stolen illegally, leading to personal privacy breaches. The development of decentralized identity is in

Z. Yu et al. (Eds.): ICPCSEE 2023, CCIS 1879, pp. 372–388, 2023.
https://doi.org/10.1007/978-981-99-5968-6_27

the early stage. There are still some areas to be improved. It is necessary to research decentralized identities constantly.

In our work, we propose a decentralized identity management scheme DIURS that can realize identity update and revocation. We make the following contributions.

First, we define the DIURS scheme. We define some algorithms in DIURS and the security model of DIURS.

Second, we design the DIURS scheme. We realize the update and revocation of DID by constructing a DID management tree and using a chameleon private key to find hash collisions. We give the detailed algorithms in the initialization phase, identity creation phase, identity update and revocation phase, and identity authentication phase.

Third, we analyse the DIURS scheme. It can be confirmed that DIURS has structural stability and irreversibility through safety demonstration. After that, we analyse its performance efficiency and compare it with other DID schemes.

2 Related Work

Self-sovereign identity management is the latest management model for digital identity that emphasizes that personal information is completely owned and controlled by users themselves and that a third party is no longer needed to manage identity. Decentralized identity, also known as distributed digital identity, or DID for short, conforms to the principle of self-sovereign identity [5]. It was proposed by the W3C in 2017, and the latest DID standard specification was released in 2022 [6].

DID consists of two parts: the base layer and the application layer. The base layer consists of the DID identifier used to uniquely identify the DID subject globally and the DID document resolved by the DID resolver based on the identifier to describe the DID subject. The DID document stored on the blockchain contains some attribute information, such as the DID identifier, user's DID public key, encryption protocol, service endpoint, time stamp, and signature. W3C provides specifications [7]. The application layer primarily refers to the user's verifiable credential. It is a collection of many verifiable claims that contain user attribute information, so it can be used to verify the legitimacy of DID. The W3C published the latest Verifiable Credential Data Model specification in March 2022 [8] (Fig. 1).

Currently, the proposed self-sovereign identity schemes include Sovrin [9], uPort [10], ShoCard [11, 12], Bitnation [13], Civic [14], EverID [15], LifeID [16] and decentralized identity schemes such as WeIdentity [17] and BaiduDID [18]. See Table 1.

Sovrin [9] is an open source decentralized identity network built on an authorized distributed ledger. It does not provide any guarantee about the correct running of the network agent. uPort [10] is an open source scheme based on Ethereum aiming to provide a decentralized identity for all people. However, it lacks portability and interoperability. The safety of the recovery mechanism still needs to be considered. ShoCard [11, 12] is able to correlate user identifiers, trusted credentials and other information using blockchain technology and hash functions. WeIdentity [17] is a blockchain-based decentralized identity scheme developed by Webank independently. DID documents can be updated and read through the Solidity event mechanism, which requires an index to

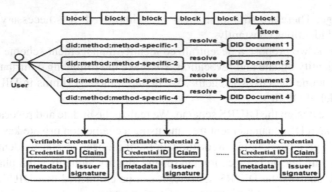

Fig. 1. Structure of DID

Table 1. Some basic information of the SSI scheme.

SSI scheme	Proposed time	Published ledger	Blockchain type
Sovrin	2016	HyperLedger Indy	permissioned
uPort	2016	Ethereum	permissionless
ShoCard	2015	Bitcoin	permission(less, ed)
Bitnation	2014	NEO	-
Civic	2015	Ethereum	-
EverID	2016	Ethereum	permissioned
LifeID	2017	LIFEID	permissionless
WeIdentity	2018	FISCO BCOS	permissioned
BaiduDID	2019	Ethereum/Quorum	-

the original document constantly for integration to obtain the latest DID document version. The convenience needs to be improved. BaiduDID [18] is a decentralized identity scheme developed by Baidu Company. The update of identity is based on the event mechanism. BaiduDID supports path query, so you can add a specific version number after the identifier to search the specified DID document. You can revoke DID by changing the value of the "*operation*" attribute to "*delete*" in the DID document. BaiduDID also saves many historical identity versions that are no longer needed, resulting in a waste of resources. The specific functional evaluation is summarized in Table 2, where "*1*" indicates satisfaction and "*0*" indicates room for improvement.

Rundong Wang designed a decentralized identity authentication system based on blockchain [19]. When the identity is updated, the information of all versions is recorded in the chain. We can perform a complete identity search through index linking to the previous identity state. It is similar to WeIdentity.

Table 2. Some functional evaluation of the SSI scheme.

SSI scheme	Control	Access	Transparency	Persistence	Portability	Interoperability	User authorized share	Minimal disclosure	Protection	Provable
Sovrin	1	1	1	1	1	1	0	1	0	0
uPort	1	1	1	1	0	0	0	1	1	1
ShoCard	1	1	1	0	0	0	1	0	1	1
Bitnation	1	1	1	1	1	0	1	0	1	0
Civic	1	1	0	0	1	0	1	1	1	0
EverID	1	1	0	1	1	1	1	0	1	0
LifeID	1	1	0	1	1	1	1	1	1	0
WeIdentity	1	1	1	1	1	1	1	0	1	1
BaiduDID	1	1	1	1	1	0	1	0	1	1

In summary, there are few relevant studies on decentralized identity. It is at an early stage of development. This paper is an innovative study on the update and revocation of decentralized identity on blockchain.

3 Preparatory Knowledge

3.1 Chameleon Hash Function

The chameleon hash function is a nonstandard collusion-resistant hash function that contains the public key cpk and the private key csk (also known as the trap gate). It can be divided into three parts.

Key generation algorithm $Ch\_Key(1^\lambda) = (cpk, csk)$: The input is constant λ. The output is chameleon public key cpk and private key csk.

Hash generation algorithm $Ch\_Hash(cpk, m, r) = h$: The input is cpk, binary message m and binary random number r. The output is the chameleon hash value h.

Hash collision algorithm $Ch\_Col(csk, m, r, m') = r'$: The input is csk, message m, random number r and new binary message m'. The output is a random number r'. They can satisfy the equation $Ch\_Hash(cpk, m, r) = Ch\_Hash(cpk, m', r') = h$.

Common chameleon hash functions include the chameleon hash function based on discrete logarithm [20], the chameleon hash function based on simple factorization or advanced factorization [21] and the chameleon hash function based on strategy.

3.2 Chameleon Hash Function Properties

A safe chameleon hash function needs to satisfy the following properties.

Collision resistance [22, 23]: In a deterministic polynomial complexity algorithm A, the probability of outputting a random number r' satisfying $Ch\_Hash(cpk, m, r) = Ch\_Hash(cpk, m', r')$ is negligible without knowing the trap gate csk when we input m, r and m'.

Semantic security [22, 23]: For any two messages m and m', the probability distribution of the $Ch\_Hash(cpk, m, r)$ value and $Ch\_Hash(cpk, m', r')$ value is computationally indistinguishable. In particular, for any r, no information about m can be obtained from $Ch\_Hash(cpk, m, r)$.

The chameleon hash functions above satisfy the property of collision resistance and semantic security.

4   Decentralized Identity Update and Revoke Scheme Definition

We use chameleon hash and standard hash together to construct a dynamic chameleon authentication tree named the DID management tree to realize the update and revocation of decentralized identity on blockchain. DIURS consists of four stages: the initialization stage, identity creation stage, identity update and revocation stage, and identity verification stage.

4.1   Algorithm Definition

The four stages contain the following six algorithms.

$SystemInit(\lambda) \rightarrow (csk, cpk, \rho, n, \Pi_n)$: initialization algorithm. The input is the security parameter λ. The output is chameleon private key csk, chameleon public key cpk, root node hash value ρ, the number of DIDs already stored n and the verification data Π_n.

$DIDCreate(csk, \rho, cpk, Document, n, \Pi_n) \rightarrow (n, \Pi_n)$: identity creation algorithm. The input is chameleon private key csk, root node hash value ρ, chameleon public key cpk, user's DID document $Document$, n and Π_n to complete the creation of user's DID. The output is the total number of identities n and verification data Π_n.

$DIDCreateNotFull(csk, cpk, v_{i,j}, Document, n, \Pi_n) \rightarrow (\Pi_n)$: identity creation algorithm, which is included in algorithm $DIDCreate$. We can create a new DID directly when the DID management tree does not need to expand the depth. The algorithm output is the latest verification data Π_n.

$DIDUpdRvk(csk, \rho, Documentı, index, \Pi_n, depth) \rightarrow (\Pi_n)$: identity update and revocation algorithm. The input is the chameleon private key csk, root node hash value ρ, user's new DID document $Documentı$, $index$ of the identity node, and verification data Π_n and $depth$ of the DID management tree. The output is the latest verification data Π_n.

$AuthPath(index, \Pi_n, depth) \rightarrow (ap_{index})$: authentication path generation algorithm. It is mainly used to generate the authentication path of a node. The input is $index$ of the identity node, verification data Π_n and $depth$ of the DID management tree. The output is authentication path ap_{index} of the identity node.

$DIDVerify(csk, \rho, cpk, index, Document, ap_{index}, depth) \rightarrow (trueorfalse)$: identity verification algorithm. The output is $true$ or $false$ to confirm whether the identity was created or updated successfully.

4.2   Security Model

We need to prove that the DIURS proposed in this paper satisfies structural stability and irreversibility by defining an interactive game to simulate the interaction between the adversary and the challenger.

Definition 1 Structural stability: We assume that an adversary A of any polynomial time algorithm attempts to obtain a forged identity and identity authentication path by changing the order of leaf nodes or replacing the content of leaf nodes. If the probability of adversary A winning is negligible, DIURS will be considered to satisfy structural stability.

Definition 2 Irreversibility: We assume that an adversary A of any polynomial time algorithm attempts to insert new data into the DID management tree and obtains the new data and its authentication path. If the probability of adversary A winning is negligible, DIURS will be considered to satisfy irreversibility.

Game: Adversary A can send DID document information several times and obtain the corresponding *index* and authentication path ap_{index} in response.

Initialization phase: the challenger runs $SystemInit(\lambda)$ to calculate private key csk and public key cpk and sends cpk to adversary A.

Challenging phase: We assume that the maximum number of DIDs that can be searched by adversary A is $q := q(\lambda)$. Adversary A sends data $d_i (1 \leq i \leq q(\lambda))$ to the challenger. The challenger runs $DIDCreate(csk, \rho, cpk, d_i, n, \Pi_n)$ to insert the data into the DID management tree and runs $AuthPath(index, \Pi_n, depth)$ to obtain the corresponding authentication path ap_{index}. The challenger returns $Q := \{(d_1, 1, ap_1),$ $(d_2, 2, ap_2), \ldots, (d_{q(\lambda)}, q(\lambda), ap_{q(\lambda)})\}$ to adversary A.

Guessing phase: Adversary A outputs data d^*, index i^* and path ap^* together to compose (d^*, i^*, ap^*). Case 1: If the output satisfies $1 \leq i^* \leq q(\lambda), (d^*, i^*, ap^*) \notin Q$ and $DIDVerify(csk, \rho, cpk, i^*, d^*, ap^*, depth)$ returns *true*. Then, we can consider that adversary A won and the structural stability of DIURS was destroyed. Case 2: If the output satisfies $i^* \geq q(\lambda)$ and $DIDVerify(csk, \rho, cpk, i^*, d^*, ap^*, depth)$ returns *true*. Then, we can consider that adversary A won and the irreversibility of DIURS was destroyed.

If DIURS satisfies both inequalities (1) and (2) while $\varepsilon(\lambda)$ is a negligible function:

$$\Pr\left[\begin{array}{c} (d^*, i^*, ap^* \notin Q) \\ 1 \leq i^* \leq\sim q(\lambda) \end{array} : DIDVerify(csk, \rho.cpk, d^*, i^*, ap^*, depth) \rightarrow true \right] \tag{1}$$
$$\leq \varepsilon(\lambda)$$

$$\Pr\left[\begin{array}{c} (d^*, i^*, ap^*) \\ i^* \geq q(\lambda) \end{array} : DIDVerify(csk, \rho.cpk, d^*, i^*, ap^*, depth) \rightarrow true \right] \tag{2}$$
$$\leq \varepsilon(\lambda)$$

We can consider that the probability of adversary A winning is negligible and the DIURS is safe.

5 Decentralized Identity Update and Revoke Scheme Design

In DIURS, the leaf nodes of the DID management tree store the DID documents corresponding to each DID. There are two kinds of intermediate nodes: chameleon nodes and ordinary nodes. The root node of the tree is initially constructed as a chameleon node and becomes a normal node with tree expansion. The structure of the DID management tree is shown in Fig. 2.

In this paper, we can choose the SHA256 hash function and the chameleon hash function based on the discrete logarithm to calculate the hash of the node. We assume

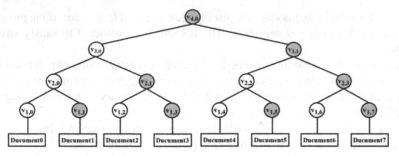

Fig. 2. Structure of DID management tree in DIURS scheme

that the prime numbers p and q satisfy $p = kq + 1$ while q is sufficiently large. Z_p^* is a group of order q, and g is the generating element of the group. First, we define the key generation function $GenCHKey() \rightarrow (csk, cpk)$ and assume $csk \in Z_p^*$, so $cpk = g^{csk} \bmod p$. Second, we define the chameleon hash generation function $GenChash(cpk, m, r) \rightarrow v$. For the message $m_1 \in Z_p^*$ and the random number $r_1 \in Z_p^*$, we can determine the chameleon hash $v = g^{m_1} cpk^{r_1} \bmod p$ of m_1. For the message $m_2 \in Z_p^*$, we can find $r_2 \in Z_p^*$ to satisfy $GenChash(cpk, m_1, r_1) = GenChash(cpk, m_2, r_2)$, namely, $GenChash(cpk, m_2, r_2) = v$. Third, we define the chameleon hash collision function $GenCol(csk, v, m_2) \rightarrow r_2$, and $r_2 = \frac{m_1 - m_2}{csk} + r_1 \bmod q$. In addition, function $CHRan(i, j) \rightarrow <x, r>$ is defined to generate a random message x and the corresponding initial random number r for chameleon node (i, j).

5.1 Initialization Stage

The system executes algorithm *SystemInit* to generate chameleon private key csk and chameleon public key cpk after receiving 128-bit security parameter λ. Then, it generates random message $x_{1,0}$ and random number $r_{1,0}$ for the initial root node. Some parameters should be initialized, such as the root node hash value ρ, the total number of DIDs stored in the tree n and the verification data Π_n. It is important to keep csk and Π_n stored by the system private and safe. cpk, ρ and n are public parameters that can be known by all users.

Algorithm SystemInit

Input : λ
Output : csk, cpk, ρ, n, Π_n
1: param $\leftarrow$ Setup(λ)
2: $<$ csk, cpk $> \leftarrow$ GenCHKey(param)
3: $< x_{1,0}, r_{1,0} > \leftarrow$ CHRan(1, 0)
4: $\rho \leftarrow$ GenChash(cpk, $x_{1,0}, r_{1,0}$)
5 : n $\leftarrow$ 0
6 : $lv_n \leftarrow \varphi, rv_n \leftarrow \varphi, rp_n \leftarrow \varphi$
7 : $\Pi_n \leftarrow \{lv_n, rv_n, rp_n\}$
8 : **return**   csk, cpk, ρ, n, Π_n

5.2 Identity Creation Stage

A DID subject can have more than one DID at a time. Each DID has a unique DID identifier that consists of a fixed character *did*, a DID method field and a DID method-specific string. We can obtain the corresponding DID document describing DID in detail after the identifier is resolved. The main fields of DID document are shown in Table 3.

Table 3. The main fields of the DID document.

Field	Description
@context	describe DID document structure
id	DID identifier
created	document creation time
updated	document update time
publicKey	DID public key list
publicKey.id	DID public key id
publicKey.type	public key verification algorithm
authentication	the public key used to verify the signature reference from publicKey
service	the service associated with the current DID
service.type	service endpoint type
service.serviceEndpoint	URI or JSON-LD object
operation	describe DID document state
signatureValue	private key signature

The total number of DIDs is 0 when system initialization is complete. We execute algorithm *DIDCreate* to construct the DID management tree by generating a random message $x_{1,0}$ and random number $r_{1,0}$ for the root node. Then, we update the verification data Π_n. When the total number of existing DIDs is within the range of $2^{d-2} < n < 2^{d-1}$, there is no need to expand the depth of the DID management tree. We find the right node (i, j) at the lowest level of the childless tree and execute algorithm *DIDCreateNotFull*. If the node is in the first layer, the system calculates the collision random number of the DID document to overwrite the contents saved by the node. The system saves the new random number and new hash. Otherwise, the system generates a leftmost right node for each layer below and creates a DID to the sibling node of the right node generated newly at the first layer. Then, the system fills in the missing left nodes at each layer bottom-up recursively. When the total number of existing DIDs is $n = 2^{d-1}$, meaning that the DID management tree is a full binary tree, it is necessary to create a new root node to expand the depth of the tree. We create node $(d, 1)$ as the right node of the new root node, while the current tree changes into a left subtree with the original root node transformed into a normal node. The system generates $< x_{d,1}, r_{d,1} >$ for node $(d, 1)$ and calculates the chameleon hash value. Then, the system executes algorithm *DIDCreateNotFull*. The total number of DIDs changed from n to $n + 1$.

Algorithm DIDCreate

Input : csk, ρ, cpk, Document, n, Π_n
Output : n, Π_n
1 : if n $==$ 0 then
2 : $r'_{1,0} \leftarrow$ GenCol(csk, ρ, Document)
3 : $rp_n \leftarrow rp_n + \{r'_{1,0}\}$
4 : else
5 :   if n $==$ 2^{d-1} then
6 : $< x_{d,1}, r_{d,1} > \leftarrow$ CHRan(d, 1)
7 : $v_{d,1} \leftarrow$ GenChash(cpk, $x_{d,1}, r_{d,1}$)
8 :   if d $==$ 1 then
9 : $v'_{d,0} \leftarrow$ Hash(Document0)
10 :   else
11 : $v'_{d,0} \leftarrow$ Hash($v_{d-1,0} \parallel v_{d-1,1}$)
12 :   end if
13 : $r_{d+1,0} \leftarrow$ GenCol(csk, $\rho, v'_{d,0} \parallel v_{d,1}$)
14 : $\Pi_n \leftarrow$ **DIDCreateNotFull**(csk, cpk, $v_{d,1}$, Document, n, Π_n)
15 : $rp_n \leftarrow rp_n - \{r_{d,0}\} + \{r_{d+1,0}\}$, $lv_n \leftarrow lv_n + \{v'_{d,0}\}$
16 : $d \leftarrow d + 1$
17:     else
18 :   search the lowest level right node (i, j) without child nodes
19 : $\Pi_n \leftarrow$ **DIDCreateNotFull**(csk, cpk, $v_{i,j}$, Document, n, Π_n)
20 :   end if
21 : end if
22 : n $\leftarrow$ n + 1
23 : **return** n, Π_n

Algorithm DIDCreateNotFull

Input : csk, cpk, $v_{i,j}$, Document, n, Π_n

Output : Π_n

1 : if i $==$ 1 then

2 : $r'_{i,j} \leftarrow$ GenCol(csk, $v_{i,j}$, Document)

3 : else

4 :   for k $\in \{i-1, \dots, 1\}$ do//generate a new left-most right child node for each layer

5 : $< x_{k,2^{i-k}\cdot j+1}, r_{k,2^{i-k}\cdot j+1} > \leftarrow$ CHRan(k, $2^{i-k} \cdot j + 1$)

6 : $v_{k,2^{i-k}\cdot j+1} \leftarrow$ GenChash(cpk, $x_{k,2^{i-k}\cdot j+1}, r_{k,2^{i-k}\cdot j+1}$)//calculate chameleon hash

7 : $rv_n \leftarrow rv_n + \{v_{k,2^{i-k}\cdot j+1}\}$//save chameleon node hash value to rv_n

8 :   end for

9 :   for k $\in \{1, \dots, i-1\}$ do//fill the missing left node at each layer bottom-up

10 :   if k $==$ 1 then

11 : $v_{1,2^{i-1}\cdot j} \leftarrow$ Hash(Document)//calculate the hash value of Document

12 :   else

13 : $v_{k,2^{i-k}\cdot j} \leftarrow$ Hash($v_{k-1,2^{i-k+1}\cdot j} \| v_{k-1,2^{i-k+1}\cdot j+1}$)//fill other nodes value

14 :   end if

15 : $lv_n \leftarrow lv_n + \{v_{k,2^{i-k}\cdot j}\}$//save node hash value to lv_n

16 :   end for

17 : $r'_{i,j} \leftarrow$ GenCol(csk, $v_{i,j}, v_{i-1,2j}\|v_{i-1,2j+1}$)//look for collision random number

18 : end if

19 : $rp_n \leftarrow rp_n + \{r'_{i,j}\}$//save chameleon node random number to rp_n

20 : **return** Π_n

5.3 Identity Update and Revocation Stage

Users need to change the value of attributes in the DID document and update their DID documents on the blockchain when they want to reset their DID keys or change other information about documents. Users can change the value of the operation attribute to "*delete*" to achieve DID revocation when they want to revoke the DID. The system executes the algorithm *DIDUpdRvk* to change the leaf node content from *Document* to *Document*/ and updates the hash value upwards until the chameleon node is encountered. The system keeps the chameleon node hash value unchanged and updates the random number. If it is the root node, the system keeps the chameleon hash unchanged and updates the random number. The algorithm returns updated verification data Π_n. We should note the signature in *Document*/ also needs to be reset if the DID document is changed.

Algorithm DIDUpdRvk

Input : csk, ρ, Document$'$, index, Π_n, depth
Output : Π_n
1 : $d \leftarrow 1$, $i \leftarrow$ index
2 : repeat
3 :   if $i \bmod 2 == 0$ then
4 :   if $d ==$ depth then
5 :   if $d == 1$ then
6 : $r'_{d,0} \leftarrow$ GenCol(csk, ρ, Document$'$)
7 :   else
8 : $r'_{d,0} \leftarrow$ GenCol(csk, ρ, leftchild $\|$ rightchild)
9 :   end if
10 : $\Pi_n \leftarrow \Pi_n - \{r_{d,0}\} + \{r'_{d,0}\}$
11 :   else
12 :   if $d == 1$ then
13 : $v'_{1,i} \leftarrow$ Hash(Document$'$)
14 :   else
15 : $v'_{d,i} \leftarrow$ Hash(leftchild $\|$ rightchild)
16 :   end if
17 : $\Pi_n \leftarrow \Pi_n - \{v_{d,i}\} + \{v'_{d,i}\}$
18 :   leftchild $\leftarrow v'_{d,i}$, rightchild $\leftarrow v_{d,i+1}$, $i \leftarrow i / 2$
19 :   end if
20 :   else
21 :   if $d == 1$ then
22 : $r'_{1,i} \leftarrow$ GenCol(csk, $v_{1,i}$, Document$'$)
23 :   else
24 : $r'_{d,i} \leftarrow$ GenCol(csk, $v_{d,i}$, leftchild $\|$ rightchild)
25 :   end if
26 : $\Pi_n \leftarrow \Pi_n - \{r_{d,i}\} + \{r'_{d,i}\}$
27 :   break
28 :   end if
29 : $d \leftarrow d + 1$
30 : until $d \leq$ depth
31 : return Π_n

5.4   Identity Verification Stage

Each DID document stored in a leaf node has its own unique authentication path. The system executes algorithm *AuthPath* and saves the hash value of sibling nodes in each layer on the path. If the node itself is a chameleon node, the system also needs to save its own random number. The authentication path ap_{index} consisting of hash value and random numbers saved will be returned.

The DID verifier needs to check the legitimacy and validity of the user's DID when the user applies for a certain service. The system executes algorithm *AuthPath* to obtain the authentication path ap_{index}. Then, the system executes algorithm *DIDVerify* to calculate the node hash or chameleon hash bottom-up to confirm the user's identity. We compare the calculated root node chameleon hash to the root node hash ρ saved before. The system returns *true* if they are consistent, which indicates that the document is valid and the user's DID is legal. Otherwise, the system returns *false*, which indicates that the document is invalid and the user's DID is illegal.

Algorithm AuthPath

Input : index, Π_n, depth
Output : ap_{index}
1 : $d \leftarrow 1, i \leftarrow$ index
2 : $lv_{index} \leftarrow \varphi, rv_{index} \leftarrow \varphi, rp_{index} \leftarrow \varphi$
3 : $ap_{index} \leftarrow \{lv_{index}, rv_{index}, rp_{index}\}$
4 : repeat
5 :   if i mod 2 $==$ 0 then//left node
6 : $rv_{index} \leftarrow rv_{index} + \{v_{d,i+1}\}$//save sibling node chameleon hash value
7 :   i $\leftarrow$ i / 2//compute the parent node index
8 :   else//right node
9 : $lv_{index} \leftarrow lv_{index} + \{v_{d,i-1}\}$//save sibling node hash value
10 : $rp_{index} \leftarrow rp_{index} + \{r'_{d,i}\}$//save chameleon node random number
11 :   i $\leftarrow$ (i − 1) / 2//compute the parent node index
12 :   end if
13 : $d \leftarrow d + 1$
14 : until d $<$ depth
15 : **return** ap_{index}

Algorithm DIDVerify

Input : csk, ρ, cpk, index, Document, ap_{index}, depth
Output : true or false
1 : $d \leftarrow 1, i \leftarrow$ index
2 : repeat
3 :   if $i \bmod 2 == 0$ then
4 :   if $d == $ depth then
5 :   if $d == 1$ then
6 : $v_{d,0} \leftarrow$ GenChash(cpk, Document, $r_{d,0}$)
7 :   else
8 : $v_{d,0} \leftarrow$ GenChash(cpk, $v_{d-1,0} \parallel v_{d-1,1}$, $r_{d,0}$)
9 :   end if
10 :   if $v_{d,0} == \rho$ then
11 :   **return** true
12 :   else
13 :   **return** false
14 :   end if
15 :   else
16 :   if $d == 1$ then
17 : $v_{1,i} \leftarrow$ Hash(Document)
18 :   else
19 : $v_{d,i} \leftarrow$ Hash($v_{d-1,2i} \parallel v_{d-1,2i+1}$)
20 :   end if
21 : $i \leftarrow i / 2$
22 :   end if
23 : else
24 :   if $d == 1$ then
25 : $v_{1,i} \leftarrow$ GenChash(cpk, Document, $r_{1,i}$)
26 :   else
27 : $v_{d,i} \leftarrow$ GenChash(cpk, $v_{d-1,2i} \parallel v_{d-1,2i+1}$, $r_{d,i}$)
28 :   end if
29 : $i \leftarrow (i-1) / 2$
30 : end if
31 : $d \leftarrow d + 1$
32 : until $d \leq$ depth

---

6 Decentralized Identity Update and Revoke Scheme Analysis

6.1 Security Analysis

Theorem: If the chameleon hash function can find the unique hash collision when the private key is known and the standard hash function is collision-resistant, then DIURS can satisfy the structural stability and irreversibility.

Proof: We assume that there is a valid algorithm *Algorithm* that makes algorithm *DIDVerify* return *true*. The following two situations may occur.

Situation 1: When $1 \leq i^* \leq q(\lambda)$, we can consider $(d_{i^*}, i^*, ap_{i^*}) \in Q$ because $(d^*, i^*, ap^*) \notin Q$. If $ap^* = ap_{i^*}$, then $d^* \neq d_{i^*}$. It has to satisfy $Hash(d^*) = Hash(d_{i^*})$ if i^* is an even number. Otherwise, algorithm $DIDVerify$ cannot return $true$ in the case of $ap^* = ap_{i^*}$. That means we found a collision of the hash function. Similarly, it has to satisfy $GenChash(cpk, d^*, r) = GenChash(cpk, d_{i^*}, r)$ if i^* is an odd number to make algorithm $DIDVerify$ return $true$ in the case of $ap^* = ap_{i^*}$. That means we found a collision of the chameleon hash function when the private key is unknown. If $ap^* \neq ap_{i^*}$, it will end up at the same node in the same layer of the DID management tree even though the authentication path is different because the index is the same. We assume that the tree depth is D; for $depth \in \{1, 2, \ldots, D\}$ and j, there is either $< v_{depth-1, 2j+1}{}^*, r\prime_{depth-1, 2j+1}{}^* > \neq < v_{depth-1, 2j+1}, r\prime_{depth-1, 2j+1} >$ or $v_{depth-1, 2j}{}^* \neq v_{depth-1, 2j}$ when $v_{depth,j}{}^* = v_{depth,j}$. When j corresponds to a normal node, satisfying $v_{depth,j}{}^* = v_{depth,j}$ means satisfying equation $Hash\big((v_{depth-1, 2j} || v_{depth-1, 2j+1})^*\big) = Hash\big(v_{depth-1, 2j} || v_{depth-1, 2j+1}\big)$, and it is necessary to have $(v_{depth-1, 2j} || v_{depth-1, 2j+1})^* = v_{depth-1, 2j} || v_{depth-1, 2j+1}$; otherwise, a collision of the hash function has been found. When j corresponds to a chameleon node, satisfying $v_{depth,j}{}^* = v_{depth,j}$ means satisfying equation $GenChash\big(cpk, (v_{depth-1, 2j} || v_{depth-1, 2j+1})^*, r_{depth,j}\big) =$ $GenChash(cpk, (v_{depth-1, 2j} || v_{depth-1, 2j+1}), r_{depth,j})$, and it is necessary to have equation $(v_{depth-1, 2j} || v_{depth-1, 2j+1})^* = v_{depth-1, 2j} || v_{depth-1, 2j+1}$, so we can consider $v_{depth-1, 2j}{}^* = v_{depth-1, 2j}$ and $v_{depth-1, 2j+1}{}^* = v_{depth-1, 2j+1}$. Because there is either $< v_{depth-1, 2j+1}{}^*, r\prime_{depth-1, 2j+1}{}^* > \neq < v_{depth-1, 2j+1}, r\prime_{depth-1, 2j+1} >$ or $v_{depth-1, 2j}{}^* \neq v_{depth-1, 2j}$, we can derive $r_{depth-1, 2j+1}{}^* \neq r_{depth-1, 2j+1}$. We know $v_{depth-1, 2j+1}{}^* = GenChash\big(cpk, x^*, r_{depth-1, 2j+1}{}^*\big), v_{depth-1, 2j+1} = GenChash(cpk, x, r_{depth-1, 2j+1})$. We want to make $v_{depth-1, 2j+1}{}^* = v_{depth-1, 2j+1}$, which needs us to find a collision $x^* \neq x$ of the chameleon hash function when the private key is unknown. Therefore, we can derive that inequality (1) in Sect. 4.2 can be satisfied.

Situation 2: When $i^* > q(\lambda)$, there must be a sibling node of the current root node in ap^*. The function $CHRan()$ generates a random message and random number $< x, r >$ for the node, and function $GenChash(cpk, x, r)$ outputs the chameleon hash value of the node. If (d^*, i^*, ap^*) from adversary A can pass identity verification, it means $GenChash(cpk, x^*, r^*) = GenChash(cpk, x, r)$. x^* is computed from i^* by calculating the child nodes upwards and $x^* \neq x$. This means that we find a collision of the chameleon hash function without knowing the chameleon private key. Therefore, we can derive that inequality (2) in Sect. 4.2 can be satisfied.

In summary, it is clear that the probability of adversary A winning is negligible. DIURS satisfies structural stability and irreversibility. DIURS is safe enough.

6.2 Performance Analysis

The static chameleon authentication tree needs to determine the depth of the tree during the initialization stage. DIURS constructs a dynamic chameleon authentication tree without determining the depth of the tree at the beginning, which saves computational overhead in the initialization operation and the DID identity insertion operation. It is more flexible and practical. When the depth of the DID management tree is d, the maximum number of DID documents that can be managed is 2^{d-1}, and the maximum number of total nodes in the tree is $2^d - 1$. The changes in the maximum number of total nodes and maximum number of DID documents in the DID management tree at different tree depths are shown in Fig. 3. In addition, the operations of identity updating and revocation stop at the first chameleon node. It is short in time and low in computation overhead. In the identity verification operation, the length of the authentication path is the tree depth, and the node number in the path is logarithmic to the number of DID documents stored in the tree. The operation takes milliseconds. Overall, DIURS performs well.

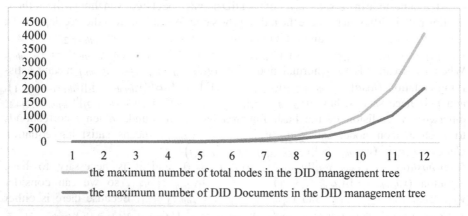

Fig. 3. The maximum number of nodes in the tree at different tree depths

6.3 Comparative Analysis

We compare the scheme proposed in this paper with other decentralized identity schemes in Table 4. We can observe that DIURS can update and revoke the initial DID document on the blockchain without the production of many historical redundant documents. DIURS does not need to reassign the blockchain address to rewrite the latest DID document, which is meaningful.

Table 4. Comparison of the update and revocation mechanisms of DID schemes.

DID scheme	Update and revocation mechanism	Advantages and disadvantages
WeIdentity [17]	It stores, reads, and updates DID Documents based on Solidity's linked event mechanism	It requires indexed continuously for integration to obtain initial documents. Convenience needs to be improved

(continued)

Table 4. (*continued*)

DID scheme	Update and revocation mechanism	Advantages and disadvantages
BaiduDID [18]	It supports DID path query to get specific document version and changes the attribute value to revoke	It saves many historical versions of DID Documents which results in a waste of resource
Scheme [19]	It links to previous state via the index like WeIdentity. New state does not overwrite the existed	Every update information is recorded. It requires continuous index to get a complete query which is tedious
DIURS	It constructs a DID management tree to edit on the initial DID document and realizes the update and revocation of DID finally	There is no need to reallocate the blockchain address for rewriting the latest version of the document, and no large amount of historical redundant data is generated. It is convenient and saves resource

7 Conclusion

In this paper, we propose a decentralized identity management scheme based on blockchain DIURS. We achieve the improvement and optimization of the DID update and revocation mechanism on blockchain by using the chameleon hash function to construct the DID management tree. Relevant analysis shows that DIURS is sufficiently safe and reliable. It has good performance efficiency. It performs better than other DID schemes.

References

1. Khalil, M.M., Lamison, M.R., Dubagunta, S.: Identity management via a centralized identity management server device. Verizon Patent and Licensing Inc. (2020)
2. Kumar, M.N., Suganthi, S., Honnavalli, P.B.: Dynamic Federation in Federated Identity Management. SSRN Electronic Journal (2020)
3. Kubach, M., Schunck, C.H., Sellung, R., Roßnagel, H.: Self-sovereign and Decentralized identity as the future of identity management?. Gesellschaft für Informatik, Bonn (2020)
4. Allen, C.: The Path to Self-Sovereign Identity. https://www.codeb.io/post/the-path-to-self-sovereign-identity-by-christopher-allen. Accessed 16 Jan 2023
5. Omar, D., Khalifa, T.: Decentralized Identity Systems: Architecture, Challenges, Solutions and Future Directions. Annals of Emerging Technologies in Computing, pp. 19–40 (2020)
6. Reed, D., Sporny, M., Longley, D., Allen, C., Grant, R., Sabadello, M.: Decentralized Identifiers (DIDs) v1.0 - Data Model and Syntaxes for Decentralized Identifiers. https://www.w3.org/TR/did-core. Accessed 26 Feb 2023
7. World Wide Web Consortium (W3C): Decentralized Identifiers v1.0: Core Data Model and Syntaxes. https://w3c.github.io/did-core/#did-document. Accessed 24 Feb 2023
8. Sporny, M., Longley, D., Chadwick, D.: Verifiable Credentials Data Model 1.1 - Expressing verifiable information on the Web. https://www.w3.org/TR/vc-data-model. Accessed 2 Mar 2023

9. Sovrin Foundation: Sovrin: A Protocol and Token for Self-Sovereign Identity and Decentralized Trust. https://sovrin.org/wp-content/uploads/Sovrin-Protocol-and-Token-White-Paper.pdf. Accessed 16 Jan 2023

10. Panait, A.E., Olimid, R.F., Stefanescu, A.: Analysis of uPort Open, an identity management blockchain-based solution. In International Conference on Trust and Privacy in Digital Business, pp. 3–13. Springer, Cham (2020)

11. Ebrahimi, A.: ShoCard, Inc. BLOCKCHAIN ID CONNECT. U.S. (2020)

12. https://coinpaprika.com/storage/cdn/whitepapers/448345.pdf. Accessed 16 Jan 2023

13. https://www.bitnations.com. Accessed Jan 18 2023

14. https://www.civic.com. Accessed 17 Jan 2023

15. https://github.com/everid. Accessed 15 Jan 2023

16. https://lifeid.health. Accessed 19 Jan 2023

17. https://weidentity.readthedocs.io. Accessed 4 Mar 2023

18. http://did.baidu.com. Accessed 4 Mar 2023

19. Rundong, W.: Research and Design of Distributed Identity Authentication System Based on Blockchain. University of Electronic Science and Technology of China, Degree (2022)

20. Krawczyk, H., Rabin, T.: Chameleon signatures. In: Proceedings of the Network and Distributed Systems Security Symposium, pp. 143–154. San Diego USA (2000)

21. Shamir, A., Tauman, Y.: Improved online/offline signature schemes. In: Proceedings of the 21st Annual International Cryptology Conference, pp. 355–367. Springer, Santa Barbara USA (2001)

22. Krawczyk, H., Rabin, T.: Chameleon hashing and signatures. IACR Cryptology ePrint Archive, (1998)

23. Chen, X., Zhang, F., Kim, K.: Chameleon hashing without key exposure. International Conference on Information Security, pp. 87–98. Springer, Heidelberg (2004). Doi: https://doi.org/10.1007/978-3-540-30144-8_8

Cloud-Edge Intelligent Collaborative Computing Model Based on Transfer Learning in IoT

Yang Long[1,2] and Zhixin Li[2(✉)] (iD)

[1] School of Information and Control Engineering, Jilin Institute of Chemical Technology,
Jilin 132022, China
[2] School of Computer Technology and Engineering, Changchun Institute of Technology,
Changchun 130012, China
rj_lzx@ccit.edu.cn

Abstract. The deep neural network is a reliable technical support for cloud computing and edge computing. It has excellent nonlinear approximation and generalization capabilities, making it suitable for classifying and predicting Internet of Things data in cloud computing and edge computing fields. However, the increasing size of neural networks poses a challenge for their deployment on devices with limited computing and storage resources. Traditional cloud computing services also suffer from high latency, which hinders real-time tasks. To address these challenges, this paper proposes a cloud-side cooperation model for deep learning based on migration learning technology. This model used migration learning technology to reduce the size of deep neural networks. Specifically, it deployed the deep neural network model (CDLM) in the cloud and the shallow neural network model (EDLM) at the edge. CDLM is used to help train EDLM and improve its performance, enabling it to run independently on edge devices with high accuracy and respond to real-time tasks. This approach reduced the amount of user data transmitted to the cloud, alleviated bandwidth pressure, and protected user privacy. Experimental results show that the proposed model improved the accuracy of EDLM by 19.58% compared with traditional neural network models. These findings provide a theoretical and experimental foundation for the study of cloud-edge collaborative models.

Keywords: Deep Learning · Cloud Computing · Edge Computing · Transfer learning · Cloud-edge collaboration

1 Introduction

With the rapid advancement of Internet of Things (IoT) technology, the number of IoT devices is increasing exponentially. As a result, the amount of structured, unstructured, or semistructured data generated by sensors and embedded devices is also growing rapidly, constituting a massive volume of data. Cloud computing and edge computing have become key technologies for the development of the IoT ecosystem. Cloud computing offers high-performance computing and storage resources, while edge computing

© The Author(s), under exclusive license to Springer Nature Singapore Pte Ltd. 2023
Z. Yu et al. (Eds.): ICPCSEE 2023, CCIS 1879, pp. 389–403, 2023.
https://doi.org/10.1007/978-981-99-5968-6_28

provides low-latency computing resources in close proximity to users. However, current IoT applications demand a more integrated approach, where the separate use of cloud computing and edge computing can no longer suffice.

In the realm of Internet of Things applications, deep learning has become a crucial component in providing intelligent services by learning and predicting large amounts of data. However, deep learning necessitates significant computing and storage resources. Although cloud computing has numerous advantages, traditional cloud computing requires transmitting all data to a cloud server for processing, which can result in delays and bandwidth issues. Although edge computing is situated in close proximity to data sources, its computing and storage capabilities are limited and may not be able to handle complex tasks [1]. Therefore, combining the strengths of cloud computing and edge computing has become an imperative concern. Zhi Zhou [2] and others have explored the integration of edge computing and artificial intelligence and how it can help tackle the "last mile problem" in artificial intelligence applications.

To enable collaboration between deep learning in the cloud and edge, the first step is to address the challenge of deploying deep neural networks at the edge. Mário P. Véstias [3] and others summarized the advantages of deploying deep learning models at the edge, as well as their key points and difficulties, and looked forward to future research directions. Xiaofei Wang [4] and others comprehensively introduced and discussed various application scenarios and basic empowerment technologies of edge intelligence and intelligent edges and proposed a key problem to extend the deep neural network model to the edge: how to design and develop the edge computing architecture under the multiple constraints of network, communication, computing power and energy consumption to achieve the best performance of deep neural network model training and reasoning. Chuntao Ding [5] and other researchers have proposed an intelligent service collaboration framework based on cloud computing and edge computing, which offers a way for cloud and edge devices to work together and overcome the challenges of intelligent service processing.

To leverage the benefits of both cloud computing and edge computing, this paper addresses the challenge of deploying deep neural networks on edge devices. The proposed solution is a cloud-edge collaboration model. Initially, a large-scale deep neural network model is deployed in the cloud, leveraging its significant computing power and storage capacity to train an accurate model. Migration learning technology is then employed to generate a small-scale deep neural network model, which is distributed to the edge for inference tasks. During the inference task, an information entropy [6] threshold is set at the edge exit, with smaller entropy indicating higher reliability of prediction results and larger entropy indicating lower reliability. After the edge reasoning task is completed, the entropy value of the result is calculated, and if it exceeds the threshold, the task is sent to the cloud. If the entropy value falls below the threshold, the result is output directly, ensuring the prediction accuracy of edge devices while saving bandwidth, reducing the transmission of original data, and protecting users' privacy.

2 Related Work

Currently, both domestic and international research has extensively explored cloud-edge collaboration in deep learning. This collaboration is widely recognized for its ability to enhance the efficiency and flexibility of the Internet of Things. Research in cloud-edge collaboration mainly encompasses two approaches: collaboration mode based on cloud computing and collaboration mode based on edge computing.

Primarily based on cloud computing, this approach delegates most computing tasks to the cloud, addressing issues such as task delay, network bandwidth overload, and privacy protection. Yamin Sepehri and colleagues [7] proposed a hierarchical training method for a deep neural network (DNN) in the edge cloud scenario. This method incorporates an early exit structure between the edge and the cloud, eliminating the need for the edge and cloud ends to share the original input data or communicate during reverse propagation. This approach reduces training time, protects users' privacy, and alleviates bandwidth pressure. Surat Teerapittayanon and colleagues [8] proposed a distributed neural network (DDNN) comprising cloud, edge, and terminal devices. The network allows fast and local inference at the edge and terminal devices, utilizing the shallow part of the neural network. This approach improves accuracy and reduces communication costs. While this collaboration method is highly reliable and related research has made efforts to reduce communication costs and protect privacy, it cannot achieve a rapid response.

The edge-based collaboration mode delegates most computing tasks to the edge to address issues such as insufficient computing storage capacity and low deep neural network training accuracy. Yoshitomo Matsubara and colleagues [9] used knowledge distillation technology to propose a structure and training process to modify the DNN model of complex image classification tasks to achieve early intra network compression in the network layer. This method reduces the reasoning time of edge devices and improves accuracy. Junhao Zhou and colleagues [10] designed a lightweight deep learning model and proposed a stacked convolutional neural network. The model significantly reduces the number of parameters while maintaining high accuracy, making it easier to deploy and run on edge devices. Although this collaboration approach can achieve a rapid response, its reliability is not guaranteed, and the powerful computing and storage capacity of the cloud has not played its due role.

To build an ideal cloud-edge collaboration model for deep learning, we need to combine their advantages while protecting users' privacy, reducing network bandwidth pressure, handling real-time tasks, and ensuring reasoning accuracy. To achieve these requirements, we need to solve the following problems:

To protect users' privacy, reduce bandwidth pressure, and handle real-time tasks, data should be limited to the edge as much as possible, enabling reasoning tasks to be performed at the edge and reducing the amount of data uploaded to the cloud.

To ensure accuracy, we need to optimize the edge neural network model. Techniques such as model pruning and model distillation are often used in edge deep learning to compress neural networks [11], ensuring accuracy. Transfer learning provides another way to compress the neural network model at the edge.

Transfer learning aims to provide a framework to solve new but similar problems faster and more effectively by using previously acquired knowledge [12]. Yosinski [13] and others have proven that the neural network model trained by transfer learning technology has better initial performance, better generalization ability and faster convergence speed than the randomly initialized neural network model.

In cloud-side collaboration, cloud computing and edge computing are complementary, and the cloud has strong computing and storage capacity and can learn more knowledge. The edge side has the advantage of being close to the data source and can provide faster response services. We hope to combine their advantages, transfer learning technology and cloud-side collaboration so that the edge can use the knowledge learned in the cloud. Therefore, this paper proposes a cloud-edge collaboration model based on transfer learning technology to address the aforementioned challenges.

3 Methodology

3.1 Model

To deploy a deep neural network model at the edge, it is necessary to reduce the scale of the neural network on the premise of ensuring the accuracy of model reasoning, enabling execution on edge devices. There are several methods to reduce the neural network scale, including network pruning, knowledge distillation, and parameter quantification. However, with the rapid growth of the Internet of Things, the number of devices has surged, making it challenging to handle multiple compressed models. To overcome this challenge, we propose using transfer learning technology to reduce the neural network model's size.

Transfer learning refers to the process of leveraging learned knowledge and experience to solve new tasks. It is a deep neural network learning method where the model developed for task A is used as a starting point in developing the model for task B. For example, the model used for car identification can be used to enhance the ability to identify trucks.

The typical definition of transfer learning involves a source domain D_s and a target domain D_t, where the source domain uses feature $D_s = \{x_i, y_i\}_i^{N_s}$ representation and the target domain uses feature $D_t = \{x_i, y_i\}_i^{N_t}$ representation. Given a source domain D_s and a learning task T_s, as well as a target domain D_t and a learning task T_t, the goal of transfer learning is to leverage knowledge from the source domain D_s and learning task T_s to improve the learning of the prediction function f_t in the target domain, where $D_s \neq D_t$ or $T_s \neq T_t$ is satisfied.

First, the model trains a large-scale neural network in the cloud by leveraging a vast amount of data and computing power, resulting in high accuracy. Once training is complete, the first n layers of the network (here, the first n layers refer to the bottom layer, which is close to the input.) are extracted to construct a smaller neural network.

To enable the cloud to aid in training the network at the edge, the top n layers of the large network (CDLM) are extracted to initialize the small network. Yosinski [13] and other researchers have verified that the features near the input layer are more universal and can serve as general image descriptors. Therefore, the deep convolutional neural network shares its first n layers to facilitate the training of the small network. The resulting edge small neural network (EDLM) comprises the first n layers of the large-scale network and the last m layers, which are randomly initialized. EDLM is a n + m layer network (Fig. 1).

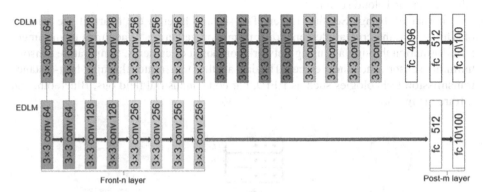

Fig. 1. Structure of CDLM and EDLM.

This study utilized transfer learning technology to construct a small neural network model, but it was apparent that even with higher accuracy at the edge, it cannot match that of the cloud. To address this, this paper incorporated a branching structure into the model. At the end of the reasoning work at the edge, a judgment is added, and the concept of information entropy is introduced. The classification results become more disorderly with increased information entropy, which diminishes their reliability. Conversely, lower information entropy produces more dependable classification results. In this paper, the edge will calculate the information entropy and set an appropriate information entropy threshold at the edge exit. If the information entropy exceeds this threshold, the classification result is deemed unreliable, and the original data are transmitted to the cloud, outputting results from the cloud. This method not only meets task delay requirements but also improves accuracy. The threshold size should be determined according to the specific task in practice.

When the edge-end neural network processes data, it generates an output result z, given an input sample x. We normalize the output result using the softmax function to ensure that the final output result falls within the range of 0 to 1. We assume that the prediction probability of each label is represented by y. The predicted output y can be expressed as:

$$y = soft\max(z) = \frac{\exp(z)}{\sum_{s \in S} \exp(z_s)} \tag{1}$$

The set of all possible class label sets for the predicted output is denoted as S. The sample output's information entropy at the exit point is defined as:

$$entropy(y) = \sum_{s \in S} y_s \log y_s \tag{2}$$

3.2 Model Structure

The entire model comprises three components: the Internet of Things perception layer, edge server, and cloud server.

Internet of Things Sensing Layer: The sensing layer is the heart of the Internet of Things and the crucial element for information gathering. It acquires environmental information through a sensing network composed of cameras, temperature sensors, humidity sensors, and other devices. The data are then transmitted through short-distance transmission technologies such as RFID, bar code, industrial field bus, Bluetooth, and infrared (Fig. 2).

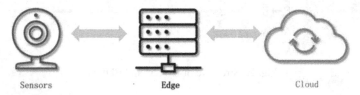

Sensors   Edge   Cloud

Fig. 2. Structure of the proposed model.

Edge Server: An edge server is located at the network's edge, such as a home gateway and micro server, typically in close proximity to end-users and data sources. Edge servers can process end users' requests and provide some computing power, storage capacity, and network connectivity functions. In this model, the edge server handles data preprocessing and some basic predictive tasks.

Cloud Server: A cloud server is a virtualization server that relies on cloud computing technology and shares computing and storage resources through the internet. Popular cloud servers include Google Cloud Platform (GCP) and Alibaba Cloud Server. Cloud servers typically have robust computing and storage capabilities. In this model, the cloud server is mainly responsible for training the deep neural network model.

3.3 Workflow

This study focuses on addressing the following issues: how to safeguard users' privacy, respond promptly to real-time tasks, reduce bandwidth pressure, and ensure the accuracy of reasoning.

To protect users' privacy and respond to delay-sensitive tasks in time, the whole reasoning process should be completed on the edge nodes as much as possible. Edge servers are typically only one hop away from users, allowing for quick response times. Furthermore, since user data do not need to be transmitted to the cloud, this approach can safeguard privacy, reduce data transmission, and minimize bandwidth pressure.

To ensure the accuracy of reasoning on edge nodes, we need to improve the accuracy of deep neural networks on a finite scale. We employ transfer learning to improve the accuracy of smaller networks. Specifically, we train a high-precision deep neural network model, called the cloud deep learning model (CDLM), on a cloud server and run a shallow neural network model, called the edge deep learning model (EDLM), on an edge server.

EDLM is composed of the first n layers of CDLM and the last m layers randomly initialized. In this way, we leverage the knowledge of CDLM to assist in initializing EDLM. During the training stage of EDLM, the edge server only needs to freeze the first n layers of EDLM and fine-tune the m layers with local or packaged data, enabling it to achieve deep neural network accuracy comparable to that of CDLM (Fig. 3).

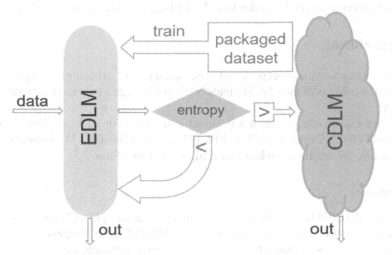

Fig. 3. The workflow of the proposed model.

The model works through the following stages:

Initialization Stage:
The cloud trains a large-scale deep neural network model, CDLM, using a significant amount of data. We extract the first n layers (near the input end) and combine them with the last m layers to generate a smaller deep neural network model, EDLM, which is deployed at the edge. The edge uses EDLM to complete local reasoning tasks, fine-tuning the last m layers with local data or cloud-packaged data to improve accuracy when the server is not busy or under maintenance. Freezing the first n layers reduces computing pressure and shortens training time.

Working Stage:
The edge side preprocesses data, runs EDLM for prediction work, and conducts small-scale training in spare time. To improve accuracy, we set an information entropy judgment at the edge exit. Information entropy reflects the degree of disorder in prediction information. We set an entropy threshold at the edge exit, and if the entropy value exceeds the threshold, data are uploaded to the cloud, and if it is lower, the results are directly output from the edge.

Update Stage:
The cloud server uses large-scale neural networks to reason the data transmitted at the edge. With the powerful computing and storage capabilities of the cloud, the accuracy of CDLM training will be very high, and the data collected in the cloud and predicted results will be packaged as training data. When the edge server is under low load or maintenance, training data are sent to the edge node to train the EDLM, improving accuracy. This uses the cloud server to help edge nodes train EDLMs and realize dynamic updates.

4   Experiment

In this section, a simulation experiment is designed to verify that initializing the EDLM with the first n layers of the CDLM yields better performance than randomly initializing the DLM.

For the experiment, we used a personal computer with an Intel Core i5-10300H CPU, an NVIDIA GeForce GTX-1650 GPU, and 16 GB of RAM to simulate a real environment. We ran the deep learning framework TensorFlow 2.3.

4.1   Datasets

We selected the CIFAR-10 and CIFAR-100 universal datasets [14]. CIFAR-10 is a widely used image classification dataset comprising 60,000 32x32 color images categorized into 10 classes. Each class contains 6,000 images, including airplanes, cars, birds, cats, deer, dogs, frogs, horses, boats, and trucks. The CIFAR-100 dataset comprises 100 classes, with each class having 600 color images sized 32x32. Of these, 500 are used as training sets, and 100 are used as test sets (Table 1).

Table 1. Datasets.

Dataset	Training set	Test set	Labers
CIFAR-10	50000	10000	10
CIFAR-100	50000	10000	100

4.2   VGG19 and Transfer Learning

We have chosen to use the VGG19 pretraining model [15] to simulate CDLM. VGG19 is a deep convolutional neural network that was proposed by researchers at Oxford University in 2014. The model has been trained on the ImageNet dataset, which consists of over 1.4 million images across more than 1,000 categories.

Because VGG19 has been pretrained on ImageNet [16], it can easily be transferred to other computer vision tasks, such as image classification, object detection, and image segmentation. These pretrained models can be utilized in various deep learning frameworks, including TensorFlow, PyTorch, and Keras, and integrated into their respective applications.

The pretraining and transfer learning process can be expressed using the following formula:

$$\theta_f = \arg\min_{\theta} \frac{1}{n} \sum_{i=1}^{n} L(y_i, f_\theta(x_i)) + \lambda\Omega(\theta) \tag{3}$$

In this context, θ represents the parameters of the model, L represents the loss function, $f_\theta(x_i)$ represents the prediction results of the model on the input x_i with the parameter θ, y_i represents the true label corresponding to input x_i, $\Omega(\theta)$ is the regularization term, and λ is the regularization coefficient.

Next, we will fine-tune the EDLM. The formula for fine-tuning can be expressed as:

$$\theta^* = \arg\min_{\theta} L_D(\theta) + \lambda L_s(\theta, \theta_0) \tag{4}$$

In this context, θ represents the model parameters to be fine-tuned, θ_0 denotes the pretrained model parameters, $L_D(\theta)$ represents the loss function on the target domain data D, $L_s(\theta, \theta_0)$ indicates the similarity between the model parameters to be fine-tuned θ and the pretrained model parameters θ_0, and λ is the hyperparameter controlling the weight between the two.

Fine-tuning can be achieved through the backpropagation algorithm, which computes the gradients of θ with respect to the model parameters to be fine-tuned L_D and the pretrained model parameters L_s and updates them accordingly.

VGG19 is capable of emulating CDLM, a deep neural network model trained on a large scale in the cloud. ImageNet is utilized to replicate the extensive and intricate data in the cloud, whereas the CIFAR-10 and CIFAR-100 datasets are used to mimic local data at the edge.

VGG19 is composed of 16 convolutional layers and three fully connected layers, and the maximum pooling operation is carried out after the second, fourth, eighth, twelfth and sixteenth convolution layers. To construct the EDLM, we extract the first four, six, and eight convolutional layers from VGG19 and combine the last two fully connected layers, which are initialized randomly, to form three EDLMs. The first EDLM, called VGG19 6, consists of four convolutional layers and two fully connected layers. It represents a six-layer neural network model trained with the assistance of VGG19. The second EDLM, called VGG19–8, consists of six convolutional layers and two fully connected layers. It represents an eight-layer neural network model trained using VGG19. The third EDLM, called VGG19–10, consists of eight convolutional layers and two fully connected layers, representing a ten-layer neural network model trained using VGG19. Additionally, VGG6, VGG8, and VGG10 represent 6-layer, 8-layer, and 10-layer deep neural network models, respectively, with all layers initialized randomly. To be consistent with VGG 19, all the models in the experiment were subjected to the maximum pooling operation with a pooling kernel of 2×2 after the second, fourth and eighth convolution layers (Fig. 4).

4.3 Performance of the EDLM

EDLM'S Accuracy on CIFAR-10. First, VGG19 is fine-tuned to obtain a model with 88.35% accuracy on the CIFAR10 test set. This model will be used to simulate CDLM.

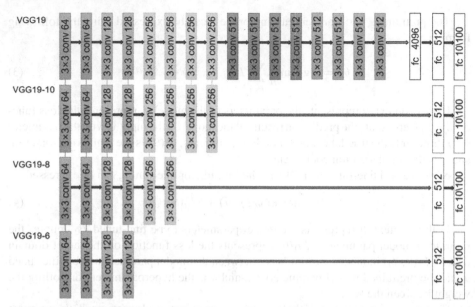

Fig. 4. The network architecture of CDLM (VGG19) and EDLM (VGG19–6, VGG19–8, VGG19-10). Note: VGG19–6 refers to the 6-layer EDLM, which is initialized with the help of the first 4 convolution layers of VGG19, and the same is true for VGG19–8 and VGG19–10.

We initialize VGG19–6, VGG19–8, and VGG19–10 using the first four layers, first six layers, and first eight layers of the VGG19 model, respectively. Then, we freeze the corresponding layers and fine-tune the last two convolution layers with the CIFAR10 training set. After that, we compare VGG19–6, VGG19–8, and VGG19–10 with randomly initialized VGG6, VGG8, and VGG10 and obtain the following data (Table 2)

Table 2. Model's test accuracy on CIFAR-10. (%)

Model\Epoch	1	5	10
VGG6	34.09	56.40	64.60
VGG19–6	73.34	78.51	79.08
VGG8	29.57	53.32	62.11
VGG19–8	80.41	83.40	84.45
VGG10	22.48	52.61	65.63
VGG19–10	83.96	85.49	85.80

The table compares the test accuracy of the VGG6, VGG8, and VGG10 models with VGG19–6, VGG19–8, and VGG19–10 after 1, 5, and 10 rounds of training on the CIFAR10 dataset. The VGG19–6, VGG19–8, and VGG19–10 models are responsible only for training the last two fully connected layers. The first n layers of the CDLM are used to help the EDLM initialize the model, and the EDLM is responsible only for fine-tuning the last m layers. When compared with a small neural network that is randomly initialized, EDLM achieves a training effect close to CDLM with fewer training rounds and less training time. After 10 epochs, the accuracy of VGG19–6 reaches 79.08%, that of VGG19–8 reaches 84.45%, and that of VGG19–10 reaches 85.8%. EDLM shows better performance than VGG6, VGG8, and VGG10.

Moreover, when the training rounds are few and the training data are limited, the gap in performance between EDLM and the other models becomes even more significant. After the first epoch of training, the accuracy of VGG6 is only 34.09%, while that of VGG19–6 is 73.34%. Similarly, VGG8 achieves an accuracy of only 29.57%, whereas VGG19–8 reaches 80.41%. VGG10's accuracy is only 22.48%, whereas VGG19–10 achieves an accuracy of 83.96%.

EDLM'S Accuracy on CIFAR-100. This study also performed experiments on the CIFAR100 dataset using the same model and obtained the following data (Figs. 5, 6 and 7):

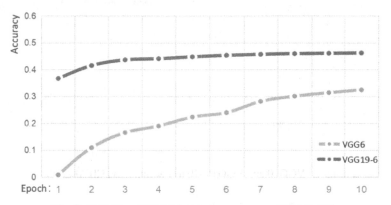

Fig. 5. VGG6 and VGG19–6 test accuracy on CIFAR-100.

The three images above compare the test accuracy of VGG6 and VGG19–6, VGG8 and VGG19–8, and VGG10 and VGG19–10 after ten rounds of training on the CIFAR100 dataset. Currently, the test accuracy of CIFAR100 on VGG19 is 61.17%, and the conclusion is similar to the experiment conducted on the CIFAR10 dataset. After ten rounds of training, the accuracy of the CIFAR100 test set is improved. Compared with VGG6, VGG8 and VGG10, VGG19–6, VGG19–8 and VGG19–10 are improved by 13.74%, 22.04% and 24.11%, respectively.

Using VGG19 to initialize smaller networks such as VGG19–6, VGG19–8, and VGG19–10 has resulted in significantly better performance than randomly initialized small networks. This performance gap is particularly evident when training rounds and time are limited.

Therefore, it is not difficult to conclude that initializing EDLM with the first n layers of CDLM can result in a more accurate neural network model. This improvement is more significant when the data volume, training times, and training time are limited.

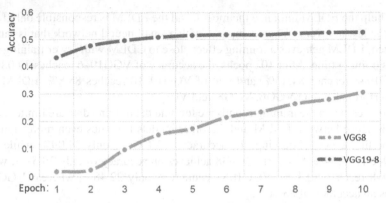

Fig. 6. VGG8 and VGG19–8 test accuracy on CIFAR-100.

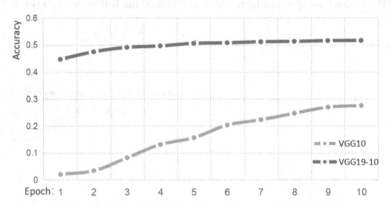

Fig. 7. VGG10 and VGG19–10 test accuracy on CIFAR-100.

4.4 Enhanced Training

This study aims to enhance the accuracy of EDLM when maintaining the server. To achieve this, we allow EDLM to train the entire m + n layer of the neural network model, resulting in the following results (Figs. 8 and 9):

Among them, VGG19-6T, VGG19-8T and VGG19-10T represent all trained neural networks.

It can be observed that training EDLM from scratch using the provided dataset can lead to further improvements in accuracy. On the CIFAR10 dataset, the accuracy of VGG19–6 increased from 79.08% to 79.6%, VGG19–8 increased from 84.45% to 85.8%, VGG19–10 increased from 85.09% to 86.13%, and the improvement range, although limited, is a viable method to enhance edge reasoning accuracy.

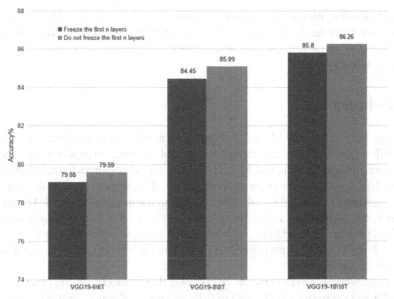

Fig. 8. EDLM test accuracy after 10 rounds of training on the CIFAR-10 dataset.

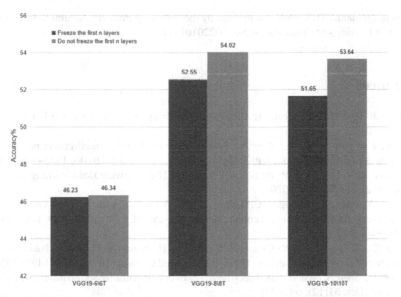

Fig. 9. EDLM test accuracy after 10 rounds of training on the CIFAR-100 dataset.

4.5   Summary

In conclusion, transfer learning techniques using CDLM to initialize EDLM can produce more accurate neural network models. Choosing to deploy CDLM and EDLM should depend on the specific task at hand. This method can bring EDLM performance closer to CDLM while significantly reducing the scale of the deep neural network, enabling the edge to independently perform most tasks. In practical situations, deploying EDLM should be based on the computing power and storage capacity of the edge, as well as the specific task requirements.

5   Conclusion

This paper proposes a cloud-based collaboration model for deep neural networks. The model utilizes transfer learning techniques to enable the edge to deploy a high-precision small neural network model (EDLM) and leverages cloud resources to facilitate dynamic updates. This approach allows for the completion of most local tasks at the edge while preventing a large volume of user data from being transmitted to the cloud. The approach also meets the requirements of time-sensitive tasks, protects user privacy, and reduces bandwidth pressure. The addition of 'information entropy' at the edge further enhances accuracy. Finally, experiments prove that EDLM outperforms traditional neural network models of the same scale.

In future work, we plan to use more complex datasets to verify the practicability of this model and study the actual effect of different types of neural networks deployed in the cloud and edge under the model proposed in this paper.

Acknowledgement. This work is supported by the following projects: Natural Science Foundation of Jilin Province of China (Grant No. 20220101136JC).

References

1. Chen, J., Ran, X.: Deep learning with edge computing: a review. Proc. IEEE **107**(8), 1655–1674 (2019)
2. Zhou, Z., Chen, X., Li, E., Zeng, L., Luo, K., Zhang, J.: Edge intelligence: paving the last mile of artificial intelligence with edge computing. Proc. IEEE **107**(8), 1738–1762 (2019)
3. Véstias, M.P., Duarte, R.P., de Sousa, J.T., Neto, H.C.: Moving deep learning to the edge. Algorithms **13**(5), 125 (2020)
4. Wang, X., Han, Y., Leung, V.C.M., Niyato, D., Yan, X., Chen, X.: Convergence of edge computing and deep learning: a comprehensive survey. IEEE Commun. Surv. Tutorials **22**(2), 869–904 (2020)
5. Ding, C., Zhou, A., Liu, Y., Chang, R.N., Hsu, C.-H., Wang, S.: A cloud-edge collaboration framework for cognitive service. IEEE Trans. Cloud Comput. **10**(3), 1489–1499 (2022)
6. Wang, C., Shen, S.: A distributed deep neural network algorithm based on multiagent. Comput. Technol. Dev. **31**(12), 6 (2021)
7. Sepehri, Y., Pad, P., Yüzügüler, A.C., Frossard, P., Dunbar, L.A.: Hierarchical Training of Deep Neural Networks Using Early Exiting. arXiv:2303.02384 (2023)
8. Teerapittayanon, S., McDanel, B., Kung, H.T.: Distributed deep neural networks over the cloud, the edge and end devices. In: IEEE 37th International Conference on Distributed Computing Systems (ICDCS), pp. 328–339. Atlanta, GA, USA, (2017)
9. Matsubara, Y., Callegaro, D., Baidya, S., Levorato, M., Singh, S.: Head network distillation: splitting distilled deep neural networks for resource-constrained edge computing systems. IEEE Access **8**, 212177–212193 (2020)

10. Zhou, J., Dai, H.-N., Wang, H.: Lightweight convolution neural networks for mobile edge computing in transportation cyber physical systems. ACM Trans. **10**(6), 1–20 (2019)
11. Kim, J., Chang, S., Kwak, N.: PQK: Model Compression via Pruning, Quantization, and Knowledge Distillation. arXiv:2106.14681 (2021)
12. Lu, J., Behbood, V., Hao, P., Zuo, H., Xue, S., Zhang, G.: Transfer learning using computational intelligence: a survey. Knowl. -Based Syst. **80**, 14–23 (2015)
13. Yosinski, J., Clune, J., Bengio, Y., Lipson, H.: How transferable are features in deep neural networks? Adv. Neural Inf. Process. Syst. **27**, 3320–3328(2014)
14. Krizhevsky, A., Hinton, G.: Learning multiple layers of features from tiny images. Handbook of Systemic Autoimmune Diseases, vol. 1, no. 4 (2009)
15. Karen, S., Zisserman, A.: Very Deep Convolutional Networks for Large-Scale Image Recognition. arXiv:1409.1556 (2014)
16. Krizhevsky, A., Sutskever, I., Hinton, G.E.: ImageNet classification with deep convolutional neural networks. Commun. ACM **60**(6), 84–90 (2017)

Design and Validation of a Hardware-In-The-Loop Based Automated Driving Simulation Test Platform

Kaichao Zheng[✉], Xianbin Xue, He Li, and Guoliang Cheng

School of Automation, Guangdong University of Technology, Guangzhou, China
1335936835@qq.com

Abstract. With the iterative development of autonomous driving technology, self-driving cars will be one of the most competitive areas in the future. In order to provide students with a better understanding and more comprehensive grasp of autonomous driving technology, a hardware-in-the-loop based autonomous driving simulation test platform has been built. The hardware-in-the-loop system integrates MATLAB/Simulink to build the core control algorithm model, CarMaker simulation software to provide a virtual display interface and vehicle dynamics model, and NVIDIA Jetson to deploy the ECU (Electronic Control Unit) to improve the algorithm power and Logitech G29 series driving simulators providing signal input. It provides a simulation test platform for the development and testing of advanced driver assistance systems, the development and testing of upper layer control algorithms and underlying actuators for autonomous driving, and can provide a teaching and experimental platform for undergraduate students and a development foundation for postgraduate practice.

Keywords: Autonomous Driving · Hardware In The Loop · Simulation Testing · Experimental Platform

1 Introduction

A new information technology revolution triggered by the rapid development of artificial intelligence, big data, the interconnectivity of transport and edge computing [1, 2], the self-driving car industry, which incorporates many new technologies, is also gaining momentum [3, 4]. The development of algorithm perception, decision control and artificial intelligence as the core technology of advanced driving assistance system, the development of this direction must be a breakthrough point for each car company or even each country to focus on investment research [5, 6].

Hardware-in-the-loop simulation test systems have become an important platform for testing and validating autonomous vehicle technology because of the significant advantages of short simulation cycles, high efficiency and low cost. Because of the advantages of hardware-in-the-loop simulation technology for autonomous driving, a large number of scholars at home and abroad have conducted in-depth research on it.

Zhao et al. from Chang'an University have developed an autonomous driving simulation platform based on whole-vehicle in-loop testing. The platform consists of a driving simulator, a virtual scene automatic generation subsystem, a virtual sensor simulation subsystem and a vehicle driving resistance simulation subsystem, which successfully simulates the real-time resistance to the vehicle while driving [7]. Chen et al. from Xi'an Jiaotong University have developed an emerging hardware-in-the-loop simulation test platform, which contains a virtual environment scene layer, a virtual sensor layer, an electronic control layer and a vehicle simulation test layer in its architecture; the platform is able to provide vehicle dynamics models, sensor models and the construction of constituted simulation test scenarios, etc., and is capable of closed-loop control at the sensing and decision-making layers, thus enabling the construction of hardware-in-the-loop [8]. A hardware-in-the-loop simulator for the verification of autonomous driving algorithms has been developed by Gelbal et al. from Ohio State University, USA. The advantage of this simulator is that it can support the simulation and verification of autonomous driving algorithms in highway scenarios up to L4 level [9]. Ma et al. from University of Cincinnati, USA, have built a hardware-in-the-loop test system for networked autonomous vehicles, which mainly integrates the vehicle model into a virtual traffic simulation environment, thus providing a rich simulation test scenario for testing autonomous driving algorithms and ensuring their accuracy and validity to the greatest extent possible [10].

With the rapid development of autonomous vehicle technology, it is natural that new and higher requirements are placed on vehicle control, placing more stringent demands on traditional performance such as vehicle safety and stability [11, 12]. However, most scholars currently studying autonomous vehicle technology focus mainly on the testing of control algorithms and sensors, which clearly lacks direct testing of the safety and stability of the vehicle by the underlying hardware inputs. To address these issues, this paper proposes a hardware-in-the-loop-based autonomous driving simulation and testing system, which can provide a direct verification platform for the development and testing of autonomous driving upper layer control algorithms and underlying actuators.

2   Overall System Framework Design

Figure 1 shows the General framework design diagram. The hardware part of the system framework mainly consists of the host computer, driving simulator, electronic control unit, display, etc. The upper computer mainly refers to the host computer; the driving simulator includes the steering wheel, accelerator pedal and brake pedal of Logitech G29 series; the electronic control unit uses NVIDIA-Jetson of NVIDIA series; the display includes the scene display and the manoeuvring driving display.

The upper computer mainly runs the CarMaker simulation software and the human-machine operation interface. CarMaker not only displays a virtual reality simulation interface and provide simulation test scenarios that tend to be realistic for autonomous driving simulation; it can also provide sensor signals required by the perception algorithm module for autonomous driving simulation tests, including cameras and various millimetre wave radars, etc. Through the human manipulated pedal to the vehicle acceleration behaviour and braking behaviour, CarMaker receives the electronic signal and then sends control commands such as braking, acceleration and steering to the electronic control unit AGX through Veristand software via UDP communication, and sends the command to the AGX hardware which deploys ACC algorithm in advance through UDP communication to make The lower computer runs the CarMaker vehicle dynamics model and realises closed-loop control of the braking, steering and drive system of the simulated vehicle based on the vehicle control signals and Object List sent by the electronic control unit AGX, achieving smoothness indicators such as control of the vehicle by human input while ensuring the completion of ADAS functions such as adaptive cruise control. The hardware-in-the-loop system human control physical diagram is shown in Fig. 2.

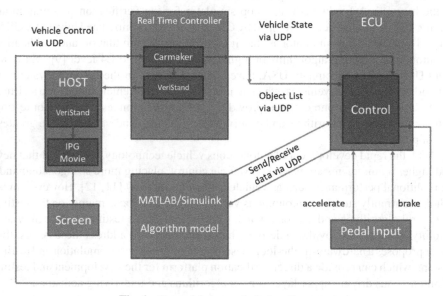

Fig. 1. General framework design diagram.

3   Software and Hardware Selection

3.1   Hardware Selection

The hardware is the NVIDIA Jetson AGX Xavier Developer Kit, shown in Fig. 3, which is powered by the NVIDIA JetPack and DeepStream SDKs, as well as CUDA, cuDNN and TensorRT software libraries. The NVIDIA Xavier processor used offers more than 20 times the performance and up to 10 times the energy efficiency improvement over the NVIDIA Jetson Tx2. Its network module uses Gigabit Ethernet.

3.2   Software Selection

When the system software and hardware are in communication, the input signal of the underlying load and the processing signal of the actuator ECU interact with the lower

Fig. 2. Physical view of human manipulation.

computer through the EDAC interface, the communication between the lower computer and the upper computer is through the network cable and the UDP protocol, and the communication between the lower computer, the upper computer and the controller is through the UDP module. The block diagram of the system software composition is shown in Fig. 4. The specific selection of each software is as follows:

(1) controller, MATLAB/Simulink software is used to build control algorithm models such as ACC for autonomous driving, and deploy the code into the hardware and send the corresponding signals to the lower computer through UDP communication.
(2) upper computer, the scene editor in CarMaker software builds simulation test scenarios such as ACC, and provides the algorithm model with relevant sensory sensor information.
(3) the lower computer, the vehicle dynamics model is provided by CarMaker software, based on the communication link of Veristand software, to realize the joint simulation between CarMaker software and Simulink algorithm model, providing real-time hardware input signals for it, and Veristand software is used to output control commands and collect the underlying signals.

Fig. 3. Physical view of the AGX.

4 Main Module Design

4.1 Algorithm Modules

The simulation test scenario constructed for this experiment is an ACC following scenario, and the main body in the algorithm module is the ACC control module, as shown in Fig. 5. Firstly, the reference speed, relative distance and Leader Speed are set respectively; secondly, they are assigned to the corresponding control inputs in the controller; finally, the closed-loop control formed between the controller and the vehicle model will iterate continuously, and when the previous vehicle is detected, the dynamic following will be achieved according to the following distance set in advance, and the information such as speed and position will be output to the upper computer and the lower computer.

4.2 Communication Modules

The communication module is mainly divided into two parts: communication sender and communication receiver, the communication sender is shown in Fig. 6 and the communication receiver is shown in Fig. 7. Firstly, the vehicle model of the front vehicle in the ACC scene outputs the real-time speed and displacement information of the vehicle model under the preset reference speed, packages the relevant data through byte packets and sends the byte packets through UDP communication protocol; then, the rear vehicle

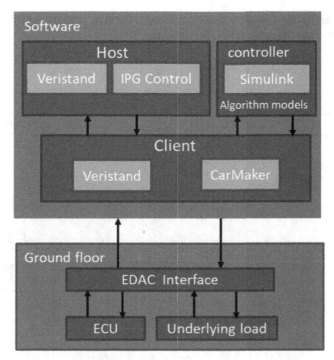

Fig. 4. Block diagram of the system software components.

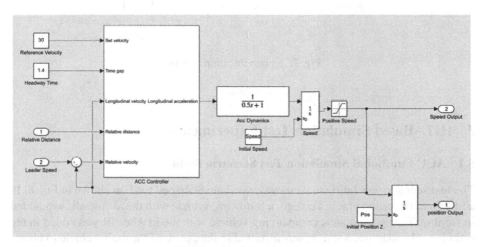

Fig. 5. Algorithm module.

must receive the byte packets from the front vehicle through UDP communication in order to realise the function of following the vehicle, and parses the data through byte unpacking and inputs them into the vehicle model of the rear vehicle. Finally, the speed and displacement of the rear vehicle are controlled within a certain range by the algorithm model to realise the dynamic following function.

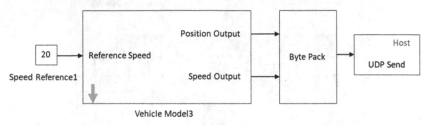

Fig. 6. Communication sender.

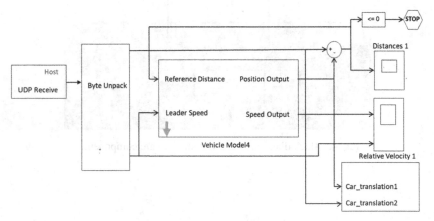

Fig. 7. Communication receiver.

5   HIL–Based Simulation Test Experiments

5.1   ACC Functional Simulation Test Scenario Build

The test scenario was built on a one-way two-lane S-shaped road, as shown in Fig. 8. In addition to the main vehicle TestEgo, a following vehicle with the id AheadL was added in front of TestEgo and an accompanying vehicle with the id AheadR was added in the right lane. In addition, a sensor was installed in the centre of the main vehicle to obtain real-time information from the preceding vehicle AheadL.

5.2   ACC Function Simulation Test Experiment

(1) Experimental description: A hardware-in-the-loop comparison experiment was done for this simulation test. The throttle variation curve and speed variation curve of the main TestEgo when there is no external throttle input are shown in Fig. 9, and the

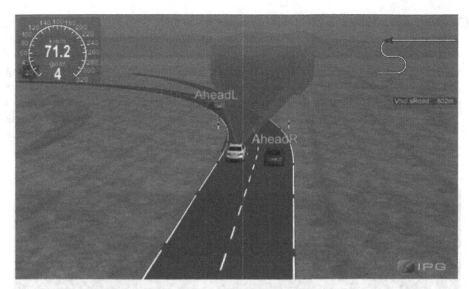

Fig. 8. 3D view of ACC scene.

throttle variation curve and speed variation curve of the main TestEgo when there is external throttle input are shown in Fig. 10. The red and blue curves in the graph correspond to the throttle and speed curves of the main TestEgo, respectively, and the horizontal axis corresponds to the simulation test time.

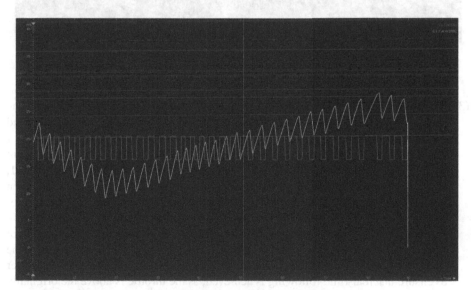

Fig. 9. Without external throttle input.

(2) Experiment without external throttle input: When there is no external throttle input, i.e. the test vehicle follows the set reference speed and reference distance for the test simulation, it can be seen from Fig. 9 that the trend of the speed value change of the main vehicle TestEgo is first gradually becoming smaller to gradually becoming larger, and finally to the speed of 0, i.e. the end of the simulation, the reason for this speed change is because the simulation test road is an S-shaped road. In order to achieve the purpose of keeping the vehicle following, the throttle of the main vehicle TestEgo is always controlled within a threshold range to ensure that the main vehicle TestEgo speed makes timely adjustment with the change of the vehicle in front.

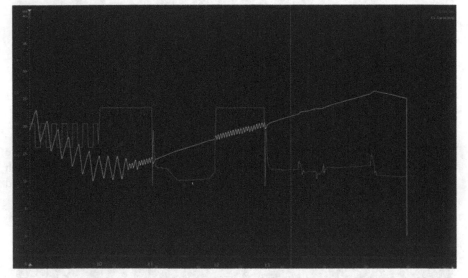

Fig. 10. With external throttle input.

(3) Experiment with external throttle input: When the external signal input is given to the simulated vehicle by manually pressing the pedal, the throttle threshold and speed change value of the main vehicle will change correspondingly with the increase of the throttle. As shown in Fig. 10, the throttle input is given via the pedal at moment t0, and the throttle is always kept at its maximum value for the time period t0 - t1, with the throttle input cancelled at moment t1; the throttle input continues to be given via the pedal at moment t2, and the throttle is always kept at its maximum value for the time period t2 - t3, with the throttle input cancelled at moment t3. The throttle input is cancelled. In contrast, when the throttle input is maintained for a period of time above the set throttle threshold, and the throttle input is suddenly cancelled, the ACC algorithm model will take a decision in response to this sudden change in order to ensure that real-time following is achieved, as the throttle is above the originally set threshold, the throttle input is suppressed and the power is provided entirely by the throttle of the previous time period, so the speed curve shows a This results in a straight-line increase in the speed profile from t1 to t2 and after t3.

(4) Experiment Summary: The experiment simulates the ACC control algorithm making the corresponding speed adjustment to ensure safe driving and automatic following of the vehicle when operating under the advanced driver assistance system, due to the driver accidentally stepping on the wrong throttle.

6 Summary

In order to shorten the development cycle and save R\&D costs, and also to keep up with the rapid development of autonomous driving technology in real time, a hardware-in-the-loop based autonomous driving simulation test system is built. This paper introduces the overall design scheme, hardware and software selection and main module design scheme of the system. By building ACC simulation scenarios and conducting simulation experiments on advanced driver assistance ACC functions based on the system, the effectiveness and safety of the system are verified. As a distinctive hardware and software teaching and engineering dual innovation practice platform, giving full play to the advantages of mixed reality, building experimental scenarios and testing functional scenarios will provide a better interactive test platform for students, which will help enhance students' innovation and practice ability and further promote the cultivation of top innovative talents in autonomous driving technology.

References

1. Yong, J., Gao, F., Ding, N.: Design and validation of an electro-hydraulic brake system using hardware-in-the-loop real-time simulation. Int. J. Autom. Technol. **18**(4), 603–612(2017)
2. Feng, N., Yong, J., Zhan, Z.: A direct multiple shooting method to improve vehicle handling and stability for four hub-wheel-drive electric vehicle during regenerative braking. Proc. Ins. Mech. Eng. Part D: J. Autom. Eng. **234**(4), 1047–1056(2020)
3. Rassolkin, A., Sell, R., Leier, M.: Development case study of first Estonian Self-driving car ISEAUTO. Sci. J. Riga Tech. Univ.-Electr. Control Commun. Eng. **14**(1), 81–88 (2018)
4. Mlorner, J., Trippl, M.: Embracing the future: Path transformation and system reconfiguration for self-driving cars in West Sweden. Eur. Plann. Stud. **27**(11), 2144–2162 (2019)
5. Luk, J.M., Kim, H.C., De Kleine, R.D.: Greenhouse gas emission benefits of vehicle light weighting: Monte Carlo probabilistic analysis of the multi material lightweight vehicle glider. Transp. Res. Part D: Transp. Environ. **62**(1), 1–10 (2018)
6. Lv, C., Hu, X., Sangiovanni-Vincentelli, A.: Driving-style-based code sign optimization of an automated electric vehicle: a cyber-physical system approach. IEEE Trans. Ind. Electron. **66**(4), 2965–2975 (2018)
7. Cicchino, J.B.: Effectiveness of forward collision warning and autonomous emergency braking systems in reducing front-to-rear crash rates. Accid. Anal. Prev. **99**(1), 142–152 (2017)
8. Chen, Y., Chen, S., Zhang, T.: Autonomous vehicle testing and validation platform: integrated simulation system with hardware in the loop, pp. 949–956 (2018)
9. Gelbal, S.Y., Tamilarasan, S., Cantas, M.R.: A connected and autonomous vehicle hardware-in-the-loop simulator for developing automated driving algorithms, pp. 3397–3402 (2017)
10. Ma, J., Zhou, F., Huang, Z.: Hardware-in-the-loop testing of connected and automated vehicle applications: a use case for queue-aware signalized intersection approach and departure. Transp. Res. Rec. **2672**(22), 36–46 (2018)

11. Nees, M.A.: Safer than the average human driver (who is less safe than me)? Examining a popular safety benchmark for self-driving cars. J. Saf. Res. **69**(1), 61–68 (2019)
12. Howard, D., Dai, D.: Public perceptions of self-driving cars: the case of Berkeley, California **14**(4502), 1–16 (2014)

Machine Learning for Data Science

Improving Transferability Reversible Adversarial Examples Based on Flipping Transformation

Youqing Fang, Jingwen Jia, Yuhai Yang, and Wanli Lyu[✉]

Key Lab of Intelligent Computing and Signal Processing of Ministry of Education,
School of Computer Science and Technology, Anhui University, Hefei 230601, China
lwl@ahu.edu.com

Abstract. Adding subtle perturbations to an image can cause the classification model to misclassify, and such images are called adversarial examples. Adversarial examples threaten the safe use of deep neural networks, but when combined with reversible data hiding (RDH) technology, they can protect images from being correctly identified by unauthorized models and recover the image lossless under authorized models. Based on this, the reversible adversarial example (RAE) is rising. However, existing RAE technology focuses on feasibility, attack success rate and image quality, but ignores transferability and time complexity. In this paper, we optimize the data hiding structure and combine data augmentation technology, which flips the input image in probability to avoid overfitting phenomenon on the dataset. On the premise of maintaining a high success rate of white-box attacks and the image's visual quality, the proposed method improves the transferability of reversible adversarial examples by approximately 16% and reduces the computational cost by approximately 43% compared to the state-of-the-art method. In addition, the appropriate flip probability can be selected for different application scenarios.

Keywords: reversible adversarial example · black-box attack · transferability · complexity

1 Introduction

In many fields, deep learning is prominent, such as autonomous driving [1], face recognition [2], semantic segmentation [3] and so on. As an important technology of artificial intelligence, it has also been challenged by various attacks. In 2013, Szegedy et al. [4] found that adding perturbations that cannot be distinguished by the human eye in the image can mislead the neural network to misclassify with high confidence. These images with specific perturbation are called adversarial examples. As shown in Fig. 1,

---

This research work is partly supported by the National Natural Science Foundation of China (62172001), the Provincial Colleges Quality Project of Anhui Province (2020xsxxkc047) and the National Undergraduate Innovation and Entrepreneurship Training Program (202210357077).

Z. Yu et al. (Eds.): ICPCSEE 2023, CCIS 1879, pp. 417–432, 2023.
https://doi.org/10.1007/978-981-99-5968-6_30

patas is misclassified as baboon when adding adversarial perturbation. The existence of image adversarial examples not only dramatically affects the application accuracy of image recognition and classification, but also seriously threatens people's personal and property safety. For example, in the autonomous driving scene, the attacker modifies the road sign into the corresponding adversarial examples [5], which causes the automatic driving system to misjudge the road sign, resulting in a traffic accident.

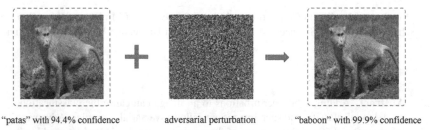

"patas" with 94.4% confidence   adversarial perturbation   "baboon" with 99.9% confidence

Fig. 1. The representation of adversarial example generation. The left image is the original image, and the right image is the adversarial example.

As a fatal attack in artificial intelligence security, if the adversarial example is offensive and reversible at the same time, it will undoubtedly have important application value. That is, attack the unauthorized model and recover the authorized model losslessly. The purpose of reversible adversarial attacks is to add adversarial perturbations to the image reversibly to generate adversarial examples. As an evolution of adversarial examples, reversible adversarial examples are offensive to models and can protect unauthorized or illegal image data processing [6]. In addition, under the legally authorized model, the original image can be restored without loss, which is convenient for image recognition processing. The applications are shown in Fig. 2. In summary, RAE is significant in further expanding artificial intelligence attack and defense technology and applications.

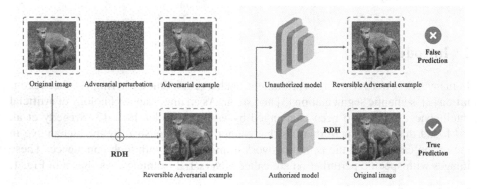

Original image   Adversarial perturbation   Adversarial example   Unauthorized model   Reversible Adversarial example   False Prediction

RDH

Reversible Adversarial example   Authorized model   Original image   True Prediction

RDH

Fig. 2. Application to reversible adversarial examples.

According to the attacker's knowledge of the target model, adversarial examples are divided into two categories: white-box attacks and black-box attacks. Under the white-box condition, the attacker can obtain all the information of the target model, and the existing techniques have achieved a good white-box attack rate. However, the attacker cannot obtain the relevant information of the target model in the black-box condition, so the attack is more difficult, but it has stronger practicability. Current reversible adversarial example generation methods mainly attack a single model, and the attack success rate on other models is poor.

The application scenario shown in Fig. 2 belongs to the scope of white-box attacks, and the authorized model, the unauthorized model and the model generating the adversarial disturbance are all unified models. Protecting image information under more unauthorized models requires reversible adversarial examples to have better black-box attack capabilities, which also means having good transferability.

To improve the attack success rate of reversible adversarial examples under black-box conditions, we improve the current attack method by introducing data augmentation technology [7, 8], which flips the color transformation of the image with a certain probability before each generation of perturbation to improve the diversity of the input. Our contributions are summarized as follows:

1. Based on data augmentation technology, we improve the transferability of reversible adversarial examples. In brief, we randomly flip the input image before adversarial perturbation generation to mitigate the overfitting of the model to the dataset.
2. We optimize the steps of data hiding to alleviate the computational complexity. The final successful adversarial perturbation is embedded, which significantly reduces the time consumption compared to the advanced method.
3. The influence of hyperparameters on the attack rate is studied through ablation experiments. By adjusting the image flipping probability, reversible adversarial examples can be generated to adapt to different scenarios, which reflects the advantages and convenience of the proposed method.

2 Related Work

2.1 Method of Generating Adversarial Examples

Adversarial examples are examples that deliberately add disturbances to induce the classification model to make incorrect judgments. At present, a variety of adversarial example generation methods have been proposed in this field, and the representative methods are introduced below.

Goodfellow et al. [9] develop the FGSM (fast gradient sign method). This method adds perturbations in the opposite direction along the gradient to increase the loss function express and reduce the computational cost of adversarial examples. To solve the problem of the low success rate of a single-step attack, Kurakin et al. [10] improved the FGSM. Through the method of iterative calculation of the gradient, I-FGSM (iterative FGSM) can effectively improve the fitting degree of adversarial examples. Dong et al. [11] introduced the momentum idea and proposed MI-FGSM (Momentum I-FGSM), which stabilizes the gradient update direction and effectively solves the local extremum problem. Subsequently, Xie et al. [7] solved the overfitting problem of the I-FGSM method based on image transformation, and proposed a diversity attack method called the diversity input method (DIM). Madry et al. [12] added randomization processing on the basis of I-FGSM and proposed the gradient projection descent method PGD (Project Gradient Descent). Carlini et al. [13] defined different objective functions, proposed the C&W method, and selected the optimal objective function to realize adversarial attacks through experimental data. Dong et al. [14] proposed the transfer invariant attack TIM (translation invariant method), which takes the set of translation-transformed images and original images as the training set.

2.2 Reversible Adversarial Example

Recently, machine learning has been diffusely applied in the field of multimedia security, especially in data hiding such as watermarking [15, 16] and steganography [17, 18]. Some researchers [19, 20] have begun to think about the similarities and differences between data hiding technology and the machine learning security field. They empirically apply the related technologies in the field of information hiding to the field of machine learning security.

The concept of reversible adversarial examples was proposed by Liu et al. [21] in this scheme, reversible data hiding (RDH) techniques are adopted, and adversarial perturbation is embedded into the image as secret information to obtain a reversible adversarial example. The framework consists of three steps: (1) generate adversarial examples; (2) use the RDH algorithm to generate reversible adversarial examples, and (3) extract the disturbance information and restore the original image. However, the size of the adversarial perturbation is limited by the ability of reversible information hiding. If the adversarial perturbation is large, the original image cannot be recovered losslessly.

To further expand the data hiding capacity, Yin et al. [22] adopted the reversible image transformation (RIT) method instead of the reversible data hiding algorithm, which can embed a large adversarial perturbation. However, there are large image distortions under small magnitude perturbations, such as the DeepFool attack. Aiming at the AdvPatch attack method generated by Brown et al., Chen et al. [23] proposed a reversible attack based on local visible adversarial perturbation, which could hide the data of adversarial patches reversibly. To obtain a good visual quality of the image, Yin et al. [24] proposed a new method, which is based on the independence of brightness and chroma channels in the YUV color space to avoid the coverage of adversarial perturbation signals by information embedding and achieved better performance than before. Table 1 lists the comparison of reversible adversarial examples. However, the previous methods are all focused on white-box attacks. When facing black-box conditions, the attack success rate of reversible adversarial examples is poor.

Table 1. Comparison of reversible adversarial examples.

RAE Method	Advantage	Disadvantage
RAE_RDH [21]	A great creative work that first generate reversible adversarial examples	The embedding capacity is limited and the visual quality of the image is poor
RAE_RIT [22]	It has a good capacity of data hiding	The optimization of DeepFool attack mode is poor
RAE_LocV [23]	The time consumption is optimized than RAE RIT	The object is single which only for AdvPatch attack
RAE_YUV [24]	The visual quality of the image is good and the white-box attack rate is high	Time consuming is serious and the transferability is limited

3 Method

To enhance the transferability of reversible adversarial examples, we combine the input diversity method [8] during perturbation generation. As shown in Fig. 3, in each iteration of the adversarial perturbation generation process, the input image is flipped randomly. Then, the gradient of the transformed images is calculated. Finally the final perturbation is generated to alleviate the overfitting in the adversarial perturbation generation process and improve the transferability of reversible adversarial examples.

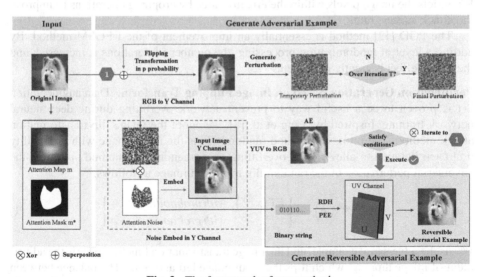

Fig. 3. The framework of our method.

In this section, we describe the implementation of our method in four steps: (1) generation of the attention map, (2) generation of adversarial examples, (3) generation of reversible adversarial examples, and (4) restoration of the original image.

3.1 Attention Map Generation

Perturbations are generally applied to the whole image, but it is difficult for too many perturbations to hide information, and the image's visual quality is affected. Study [25] has shown that convolutional neural networks pay more attention to the foreground part of the image, so adding perturbations to the foreground will be more effective than the background. Inspired by [24], CAM [26] is introduced to generate the attention map. We use the ResNet50 network to extract the key regions of the image as constraints, which can reduce the adversarial perturbation region and the quantity of hiding data.

3.2 Adversarial Example Generation

We use the traditional iterative attack algorithm to generate adversarial perturbations, and combine the image flipping transformation to improve the transferability. The components of the perturbation on the Y channel are extracted and combined with the binarized attention map generated by 3.1 to obtain the attention perturbation. The attention perturbation is embedded into the y-channel of the original image, and the adversarial example is obtained iteratively. Each step is described in detail below.

Adversarial Attack. To mislead the unauthorized model, we use adversarial attack techniques to make adversarial perturbations. Here, we introduce some common iterative attack algorithms.

I-FGSM [10] was proposed as an updated version of FGSM [9]. The method divides the one-step perturbation calculation process in the FGSM method into multiple steps. It restricts the image pixels within the effective area by cropping operations to improve the attack success rate.

The PGD [12] method is essentially an improvement of the I-FGSM method. By adding a layer of randomization processing, the number of iterations is increased, and the attack effect is greatly improved.

Perturbation Generation Based on Image Flipping Transform. Data augmentation [7] is proven to be a useful way to prevent network overfitting during deep neural network training. Inspired by Yang et al. [8], we adopt the image flip transformation attack method. The proposed method randomly flips the input image with probability p at each iteration to alleviate the overfitting phenomenon. The method optimizes the adversarial perturbation of the random flip transform image as follows.

$$FT\left(x^{adv}; p\right) = \begin{cases} FT\left(x^{adv}\right), & with\, probability\, p \\ x^{adv}, & with\, probability\, 1 - p \end{cases} \tag{1}$$

x^{adv} is the adversarial example, and the image transformation function FT() is the adjustment of image flipping, which flips the input image left and right. The balance between the transformed image and the original input image is controlled by the probability p. In this way, an effective attack is achieved through data augmentation, which can avoid the "overfitting" attack of the white-box model and improve the transferability of adversarial examples.

YUV Channel Transform. The main distinctive features for a class of objects can be obtained from luminance, and adversarial noise added to the luminance channel is expected to be more detrimental to the performance of the network than noise in the color channels.

Therefore, we convert the RGB color space into the YUV color space and add disturbance to the Y channel. It can greatly retain the aggressiveness of adversarial noise, and the vacated U and V channels can be used for subsequent data hiding.

According to the standard [27], RGB to YUV color space conversion is formulated as follows:

$$
\begin{aligned}
Y &= 0.299R + 0.587G + 0.114B, \\
U &= -0.169R - 0.331G + 0.500B, \\
V &= 0.500R - 0.419G - 0.081B.
\end{aligned}
\tag{2}
$$

Then, the reversed conversion is formulated as follows:

$$
\begin{aligned}
R &= Y + 1.402(V - 128), \\
G &= Y - 0.344(U - 128) - 0.714(V - 128), \\
B &= Y + 1.772(U - 128).
\end{aligned}
\tag{3}
$$

Perturbed Y Channel Iteratively. We extract the adversarial perturbation generated based on the image flip transformation in the Y channel and combine it with the attention mask generated in 3.1 to obtain the attention perturbation. This perturbation is embedded into the original image Y channel to obtain a temporary adversarial example. Since the perturbation on the brightness channel has a large interference on the neural network, this temporary adversarial example is also very confusing. Here, we set a judgment condition: when the temporary adversarial example can misclassify the current classifier or reach the maximum number of iterations we set, the final adversarial example is generated. The next work is to combine data hiding to change the adversarial example into a reversible adversarial example.

3.3 Reversible Adversarial Examples Generation

We use the prediction error expansion (PEE) [28] embedding algorithm to achieve lossless reversible perturbation hiding.

The embedding process can be summarized in the following three steps:

Step 1: Computing the prediction value $\hat{a}$. The predicted value $\hat{a}$ is obtained from pixel a based on its right, bottom and diagonal pixel values.

Step 2: Computing the prediction error. The prediction error can be calculated as:

$$
e = a - \hat{a}
\tag{4}
$$

Step 3, data embedding. The prediction error after embedding a bit m can be calculated as:

$$
e' = 2e + m
\tag{5}
$$

Then, the pixel is calculated by:

$$a' = \hat{a} + e' \tag{6}$$

In the recovery process, we obtain m by extracting the least significant bit (LSB) in a' and recover e divided by two. Then, obtain a:

$$a = e + \hat{a} \tag{7}$$

In this paper, we regard the latest attention noise as secret information and embed it into the UV channel of the adversarial example by PEE. Finally, the image is converted from the YUV color space to the RGB color space to obtain a reversible adversarial example.

3.4 Original Image Restoration

Under the authorization model, the reversible adversarial examples can be restored to the original image without loss. The process is as follows: the reversible adversarial example is converted into the YUV channel, and the secret information (attention noise) in the UV channel is extracted by PEE. Subtract the noise in the Y channel, and then obtain the original image by converting to the RGB channel.

4 Experiment

To verify the proposed method's effectiveness and superiority, we designed the experimental section as follows.

4.1 Experiment Settings

- **Datasets and Models.** ImageNet [29] is used in the experiment. We randomly choose 1000 images that can be correctly classified by the model for experiments. In addition, InceptionV3 [30], ResNet50 [31], ResNet101 [31], ResNet152 [31] and VGG16 [32] are chosen as threat models.
- **Evaluation Metric.** The structural similarity (SSIM), peak signal to noise ratio (PSNR) and attack success rate (ASR) are used as evaluation metrics. A larger value of the ASR shows better attack performance, and larger values of the PSNR and SSIM indicate higher image visual quality.

4.2 Attacking Performance

Table 2 shows the attacking performance of different methods with different noise percentages. The first row in each patch represents the original attack method PGD's ASR of adversarial example generated by the literature [12]. RAE_RDH, RAE_RIT, RAE_YUV and RAE Ours represent the attack success rate of reversible adversarial examples generated by Liu et al. [21], Yin et al. [22], Yin et al. [24] and our method. Inspired by RAE_YUV, which iteratively generates reversible adversarial examples together with

Table 2. The ASR of reversible adversarial examples in PGD. * indicates a white-box setting, and the highest success rate is in bold.

Attack Method	classifier	RAE Method	Inc-v3	resnet50	resnet101	resnet152	vgg16
PGD	Inc-v3	PGD	94.8*	67.5	60.7	60.3	74.6
		RAE_RDH	84.2*	66.3	55.3	59.0	65.6
		RAE_RIT	86.2*	60.5	53.4	51.3	75.5
		RAE_YUV	**99.5***	65.2	58.6	55.3	70.2
		ours	98.1*	**81.7**	**75.2**	**73.6**	**83.9**
	resnet50	PGD	38.3	100.0*	68.8	64.8	76.4
		RAE_RDH	33.1	89.4*	46.0	47.8	60.4
		RAE_RIT	37.9	89.9*	51.7	48.1	67.7
		RAE_YUV	36.7	**99.4***	55.6	50.6	69.9
		ours	**39.6**	98.8*	**62.0**	**57.9**	**73.3**
	resnet101	PGD	39.8	80.5	100.0*	70.3	74.9
		RAE_RDH	31.1	70.9	89.8*	53.0	66.4
		RAE_RIT	36.1	73.6	88.9*	52.9	68.1
		RAE_YUV	39.6	68.3	**99.5***	55.5	71.9
		ours	**42.1**	**73.5**	99.2*	**61.9**	**73.2**
	resnet152	PGD	40.3	80.5	73.3	99.5*	75.2
		RAE_RDH	30.2	61.5	63.0	88.3*	68.9
		RAE_RIT	36.1	68.8	64.8	97.7*	67.3
		RAE_YUV	39.2	68.3	63.7	**99.2***	72.8
		ours	**43.7**	**77.0**	**69.1**	98.6*	**75.5**

data hiding, we iteratively generate adversarial examples multiple times. As shown in Table 2, the ASRs of RAE_YUV and RAE ours are higher than those of RAE_RDH and RAE_RIT.

Compared to RAE_YUV, the ASR of our method is higher in the black-box condition because the input image is probabilistically flipped when generating the adversarial perturbation in each iteration, which enhances the diversity of the dataset to alleviate the overfitting phenomenon. For example, the black-box attack rates of the reversible adversarial examples generated by our method in the Inc-v3 discriminator attacking the ResNet50, ResNet101, ResNet152, and VGG16 models are 81.7%, 75.2%, 73.6%, and 83.9%, respectively. The methods of RAE_YUV are 65.2%, 58.6%, 55.3%, and 70.2%, respectively, with an average increase of 16%. At the same time, the ASR of RAE_YUV under white-box conditions is equivalent to that of RAE_YUV, and the difference is only 1%. Table 3 shows the ASR of our method and RAE_YUV on reversible adversarial examples under the I-FGSM attack method, and the trend is the same as in Table 2. The ASR is improved under the black-box condition.

Table 3. The ASR of reversible adversarial examples in I-FGSM. * indicates a white-box setting, and the highest success rate is in bold.

Attack Method	classifier	RAE Method	Inc-v3	resnet50	resnet101	resnet152	vgg16
I-FGSM	Inc-v3	RAE_YUV	99.1^*	64.8	57.1	53.9	70.3
		ours	98.0^*	**69.6**	**62.7**	**58.5**	**75.6**
	resnet50	RAE_YUV	37.8	99.1^*	55.2	50.9	70.4
		ours	**40.8**	99.1^*	**61.9**	**58.2**	**73.2**
	resnet101	RAE_YUV	37.9	66.4	98.9^*	52.9	69.7
		ours	**42.7**	**74.0**	98.8^*	**61.8**	**74.0**
	resnet152	RAE_YUV	39.2	68.3	60.9	99.1^*	71.1
		ours	**42.6**	**76.8**	**70.5**	99.3^*	**75.8**

It is a valuable boost for reversible adversarial examples, which can protect images from being correctly classified in multiple unauthorized models.

4.3 Image Visual Quality

PSNR is one of the most widely used image similarity assessment metrics in the field of image processing. The mean squared error (MSE) and the PSNR can be defined as follows:

$$MSE = \frac{1}{mn} \sum_{x=0}^{m-1} \sum_{y=0}^{n-1} \left[I(x, y) - J(x, y) \right]^2, \tag{8}$$

$$PSNR = 10 \times \log_{10} \left(\frac{MAX_1^2}{MSE} \right), \tag{9}$$

where I and J represent the reference image and test image with size $m * n$, respectively, and MAX_1 represents the maximum pixel value of the image, which equals 255. In contrast to PSNR, SSIM measures image similarity from luminance l, contrast c, and structure s. The SSIM is formulated as follows:

$$SSIM(X, Y) = l(X, Y) * c(X, Y) * s(X, Y), \tag{10}$$

where:

$$\begin{cases} l(X, Y) = \dfrac{2\mu_X \mu_Y + C_1}{\mu_X^2 + \mu_Y^2 + C_1} \\ c(X, Y) = \dfrac{2\sigma_X \sigma_Y + C_2}{\sigma_X^2 + \sigma_Y^2 + C_2} \\ s(X, Y) = \dfrac{\sigma_{XY} + C_3}{\sigma_X \sigma_Y + C_3} \end{cases} \tag{11}$$

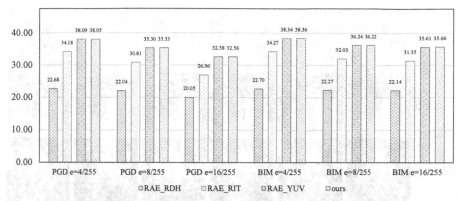

Fig. 4. Comparison of the results of image quality between adversarial examples and reversible adversarial examples with PSNR (dB).

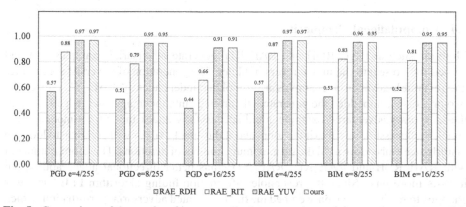

Fig. 5. Comparison of the results of image quality between adversarial examples and reversible adversarial examples with SSIM.

where μ_X and σ_X^2 are the mean and variance, respectively, of image X. σ_{XY} is the covariance between X and Y. C_1, C_2 and C_3 are positive constants to avoid a null denominator.

Figure 4 and Fig. 5 show the visual quality of reversible adversarial examples generated by different methods in different perturbations.

It can clearly be seen that our method is equal to RAE_YUV in image visual quality evaluation and much better than the RAE_RDH and RAE_RIT methods. This is due to using the attention map CAM to limit the adversarial disturbance to the key area. We only retain the disturbance on the luminance channel and embed them into the chrominance and saturation channels, which significantly maintains the original features of the image. However, RAE_RDH and RAE_RIT both add disturbances in all regions of the RGB channel, so they show poor visual effects. Reversible adversarial examples with our method under different conditions are shown in Fig. 6.

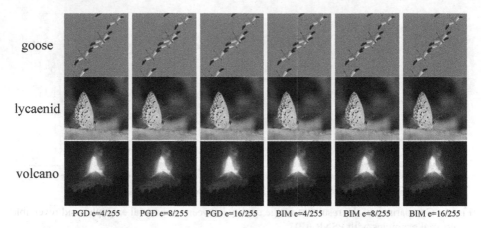

goose

lycaenid

volcano

PGD e=4/255   PGD e=8/255   PGD e=16/255   BIM e=4/255   BIM e=8/255   BIM e=16/255

Fig. 6. Reversible adversarial examples with our method under different conditions.

4.4 Computational Complexity

Methods such as RAE_RDH and RAE_RIT generate adversarial examples first and embed the noise into the image reversibly later. The generation of adversarial examples and reversible adversarial examples are completed independently. RAE_YUV fused the two processes, embedded the generated disturbance into the image, and judged whether the temporary adversarial example could successfully attack the current model. If the attack is successful, it will exit the loop, and if the attack fails, it will continue to loop the disturbance embedding until the set maximum number of iterations. Therefore, compared with RAE_RDH and RAE_RIT, RAE_YUV has better directional attack ability, but it costs more time. Since the reversible information hiding algorithm PEE has fewer changes to the image, we only embed the final generated adversarial perturbation, which saves the time-consuming embedding in each iteration. Due to the improved reversible adversarial example generation mode, the time to produce 1000 reversible adversarial examples in the experimental environment of this chapter is greatly reduced compared with RAE_YUV. Taking I-FGSM as the attack method as an example, when ResNet152 is used as the discriminator, RAE_YUV takes 14001 time units, while our method only needs 7976 time units, which is 43% shorter. Therefore, the proposed method has more practical value (Fig. 7).

4.5 Hyperparameter Research

Here, we study the effect of the hyperparameter flip probability p on the attack success rate under both white-box conditions and black-box conditions. The range of p is from 0 to 1, growing in steps of 0.1. Figure 8(a) shows that under the PGD attack method, as the flip probability p rises, the attack success rate of white-box attacks decreases while the attack success rate of black-box attacks increases. In the beginning, the attack success rate of white-box attacks decreases slowly and then decreases rapidly when p is greater than 0.7. Currently, the attack success rate of black-box attacks is increasing

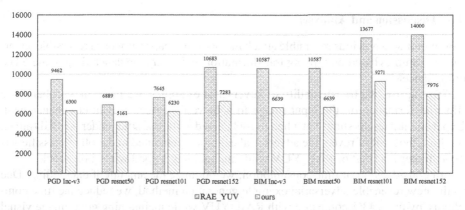

Fig. 7. Comparison results of computational complexity of reversible adversarial example generation.

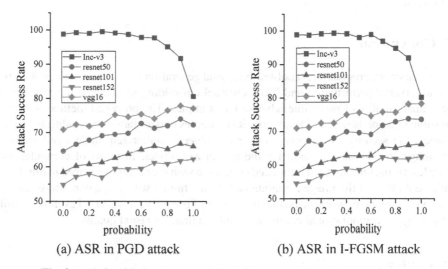

(a) ASR in PGD attack   (b) ASR in I-FGSM attack

Fig. 8. Relationship between translation probability and attack success rate.

gradually. The trend is the same under the I-FGSM attack method, and the turning point is approximately $p = 0.6$.

The trend also provides an effective adversarial attack method: if we want to achieve a high black-box attack rate, we can increase the probability of image enhancement. If we want to maintain a high white-box attack rate, we can set the probability p to an intermediate value. Considering the calculation time, the probability p of this paper is chosen to be 0.5 in the previous test.

4.6 Discussion and Analysis

Compared with previous reversible attack methods, our paper proposed reversible adversarial examples based on flipping transformation and reversible data hiding in the YUV color space.

To improve the transferability of adversarial examples, we perform a probabilistic flip transformation on the input image to achieve the effect of dataset enhancement. Experimental results show that when Inc-v3 is used as the classifier under the PGD attack method, the ASR of reversible adversarial examples generated under other classifiers is 16% higher than that of RAE_YUV while maintaining a good white-box attack.

In consideration of time consumption, we optimize the amount of data hiding. Due to the new reversible adversarial example generation method, we reduce the time complexity by up to 43% compared with RAE_YUV while maintaining good image visual quality.

In addition, the appropriate flip probability can be selected for different application scenarios.

5 Conclusion

In this paper, we propose a reversible adversarial generation method based on probability flip adversarial perturbation and YUV channel embedding to alleviate the problem of low transferability of reversible adversarial examples. The proposed method improves the robustness of the reversible adversarial examples under black-box conditions, which protects the security of the image under the other unauthorized models and recovers the image in a lossless fashion for the authorized model. In terms of computational complexity, the running time is reduced compared with the current advanced method, and the visual effect of the image is maintained. In the future, we will explore more suitable methods to improve the black-box attack ability of reversible adversarial examples while balancing the computational complexity and the image's visual quality.

References

1. Feng, D., Harakeh, A., Waslander, S.L., et al.: A review and comparative study on probabilistic object detection in autonomous driving. IEEE Trans. Intell. Transp. Syst. **23**(8), 9961–9980 (2021)
2. Rivero-Hernandez, J., Morales-Gonzalez, A., Denis, L.G., et al.: Ordered weighted aggregation networks for video face recognition. Pattern Recogn. Lett. **146**, 237–243 (2021)
3. Santos, T.I., Abel, A., Wilson, N., et al.: Speaker-independent visual speech recognition with the Inception V3 model. In: IEEE Spoken Language Technology Workshop (SLT), pp. 613–620. IEEE (2021)
4. Szegedy, C., Zaremba, W., Sutskever, I., et al.: Intriguing properties of neural networks. In: arXiv preprint arXiv:1312.6199 (2013)
5. Eykholt, K., Evtimov, I., Fernandes, E., et al.: Robust physical-world attacks on deep learning visual classification. In: Proceedings of the IEEE Conference on Computer Vision and Pattern recognition, pp. 1625–1634 (2018)

6. Hou, D., Zhang, W., Liu, J., et al.: Emerging applications of reversible data hiding. In: Proceedings of the 2nd International Conference on Image and Graphics Processing, pp. 105–109 (2019)
7. Xie, C., Zhang, Z., Zhou, Y., et al.: Improving transferability of adversarial examples with input diversity. In: Proceedings of the IEEE/CVF Conference on Computer Vision and Pattern Recognition, pp. 2730–2739 (2019)
8. Yang, B., Zhang, H., Li, Z., et al.: Adversarial example generation method based on image flipping transform. J. Comput. Appl. **42**(8), 2319 (2022)
9. Goodfellow, I.J., Shlens, J., Szegedy, C.: Explaining and harnessing adversarial examples. In: arXiv preprint arXiv:1412.6572 (2014)
10. Kurakin, A., Goodfellow, I.J., Bengio, S.: Adversarial Examples in the Physical World. In: Artificial Intelligence Safety and Security. Chapman and Hall/CRC, pp. 99–112 (2018)
11. Dong, Y., Liao, F., Pang, T., et al.: Boosting adversarial attacks with momentum. In: Proceedings of the IEEE Conference on Computer Vision and Pattern Recognition, pp. 9185–9193 (2018)
12. Madry, A., Makelov, A., Schmidt, L., et al.: Towards deep learning models resistant to adversarial attacks. In: arXiv preprint arXiv:1706.06083 (2017)
13. Carlini, N., Wagner, D.: Towards evaluating the robustness of neural networks. In: IEEE Symposium on Security and Privacy (sp). 2017, pp. 39–57. IEEE (2017)
14. Dong, Y., et al.: Evading defenses to transferable adversarial examples by translation-invariant attacks. In: Proceedings of the IEEE/CVF Conference on Computer Vision and Pattern Recognition (2019)
15. Xiong, L., Han, X., Yang, C.N., et al.: Robust reversible watermarking in encrypted image with secure multiparty based on lightweight cryptography. IEEE Trans. Circ. Syst. Video Technol. **32**(1), 75–91(2021)
16. Zhang, X., Sun, X., Sun, X., et al.: Robust reversible audio watermarking scheme for telemedicine and privacy protection. CMC-Comput. Mater. Continua, **71**(2), 3035–3050 (2022)
17. Yin, Z., Longfei, K.: Robust adaptive steganography based on dither modulation and modification with re-compression. IEEE Trans. Sig. Inf. Process. Over Netw **7**, 336–345 (2021)
18. Ke, L., Yin, Z.: On the security and robustness of "Keyless dynamic optimal multi-bit image steganography using energetic pixels." Multimedia Tools Appl. **80**, 3997–4005 (2021)
19. Schöttle, P., Schlögl, A., Pasquini, C., et al.: Detecting adversarial examples-a lesson from multimedia security. In: 26th European Signal Processing Conference (EUSIPCO). IEEE, 2018, pp. 947–951 (2018)
20. Quiring, E., Arp, D., Rieck, K.: Forgotten siblings: unifying attacks on machine learning and digital watermarking. In: 2018 IEEE European symposium on security and privacy (EuroS&P), pp. 488–502. IEEE (2018)
21. Liu, J., Hou, D., Zhang, W., Yu, N.: Reversible adversarial examples. In: arXiv preprint arXiv:1811.00189 (2018)
22. Yin, Z., Wang, H., Chen, L., et al.: Reversible adversarial attack based on reversible image transformation. In: arXiv preprint arXiv:1911.02360 (2019)
23. Chen, L., Zhu, S., Yin, Z.: Reversible Attack based on Local Visual Adversarial Perturbation. In: arXiv preprint arXiv:2110.02700 (2021)
24. Yin, Z., Chen, L., Lyu, W., et al.: Reversible attack based on adversarial perturbation and reversible data hiding in YUV colorspace. Pattern Recogn. Lett. **166**, 1–7 (2023)
25. Dong, X., Han, J., Chen, D., et al.: Robust superpixel-guided attentional adversarial attack. In: Proceedings of the IEEE/CVF Conference on Computer Vision and Pattern Recognition, pp. 12895–12904 (2020)

26. Hou, Q., Zhou, D., Feng, J.: Coordinate attention for efficient mobile network design. Proceedings of the IEEE/CVF Conference on Computer Vision and Pattern Recognition (2021)

27. Jack, K.: Video Demystified: a Handbook for the Digital Engineer. In: Elsevier (2011)

28. Wenguang, H., Cai, Z.: Reversible data hiding based on dual pairwise prediction-error expansion. IEEE Trans. Image Process. **30**, 5045–5055 (2021)

29. Russakovsky, O., Deng, J., Su, H., et al.: ImageNet large scale visual recognition challenge. Int. J. Comput. Vis. **115**, 211–252 (2015)

30. Szegedy, C., Vanhoucke, V., Ioffe, S., et al.: Rethinking the inception architecture for computer vision. In: Proceedings of the IEEE Conference on Computer Vision and Pattern Recognition, pp. 2818–2826 (2016)

31. He, K., Zhang, X., Ren, S., et al.: Deep residual learning for image recognition. In: Proceedings of the IEEE Conference on Computer Vision and Pattern Recognition, pp. 770–778 (2016)

32. Simonyan, K., Zisserman, A.: Very deep convolutional networks for large-scale image recognition. In: arXiv preprint arXiv:1409.1556 (2014)

Rolling Iterative Prediction for Correlated Multivariate Time Series

Peng Liu, Qilong Han[✉], and Xiao Yang

College of Computer Science and Technology, Harbin Engineering University, Harbin 150001, China
hanqilong@hrbeu.edu.cn

Abstract. Correlated multivariate time series prediction is an effective tool for discovering the chang rules of temporal data, but it is challenging to find these rules. Recently, deep learning methods have made it possible to predict high-dimensional and complex multivariate time series data. However, these methods cannot capture or predict potential mutation signals of time series, leading to a lag in data prediction trends and large errors. Moreover, it is difficult to capture dependencies of the data, especially when the data is sparse and the time intervals are large. In this paper, we proposed a prediction approach that leverages both propagation dynamics and deep learning, called Rolling Iterative Prediction (RIP). In RIP method, the Time-Delay Moving Average (TDMA) is used to carry out maximum likelihood reduction on the raw data, and the propagation dynamics model is applied to obtain the potential propagation parameters data, and dynamic properties of the correlated multivariate time series are clearly established. Long Short-Term Memory (LSTM) is applied to capture the time dependencies of data, and the medium and long-term Rolling Iterative Prediction method is established by alternately estimating parameters and predicting time series. Experiments are performed on the data of the Corona Virus Disease 2019 (COVID-19) in China, France, and South Korea. Experimental results show that the real distribution of the epidemic data is well restored, the prediction accuracy is better than baseline methods.

Keywords: Time Series Prediction · Correlated Multivariate Time Series · Trend Prediction of Infectious Disease · Rolling Circulation

1 Introduction

A time series refers to the series of observed values under the same index that are arranged strictly in chronological order, which is used to describe the change of state of objects over time [1]. Correlated multivariate time series simultaneously records multiple time series, containing more target information, and there may be complex dependency relationships between variables [2]. As time series data are a natural pattern, correlated multivariate time series data exist in most complex tasks that require human cognition [3]. Therefore, effective analysis and prediction of such data is helpful for people to grasp the direction of changes in things, and has important reference value for the arrangement and decision-making in production and life.

© The Author(s), under exclusive license to Springer Nature Singapore Pte Ltd. 2023
Z. Yu et al. (Eds.): ICPCSEE 2023, CCIS 1879, pp. 433–452, 2023.
https://doi.org/10.1007/978-981-99-5968-6_31

Correlated Multivariate Time Series (CMTS) contains a lot of useful information. In addition to intuitive observed values, many implicit intrinsic information can be obtained through data mining. The goal of CMTS prediction is to capture the change rule of data from the change of historical data, combine the factors that may affect the results, and establish a suitable model to predict the most likely time series values on the future time node [4]. The prediction of CMTS can mine the change rule of time series data and predict the development trend of things. For example, in the medical field, it can prevent and suppress the spread of diseases by predicting the incidence of infectious diseases [5].

In the application of CMTS prediction, people usually predict new trends or potential hazardous events based on knowledge in historical sequences. In the field of medicine, the outbreak and spread of epidemics have caused great damage for society and finance [6]. Therefore, epidemics need to be intervened and controlled in a timely manner, but successfully predicting the inflection point and trend of epidemic transmission is a complex problem. Multivariate time series of epidemic diseases are very typical CMTS, they have strong time correlation and variable dependency relationship, such as Corona Virus Disease 2019 (COVID-19). The epidemic data is very sparse, making it more important to mine potential knowledge from data than to observe the observed sequence of data. Accurately predicting the inflection point of the infectious diseases and other transmission trends can provide abundant decision-making information for the epidemic prevention department.

In recent years, there have been many classical time series prediction methods. Traditional methods use statistical knowledge to model and predict the development process and trends of time series, including Vector Autoregressive Model (VAR), AutoRegressive Moving Average Mode (ARMA), Support Vector Machine (SVM) and Random Forest (RF). These methods are efficient, easy to use and interpretable. However, these models usually assume that the time series has a certain distribution or function form, requiring that the predicted values have a linear correlation with the historical sequence data. They may not be able to capture the potential complex nonlinear relationship between time series. For the complex transmission process of epidemics, existing mathematical theories find appropriate models to describe the dynamic transmission process of epidemics. For example, Susceptible Infected Recovered (SIR) model is characterized by the susceptible, the infective and the removal to describe the spread of infection. Susceptible-Exposed-Infectious-Removed (SEIR) model can clarify the relationships among people in different states. However, in actual scenarios, there are often very complex association relationships and high dimensions between data. As a result, SIR and SEIR cannot achieve the expected prediction accuracy.

Most of the traditional time series prediction methods only model the predicted objects from the time dimension, which may not capture the potential complex nonlinear relationships between time series. The method based on neural network belongs to the data-driven prediction method. The suitable model parameters are sought through the data, which can effectively establish the association relationship of data in time and the dependency relationship between multivariate variables. However, most neural network-based prediction methods can not effectively capture the time series' long-term dependencies, and can only use fixed-length data series for prediction, making it difficult to achieve the best prediction effect. Recurrent Neural Network (RNN) can flexibly capture the nonlinear relationships between variables and output sequences by relying on the iteration of its own hidden layer state and the input of each time step. However, conventional RNNs cannot capture the long-term dependences. Long short term memory (LSTM) adds memory and gating mechanisms to RNN, it can capture long-term dependencies in time series well and has strong nonlinear fitting ability of variables. The long-term and short-term relationships between sequences are used to improve model performance, but the interpretability of LSTM is low.

This has prompted us to study how to accurately predict the long-term trend of sparse correlated multivariate data (e.g., infectious disease transmission data), and many practical factors have brought difficulties to the research, including (1) There are data missing and data sparsity in CMTS, the model is required to eliminate noise interference; (2) Most of the existing methods model the predicted object from the time dimension, and there is no effect of the internal mechanism of the object on the object; (3) The epidemics trend prediction is a long-term and continuous task. Most existing models simplify this propagation process into a constant attenuation coefficient or a unified functional form. The model is required to balance long-term historical information and short-term mutation information to ensure the performance of the model. Therefore, a Rolling Iterative Prediction (RIP) method is proposed to learn the variable dependence and temporal propagation parameters of CMTS in this article. The prediction direction of LSTM can be guided by clear potential information to ensure that the prediction results conform to the long-term trend of the sequence. Overall, the main contributions of our work are:

- The effective distribution of epidemic data is restored by using Time-delay Moving Average. The correlation of sparse data in time order is enhanced and the influence of time delay on data structure is reduced.
- Transmission dynamics model is established to describe clear transmission mechanisms of infectious diseases. The SEIR model is optimized in this paper, the model adds incubation infectivity and asymptomatic carriers and studies more complex forms of transmission.
- Different nodes are clustered predict node trends in a targeted manner. Through explicit data modeling of infectious disease transmission process combined with LSTM, the implicit propagation parameters of the data are iteratively calculated and explicit time series are predicted to improve the accuracy of trend prediction of infectious diseases.

The remainder of this article is structured as follows: In Sect. 2, a review of related works is provided. Section 3 outlines the methods utilized in this study. The experimental results are presented in Sect. 4, and finally, Sect. 5 concludes this paper and discusses possible future works.

2 Related Works

The transmission data of epidemic diseases have strong variable correlation and sparsity. Since our approach relies on propagation dynamic model and the LSTM to predict the transmission trend of epidemics, we summarize the related works in these two directions in this paper.

Transmission Dynamics Model. The emergence of infectious disease dynamics provides a solid foundation for the theoretical research on infectious diseases. A mathematical model corresponding to the characteristics of epidemic spreading is constructed by reasonable assumptions [7]. Britton et al. analyzed the causes and influencing factors of infectious diseases, and theoretical basis and parameter basis are provided for decision-making of epidemic prevention department [8]. Zhang et al. integrated government intervention into the SEIR model to dynamically estimate all parameters of the model to simulate the transmission process of the virus [9]. Traditional SEIR model was optimized by Nanshan Zhong et al., and the population flow information was added to the SEIR model. Trends and key nodes of dissemination of COVID-19 in China are effective predicted by adding inflow and outflow coefficients of lurkers [10]. Read et al. added the latent phase of the epidemic and asymptomatic individuals to the SEIR model, which was able to accurately separate infected individuals from the population, the difficulty of collecting accurate individual data makes it difficult to obtain optimal fitting parameters [11]. Wu et al. analyzed the epidemic transmission process at the city level using the SEIR model and estimated the basic reproductive number using the Monte Carlo method [12]. Due to the limited factors included in the commonly used transmission dynamics models, it is difficult to effectively capture some important factors, such as asymptomatic cases, isolated cases, etc.. This leads to very complicated calculations. At the same time, in the immediate aftermath of the outbreak, there are various degrees of lack and noise in various data collections, this may have a significant influence on the accuracy of the transmission dynamics method.

Deep Learning Methods. Recently, deep learning methods have been extensively studied and applied in CMTS prediction, especially LSTM, it is used to predict epidemic trends of infectious diseases. Pathan et al. analyzed the nucleotide mutation rate and pattern in the codon mutation set, and the LSTM was used to estimate the mutation rate of the virus in humans and to predict the increasing number of cases per day [13]. Jana et al. converted epidemic transmission data into two-dimensional image sets without time components and extracted spatial causality of data using convolutional neural networks [14]. To address the problems caused by the limited data of COVID-19 for modeling and prediction, Kırbaş et al. utilized ARIMA, nonlinear autoregressive neural network (NARNN) and LSTM methods to model cumulative confirmed case data of European countries, it is verified that LSTM has higher accuracy [15]. Arora et al. tested LSTM variants such as Convolutional LSTM, Deep LSTM, and Bi-directional LSTM models. Absolute errors of the prediction results of models show that the bidirectional LSTM has the best prediction results, while the convolutional LSTM has the worst prediction results [16]. However, LSTM models tend to prioritize the quantity of infections, but a significant limitation of these models is their failure to account for external factors such as isolated cases, asymptomatic cases, and protected populations. These factors are critical in assessing the impact of COVID-19.

3 Methods

3.1 Problem Statement

A single variable time series is a sequence of observed data arranged according to the time order, denoted by.

$$X = (x_1, x_2, \ldots x_j \ldots, x_T),$$

At the time point $t \in (1,2,3,\ldots,T)$, there is a unique observation result. The multivariate time series contains M univariate time series, denoted by $MTS_{train} = \left(X_1^{T(train)}, X_2^{T(train)}, \ldots X_i^{T(train)} \ldots, X_M^{T(train)}\right)$, MTS_{train} obeys the distribution $f\left(MTS^{T(train)}\right)$. At the time point t, the value $x^t \in \mathbb{R}^{M \times T}$ of each variable forms an M-dimensional vector. There may be correlation relationships among single variable time series in MTS, denoted by

$$f_k(X_k^t) = \left(\left(f_i(X_i^t) \cdots f_j(X_j^t)\right), \Phi\right) \tag{1}$$

where Φ is a set of single variable fitting parameters, and a single variable time series in MTS can be linearly represented by other single variable time series.

$$x_{ij} = \sum_{k=1}^{M} \phi_k x_{kj} \tag{2}$$

where $\varphi \in \Phi$, $j \in (1,2,3,\ldots,T)$, $k \in (1,2,3,\ldots,N)$, $i \in (1,2,3,\ldots,M)$, when $k = i$, $\varphi_k = 0$.

If any one single variable time series in the MTS has the time dependently, we define $\Omega_i = \{x_{ij} | x_{t-K}, x_{t-K+1}, \ldots, x_{t-1}\}$. The observed value of the single variable X_i at time j in the time period $[p, q]$ can be represented as

$$x_{ij} = \frac{1}{j - p} \sum_{k=p}^{j-1} \lambda_k x_{ik} \tag{3}$$

where λ is the weight vector of x_{ij} represented by data in the time period $[p, q]$.

Our goal is to forecast the patterns in a specific future through the acquisition of trends of historical data MTS_{train}, and output the time series with the same variables as MTS_{train}, denoted by $MTS_{test} = \left(X_1^{T(test)}, X_2^{T(test)}, \ldots X_i^{T(test)} \ldots, X_M^{T(test)}\right)$, MTS_{test} approximately obeys the distribution $f\left(MTS^{T(train)}\right)$. The description of the development process of the events needs to express not only its characteristic change process, but also the nonlinear change process of its internal mechanism and complex relationships. In order to clearly obtain the correlation Φ between variables, we establish the time series model S of CMTS. S maps the explicit observation sequence MTS_{train} to the potential propagation parameter sequence *m* of multivariate time series, and m can capture the potential development trend or mutation signal of time series. We use LSTM to characterize the long and short term dependence of time series.

Due to the infectious disease of long duration of the epidemic and the sparse data, it is difficult to collect more meaningful original information. Therefore, we add analytical data containing more statistical information to the LSTM. The sparse CMTS with strong correlation and long time interval can be predicted by alternating work and iteration of S and LSTM.

3.2 Overview

To accurately predict long-term trends of CMTS of epidemics, a rolling iterative prediction method for predicting epidemiological parameters and long time trends of epidemics is proposed, illustrates the framework of the method (Fig. 1).

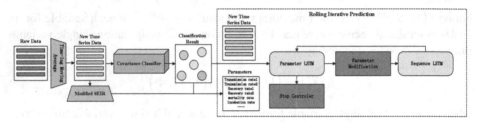

Fig. 1. The proposed method in the paper.

(1) **Time-Delay Moving Average (TDMA) method**, is adopted to the raw data preprocessing, and the raw data are reversely traversed by a time window that the width is the length of the latent period. The data are regenerated that are consistent with the real data distribution of epidemic, and data missing and noise interference are reduced.

(2) **SEIR Model**, the optimized SEIR model is constructed according to epidemiological characteristics of epidemic, and the epidemiological parameters consistent with COVID-19 are calculated.

(3) **Nodes Classification**, the high-dimensional data of different lengths of each transmission node is aligned through the covariance matrix, and the matrix feature vectors are extracted. The transmission nodes are clustered for personalized prediction of cities with different epidemiological manifestations.

(4) **The Rolling Iterative prediction**, epidemiological parameters and regenerated data are input into the LSTM. The models of different clusters are pre-trained to guide the LSTM to predict the spread trend of epidemics effectively by the dynamic factors of epidemic spreading. Throughout the two incubation periods, the controller halts the prediction process until the count of infections reaches 0. Bias that our data processing method brings to the prediction results are corrected.

3.3   Time-Delay Moving Average

The observational data of epidemics are generally collected manually. The collected data general have some quality problems, particularly during the initial phases of an infectious disease outbreak, a large number of cases and transmission information are incomplete and noisy, but they still have potential data distribution characteristics. There is still some inherent law in the dirty data that makes the overall trend similar to the real world. Therefore, it cannot simply be replaced by deletion or average value. Since infectious diseases usually have an incubation period, the lag of the onset caused by the incubation period will affect the true distribution of time series. We use TDMA to smooth the raw data. The smoothed time series is regenerate on the basis of the influence of the historical residual trend of data on the current node. The temporal structure of data is enhanced to make the processed data more in line with the spread trend of epidemics.

The mean value in the window is calculated through a window of length ω that slides on the time series in reverse. All calculated mean values are arranged in chronological order in the raw data to obtain a new series.

Definition 1 (Time-Delay Moving Average, TDMA). Suppose the time series $X = \{x_1, x_2, \ldots, x_M\}$ be represented by another series $X' = \{x'_1, x'_2, \ldots, x'_M\}$, the i-th element in X' can be calculated by formula (4).

$$X'_i = \frac{1}{\omega} \sum_{j=i-\omega+1}^{i} x_j \tag{4}$$

where ω is length of the window.

New data generated by TDMA is always based on the residual influence of the raw data. The sum result of the processed data is as follows:

$$
\begin{aligned}
SUM &= \sum_{i=1}^{T} \frac{x_{i-6} + x_{i-5} + x_{i-4} + x_{i-3} + x_{i-2} + x_{i-1} + x_i}{7} \\
&= x_7 + x_8 + \cdots + x_{T-6} + \frac{6 \cdot x_{T-5} + 5 \cdot x_{T-4} + 4 \cdot x_{T-3} + 3 \cdot x_{T-2} + 2 \cdot x_{T-1} + x_{T-4}}{7} \\
&\approx \sum_{i=1}^{T} x_i
\end{aligned}
\tag{5}
$$

where x_i is the element in the regenerated epidemic time series, and T represents the overall length of series.

To judge the stage of the end of the epidemic at the current node, it is necessary to collect zero new cases for 14 consecutive days. The new data collected in the last 14 days of the epidemic are all 0 values. Therefore, according to formula 5, the people counting in the dimensions of the new dataset processed after using TDMA is the same as the raw data. Therefore, our TDMA method has no effect on the sum of the daily new data.

As per the findings of the classical SEIR model, the daily new increased in the count of people in the process of infectious disease transmission conforms to the normal distribution. When the data fully conforms to the normal distribution, as is shown in

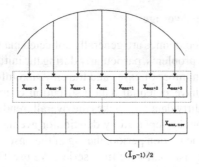

Fig. 2. The inflection point of time-delay moving average.

Fig. 2. In the data processed by our TDMA method, the real date of each point is pushed back by $(l_p-1)/2$ days.

In reality, the distribution of the epidemic does not strictly conform to the normal distribution. The date delay for each point is actually $[0, (l_p-1)/2]$ days that combines with the characteristics of change-points in practice, and the time migration is corrected for $(l_p-1)/2 \pm [(l_p-1)/4]$.

3.4 Optimized SEIR Model

In the process of epidemic spread observation, the patients in incubation period also have infectivity. There are also asymptomatic carriers appear, and they have some infection risk. However, they are not brought into confirmed cases and most of them will cure by themsleves. To sum up, the traditional SEIR model is optimized in this paper by considering the infectivity and asymptomatic self-cure conditions in incubation period. The optimized model is depicted in Fig. 3.

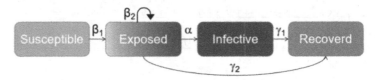

Fig. 3. Our optimized SEIR model.

Base model is as follows:

$$\begin{cases} N(t) = S(t) + E(t) + I(t) + R(t) \\ dS/dt = -r_1\beta_1 IS/N - r_2\beta_2 ES/N \\ dE/dt = r_1\beta_1 IS/N + r_2\beta_2 ES/N - \alpha E - \gamma_2 E \\ dI/dt = \alpha E - \gamma_1 I \\ dR/dt = \gamma_1 I + \gamma_2 E \end{cases} \quad (6)$$

Here, the people counting is assumed as N, including Susceptible(S), Exposed(E), Infective(I), Recoverd(R). S has β_1 probablility to be infected per day, while S has β_2 probablility to be exposed per day. E is converted to I with probability α per day, and converted to R with probability γ_2 (cure or death). I is converted to R with probability γ_1. The optimized model is as follows:

$$
\begin{cases}
N = S_n + E_n + I_n + R_n \\
S_n = S_{n-1} - r_1\beta_1 I_{n-1} S_{n-1}/N_{n-1} - r_2\beta_2 E_{n-1} S_{n-1}/N_{n-1} \\
E_n = E_{n-1} + r_1\beta_1 I_{n-1} S_{n-1}/N_{n-1} + r_2\beta_2 E_{n-1} S_{n-1}/N_{n-1} - \alpha E_{n-1} - \gamma_2 E_{n-1} \quad (7) \\
I_n = I_{n-1} + \alpha E_{n-1} - \gamma_1 I_{n-1} \\
R_n = R_{n-1} + \gamma_1 I_{n-1} + \gamma_2 E_{n-1}
\end{cases}
$$

where:

r_1: Number of contacts per day per lurker;

r_2: Number of contacts per person per day for diagnosed persons;

β_1: Infection rate for susceptible individuals;

β_2: The rate of transmission for the susceptible to exposed;

γ_1: The probability of recovery for diagnosed persons, the estimated value is 0.1024;

γ_2: The probability of recovery for asymptomatic carriers, the estimated value is 0.2978;

α: The incubation rate, the estimated value is 1/5.

In this paper, COVID-19 is taken as an example. The following parameters are estimated. As shown in Table 1.

Table 1. Parameter estimation.

Incubation period	2–14 days, the middle value (7 days) is selected in this paper
Number of daily contacts	3–15 persons/day, before Jan 23, $r = 15$, after Jan 23, $r = 3$, and after March 1, $r = 10$
Detection time	Average time from infection to discovery of COVID-19 is 10 days
Cure time	The average cure time of Hubei is 9 days. The average cure time beyond Hubei is 20 days. The average cure time of the whole is 14 days
The lowest infection rate[①]	The lower limit of modified model for infection rate is 2%. The result of infection rate comes from the infection rate of medical workers in isolation region
Estimate the number of lurkers at the beginning of epidemic	13118

(continued)

Table 1. (*continued*)

The cardinal number of lurkers	Suppose that all the nodes are from Wuhan after the epidemic, $E_{node} = \frac{E_{wuhan} M_{wuhan} m_{node}}{S_{wuhan}}$ ②

① Infection rates among medical workers during COVID-19 transmission. Due to the protection and disinfection of medical workers is the highest standard during epidemic transmission, the infection rate is the lowest infection rate [17].

② The transmission of lurkers from Wuhan to other nodes. E_{wuhan}/S_{wuhan} is the proportion of the lurkers in Wuhan. M_{wuhan} is the number of outflow population before Wuhan is on lockdown, and m node is the Baidu migration coefficient of the nodes are input by Wuhan [18].

Formula 7 is employed to estimate parameters of the model, then the sequence $S_p \in \mathbb{R}^{sp}$ of the parameters of the model changing with time is obtained. The value ranges of β_1 and β_2 are $\beta_1 \in [0.1371, 0.2763]$, $\beta_2 \in [0.0012, 0.0881]$.

3.5 Nodes Classification

The starting and ending time of epidemic observation data reported at each node are different, resulting in different length of the collected time sequence data. In order to achieve effective clustering, the covariance matrix is used to process high-dimensional unequal length time series at each node. Then eigenvalue vectors of each matrix are calculated. Finally, eigenvalue vectors are clustered through K-means.

The high-dimensional time series of each epidemic transmission node (city as a unit) is regenerated by using the TDMA method. The covariance matrix C is as follows,

$$C_{M \times M} = \begin{pmatrix} c_{11} & c_{12} & \cdots & c_{1M} \\ c_{21} & c_{22} & & c_{2M} \\ \vdots & \vdots & \ddots & \vdots \\ c_{M1} & c_{M2} & \cdots & c_{MM} \end{pmatrix} \tag{8}$$

where $C_{M \times NM}$ represents the covariance between M variables within the node.

Eigenvalues of covariance matrices are clustered through K-means, it achieves personalized prediction of different propagation nodes.

3.6 The Rolling Iterative Prediction

Most of the existing methods assume that the time series have a certain distribution or function form, so they cannot capture different forms of data under nonlinearity. Whether it is a traditional method or a neural network method, it is essentially a linear fitting of the data through the internal distribution of the data. They don't learn the structural relationships between variables explicitly, nor do they utilize such structural relationship to forecast the expected behavior of the time series, especially for the nonlinear transmission process of infectious diseases. In this paper, the optimized SEIR model is utilized to determine the system's own development information. LSTM is used to capture more potential characteristics of the epidemic development process, including the impact of the external environment on the system. We eliminate the degree of uncertainty in the spread of the epidemic by alternating rolling predictions with two predictors.

At present, LSTM has been able to accurately predict nonlinear high-dimensional time series. However, this method is not effective for the changes in the transmission process of infectious diseases. Therefore, the important parameters learned from the improved SEIR model is added to the prediction process to guide the prediction process. Meanwhile, predicting the inflection point of infectious diseases is particularly crucial in the process of infectious disease trend prediction. The inflection point is an important time node, and its appearance indicates the improvement of herd immunity. It shows that the intervention measures have effectively curbed the spread of infectious diseases, which will have an impact on health policy formulation, disease control scheme, and even the daily life of the public. According to the event dynamics model, it can be seen that $r\beta$ potentially affects the transmission trend of infectious diseases. When $r\beta > 1$, the pathophoresis is positive; otherwise, the pathophoresis is inhibited. Therefore, the value of $r\beta$ can be used to determine whether the inflection point occurs.

The LSTM model's cell structure is depicted in Fig. 4, denoted by:

$$i_t = \sigma(W_{xi}x_t + W_{hi}h_{t-1} + W_{ci}c_{t-1} + b_i) \tag{9}$$

$$f_t = \sigma\left(W_{xf}x_t + W_{hf}h_{t-1} + W_{cf}c_{t-1} + b_f\right) \tag{10}$$

$$c_t = f_t c_{t-1} + i_t \tanh(W_{xc}x_t + W_{hc}h_{t-1} + b_c) \tag{11}$$

$$o_t = \sigma(W_{xo}x_t + W_{ho}h_{t-1} + W_{co}c_t + b_o) \tag{12}$$

$$h_t = o_t \tanh(c_t) \tag{13}$$

where i is input gate, f is forgetting gate, c is cell state, o is output gate; W and b are the corresponding weight coefficient matrix and bias term respectively; σ and tanh are sigmoid and hyperbolic tangent activation function respectively.

The paper proposes a rolling iterative prediction method, the long and whole time series is decomposed into multi-short time future prediction. The $r\beta$ is used to guide the segmented prediction of the LSTM, and correct the cumulative error of LSTM. The prediction model of internal dynamic dependence in the development process of things is established. The long and whole time series is predicted segmentally, and the sub-prediction sequences are normalized, so as to the accumulation of errors is weakened. The trend of time series data is predicted incrementally based on the dynamic characteristics of event development.

Our method is not to predict the high-dimensional time series data directly through historical observations, it predicts the next period in the time series based on the historical background of the current moment. Therefore, instead of predicting $x_t, x_{t+1}, \ldots,$

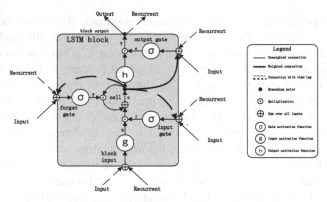

Fig. 4. LSTM cell structure.

x_{t+n}, we predict a short subsequence with the length is ω in multiple times. Where ω is much smaller than the single prediction sequence length T. It is connected to a high-dimensional prediction vector for n time points in the future through $x_{t:t+\omega}$, $x_{t+\omega:t+2\omega} \cdots x_{n-\omega:n}$, which is called the precursory prediction result x_{tm}. The prediction process is alternating, and we repeatedly predict dynamic parameters through historical data. The dynamic parameters are used to guide the network to predict the iteration process of short-term future sequence. These sub-sequences provide interpretable information and internal dynamic changes of events for each observation period. For example, during the development process of the COVID-19, a period of observation may show that the epidemic continues to spread and intensify. The potential dynamics of whether the epidemic will be accelerated propagation or contained within a short window can be predicted through the dynamic parameter changes in each prediction sub-sequence, so that decision makers can have a more comprehensive understanding of the development of the epidemic.

Therefore, a prediction model is established that accurately reflects the dynamic interdependent relationship that contained in the time series. The dynamic model of the epidemic is obtained according to the SEIR model, as shown in formula (14).

$$I = ((k_1\beta_1 - \gamma_1)I_0 + (k_2\beta_2 - \gamma_2)E_0)t^2 \tag{14}$$

The estimated epidemiological parameters and the processed time series are input into the LSTM to predict the epidemic trend of the epidemics. Parameter LSTM interrupts the prediction by using $r\beta$ and the population growth to control the controller. The sum of the data of the last $2*l_{in}$ time points (the isolation observation period of epidemic) are calculated whether it is zero. The results of {STOP = 1, CONTINUE = 0} are sent to the Sequence LSTM and the model is stop when the epidemiological parameters $r\beta <$ 1 and $\sum_{i=1}^{2l_{in}} S_i = 0$.

$$p\left(\text{Stop}|S'_p\right) = \begin{cases} 1, & \text{if } r\beta < 1 \cap \sum_{i=0}^{2l_{in}} S_i = 0 \\ \\ 0, & \text{if } r\beta > 1 \end{cases} \tag{15}$$

where the modified SEIR model shows that the virus can only spread when the daily growth rate of E, I is greater than 1.

$$\begin{cases} dE/dt = r_1\beta_1 IS/N + r_2\beta_2 ES/N - \alpha E - \gamma_2 E \\ dI/dt = \alpha E - \gamma_1 I \end{cases} \tag{16}$$

When the main influence parameters are satisfied $\begin{cases} r_1\beta_1 > 1 \\ r_2\beta_2 > 1 \end{cases}$ after simplification, the epidemic continues to spread. When the termination condition is reached, the Parameter LSTM will stop producing new parameter vectors, and the Sequence LSTM will also stop producing new prediction result.

The dataset used for model training and prediction is the time series newly generated by TDMA. The mean function will lead to a smaller value of the prediction results than the data collected in the middle-later period of the epidemic (when the confirmed detection is more stable and the kit is more effective). Therefore, the final prediction results are inversely transformed based on the TDMA method to describe the transmission trend of the epidemic in the real world.

4 Experiments

Datasets: Historical data of COVID-19 in China from 20 January, 2020 to 22 April, 2020 (not including foreign input), including the daily new addition and historical accumulation of the confirmed cases, suspected cases, cured cases and death cases in China, as well as the information on various transmission nodes. Historical data of COVID-19 in the world from 28 January, 2020 to 15 May, 2020, including new confirmed diagnosis, cumulative confirmed diagnosis, cumulative cures and cumulative deaths in countries around the world is used as testing data. Among them, the first 80% of the dataset is selected as the training set and the last 20% as the verification set.

Experimental Environment: The rolling iterative prediction method in this paper is implemented with the Keras framework. In order to train the deep learning network, an i7-9750H CPU and a NVIDIA GeForce RTX 2060 GPU is used. The cuda used V10.1, the cudnn used 9.0, and TensorFlow used 2.1.

Evaluation Metrics:
The measurement expression of LB is:

$$LB = n(n+2)\sum_{k=1}^{m}\widehat{\rho}_k^2/(n-k) \tag{17}$$

Here, n represents the number of samples and $\widehat{\rho}_k^2$ is the correlation coefficient of the sample k-order lags.

The calculation formula of the root mean square error is:

$$RMSE(X, h) = \sqrt{\frac{1}{m}\sum_{i=1}^{m}(h(x_i) - y_i)^2} \tag{18}$$

where m is the sequence length, $h(x_i)$ is the i-th observed value of the original sequence, and y_i is the i-th predicted value.

Experiment 1: The Results of Data Preprocessing

National daily new addition and accumulated data in raw data are generated by TDMA. Abnormal cycles in the generated results are removed and then the data are filled. Figure 5 displays the results.

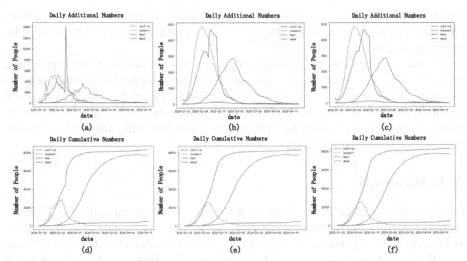

Fig. 5. The generation results of raw data. (a) original daily new additions; (b) generated daily new additions; (c) daily new additions after removing abnormal cycles; (d) original daily accumulation; (e) generated daily accumulation; (f) daily accumulation after removing abnormal cycles.

Figure 5 shows that the generated time series is smoother than the raw data and conforms to the data trend of SEIR model. Especially in the generation of daily new addition data, TDMA does a good job of eliminating the effect of change-points and repairing the real-world distribution of the data of COVID-19. However, abnormal period still exists in the generated data. Therefore, our approach involves removing any abnormal periods present in the generated data and filling in the missing values to ensure that the length of the data remains unchanged.

Experiment 2: The Stationarity of the Generated Data

In this experiment, the stationarity and availability of the generated time series data are tested through descriptive test and metrological test, respectively.

(1) ACF graph test

Time series usually has strong autocorrelation. The stationarity of time series is described by using autocorrelation function (ACF). The calculation method of ACF: Assuming that the time series X_t has a total of T phase observed values, and T-k arrays (x_i, x_{i+k}) are constructed. The dataframe composed of T-k arrays is calculated by the sample covariance. The point estimation of the autocorrelation function is obtained by dividing the covariance by the point estimation of sequential variance.

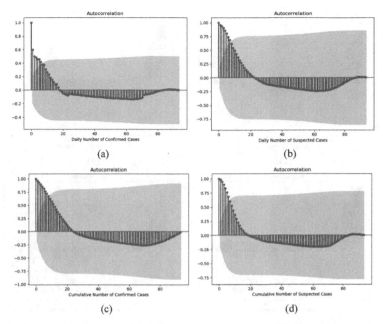

Fig. 6. The ACF test of raw data. (a) daily new addition confirmed diagnosis cases in raw data; (b) daily new addition suspected in raw data; (c) the cumulative confirmed in raw data; (d) the cumulative suspected in raw data.

The stationary time series usually has only short-term autocorrelation, and the long-term ACF will oscillate and approach to 0 randomly. In Fig. 6 and Fig. 7, each ACF graph shows the shape of an inverse triangle, which means that there is an obvious trend in the time series of COVID-19, which is an unstable time series. The final curves are all close to 0 which prove that the data of COVID-19 presented an obvious and stable time sequence in the later stage. At the same time, the oscillation trend of ACF graph of the generated series data is smoother. It proves that there are no abnormal change points in the generated data.

(2) LB test

Descriptive test of the generated data has been verified in ACF experiment. To further test the effectiveness of generated data, LB method which is more suitable for small samples is selected according to the characteristics of the data generated results. The measurement test of the generated data through LB in this experiment. The objective statistics is introduced and LB is used to conduct a pure random test on time series to determine the overall significance of the time series model. It is proved that the generated time series data is not the pure-noise, but contains the information that can be described by the model.

Table 2 indicate that the value of P in the generated sequence is less than the significance level (0.05). It is proved that the generated sequence is not white noise.

Experiment 3: The Cluster Results

The data of COVID-19 from January 20, 2020 to April 22, 2020 in 34 provinces of China

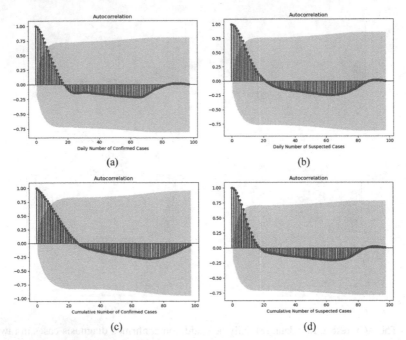

Fig. 7. ACF test for the generated time series. (a) daily new addition confirmed diagnosis cases of the generated data; (b) daily new addition suspected of the generated data; (c) the cumulative confirmed of the generated data; (d) the cumulative suspected of the generated data.

Table 2. The results on LP of the data generation.

Items	LB	P
Daily Confirmed Case Count	[471.51,631.56]	$[1.15 \times 10^{-8}, 1.92 \times 10^{-7}]$
Daily Suspected Case Count	[503.04,704.32]	$[1.86 \times 10^{-5}, 5.25 \times 10^{-4}]$
Cumulative Confirmed Case Count	[502.42,791.62]	$[2.53 \times 10^{-5}, 1.04 \times 10^{-6}]$
Cumulative Suspected Case Count	[485.03,628.76]	$[1.41 \times 10^{-4}, 7.63 \times 10^{-7}]$

are clustered by using the method in this paper. The obtained cluster results are shown in Table 3. The classification results are close to the risk level of each province issued by National Health Commission of the People's Republic of China during the epidemic. The visualization result of cluster is shown in Fig. 8:

Experiment 4: The Loss Trained by the Model

The model is trained and tested for 20 times in this experiment. The training and test set losses are calculated and visualized in the training process, while RMSE is also calculated throughout this stage. The final average RMSE of LSTM is 1.955. The loss with the lowest RMSE of training set and test set in multiple training is shown in Fig. 9. Our model achieves a much lower average RMSE of 1.955, compared to the RMSE

Table 3. The table of cluster results of COVID-19 of each province in China.

Province	Classes	Province	Classes	Province	Classes	Province	Classes
Hubei	5	Chongqing	3	Yunnan	2	Xianggang	2
Guangdong	4	Heilongjiang	3	Hainan	2	Neimenggu	2
Henan	4	Sichuan	3	Guizhou	2	Ningxia	2
Zhejiang	4	Beijing	3	Shanxi	2	Taiwan	1
Hunan	4	Shanghai	3	Tianjin	2	Aomen	1
Anhui	4	Hebei	3	Liaoning	2	Xizang	1
Jiangxi	4	Fujian	3	Jilin	2	Qinghai	1
Shandong	4	Shaanxi	3	Gansu	2		
Jiangsu	3	Guangxi	3	Xinjiang	2		

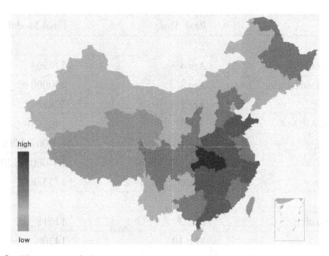

Fig. 8. The graph of cluster results of COVID-19 of each province in China.

of 54.26 from the mathematical model. The mathematical model is more suitable for predicting infectious diseases at the initial stage of an epidemic. In contrast, our model is more suitable for long and whole time series prediction of epidemics.

Experiment 5: Comparative Experiments on COVID-19 Prediction in France and South Korea

Data from France and South Korea are applied to verify the accuracy of the model. Our model is pre-trained through the clustering results and historical data of provinces in China. The classification of France and South Korea is obtained by using Covariance Classifier, that is, France is similar to Hubei and South Korea is similar to Guangdong. Models trained with different types of data are used to predict the epidemic in France and Hubei, and the accuracy of models is verified.

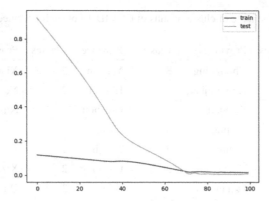

Fig. 9. Loss of training set and test set for model training.

Table 4. The prediction results of COVID-19 in France and South Korea.

Methods	Peak Time	Final Model
France		
The Actual Situation	Apr.4	178994
Segmented Poisson Model	Apr.8	750000
ARIMA	Apr.7	378253
The Optimized ARIMA	Apr.5	140,320
SIRU [19]	Apr.16	470000–700000
SIRU [20]	Mar. 30-Apr. 9	104000–212000
RIP	Apr.6	175538
South Korea		
The Actual Situation	Mar.3	11018
ARIMA	Mar.10	14708
SIRU [19]	Feb.30	8800–10000
SIRU [20]	Feb. 29 − Mar. 1	9000–9400
RIP	Mar.2	10883

SEIR characterizes the dynamics of infectious diseases by dividing the population into four states: susceptible, exposed, infectious, and removed. The rate of change between each state determines the proportion of the population in each state. The epidemiological information of COVID-19, such as transmission probability, incubation period, and recovery rate, is incorporated into the LSTM model to modify the prediction. Models are trained for similar transmission trends in France and South Korea by classifying the actual reported data from China. LSTM is then used to predict the inflection points of the infectious disease and the final number of cases. Based on the COVID-19 report, our model is applied to the French dataset. It predicts an inflection point on April 6, and the final total number of cases will reach 175,538. By applying our model to the Korean dataset, an inflection point on March 2 is predicted, and the final total number of cases will reach 10,883. After the inflection point, although the number of infected people is still increasing, the growth rate has declined. This is because when $r\beta < 1$, the infectious disease gradually decreases from continuous prevalence until it finally disappears. This proves that the epidemic prevention and control measures are effective. The chance of spreading infectious diseases is reduced, and thus the epidemic is controlled.

As can be seen from Table 4 that the existing methods are relatively accurate in predicting the inflection points of the epidemic, while our method is much closer to the actual situation. Although the inflection point prediction of France is a day behind the optimized ARIMA method, our method is more accurate in predicting the eventual size of the epidemic. Before the TDMA inverse transformation of the prediction result, the predicted result of using our method is 173617 in France epidemic prediction, which is only 5,377 fewer than the real data. The predicted result of our method is 10883 in South Korea epidemic prediction, which is only 135 fewer than the real data. However, after the inverse transformation of TDMA, the predicted result of our method is 175538 in France epidemic prediction, which is only 3456 fewer than the real data. The predicted result of our method is 10911 in South Korea epidemic prediction, which is only 107 fewer than the real data.

5 Conclusion

A rolling iterative prediction (RIP) method is proposed in this paper. This method learns the variable correlation relationships and time series dependency relationships in the correlated multivariate time series. The method utilizes the propagation dynamics model and Long-Short-Term Memory to iteratively calculate the propagation parameters and predict the time series. The experiments conducted on two real COVID-19 datasets demonstrate that the RIP method outperforms baseline methods and can obtain short-term mutation points of time series. In our future studies, additional architectures, hyper-parameter selection and online training methods are considered to further enhance the practicability of the method.

Acknowledgements. This work was supported by the National Key R&D Program of China under Grant No. 2020YFB1710200.

References

1. Pape, R.A., Price, C.: A slow-rolling disaster: assessing the impact of the Covid-19 pandemic on militant violence. J. Conflict Resolut. , 002200272311801 (2023). https://doi.org/10.1177/00220027231180101
2. Shang, B., Shang, P.: Directed vector visibility graph from multivariate time series: a new method to measure time series irreversibility. Nonlinear Dyn. **104**(2), 1737–1751 (2021). https://doi.org/10.1007/s11071-021-06340-3

3. Längkvist, M., Karlsson, L., Loutfi, A.: A review of unsupervised feature learning and deep learning for time-series modeling. Pattern Recogn. Lett. **42**, 11–24 (2014). https://doi.org/10.1016/j.patrec.2014.01.008
4. Das, M., Ghosh, S.K.: BESTED: an exponentially smoothed spatial bayesian analysis model for spatio-temporal prediction of daily precipitation. In: International Conference on Advances in Geographic Information Systems, ACM, pp. 1–4 (2017)
5. Patil, S., Khule, S., Toshniwal, S.: Role of D-Dimer in assessing severity, monitoring, and predicating outcome in COVID-19 pneumonia: a single center study. Glob. J. Health Sci. Res. **1**, 31–37 (2023)
6. Mishra, P., Verma, S., Arya, D., et al.: Early predication of Covid-19 by machine learning algorithms. J. Pharm. Negative Results, 2907–2914 (2022)
7. Dietz, K., Schenzle, D.: Mathematical models for infectious disease statistics. In: Atkinson, A.C., Fienberg, S.E. (eds.) A Celebration of Statistics: The ISI Centenary Volume A Volume to Celebrate the Founding of the International Statistical Institute in 1885, pp. 167–204. Springer New York, New York, NY (1985). https://doi.org/10.1007/978-1-4613-8560-8_8
8. Britton, T.: Stochastic epidemic models: a survey. Math. Biosci. **225**(1), 24–35 (2010)
9. Zhang, G., Pang, H., Xue, Y., Zhou, Y., Wang, R.: forecasting and analysis of time variation of parameters of COVID-19 infection in China using an improved SEIR model (2020)
10. Yang, Z., et al.: Modified SEIR and AI prediction of the epidemics trend of COVID-19 in China under public health interventions. J. Thorac. Dis. **12**(3), 165 (2020)
11. Read, J.M., Bridgen, J.R., Cummings, D.A., Ho, A., Jewell, C.P.: Novel coronavirus 2019-nCoV: early estimation of epidemiological parameters and epidemic predictions. MedRxiv (2020)
12. Wu, J.T., Leung, K., Leung, G.M.: Nowcasting and forecasting the potential domestic and international spread of the 2019-nCoV outbreak originating in Wuhan, China: a modelling study. Lancet **395**(10225), 689–697 (2020)
13. Pathan, R.K., Biswas, M., Khandaker, M.U.: Time series prediction of COVID-19 by mutation rate analysis using recurrent neural network-based LSTM model. Chaos, Solitons Fractals **138**, 110018 (2020)
14. Jana, S., Bhaumik, P.: A multivariate spatiotemporal spread model of COVID-19 using ensemble of ConvLSTM networks. MedRxiv (2020)
15. Kırbaş, İ, Sözen, A., Tuncer, A.D., Kazancıoğlu, F.Ş: Comparative analysis and forecasting of COVID-19 cases in various European countries with ARIMA, NARNN and LSTM approaches. Chaos, Solitons Fractals **138**, 110015 (2020)
16. Arora, P., Kumar, H., Panigrahi, B.K.: Prediction and analysis of COVID-19 positive cases using deep learning models: a descriptive case study of India. Chaos, Solitons Fractals **139**, 110017 (2020)
17. The lowest infection rate. https://new.qq.com/rain/a/20200210A0JLX100 (2020)
18. Li, R., et al.: Substantial undocumented infection facilitates the rapid dissemination of novel coronavirus (SARS-CoV-2). Science **368**(6490), 489–493 (2020)
19. Magal, P., Webb, G.: Predicting the number of reported and unreported cases for the COVID-19 epidemic in South Korea, Italy, France and Germany. Italy, France and Germany (2020)
20. Liu, Z., Magal, P., Webb, G.: Predicting the number of reported and unreported cases for the COVID-19 epidemics in China, South Korea, Italy, France, Germany and United Kingdom. J. Theor. Biol. **509**, 110501 (2021)

Multimedia Data Management
and Analysis

Video Popularity Prediction Based on Knowledge Graph and LSTM Network

Pingshan Liu[1], Zhongshu Yu[2]([✉]), Yemin Sun[2], and Mingjun Xi[2]

[1] Business School, Guilin University of Electronic Technology, Guilin, China
[2] Guangxi Key Laboratory of Trusted Software, Guilin University of Electronic Technology, Guilin, China
1450424268@qq.com

Abstract. The prediction of the popularity of online content, particularly videos, has recently gained significant attention as successful popularity prediction can assist many practical applications such as recommendation systems and proactive caching, as well as aid in optimizing advertising strategies or balancing network throughput. Despite much work being done on predicting the popularity of online videos, there are still challenges to be overcome: (1) popularity is greatly influenced by various external factors, resulting in significant fluctuations that are difficult to capture and track; (2) online video content and metadata information are typically diverse, sparse, and noisy, making the prediction task complex and unstable; (3) some data have temporal relevance, and the impact on popularity varies at different times.

In this paper, we propose an Adaptive Temporal Knowledge Graph Network (ATKN) video popularity prediction model to address the issues surrounding video popularity prediction. First, we employ the attention-based Long Short-Term Memory (ALSTM) network to capture the trend of popularity change. Then, we introduce an Attention-based Factorization Machine (AFM) with attention mechanism to model the feature cross of video content, thereby enhancing the distinction of importance after different feature crosses. Next, we use a Relational Graph Convolutional Network (RGCN) to extract the associated features between entities in the knowledge graph. Finally, we propose a dynamic feature fusion method that adaptively assigns the weights of temporal features and content features at different time intervals by constructing an exponential decay function, thereby obtaining an effective and stable feature fusion module. Experimental results demonstrate the superiority and interpretability of ATKN on the MovieLens-20M dataset and the Microsoft Satori-built movie knowledge graph.

Keywords: Popularity Prediction · Knowledge Graph · Neural Network · Feature Fusion

1 Introduction

With the rapid development of the internet and online platforms, an increasing number of users are joining online video platforms such as YouTube, Netflix, and Tencent Video for sharing and viewing. Therefore, the prediction of content popularity has attracted

© The Author(s), under exclusive license to Springer Nature Singapore Pte Ltd. 2023
Z. Yu et al. (Eds.): ICPCSEE 2023, CCIS 1879, pp. 455–474, 2023.
https://doi.org/10.1007/978-981-99-5968-6_32

wide attention from researchers and practitioners and has become an important task for various applications. One of the goals is to infer the cumulative view count of a video at some future time. For users, this task can help filter information. For video platforms, accurate prediction of video popularity can help them better recommend and display valuable videos, improve user experience and retention rate, and greatly assist in related business operations such as precision advertising and edge caching strategies.

Traditional methods for predicting video popularity attempt to build classification or regression models on historical time series data [1, 16]. While these methods can obtain the general trend of popularity changes, they have disadvantages such as high data requirements and limited factors considered. Therefore, many studies have further improved the models by utilizing different features of videos, including user features [3], content features [4], time features [1], and structural features [6]. These feature-based methods use time series and contextual data to learn better prediction models. However, the features involved in these methods are often extracted through manual or heuristic methods, making the performance of the model heavily dependent on the extracted features. Hand-crafted features also face many challenges such as difficulty in design and measurement, strong platform-relatedness, low scalability, and ineffective use of contextual information.

In recent years, the popularity of neural networks has led to the development of many deep learning-based models for predicting content popularity [5, 7, 17, 18]. However, for predicting video popularity, few works can simultaneously address the following three issues: (1) popularity is influenced by various external factors, resulting in large fluctuations that are difficult to predict and to capture the dynamic trends of popularity; (2) the content and metadata information of online videos are often diverse, sparse, and noisy, making the prediction task complex and unreliable. Therefore, a reasonably approach is needed to effectively utilize the content and metadata information of videos. (3) Some feature data are time-sensitive. As the video ages, temporal features are often more important than content features. Therefore, it is necessary to model temporal and content features reasonably and design a feature fusion scheme to flexibly capture the popularity of videos in different lifecycles.

To better extract useful information from diverse and sparse content and metadata and to learn underlying knowledge cascades, introducing knowledge graphs into video popularity prediction is a new research approach. A knowledge graph is a graphical model that represents knowledge, including various entities and their relationships. By incorporating entities and relationships from the knowledge graph into video popularity prediction, we can comprehensively understand the connections between video content and related information, thus improving the accuracy and prediction performance of the model.

In response to the challenges faced by video popularity prediction, we drew inspiration from the successful experiences of deep learning and knowledge graphs in different fields.

In this paper, we propose an Adaptive Temporal Knowledge Graph Network (ATKN) model that integrates temporal features and knowledge graph representations for video popularity prediction. The key to this model lies in the effective fusion of historical popularity-related time series and knowledge graphs. For time series data, inspired by the efficiency of modelling popularity dynamics using LSTM networks in recent popularity prediction works based on LSTM networks [5, 14, 18], we found that modelling popularity dynamics using LSTM networks is efficient. In this paper, we use attention-based LSTM (ALSTM) [20] to obtain dynamic temporal features. In addition, we introduce video metadata to modify the model. In this paper, we use deep attention-based factorization machine (AFM) to model video content features. The attention mechanism assigns higher weights to useful features for the target, thus distinguishing the impact of different features on the prediction result. Considering that video metadata features cannot accurately reflect the relationships between features, we use a Relational Graph Convolution Network (RGCN) [22] to extract the relational features of the knowledge graph. By initializing node features and constructing a relationship matrix, not only can we learn the features and relationships of the knowledge graph itself but also in the process of convolutional layer propagation, the node will receive information from neighboring nodes and constantly update itself, integrating feature information and structural information through self-learning to increase the model's expression ability. Meanwhile, previous research [5] has shown that the impact of different features on the video lifecycle in different stages is different. Therefore, we propose a dynamic feature fusion method. This method adaptively assigns weights to time series features, metadata features, and knowledge graph-related features at different time periods by constructing an exponential decay function, obtaining an effective and stable feature fusion module.

We summarize the main contributions of this paper as follows. First, we introduce knowledge graphs into video popularity prediction, and by using relational graph convolution networks, we not only learn diverse and sparse feature information of entities and relationships but also learn the underlying semantic information between entities. Second, we propose a novel prediction model that utilizes global attention mechanism to dynamically combine the time-series features captured by ALSTM, metadata cross-features mined by AFM, and knowledge graph relationship features extracted by RGCN. Third, we conduct extensive experiments on real-world datasets and compare our model with several baseline methods, demonstrating its effectiveness and superiority.

2 Related Work

In this section, we review the related research on popularity prediction.

The traditional approach to popularity prediction is to build regression or classification models based on historical popularity statistics [1, 16]. However, these simple models based on time-series data cannot effectively capture complex content features. With the emergence of rich Internet contexts, many studies have further improved models by utilizing different feature information, including user features [3], content features [4], time features [1], and structural features [6]. Keneshloo et al. extracted metadata and time features and used tree regression for news popularity prediction. Tang et al. [10] and Rizoiu et al. [11] predicted video popularity based on Hawkes processes. Although this method does not require too much feature engineering, it usually makes assumptions about fixed parameters, which limits the expression of the model [12, 13]. Therefore, influenced by the great success of deep learning in many fields, we use neural networks to model content features and time series, reducing the inherent problems of feature engineering.

In recent years, the popularity of neural networks has led to the development of many deep learning-based prediction models. A typical approach is to use time-series data of historical popularity statistics as input to build regression or classification models [1, 2]. This deep learning-based approach typically employs Recurrent Neural Networks (RNN) to capture time dependencies [7–9, 14]. Cao et al. [1] combined Hawkes processes with deep learning methods for popularity prediction to overcome the limitations of simple parameterized point process models. Liao et al. [9] proposed a prediction network that deeply fused temporal and content features, modelling multimodal data to predict the popularity of articles. To better extract useful information from diverse and sparse content and metadata and to learn latent knowledge cascades, introducing knowledge graphs into video popularity prediction is a new research approach. Recent studies have successfully applied knowledge graphs to recommendation systems [22], article popularity prediction [5], and conversational interaction [23]. Dou et al. [18] used embeddings of knowledge graph entities and their neighbors to enhance LSTM-based video popularity prediction. Zhang et al. [15] constructed a multichannel popularity prediction model with an attention mechanism utilizing visual, textual, and user representations. They simultaneously utilized time-series and content features, learning a better popularity prediction model through machine learning.

3   Method

In this section, we introduce the ATKN model, whose framework is shown in Fig. 1. We first describe the problem of predicting video popularity based on knowledge graph and LSTM network. Then, we introduce the attention-based LSTM network, AFN model, and RGCN model. Finally, we present the ATKN model proposed in this paper and introduce the dynamic feature fusion method.

3.1   Attention-Based LSTM

According to the leftmost part of the model architecture in Fig. 1, we process the time-series data by simulating the evolution of popularity over time using the Attention-based Long Short-Term Memory (ALSTM) network.

We have two considerations. First, as the most widely used Recurrent Neural Networks (RNN) structure, LSTM is efficient in modelling long-term historical information of time series. Second, introducing an attention mechanism can reduce the interference of external factors and help the model focus on the relevant parts of the input data for the current task. When processing sequential data, especially long sequences, the LSTM model may be limited by the flow of information, making it difficult to remember earlier inputs. The attention mechanism can dynamically calculate the importance of each time step in the input sequence through weighted calculation so that LSTM can better handle long sequences while improving the interpretability of the model. The Attention-based LSTM model is shown in Fig. 2. We can construct a time series ($X = (x_1, x_2, \ldots, x_n)$) as the input of the LSTM network by analysing user feedback such as video views, likes, and comments over time. The calculation of each cell in the LSTM network is as follows:

$$i_t = \sigma(W_i x_t + U_i c_{t-1} + V_i h_{t-1} + b_i) \tag{1}$$

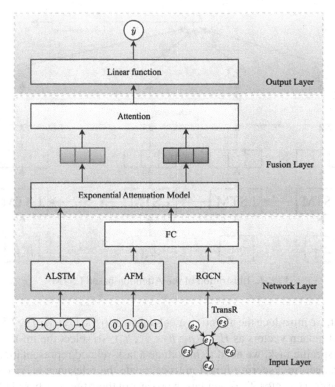

Fig. 1. Illustration of the ATKN model.

$$f_t = \sigma\left(W_f x_t + U_f c_{t-1} + V_f h_{t-1} + b_f\right) \tag{2}$$

$$c_t = f_t * c_{t-1} + i_t * tanh(W_c x_t + V_c h_{t-1} + b_c) \tag{3}$$

$$o_t = \sigma(W_o x_t + U_o c_{t-1} + V_o h_{t-1} + b_o) \tag{4}$$

$$h_t = o_t * tanh(c_t) \tag{5}$$

where i_t, f_t, o_t are the input gate, forget gate, and output gate of the LSTM network at time t, h_{t-1} is the output of the hidden layer at time $t-1$, x_t is the input at time t, and c_{t-1} is the cell state at time $t-1$. In addition, W_i, W_f, W_c, W_o and b_i, b_f, b_c and b_o are trainable weight matrices and biases, and σ represents the sigmoid activation function.

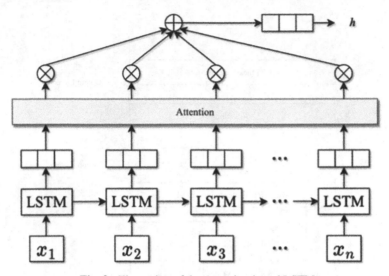

Fig. 2. Illustration of the Attention-based LSTM

After that, we introduce the attention mechanism into the LSTM. The LSTM obtains a sequence of hidden vectors as $H = (h_1, h_2, \ldots, h_n)$, To select the information that is relevant to a specific task, we need to introduce a task-related representation q, called a query vector, and use a scoring function to calculate the relevance between each input vector and the query vector. The calculation process of the Attention-based LSTM model is as follows:

$$a_i = w^T tanh(W_q * q + W_h * h_i) \tag{6}$$

$$\alpha_i = \frac{exp(a_i)}{\sum_{j=1}^{n} exp(a_j)} \tag{7}$$

$$v^{A-L} = \sum_{i=1}^{n} \alpha_i h_i \tag{8}$$

where W_q, W_h, w^T represents learnable parameters and a_i represents the attention scoring function. After passing through the softmax layer, attention weights α_i are obtained. The popularity trend output by the Attention-based LSTM model is denoted as v^{A-L}.

3.2 AFM

The information about a movie mainly includes metadata such as its genre, cast, and release date. This information can elicit vastly different feedback from users based on their preferences and largely determines the popularity of a movie, making it an essential condition for popularity prediction. In reality, different cross-features have varying impacts on the results. More important features should be assigned higher weights. Therefore, this paper adopts the attention mechanism of the factorization machine (AFM) to model the content features of movies to learn the weighted cross-features of movies. A multilayer perceptron is used to parameterize the attention score. The attention mechanism assigns higher weights to useful features, allowing FM to extract important first-order and second-order linear cross-features that are crucial for target prediction. Figure 3 shows the schematic structure of the AFM model.

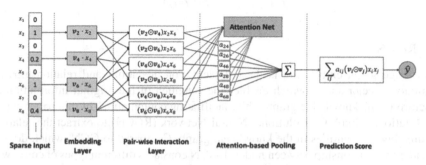

Fig. 3. Illustration of the AFM

The input layer of AFM adopts the sparse representation of input features, with items and user data encoded using one-hot encoding. The high-dimensional sparse features are then embedded into low-dimensional dense features through an embedding layer. The main function of the pairwise interaction layer is to perform second-order cross of features, generating n(n-1)/2 s-order cross-feature vectors by pairwise crossing n features. The pairwise interaction layer can be represented as $f_{PI}(\xi)$:

$$f_{PI}(\xi) = \left(v_i \odot v_j\right)x_i x_{j\,(i,j)\in R_x} \tag{9}$$

where ξ represents the output of the embedding layer, R_x represents all possible pairwise combinations of features, x_i represents the feature value, v_i represents the latent vector corresponding to feature value x_i, and the operator $\odot$ denotes elementwise multiplication of vectors.

It can be understood that different cross features should have different levels of importance and should be assigned different weights. To achieve this, AFM introduces the concept of attention into the model, assigning importance weights to each cross-feature. When summing up the feature vectors, the importance weights are used to perform a weighted sum of the second-order cross-features. This essentially involves a multilayer perceptron (MLP) with a hidden layer. It can be computed as follows:

$$a'_{i,j} = h^T Relu\left(W\left(v_i \odot v_j\right)x_i x_j + b\right) \tag{10}$$

$$a_{i,j} = \frac{exp\left(a'_{i,j}\right)}{\sum_{(i,j)\in R_x} exp\left(a'_{i,j}\right)} \tag{11}$$

where h, W and b are trainable model parameters, t represents the number of nodes in the hidden layer of the attention network, and $a_{i,j}$ represents the importance weight of the corresponding second-order cross-feature $(v_i \odot v_j)x_i x_j$. The importance weights are used to perform a weighted sum, which ultimately produces the output v^A of the AFM model. The calculation is as follows:

$$v^A = w_0 + \sum_{i=1}^{n} w_i x_i + h^T \sum_{i=1}^{n} \sum_{j=i+1}^{n} a_{i,j} (v_i \odot v_j) x_i x_j \tag{12}$$

3.3 RGCN

Knowledge graph embedding is the representation of entities and relationships in a continuous vector space, which enables the retention of the original information and structure of the knowledge graph while facilitating computation. In this paper, we use the Relational Graph Convolutional Neural Network (RGCN) to extract the relational features between entities in the knowledge graph. Compared to GCN, which does not consider the relationship between nodes, RGCN considers different types of edges when computing graph convolutions, allowing it to better capture relationship information in the knowledge graph. Through the initialization of node features and the construction of relationship matrices, RGCN not only learns the features and relationships of the knowledge graph itself but also updates its nodes continuously by accepting information from neighboring nodes during the convolutional layer propagation process. This self-learning process allows for the fusion of feature and structure information, increasing the expressiveness of the model.

3.3.1 KG Embedding Layer

Knowledge graph embedding refers to the mapping of entities and relationships in a knowledge graph to a low-dimensional vector space while preserving their semantic information. In this paper, we choose TransR as the embedding model for the knowledge graph. For each triplet (h, r, t) its represented vector consists of h, r, and t. TransR projects h and t into the vector space of relationship r and calculates their distance. The definition of this distance uses the matrix product between the projected vectors and the relationship transfer matrix M_r. Their representations in the vector space of relationship r are $h_r = M_r h$ and $t_r = M_r t$, and training is conducted in the relationship space to make $h_r + r \approx t_r$. The scoring and loss function are as follows:

$$f_r(h, t) = ||h_r + r - t_r||_2^2 \tag{13}$$

$$L = \sum_{(h,r,t) \in S} \sum_{(h\prime,r,t\prime) \in S\prime} max(0, f_r(h, t) + \gamma - f_r(h\prime, t\prime)) \tag{14}$$

γ is the margin, S is the set of correct triplets, and $S\prime$ is the set of incorrect triplets.

3.3.2 KG Structure

For a given movie i, the movie entity is represented as the head node in the knowledge graph triplet by h_i. The corresponding tail node and the relationship between them are defined as the set $\mathcal{G}_i = \{(h_i, r, t) | (h_i, t) \in \epsilon, r \in \mathcal{R}\}$ to represent the set of triplets in the knowledge graph. Where ϵ and $\mathcal{R}$ represent the entity set and relationship set of the knowledge graph, respectively. For example, the triplet (Titanic, movie.director, James Cameron) represents that James Cameron is the director of the movie Titanic. Titanic and James Cameron are entities, while director represents the relationship between them.

3.3.3 Forward Propagation Model

Both RGCN and GCN utilize graph convolutional models to simulate the propagation of information in network structures. Therefore, the framework for this section can be illustrated as follows:

$$
h_i^{(l+1)} = \sigma \left(\sum_{m \in M_i} g_m \left(h_i^{(l)}, h_j^{(l)} \right) \right) \tag{15}
$$

$g_m(\cdot, \cdot)$ refers to the aggregation of incoming messages passing through an aggregation function. M_i Refers to the set of incoming messages for node i, which is typically chosen to be the set of incoming edges. $h_i^{(l)}$ Represents the node representation of node i in the l-th layer, and $h_j^{(l)}$ represents the node representation of all neighbor nodes of node i in the l-th layer.

Aggregating information from the local neighborhood nodes to encode features is highly effective for graph classification and semisupervised node classification tasks. Based on the aforementioned approach, RGCN defines a propagation model on a relational multigraph. The updating rule for node i in the graph is as follows:

$$
h_i^{(l+1)} = \sigma \left(\sum_{r \in \mathcal{R}} \sum_{j \in \mathcal{N}_i^r} \frac{1}{C_{i,r}} W_r^{(l)} h_j^{(l)} + W_0^{(l)} h_i^{(l)} \right) \tag{16}
$$

$\mathcal{N}_i^r$ represents the set of neighboring nodes of node i with relation r, $C_{i,r}$ is a regularization constant that can be defined as $|\mathcal{N}_i^r|$ or a custom value, and $W_r^{(l)}$ is a linear transformation function that applies the same parameter matrix $W_r^{(l)}$ to transform the neighboring nodes with the same type of relation r. Figure 5 is a schematic diagram of the RGCN model structure.

From Fig. 4, we can see that RGCN needs to linearly transform the neighboring nodes connected by edges of different types. Among them, e, c, and h represent different entity nodes. The number of $W_r^{(l)}$ in formula (16) is the number of edge types. RGCN considers both the type and direction of edges, which is in line with the structural characteristics of the movie knowledge graph.

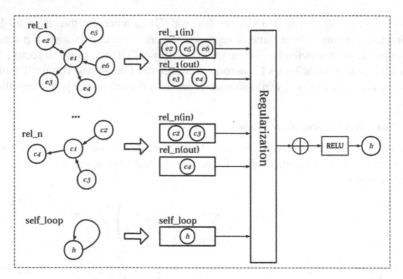

Fig. 4. Illustration of the RGCN

3.4 ATKN Model

3.4.1 Structural Description

The dynamic changes in movie popularity are difficult to capture, therefore, we use an attention-based LSTM network to extract temporal features and capture the changing trend of movie popularity. In this paper, we use num_score and avg_score to construct the time series. The addition of attention mechanism allows the model to selectively focus on the relevant parts of the input sequence during prediction. The content features of movies also have an important impact on their popularity changes. The movie content features described in this paper include the cross features of movie metadata and the association features between entities in the movie knowledge graph. The cross features of movie metadata include five categories: "category", "director", "actor", "release_date" and "runtime". We can preprocess the data through one-hot encoding technology and then use embedding technology to reduce the dimensionality of high-dimensional features and obtain low-dimensional dense features. For the movie knowledge graph, we can extract the triple set, which contains seven types of relationships, and use TransR to embed the triples. After learning the relevant feature vectors of metadata cross features and knowledge graph association features, we use a fully connected layer to aggregate the feature vectors of these two types of features as the representation of movie content features. Then we consider the dynamic fusion between the temporal features and content features.

3.4.2 Dynamic Feature Fusion

We use v^{A-L}, v^A, and v^{KG} to represent the time-series feature vectors captured by the ALSTM, the cross-feature vectors mined by the AFM model, and the knowledge graph association feature vectors extracted by the RGCN, respectively. We need to design a reasonable and effective method to fuse these feature vectors. Common fusion methods such as concatenation and weighted averaging lack specificity and flexibility and do not consider the impact of time variation on different feature vectors. As we know, the content features related to movie content do not change with time, while the content of time-series features dynamically changes with time. Therefore, considering that the importance of content features gradually decreases with time, while the importance of time-series features gradually increases, we propose a dynamic feature fusion method.

We can use Newton's law of cooling formula from physics to establish the functional relationship between popularity and time and build a learnable exponential decay model to adaptively allocate the weights of temporal and content features at different time points. This will result in an effective and stable feature fusion module. The formula is expressed as follows:

$$D(\Delta t) = D(t - t_0) = exp(-\alpha(t - t_0)) \tag{17}$$

$$v^T = (1 - D(\Delta t)) * v^{A-L} \tag{18}$$

$$v^C = D(\Delta t) * \left(W_a v^A + W_k v^{KG} \right) \tag{19}$$

where t represents the predicted time, t_0 represents the time when the video was published, $\Delta t = t - t_0$, $D(\Delta t)$ is a learnable exponential decay function, and α is a learnable parameter that controls the decay rate of $D(\Delta t)$.

With this, we obtain time feature vectors and content feature vectors with dynamically changing weights over time. The content feature vector is composed of video metadata cross-features and knowledge graph association features. Now, we need to consider how to cross-fuse the time feature and content feature. We use an attention mechanism to fuse these two types of feature representations. First, we use a fully connected layer to fuse $[v^T, v^C]$ into a global vector. Then, we calculate the dot product between the global vector and each feature vector and obtain the attention weights through softmax. Finally, we weight each feature and sum them to obtain the final result. The calculation is as follows:

$$h = W_h * \left[v^T, v^C \right] + b_h \tag{20}$$

$$\alpha_i = \frac{exp\left(h^T * v^i\right)}{\sum_{i* \in T, C} exp\left(h^T * v^i\right)} \tag{21}$$

$$F = \sum_{i* \in T, C} \alpha_i * v^i \tag{22}$$

After obtaining F we use a linear function to obtain the final prediction result. The calculation is as follows:

$$\hat{y} = \frac{1}{1 + exp(-W_f * F + b_f)} \tag{23}$$

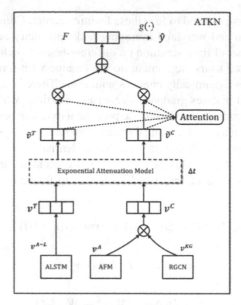

Fig. 5. Illustration of the ATKN

3.5 Model Training

We define the loss function for the video popularity prediction as follows:

$$L = \sum_{i \in \mathcal{D}} \sum_{t \in [1,n]} MSE\left(y_t^i, \hat{y}_t^i\right) \tag{24}$$

where $\mathcal{D}$ is the training set of the movie dataset, t is the predicted time, $P_{FIN}{}^i$ and $\widehat{P_{FIN}}{}^i$ are the true and predicted popularity values of movie i at time t.

4   Experiments

This section presents the experimental setup and analysis of the results in this paper.

4.1   Dataset and Preprocessing

4.1.1   Dataset

The datasets used in this experiment are the Movielens-20M dataset from GroupLens, the movie metadata dataset released on Kaggle platform, and the triple dataset extracted from the movie knowledge graph built from Microsoft Satori. The Movielens-20M dataset contains approximately 2 million ratings from 138,000 users on more than 27,000 movies, with a rating range of 0.5 to 5, with higher scores indicating that users like the movie more. Therefore, this dataset is considered to be quite dense. In addition to the basic information of movies and users, this dataset also includes some other auxiliary data, such as the IMDB number of the movie. Therefore, this dataset can be expanded by connecting to the TMBD and IMDB datasets to obtain richer metadata information for the temporal dataset and movie content dataset. The movie metadata dataset released on the Kaggle platform is extracted from the movie metadata on the TMDB website and contains rich metadata for 45,000 movies. The movie knowledge graph dataset is built using the movie knowledge graph from Microsoft Satori. Microsoft Satori is used to organize and present information from the Internet. It covers multiple domains, including movies. By using Satori, entity information and relationships can be constructed. It organizes movie information into a hierarchical structure and provides related data.

4.1.2 Data Preprocessing

In this experiment, data preprocessing was performed on three datasets. To construct the time series, an attention-based LSTM network was used to capture the changing trend of movie popularity. For the popularity time series data, the user rating dataset from Movielens-20M was used. This dataset contains approximately 2 million ratings from 138,000 users for over 27,000 movies (with ratings ranging from 0.5 to 5, where a higher score indicates a higher level of user preference). We used a uniform sampling method to construct the macrolevel time series with a time interval of one day. To ensure diversity in the training data, we sampled the number of ratings per day (num_score) and the average rating (avg_score) as operations and arranged these data in chronological order. Additionally, we converted explicit rating data into implicit feedback data by categorizing movie popularity into three levels: "hot," "normal," and "cold." Specifically, a movie is considered "hot" if it has a rating of greater than 4, "normal" if its rating is greater than 2 but less than or equal to 4, and "cold" if its rating is less than or equal to 2. Based on this categorization, we found that only 22.17% of movies were "hot" 64.68% were "normal" and 13.15% were "cold" In addition to time series data, the metadata features consist of "category", "director", "actor", "release_date" and "runtime." For the movie knowledge graph, we retained only the entity types "name", "actor", "director", "country", "genre", "language" and "production_company" as relationships in the triple dataset. The construction of the movie knowledge graph is consistent with MKR, which collects the valid movie IDs by matching the tail entity (head, film.film.name, tail) in the triple set with all movie entities named "name" and then deletes the corresponding content of unmatched movie IDs. Then, a mapping relationship between item_id and entity_id is established to obtain a valid subset. The movie IDs in the valid subset are matched with the KG triples, and entities that do not match with the KG triples are filtered out. Finally, the KG is preprocessed to transform the intuitive triple dataset into a numerical representation that can be read by computers.

4.2 Experimental Setup

In this paper, the preprocessed rating dataset from MovieLens-20M is split into training, validation, and testing sets using 85%, 5%, and 15% splits, respectively. The statistical information of the dataset is shown in Tables 1 and 2.

Table 1. Statistical Information of Datasets

Set	hot (> 4)	normal (> 2, ≤ 4)	cold (≤ 2)
Train	376, 890	1, 099, 560	223, 550
Validation	22, 170	64, 680	13, 150
Test	44, 340	129, 360	26, 300

Table 2. Knowledge Graph Information

Dataset	# movies	# entities	# relations	# triples
KG	16954	102568	7	499473

4.2.1 Evaluation Metrics

We use the following two evaluation metrics to assess the performance of our proposed ATKN.

1) C - Accuracy: Accuracy is used to measure the proportion of correctly predicted popularity, defined as:

$$C - Accuracy = \frac{1}{n} \sum_{i=1}^{n} |l(y_i) = l(\hat{y}_i)| \tag{25}$$

y_i and $\hat{y}_i$ represent the true popularity and predicted popularity of movie i, respectively, while $l(*)$ denotes the class corresponding to the computed popularity result, such as hot, normal, and cold.

2) F1-score: F1-score is a commonly used metric for evaluating the performance of classification models. In binary classification tasks, F1-score is calculated by combining precision and recall.

Precision is the ratio of true positive samples to the samples predicted as positive. The formula is as follows:

$$precision = \frac{TP}{TP+FP} \tag{26}$$

TP is the number of true positive samples, and FP is the number of samples predicted as positive but actually negative.

Recall is the proportion of truly positive samples that are correctly predicted as positive. The formula is as follows:

$$recall = \frac{TP}{TP+FN} \tag{27}$$

FN represents the number of true negative samples that are predicted as positive by mistake.

In this paper, we calculate F1-score by taking one of the "popular", "normal", and "cold" categories as the positive class, while the negative class consists of the other two categories. The F1 score is the harmonic mean of precision and recall, which can consider both metrics simultaneously. The formula for F1 score is as follows:

$$F1 = \frac{2*precision*recall}{precision+recall} \tag{28}$$

The range of F1 score is [0, 1], where 1 indicates the model makes all correct predictions, and 0 indicates complete failure of the model. In practical applications, the higher the F1 score, the better the performance of the classification model.

4.2.2 Comparison Methods

In order to demonstrate the effectiveness of our model, we compared it with the following baselines:

1) MLR: Multiple Linear Regression is a method that uses the linear combination of multiple variables as the predicted value.
2) SVR: The Support Vector Regression model uses time-series data as features to predict the popularity of movies.
3) RF: Random Forest is also a popular baseline method for popularity prediction in classification tasks.
4) Attention-based LSTM: The attention mechanism in the long short-term memory network can dynamically weight different features in a sequence, allowing for more accurate capturing of important information in the sequence.
5) AFM: The Attentional Factorization Machine combines the Factorization Machine (FM) with the attention mechanism to automatically calculate the cross-interaction between features and learn the importance of different features.
6) ALSTM$_{+AFM}$: The ATKN model used in this study does not utilize knowledge graphs. It only learns from input sequences and metadata features.
7) ALSTM$_{+KG}$: The ATKN model used in this study does not employ the AFM. It only learns from input sequences and knowledge graph-related features.
8) ATKN: The ATKN model used in this study is a complete version that combines input sequences, cross-metadata features, and knowledge graph-related features between entities.

4.2.3 Parameter Setting

To modeling the trend of movie popularity using time series data, we utilized a single-layer LSTM network with a hidden layer size of 128. When using AFM to learn the cross-metadata features of movies, we set the metadata feature embedding size to 128 and the number of neurons in the DNN hidden layer to 128. For the TransR model, we set the margin size to 4, and the entity and relation embeddings for each triple were also set to 128. For all neural network models, we initialized the other parameters using Xavier initialization and used the Adam optimizer to optimize the models. We employed ReLU as the activation function for all layers, set the batch size to 128, and used dropout with a rate of 0.2 on each FC and RNN layer for regularization to avoid overfitting.

4.3 Results and Analysis

Table 3 presents the experimental results of different methods for predicting popularity. As the datasets used by different prediction methods are not entirely the same, they often exhibit different levels of popularity. Therefore, according to common experimental analysis methods in related studies, we divided the experimental methods into two categories: traditional methods and neural network methods. For our own proposed neural network models, we prepared three variants of our model: 1) ALSTM $_{+KG}$, which uses KG embeddings; 2) ALSTM $_{+AFM}$, which uses metadata feature embeddings; and 3) ATKN, which is a complete version that uses both KG and metadata feature embeddings.

Table 3. Comparison with Baselines

Method	ACC	F1(hot)	F1(normal)	F1(cold)	WA-F1
MLR	0.380	0.550	0.755	0.426	0.663
SVR	0.419	0.583	0.753	0.432	0.673
RF	0.436	0.617	0.765	0.459	0.692
ALSTM	0.473	0.653	0.783	0.443	0.708
AFM	0.479	0.689	0.793	0.547	0.738
ALSTM$_{+AFM}$	0.498	0.693	0.817	0.574	0.758
ALSTM$_{+KG}$	0.501	0.691	0.830	0.567	0.765
ATKN	**0.513**	**0.750**	**0.863**	**0.605**	**0.804**

4.3.1 Comparison with Baselines

Table 3 shows the accuracy and F1 scores of the experiments on the test set. Classifiers based on movie content features, such as RF, MLR, and AFM, achieved relatively poor results. This is largely because these features are often difficult to design and measure and cannot capture the trend of popularity changes, resulting in poor predictive performance of the model. Attention-based LSTM models can only handle time-series features to capture the trend of popularity changes but they cannot learn directly from content features that have a direct impact on popularity results. This also limits the model's expressive power. However, from the experimental results, the ALSTM model performed better than content-based classifiers in most cases but was slightly inferior to AFM. This indicates that the AFM model has a strong ability to express cross-features, while ALSTM can effectively capture popularity trends but has strict requirements for time-series data, with limited feature expression ability.

As a regression model for handling time-series data, SVR only calculates a part of the time process and cannot capture the long-term dynamic changes in popularity, resulting in a certain gap in predictive performance compared to ALSTM. Therefore, it can be concluded that both purely time-series-based and purely content-based methods have significant limitations. Compared with other control methods, the model established in this experiment not only effectively captures the trend of video popularity changes but also models the learning of movie content features, thus improving the model's learning and expressive abilities.

From the experimental results, it can also be seen that the proposed model has higher predictive accuracy and F1 score than other models, demonstrating good popularity prediction performance. To better visualize the learning ability of each prediction model, we provide line graphs of the accuracy of each model over training time, as shown in Table 3. We also calculated the weighted average F1 scores of each model and provided line graphs of the F1 scores of some models over training time, as shown in Fig. 7.

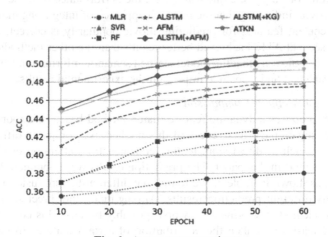

Fig. 6. Accuracy comparison

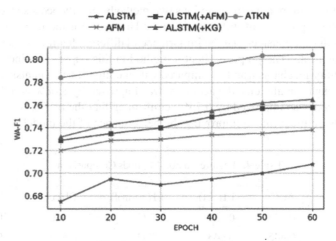

Fig. 7. F1-Score comparison

4.3.2 Ablation Analysis

1) Variant Comparison

We also conducted ablation experiments to compare our proposed model with its variants. This research method is used to determine which parts of the model have an important impact on predicting the popularity of movies. In this study, we performed ablation experiments on three methods: ALSTM $_{+AFM}$, ALSTM $_{+KG}$, and ATKN. ALSTM $_{+AFM}$ only uses the temporal features and metadata features of movies for prediction, while ALSTM $_{+KG}$ only uses the temporal features of movies and the association features between entities in movie knowledge graphs. Compared with methods that only use temporal features or only use content features, the performance of these two methods has been improved, indicating that the research approach of integrating movie temporal features and content features for predicting movie popularity is correct. At the same time, we can see that ATKN achieved better results than the other methods in the ablation experiment, indicating that the model learned semantic information from the movie knowledge graph and had a positive impact on the experimental results.

2) Feature Fusion Methods Comparison

This paper proposes a dynamic feature fusion method based on exponential decay model. To investigate the effectiveness of this method, we compared different feature fusion methods, such as direct sum, feature concatenation, attention mechanism, and the proposed method in this paper. The prediction results were obtained through ACC and F1-score, as shown in Table 4. We found that different feature fusion methods lead to significantly different prediction results. Among them, the direct sum and feature concatenation methods performed the worst, possibly because this completely mixed method could mask or weaken the contribution of certain features to the prediction results or introduce redundant information, thereby reducing the performance of the model. In addition, the introduction of attention mechanism improved the fusion of different features compared to the previous two methods, mainly because it dynamically focuses on the importance of different features. However, for the prediction of movie popularity, it did not consider that the importance of different features varies at different prediction moments. The proposed fusion method in this paper combines this feature with the attention mechanism and is compared with the attention mechanism as the fusion method. Through calculations, the accuracy was improved by 8.68% and the weighted average F1 score was improved by 6.73%, confirming that the proposed feature fusion method plays a critical role in improving the performance of the model.

Table 4. Feature Fusion Methods Comparison

Fusion	ACC	F1(hot)	F1(normal)	F1(cold)	WA-F1
Sum	0.427	0.630	0.772	0.566	0.713
Concatenation	0.445	0.643	0.753	0.572	0.705
Attention	0.472	0.687	0.815	0.594	0.757
Our	**0.513**	**0.750**	**0.863**	**0.605**	**0.808**

5 Conclusion

In this paper, we introduce knowledge graph into video popularity prediction and propose a video popularity prediction model that integrates temporal features and knowledge graph representations. The model consists of four parts: 1) for temporal data, we use an LSTM with attention mechanism to capture the trend of popularity changes, where we select the number of ratings at fixed time intervals and the average rating as input parameters; 2) we use the AFM model to explore cross-features of movie metadata, learning the impact of different features on popularity prediction results, where we select five key movie metadata categories, such as "category" and "director," for experimental research; 3) we use RGCN to extract association features from the knowledge graph. We first obtain a triple dataset of "entity-relation-entity" through the knowledge graph, and then clean and embed the data into the RGCN model to obtain new entity features with relationship information; 4) we propose a dynamic feature fusion method, modeling temporal and content features with the Newton cooling formula, fusing the modeled features through attention mechanism, and obtaining the final result through a linear function. Finally, we conduct experiments on a real dataset to validate the effectiveness and superiority of the proposed model.

References

1. Pinto, H., Almeida, J.M., Gonçalves, M.A.: Using early view patterns to predict the popularity of YouTube videos. In: Proceedings of the Sixth ACM International Conference on Web Search and Data Mining (2013)
2. Szabo, G., Huberman, B.A.: Predicting the popularity of online content. Commun. ACM 53(8), 80–88 (2010)
3. Wu, B., et al.: Unfolding temporal dynamics: predicting social media popularity using multiscale temporal decomposition. In: Proceedings of the AAAI Conference on Artificial Intelligence, vol. 30, no. 1 (2016)
4. Roy, S.D., et al.: Towards cross-domain learning for social video popularity prediction. IEEE Trans. multimedia 15(6), 1255–1267 (2013)
5. Liao, D., et al.: Popularity prediction on online articles with deep fusion of temporal process and content features. In: Proceedings of the AAAI Conference on Artificial Intelligence, vol. 33. no. 01 (2019)
6. Hong, L., Dan, O., Davison, B.D.: Predicting popular messages in twitter. In: Proceedings of the 20th International Conference Companion on Worldwide Web (2011)
7. Lai, G., et al.: Modelling long-and short-term temporal patterns with deep neural networks. In: The 41st International ACM SIGIR Conference on Research & Development in Information Retrieval (2018)
8. Mishra, S., Rizoiu, M.-A., Xie, L.: Modelling popularity in asynchronous social media streams with recurrent neural networks. In: Proceedings of the International AAAI Conference on Web and Social Media, vol. 12, no.1 (2018)
9. Qin, Y., et al.: A dual-stage attention-based recurrent neural network for time series prediction. arXiv preprint arXiv:1704.02971 (2017)
10. Tang, L., et al.: Popularity prediction of Facebook videos for higher quality streaming. In: USENIX Annual Technical Conference (2017)
11. Rizoiu, M.-A., et al.: Expecting to be hip: Hawkes intensity processes for social media popularity. In: Proceedings of the 26th International Conference on Worldwide Web (2017)

12. Du, N., et al.: Recurrent marked temporal point processes: embedding event history to vector. In: Proceedings of the 22nd ACM SIGKDD International Conference on Knowledge Discovery and Data Mining (2016)

13. Zhou, F., et al.: A survey of information cascade analysis: Models, predictions, and recent advances. ACM Comput. Surv. (CSUR) **54**(2), 1–36 (2021)

14. Jia, X., et al.: Incremental dual-memory lstm in land cover prediction. In: Proceedings of the 23rd ACM SIGKDD International Conference on Knowledge Discovery and Data Mining (2017)

15. Zhang, W., et al.: User-guided hierarchical attention network for multimodal social image popularity prediction. In: Proceedings of the 2018 Worldwide Web Conference (2018)

16. Szabo, G., Huberman, B.A.: Predicting the popularity of online content. Commun. ACM **53**(8), 80–88 (2008)

17. Cao, Q., et al.: Deephawkes: bridging the gap between prediction and understanding of information cascades. In: Proceedings of the 2017 ACM on Conference on Information and Knowledge Management (2017)

18. Dou, H., et al.: Predicting the popularity of online content with knowledge-enhanced neural networks. In: ACM KDD (2018)

19. Logan, I.V., Robert, L., et al.: Barack's wife hillary: using knowledge-graphs for fact-aware language modelling. arXiv preprint arXiv:1906.07241 (2019)

20. Wang, Y., et al.: Attention-based LSTM for aspect-level sentiment classification. In: Proceedings of the 2016 Conference on Empirical Methods in Natural Language Processing (2016)

21. Schlichtkrull, M., Kipf, T.N., Bloem, P., van den Berg, R., Titov, I., Welling, M.: Modeling relational data with graph convolutional networks. In: Gangemi, A., et al. (eds.) ESWC 2018. LNCS, vol. 10843, pp. 593–607. Springer, Cham (2018). https://doi.org/10.1007/978-3-319-93417-4_38

22. Wang, X., et al.: KGAT: knowledge graph attention network for recommendation. In: Proceedings of the 25th ACM SIGKDD International Conference on Knowledge Discovery & Data Mining (2019)

23. Ding, M., et al.: Cognitive graph for multihop reading comprehension at scale. arXiv preprint arXiv:1905.05460 (2019)

Design and Implementation of Speech Generation and Demonstration Research Based on Deep Learning

Wanyu Luo, Yanqing Wang(✉), Yujia Liu, and Yiqin Xu

Nanjing Xiaozhuang University, Nanjing, Jiangsu 211171, China
wyq0325@126.com

Abstract. Aiming at complex and changeable factors such as speech theme and environment, which make it difficult for a speaker to prepare the speech text in a short time, this paper proposes a speech generation and demonstration system based on deep learning. This system is based on the Deep Learning Development Framework (PyTorch), trained through the theory of GPT-2 and the open source pretrained model, to generate multiple speeches according to the topics given by users, and the system generates the final speech and corresponding voice demonstration audio through text modification, speech synthesis and other technologies to help users quickly obtain the target document and audio. Experiments show that the text generated by this model is smooth and easy to use, which helps shorten the preparation time of speakers and improves the confidence of the impromptu speaker. In addition, the paper explores the application prospects of text generation and has certain reference value.

Keywords: Natural Language Processing · Text Generation · Deep Learning · Speech Generation

1 Introduction

Text generation in natural language processing (NLP) is one of the most important research directions, and its extremely high research value and application value attract scholars to explore constantly. The development of text generation can drive the progress of natural language processing, which can make the whole artificial intelligence field a greater advance [1]. The method of studying text generation was originally based on rule templates and generated text by stipulating language rules such as lexical analysis and part of speech annotation. This method works well in the era of small data volume and simple rules. With the development of technology, people usually use the method of summarizing long texts to generate short texts. In the long text generation task, the machine uses a given linguistic format to fill in the already fixed "eight-part text" and then obtains the final text through simple replacement. With the development of artificial intelligence, an increasing number of scholars use deep learning to solve the problem of text generation. At present, deep learning has relatively successful products in speech

Z. Yu et al. (Eds.): ICPCSEE 2023, CCIS 1879, pp. 475–486, 2023.
https://doi.org/10.1007/978-981-99-5968-6_33

recognition, machine translation and other fields. There are also some application cases in prose generation and fairy tale [2] generation, but there is little research in the field of illustrative text generation, such as speech.

2 Related Work

With the development of deep learning, recurrent neural networks have been applied to the field of text generation. In 2010, Mikolov proposed that RNN can learn all the information before the sequence and compared the differences between the previous and subsequent texts [3]. However, the dependence on sequences has prevented RNNs from achieving good results in long text generation. In 2018, Allen NLP proposed the ELMO model and used LSTM as a feature extractor to solve the problem of overdependence on sequences in RNN [4]. Compared to traditional static word vector models, ELMO is able to solve partial polysemy problems [5], but its effectiveness is poor. Then, the feature extractor transformer architecture of natural language processing appeared. Transformers use encoders, decoders, and self-attention to solve the problems of low parallel computing efficiency [6] and the poor ability to capture long-distance features that occur during the training of recurrent neural networks. The GPT series module takes the transformer architecture as the core, and it has achieved remarkable results in the field of text generation with excellent model architecture, powerful computing power and massive training data, bringing the development of the dynamic word vector model to a new height.

Transformers aim to process sequential input data, and their internal structure mainly consists of two sets of "encoders" and "decoders". The encoder uses the self-attention layer and feedforward neural network. The self-attention layer helps the encoder view other words in the input sentence when encoding specific words to link the context and improve the relevance of the whole text. For the decoder, an attention layer is added in the middle on top of the encoder to help the decoder focus on the relevant parts of the input sentence. The self-attention mechanism is the core mechanism of Transformer, which scores all vocabulary in a sentence and obtains new vectors based on the scores. In addition, the self-attention mechanism can also enable the training of each layer to be carried out simultaneously, which will increase parallelism. The GPT series models use a unidirectional model for pretraining and fine-tuning to process lower-level tasks. The feature extractor in GPT-1 [7] uses the decoder module in Transformer. GPT-2 [8] increases the Transformer model from 12 layers to 48 layers and increases the training parameters. Downstream tasks use zero shot, which does not require any annotation information or further training of the model. The GPT-3 [9] model adopts a larger database and more network layers, dedicated to training a large amount of general data and achieving problem solving with fewer specialized data in specific fields. Since the domestic models are still in the construction and experimental stage and there are few GPT-3 language models with good effects that are open source in Chinese, this paper uses the GPT-2 open source model to train the text generative model used in this paper.

In terms of speech synthesis, with the development of deep learning, speech synthesis technology has evolved from parameter synthesis or concatenation synthesis to end-to-end speech synthesis based on deep learning. Speech synthesis technology based on deep learning is divided into three parts: front-end processing, acoustic models, and vocoders. Because most existing speech synthesis models currently use English datasets, it is necessary to preprocess Chinese text before front-end processing, such as eliminating the ambiguity of polyphonic characters, predicting prosody, and converting pinyin into phoneme sequences. The front-end processing part converts the input text sequence into the corresponding phoneme sequence and inputs the phoneme sequence again into the acoustic model to train the Mel spectrum. The vocoder generates the corresponding speech waveform based on the Mel spectrum.

There are two types of acoustic models: autoregressive models and nonautoregressive models. The autoregressive models represent the Tacotron series models with an encoder, an attention mechanism and a decoder. Tacotron [10] is the first true end-to-end TTS deep neural network model. The model can be implemented to generate the corresponding speech and sound spectrum through the input text sequence. However, the Recurrent Reural Network is used in both the encoder and decoder in Tacotron, so the model cannot be trained in parallel, and speech synthesis is not effective in real-time performance. Fastspeech [11] is a nonautoregressive acoustic model that is based on the Transformer model. The model removes the decoder part in the transformer and replaces it with the FFT block, length regulator and duration predictor modules. Compared with the autoregressive acoustic model, Fastspeech improves the speed of speech synthesis and solves the problem of instability of synthetic speech. Moreover, the time length of synthetic speech in the duration predictor module becomes more controllable.

Similar to acoustic models, vocoders are also divided into autoregressive and nonautoregressive models. WaveNet [12] is a typical autoregressive model that uses a deep neural network to model the original audio and synthesize speech combined with the Mayer spectrum features. WaveGAN [13] is a nonautoregressive model based on GAN that uses WaveNet as the generator, trains the generator through the loss combined with the multiresolution STFT loss function and the confrontation loss, and judges whether the generated audio is true through the discriminator. The WaveGAN model has significantly improved the speed of speech synthesis.

3 System Design

This project is trained and tested on a deep learning framework, which mainly includes two parts: speech text generation and speech synthesis. The text generation module preprocesses the dataset by segmenting words, represents vocabulary information through word vectors and position vectors, and utilizes multihead attention to represent the correlation between words. The speech synthesis module obtains the text features through the text front end [14], uses the acoustic model, and synthesizes the audio through the vocoder.

The process design of the speech generation system is shown in Fig. 1, which is divided into two parts: speech text generation and speech synthesis. The speech text generation section first uses crawler technology to obtain the speech text as a training dataset and then preprocesses the datasets through word segmentation. Next, the text is converted into feature vectors, and the vocabulary is transformed into a matrix-vector form that can be understood by the computer. The position encoding information of the vocabulary is added to the word embedding vector to record the position of the vocabulary in the sentence. Next, the word vector is input into the pretrained model to train the model. After several rounds of training, we obtain the speech generative model used in this paper. After the model training is completed, users can input requirements such as topic and word count. The model will process the input information and automatically generate the corresponding first draft of the speech text. Then, the first draft was divided into short sentences, and the keyword in Python was used to determine whether the text was duplicated. Users will obtain the final speech text after deleting the duplicated text. The speech synthesis section uses the FastSpeech Acoustic Feature Network and WaveGAN vocoder to process text and audio, generating audio files corresponding to the speech text.

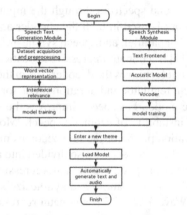

Fig. 1. System flow diagram

3.1 A Subsection Sample

The speech text generation module uses the GPT-2 model for text generation. The training process diagram of the GPT-2 model is shown in Fig. 2. The implementation of text generation in GPT-2 is mainly divided into two stages: The first step is using the language model for pretraining, and the second step is to fine-tune the pretrained model and then apply it downstream. The GPT-2 model adopts the decoder structure of the transformer model and generates text sequentially from left to right through an autoregressive mechanism, which is more in accordance with human writing and reading habits. The GPT-2 model adopts a 48-layer transformer decoder structure, with each layer containing modules such as a word vector matrix, position encoding matrix, and multihead attention mechanism. Through these modules, text generation can be achieved, and the relevance of the text can be guaranteed. The decoder structure is shown in Fig. 2.

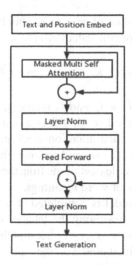

Fig. 2. The structure of a decoder

When GPT-2 generates text, the first step is to represent each input word with a word embedding vector matrix and add position embedding information to this vector to represent the position of each word in the sequence. The word embedding vector and position encoding information are input into the decoder of the first transformer as input information. The data are processed by the self-attention layer in the decoder, which calculates the correlation between each word in a text and other words and scores them. After summing all the correlation scores of a word, it is processed by a neural network to obtain a word vector for the word. The processed results are sent to the next transformer decoder, and the above calculation process is repeated continuously. Finally, the model will multiply the word vector generated by the last decoder and the word embedding vector to obtain the final output result.

The self-attention layer in the decoder allows for modelling dependencies without considering their distance in the input or output sequence and can concatenate different positions of a sequence. Therefore, the self-attention mechanism is the core module in GPT, which will be introduced in detail below. Self-attention achieves text generation by comparing the degree of correlation between a word and other words and assigning different amounts of attention to each word. Therefore, the core of self-attention is to calculate the attention value assigned to each word, which is the weight of each word. There are two steps to calculating the weight coefficient of words: calculating similarity and normalizing using the softmax function. When calculating similarity, self-attention generates three vectors for each word: query, key, and value. The query vector is multiplied by the key vectors of other words to obtain the correlation score between that word and other words. The Key vector is the number of each word, and by searching for the Key vector, the desired word can be retrieved. Value vectors are actual word representations, and once we rate the relevance of each word, we need to weigh and sum these vectors to represent the current word. The query [15] vector of each word is linearly added to the key vector of other words after calculating the weight, and the resulting vector is the representation of each word after being weighted and summed by the attention mechanism. The specific calculation formula is as follows:

$$Attention(Q, K, V) = softmax(\frac{QK^T}{\sqrt{d_k}})V \tag{1}$$

d_k is the number of columns of the Q and K matrices, namely, the vector dimension. The self-attention mechanism used in GPT-2 is a multiheaded self-attention mechanism. The purpose is to enable each attention mechanism to map to different spaces through the QKV vector to learn features and optimize the different feature parts of each word, thereby balancing the possible deviations from the same attention mechanism and allowing more meta-expressions of word meanings.

After the pretrained GPT model is completed, it will be applied to different tasks through supervised fine-tuning. Assuming the labelled dataset is C and the data structure is $(x_1, x_2, x_3 ... x_m, y)$, we enter $(x_1, x_2, x_3 ... x_m)$. Obtain the output vector h_l^m after iteratively processing the pretrained model on the input. Then, through linear layers and softmax, we obtain and guess the labels:

$$P\left(y|x^1, x^2, x^3 ... x^m\right) = softmax(h_l^m W_y) \tag{2}$$

where W_y represents the parameters of the prediction output, and the fine-tuning needs to maximize the following functions:

$$L_2(C) = \sum_{(x,y)} logP(y|x_1, x_2, x_3 ... x_m) \tag{3}$$

The GPT needs to consider the pretrained loss function when fine-tuning, and the final function that needs to be optimized is:

$$L_3(C) = L_2(C) + \lambda^* L_1(C) \tag{4}$$

The text generation adopts the general model and prose model provided by the pretrained model GPT2 Chinese. The training process is shown in Fig. 3. First, 7000 speech manuscript texts with different themes are obtained through crawler technology as the training set, the datasets are preprocessed through filtering, word segmentation and so on, and the text is transformed into a feature vector through the vocab_small dictionary of Goggle. The processed feature vectors are fed into the GPT-2 pretrained model, and vocabulary is converted into matrix vectors through word vectors and word position vectors for computer calculation. The multihead attention mechanism is used to calculate the correlation between vocabulary to achieve vocabulary prediction. The text generative model used in this paper is obtained by constantly adjusting the model training parameters. Then, input the expected speech theme, word count, and other requirements as input information into the existing model, which will automatically generate the speech text. The initial generated text has some duplication, so use the 'keyword in' method provided by python to delete repeated content. The 'keyword in' is used to determine the dependency relationship between object A and object B. By using the keyword, it can determine whether the current value already exists and remove duplicate content. Then, duplicate text is deleted to obtain the final output text. The main process of text generation is shown in Fig. 3.

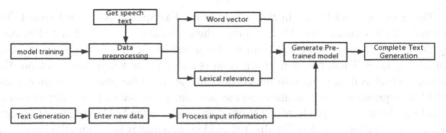

Fig. 3. Flow diagram of Text generation

3.2 Speech Synthesis Module

The speech synthesis part mainly includes three modules: text front end, acoustic model and vocoder. The text front end is responsible for realizing the text-to-character conversion. The acoustic model can convert the character features into acoustic features and finally use the vocoder to convert the acoustic features into waveforms. The flowchart of the speech synthesis module is shown in Fig. 4.

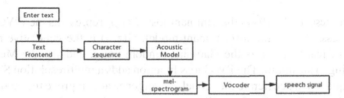

Fig. 4. Flowchart of the speech synthesis module

The input text first needs to undergo preprocessing operations, such as analysing polyphonic characters, prosodic prediction, and text normalization. In the text generation section, we have already deleted the duplication context, so the text normalization step can be omitted. The purpose of prosody prediction is to reflect the different intonations of sentences through annotations, making the synthesized speech more in line with human speech habits. According to the four common situations in Chinese conversations, this model divides prosodic annotation into four levels, represented by 1, 2, 3, and 4 [16]. Different prosodic annotations represent different pause times of sentences, and the pause time is used to determine whether the sentence ends. The shortest pause time indicated by prosody 1 and the longest pause time indicated by prosody 4 indicate the end of the sentence, and the synthesized speech of the sentence ends. The prosodic annotation is carried out at the same time as the text generation part. Bert's tokenizer is used for word segmentation, and 1 is marked after each word, 2 is marked after the words with transliteration, 3 is used for pausing punctuation marks such as comma, semicolon, colon, etc., and 4 is used for ending punctuation marks such as full stop, exclamation mark, question mark. Send the preprocessed data into the front end of the text to obtain the phoneme sequence. This article uses the library function pypinyin tool provided in Python to convert the pinyin sequence into a phoneme sequence.

The acoustic model used in this article is the FastSpeed2 pretrained model. The structure of the acoustic model is as follows: the embedding vector is obtained by passing the phoneme sequence input from the front end of the text through the phoneme embedding layer. Then, position information is added to the vector through a position encoder, which is then input into the encoder to transform the phoneme sequence into a hidden sequence through an attention mechanism. Then, the variable adapter is used to add the duration and pitch information of the audio to the hidden sequence for easy prediction of the target voice. Finally, the hidden sequence is input into the decoder to obtain the Mel spectrum.

The vocoder is used to convert the Mel spectrogram into corresponding audio. The vocoder used in this article is the parallel wave GAN. The loss function adopted by PWG retains the multiresolution STFT loss in the original vocoder and adds the confrontation loss, which speeds up the convergence speed of the model while maintaining the stability of the model. The specific loss formula is as follows [17]:

$$PG_{adv} = (1 - D(G(z, s)))^2 \tag{5}$$

$$PD_{loss} = (1 - D(x))^2 + D(G(z, s))^2 \tag{6}$$

PD_{loss} represents the PWG discriminant loss, PG_{adv} represents the PWG generator against the loss, $D(x)$ is the discriminant model, $G(z, s)$ is the generative model, x is the sample of real audio, z is the Gaussian random noise, and s is the Mel spectrum corresponding to the audio. The PWG loss function adds a multiresolution STFT loss to the above equation to accelerate the model convergence and improve the model stability. Therefore, the loss function of the PWG generator is as follows:

$$PG_{loss} = L_{aux} + \lambda \cdot PG_{adv} \tag{7}$$

PG_{loss} is the loss of the PWG generator, and λ is the proportion coefficient of antagonistic loss.

The structure of the PWG includes two parts: a discriminative network and a generative network. The Generate network is composed of dilated convolution layers and uses a non-autoregressive structure to ensure the synthesis effect while accelerating the synthesis speed. The accuracy of the distribution of each sampling point is judged by the discriminator, which consists of 10 dilated convolution layers. The dilation rates of each layer increase from 1 to 10. Finally, the score of each point is judged by outputting the one-dimensional result of the convolution layer. The Generate and Discriminative network structures of the PWG are shown in Fig. 5.

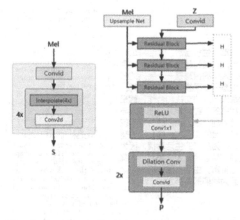

Fig. 5. Structure diagram of the PWG network

4 Experiment

4.1 Speech Generation

The dataset used in this article is a crawler technology that crawls the topic of speech notes on the "Worryfree Test." A total of 7000 speech notes were obtained as a training set, with one line for each paragraph. Due to the lack of a large number of advertisements on the website itself, the datasets do not need to be carefully filtered and checked for formatting errors. The data can be directly preprocessed. BERT's tokenizer is used to process the obtained datasets and obtain word segmentation results.

Model pretrained uses the GPT2 Chinese general model and prose model for testing and training, and the basic network settings are directly invoked by referring to the parameter settings on the official website of Transformer, resulting in the final new model. During training, the preprocessing datasets were divided into 90 groups, and the training cycle was conducted 4 times. The learning rate was 1.5c-3, and the training batch size was 8.

Enter the topic, number of words, and number of articles into the pretrained model, which preprocesses the input information and automatically generates the speech text. Perform a duplication removal operation on the first generated speech text. Divide the generated speech text into short sentences, use the keyword provided by python to determine whether it is duplicated, and then delete the repetitive short sentences to obtain the final draft of the speech. The text generation results comparison before and after and English translation (see Fig. 6).

There is currently no unified optimal standard for evaluating the credibility of text generation models. Common model evaluation indicators include BLEU, ROUGE, and PPL. BLEU evaluates the model by comparing the coincidence degree of n-grams in the generated text and the reference text. ROUGE focuses on calculating recall rates, commonly used for evaluating machine translation and article abstracts. PPL, or degree of confusion, is used to measure the degree to which a probability distribution or probability model predicts a sample. In this paper, the BLUE evaluation index was used to evaluate the model (Table 1).

老师们，同学们：
　大家好！
智慧之地、智慧之地。
中国人的智慧之地，人文荟萃，生机盎然，生活富足，
精神生活丰富，生活质量高。人类智慧是人类的精神家
园，是人类的灵魂之所在。人类的智慧是一个民族的智
慧，是人类智慧的源泉，也是一个民族的智慧的体现。
人与人的交流、沟通，就是人与人的交往和合作，就是
人与人的交际，就是人与社会的交往。智慧之地是人与
人的交互，也是一个人与社会的交往。我们的智慧不仅
是人与人的交往，更是人与社会的交流和合作。因为智
慧，所以我们的生活富足，因为我们的生活幸福，因为
我们的生活幸福。中国人的智慧之地，人文荟萃，人文
荟萃，生活富足，精神生活丰富，生活富足，精神生活
丰富，我们的生活富足；因为我们的精神生活

A land of wisdom, a land of wisdom.
The land of wisdom for the Chinese people is a place of cultural diversity, full of vitality, abundant life, rich spiritual life, and high quality of life. Human wisdom is the spiritual home of humanity and the soul of humanity. Human wisdom is the wisdom of a nation, the source of human wisdom, and the embodiment of a nation's wisdom. The communication and interaction between people is the interaction and cooperation between people, the communication between people, and the interaction between people and society. The land of wisdom is the interaction between people, as well as the interaction between a person and society. Our wisdom is not only about interpersonal communication, but also about communication and cooperation between people and society. Because of wisdom, our lives are rich, because our lives are happy, because our lives are happy. The land of wisdom for the Chinese people, where there is a wealth of human culture, a wealth of life, a rich spiritual life, a wealth of life, a rich spiritual life, and a wealth of our lives; Because of our spiritual life

老师们，同学们：
　大家好！
智慧之地
中国人的智慧之地，人文荟萃，生机盎然，生活富足，
精神生活丰富，生活质量高。
人类智慧是人类的精神家园，是人类的灵魂之所在。
人类的智慧是一个民族的智慧，是人类智慧的源泉，也
是一个民族的智慧的体现。
人与人的交流、沟通，就是人与人的交往和合作，就是
人与人的交际，就是人与社会的交往。
智慧之地是人与人的交互，也是一个人与社会的交往。
我们的智慧不仅是人与人的交往，更是人与社会的交流
和合作。
因为智慧，所以我们的生活富足，因为我们的生活幸
福，因为我们的生活幸福。
中国人的智慧之地，我们的生活富足；因为我们的精神
生活

The Land of Wisdom
The land of wisdom for the Chinese people is a place of cultural diversity, full of vitality, abundant life, rich spiritual life, and high quality of life. Human wisdom is the spiritual home of humanity and the soul of humanity.
Human wisdom is the wisdom of a nation, the source of human wisdom, and the embodiment of a nation's wisdom.
The communication and interaction between people is the interaction and cooperation between people, the communication between people, and the interaction between people and society.
The land of wisdom is the interaction between people, as well as the interaction between a person and society.
Our wisdom is not only about interpersonal communication, but also about communication and cooperation between people and society.
Because of wisdom, our lives are rich, because our lives are happy, because our lives are happy.
The land of Chinese wisdom, our lives are abundant; Because of our spiritual life

Fig. 6. Comparison before and after repetition and English translation

Table 1. Results of the model evaluation in this article.

Model	BLEU1	BLEU2	BLEU3
Speech Prediction Model	0.72	0.58	0.44
CNN-Transformer	0.42	0.39	0.32
VGG-Transformer	0.64	0.50	0.42
ResNet-Transformer	0.71	0.63	0.41

4.2　Speech Synthesis

The dataset is selected from the Chinese Standard Mandarin Speech Copus dataset, which is an open-source dataset provided by Standard Bay Technology Co., Ltd. The audio in these datasets is a professional standard Mandarin female voice. The recorded corpus covers a wide range. The datasets also perform text pronunciation and character proofreading, prosody level labelling, and voice file boundary segmentation labelling for the voice database according to the synthetic voice labelling standard. It is very suitable as a dataset for training speech synthesis models.

The model training section uses the pretrained FastSpeed2 model as the acoustic feature network and the parallel wave GAN as the neural vocoder. The basic network settings directly refer to the parameter settings of the FastSpeed2 model.

During speech synthesis, the speech text generated by the speech generation module is input, and the corresponding Mel spectrum map and waveform map of the speech are generated by calling the acoustic model and vocoder. The audio file is saved as a.wav file. Graphics about the generated results taking the topic of the speech Children's Day as an example (see Fig. 7 and Fig. 8).

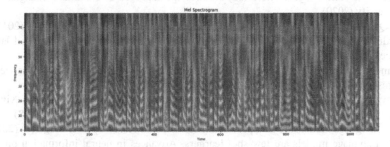

Fig. 7. Mel spectrogram

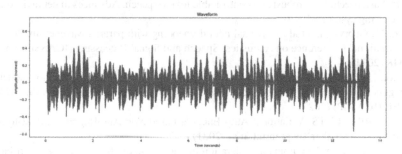

Fig. 8. Waveform plots

5   Conclusion

With the development of text generation technology, applying text generation technology to daily life is a potential topic. Aiming at the pain points of speakers' difficulty in quickly preparing speeches based on different environments, objects, topics, and other requirements, this paper proposes a demonstration system for speech generation based on deep learning.

This system is based on the Deep Learning Development Framework (PyTorch) and trained through the theory and pretrained model of GPT-2 to achieve the goal of transforming user requirements into speech. At the same time, the speech and corresponding voice demonstration audio are generated through modules such as text error detection and speech synthesis, helping users quickly obtain target documents and audio. Experiments have shown that this system works well and that the generated text has a certain degree of readability. By encapsulating the system, it is convenient to use, helping to shorten the speaker's preparation time and enhance the confidence of impromptu speech.

References

1. Hu, M.: Text generation based on generative adversarial networks. University of TC (2020)
2. Ma, Z.: Fairy tale text generation study based on the improved GPT-2 model. Shanghai Normal University (2020)
3. Le, Q.: Distributed representations of sentences and documents. Journal PMLR (2014)
4. Gardner, M., Grus, J., Neumann, M., et al.: AllenNLP: a deep semantic natural language processing platform. Journal (2018)
5. Ghanem, R., Erbay, H.: Spam detection on social networks using deep contextualized word representation. J. **82**(3), 3697–3712 (2023)
6. Vaswani, A., Shazeer, N., Parmar, N., et al.: Attention Is All You Need. Journal arXiv (2017)
7. A R. Improving language understanding by generative pretraining. Journal (2018)
8. A R. Language models are unsupervised multitask learners. J. OpenAI blog **1**(8) (2019)
9. T B. Language models are few-shot learners. Advances in neural information processing systems (2020)
10. Y W. Tacotron: Towards end-to-end speech synthesis. arXiv preprint arXi (2017)
11. Y R. Fastspeech: Fast, robust and controllable text to speech. Advances in neural information processing systems (2019)
12. Jiao, Y., Gabrys, A., et al.: Universal neural vocoding with parallel wavenet. In: 2021 IEEE International Conference on Acoustics, Speech and Signal Processing (ICASSP), pp. 6044–6048 (2021)
13. Amoto, R., Song, E., Kim, J.M.: Parallel WaveGAN: A fast waveform generation model based on generative adversarial networks with multiresolution spectrogram//ICASS, (ICASSP). IEEE (ICASSP) (2020)
14. H L. VAENAR-TTS: Variational Auto-Encoder based Non-AutoRegressive Text-to-Speech Synthesis. Journal arXiv preprint arXi (2021)
15. Sun, F.: Study on the automatic generation method of scientific consulting report based on GPT-2 model. Xidian University (2021)
16. Wang, Z.: Research and implementation of speech synthesis based on Fastspeech. Sichuan University of Light and Chemical Technology (2021)
17. Chen, F.: Multidiscriminant song synthetic vocoders based on generative adversarial networks. Zhejiang University (2021)

Testing and Improvement of OCR Recognition Technology in Export-Oriented Chinese Dictionary APP

Qingpei Yang$^{(\boxtimes)}$ (iD), Xinrui Wan (iD), Hongjun Chen (iD), Mengjiao Wu (iD), Yao Wu (iD), and Jiyun Qiao (iD)

NingboTech University, Ningbo 315100, China
2140765316@qq.com

Abstract. With the increasing number of Chinese learners each year, the international influence of the Chinese language has also grown. Consequently, Chinese auxiliary learning tools should align more closely with the needs of present-day learners. This paper focuses on a typical Chinese dictionary app and conducts tests and comparisons of the optical character recognition (OCR) function. Specifically, it compares the OCR function of a widely used Chinese learning dictionary with the more advanced and sophisticated mobile built-in OCR recognition technology available today. By identifying the technical gap, the paper proposes a design that incorporates deep learning techniques to enhance the built-in OCR function of the dictionary. This improvement aims to significantly enhance recognition accuracy in natural scenarios and ultimately enhance the user experience of the Chinese learning dictionary app.

Keywords: Extroverted Dictionary · OCR Technology · Deep Learning

1 Introduction

According to Hanban's rough estimates, the number of people learning and using Chinese worldwide, excluding China (including Hong Kong, Macao, and Taiwan), has surpassed 100 million. With the rapid development of the mobile Internet, mobile applications, represented by apps, have become a crucial method for online language learning. Among these apps, Chinese learning dictionaries and translation software play a significant role in determining user learning efficiency and interest.

This paper examines Chinese learning mobile applications available in the app market. Specifically, it focuses on evaluating the photo translation functions of the most downloaded Chinese learning software, namely, Pleco on the Apple App Store. The study reveals that Pleco employs traditional OCR technology for its photo translation functions. This technology employs computer algorithms to identify and extract characters from optical image information to obtain editable electronic text. During our research, we observed that Chinese OCR character recognition technology primarily relies on traditional image processing techniques, including template matching, manual character

Z. Yu et al. (Eds.): ICPCSEE 2023, CCIS 1879, pp. 487–494, 2023.
https://doi.org/10.1007/978-981-99-5968-6_34

template matching, and feature extraction methods. As a result, the OCR technology has achieved a high level of effectiveness and recognition accuracy in identifying black and white printed fonts. However, its performance significantly declines when dealing with complex scenes and handwritten fonts. This limitation poses a significant inconvenience to users [1].

Research on export-oriented Chinese learning dictionaries in China started relatively late compared to English dictionaries, and the investigation of app learning dictionaries combined with Internet technology is relatively uncommon. Existing research mainly focuses on optimizing the phonetic interpretation content from a linguistic perspective, improving learning methods from a pedagogical standpoint, and enhancing the development and interface function of the app itself from a design perspective.

With the rapid advancements in deep learning artificial intelligence within the computer field, there is still a lack of a comprehensive research system on how to effectively combine traditional dictionary learning tools with current science and technology to optimize the user experience in the Internet era. This paper focuses on the most popular app dictionaries as research subjects, aiming to investigate the current application status and limitations of their commonly used photo search function. It introduces deep learning techniques to test and enhance this function, with the aim of shedding light on interdisciplinary research between app dictionaries and emerging computer technologies. Additionally, it seeks to contribute to the improvement and optimization of teaching tools in the international Chinese language education domain.

2   Research Design

2.1   Background of OCR Technology

Optical character recognition (OCR) Optical character recognition (OCR), also known as optical character recognition, is an important component of computer vision that simulates human vision to extract and digitize text information from target images [2]. It efficiently and quickly performs numerous simple and repetitive text recognition tasks.

Traditional OCR technology primarily utilizes the template matching method for character recognition. This involves comparing preprocessed and segmented characters with character templates and determining the most suitable result by evaluating the similarity between them.

Initially, the feature extraction of the processed character image is performed, and the similarity value is calculated by measuring the distance between the character image feature and the template feature [3]. The output result with the highest similarity and the closest distance is extracted. During the template matching process, multiple position coordinates are matched against each template, and the nonmaximal suppression algorithm is employed to eliminate overlapping coordinates [4]. Equations A and B are the matched position coordinate areas:

$$nms = (A \cap B) \div (A \cup B) \tag{1}$$

In the postprocessing stage, the word matching method is primarily employed when the recognition background environment is complex, and issues such as recognition angle and lighting problems cause some character recognition to become blurred. In such cases, the system searches the phrase library to find the most appropriate individual word match. For instance, when recognizing an image containing the word "Olympic," there might be various results such as "four," "Kuang," and "horse" for the word "horse." However, through the use of a thesaurus search, the system retrieves the phrase "Olympic" and ultimately selects "horse" as the output.

2.2 Development of OCR Technology

OCR technology was first proposed by a German scholar named Tausheck in 1929 [5]. He utilized optical recognition of text and converted it into code that could be recognized by computers [6]. English, German, and other languages primarily consist of a few dozen alphabetic arrangements and combinations, making them relatively simple to recognize. As a result, OCR technology for these languages has rapidly developed, leading to the emergence of numerous OCR systems for various application scenarios, such as bank check recognition [7].

The development of OCR technology for Chinese characters began later. In the 1960s, Casey et al. achieved the recognition of 1,000 Chinese characters using the template matching method, marking a breakthrough in Chinese character recognition. Since then, the template matching method has become the main approach for Chinese character OCR research. In the 1990s, several Chinese OCR systems emerged that could recognize printed fonts, such as TH-OCR and BI-OCR. However, due to the complexity and diversity of Chinese character structures, the template matching method is not effective for the OCR recognition of Chinese characters.

With the rise of machine learning, OCR technology has incorporated various machine learning algorithms. Convolutional neural networks (CNNs) are used to preprocess scanned images and extract features from them. By applying CNNs in OCR, features of handwritten characters can be better extracted, resulting in higher recognition accuracy [8]. Hidden Markov models (HMM) are utilized in OCR for scanned texts to improve accuracy by modelling the probability of character sequence extracted features [9]. Recurrent neural networks (RNNs) are capable of extracting temporal and semantic information from data and perform well in processing data with sequence characteristics, but they face the issue of gradient disappearance. To address this problem, some researchers have proposed combining RNN with long short-term memory (LSTM), as LSTM is effective in correcting characters with OCR recognition errors [10].

Due to the complexity of Chinese characters and the diversity of application scenarios, there is no unified algorithm system for Chinese character OCR. Instead, it needs to be designed according to specific requirements and practical needs.

3 Existing OCR Technology Testing and Problem Analysis

3.1 Comparison Test of Text Scene and Natural Scene Recognition Accuracy

Existing real-world test data demonstrate that the performance and recognition accuracy of the popular Chinese dictionary software Pleco in recognizing black and white printed fonts have achieved a high level.

This experiment aims to test the practical application of images in various usage scenarios and introduces SamScanner, a scanning software, as a reference. SamScanner utilizes deep learning technology for OCR recognition and is widely utilized in the field of text scanning, boasting a large number of downloads (Table 1).

Table 1. OCR recognition accuracy under different conditions.

Font Category			Recognition Rate		
Picture properties	Picture scene	Font properties	SamScanner	Pleco	Train Chinese
Book text	Text	Printed	≥99%	99.4%	96.5%
Notes	Text	Handwritten	≥99%	43.3%	27.0%
Posters	Natural	Printed	94.1%	24.2%	31.8%
Placard	Natural	Printed	97.9%	67.4%	49.6%

3.2 Test Conclusions

1. Printed text only: Pleco's character template matching is designed for common typographic Chinese fonts, which may result in lower accuracy or failure to recognize handwritten characters in handwritten notes.
2. Difficulties in recognizing rare words: Pleco's template library consists of 6,763 simplified characters and 5,401 traditional characters. However, some rare words may be challenging to recognize because they are not included in the database.
3. Limited background anti-interference capability: Due to the utilization of template matching technology, Pleco may encounter difficulties in achieving accurate recognition when faced with complex background patterns, varying shooting angles, or challenging lighting conditions such as sunlight or shadows, leading to potential errors in code recognition.

4   Technical Improvement Program Design

4.1 Overview

This study aims to integrate deep learning techniques into the conventional OCR recognition technology used in the dictionary. By utilizing advanced deep learning models, it enhances the feature point extraction of character molds and improves image processing and recognition capabilities [11]. The study achieves the following objectives:

1. Expanding the application scenarios of OCR recognition from text scene recognition to natural scene recognition.
2. Enhance recognition accuracy and improve anti-interference capabilities in complex environments.

Fig. 1. a and e are test images, b and f are SamScanner results, and c, d, g and h are Pleco results.

4.2 Improvement Program

Introduction to Principle

Traditional OCR methods for text recognition typically employ template matching to "classify" each individual character. This process is often carried out through "over-segmentation - dynamic programming" techniques, where an optimal combination is obtained through dynamic programming algorithms. However, this approach overlooks the overall context of the text, focusing solely on individual character recognition. To address this limitation, this paper proposes the use of a convolutional recurrent neural network (CRNN) for OCR character recognition.

As depicted in Fig. 1, a CRNN network architecture consists of three components: a convolutional layer, a recurrent layer, and a transcription layer. The convolutional layer extracts feature sequences from the input image, while the recurrent layer predicts the label for each frame. The transcription layer converts the frame predictions into a final sequence of labels. In this study, a residual neural network (ResNet) is utilized as the convolutional layer, which offers higher accuracy compared to networks such as AlexNet and VGG. ResNet addresses issues such as information loss and transmission loss that are associated with traditional convolutional or fully connected layers. By directly bypassing input information to the output, ResNet preserves the integrity of the information, simplifies the learning objectives, and reduces complexity (Fig. 2).

The recurrent layer in this paper employs a bidirectional long short-term memory (BiLSTM) network. Compared to ordinary LSTM networks, BiLSTM networks

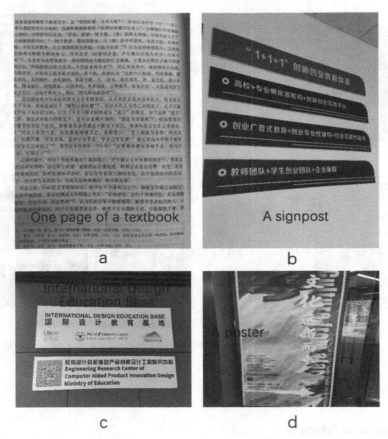

Fig. 2. a is a text scene, b, c and d are natural scenes

can effectively handle context in both forward and backwards directions, offering complementary information that is well suited for processing image sequences [12] (Fig. 3).

Dataset Used for the Experiment

The model was trained using the dataset provided by Baidu's Chinese Scene Text Recognition Technology Innovation Competition. This dataset consisted of a total of 330,000 images. The dataset was derived from Chinese Street View and comprised text line regions extracted from Street View images, including shop signs, landmarks, and other text-containing areas (Figs. 4 and 5).

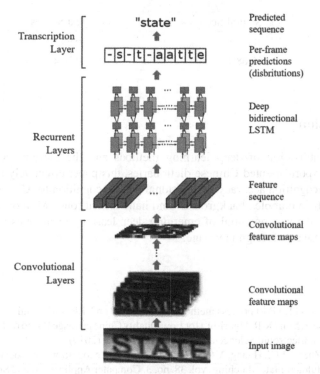

Fig. 3. CRNN-based text recognition

Fig. 4. Dataset for training: The names of some shops on the street

Fig. 5. Field test image: Some signages

Results and Analysis

Based on multiple tests, the model utilized in this study demonstrates improved training results when adjusting the base learning rate (base lr) to 0.00001, the number of epochs to 30, and the batch size to 256 during training. After two hours of training on an NVIDIA Tesla V100 GPU, the model achieves an overall accuracy of approximately 85%. During field testing, it attains an accuracy of 81.53%.

	Natural Scenes	Text Scenes
Pleco	36.52%	46.95%
Ours	85.68%	88.78%

5 Conclusion

Through the utilization of deep learning methods and the current OCR approach employed in export-oriented Chinese dictionaries, this paper effectively addresses the issue of low recognition accuracy in traditional OCR recognition for Chinese dictionaries, particularly in complex backgrounds and handwritten fonts. Moreover, it provides a glimpse into the vast potential of emerging deep learning techniques in the field of teaching Chinese as a foreign language.

References

1. Li, W.H., Luo, G.L.: Post process method of OCR based on NLP. Softw. Guide **9**(10), 2 (2010)
2. Liu, Y.: Research on OCR Algorithm for Low Quality Chinese Image Based on Deep Learning. University of Electronic Science and Technology, China (2019)
3. Cao, C.Y., Zheng, J.C., Huang, Y.Q.: A Special Symbol Recognition and Location Algorithm Based on Multitemplate Matching, vol. 38, no. 3. Computer Applications and Software, China (2021)
4. Shi, B.G., Bai, X., Yao, C.: An end-to-end trainable neural network for image-based sequence recognition and its application to scene text recognition. IEEE Trans. Pattern Anal. Mach. Intell. **39**(11), 2298–2304. IEEE (2017)
5. Mori, S.: Historical review of OCR research and development. Proc. IEEE **80**(7), 1029–1058 (1992)
6. Cao, Z.M.: Research and Implementation of Character Recognition System Based on Deep Learning. Southeast University, China (2020)
7. Gorski, N., Anisimov, V., et al.: A2i a check reader: a family of bank check recognition systems. In: International Conference on Document Analysis and Recognition. IEEE (1999)
8. Bora, M.B., et al.: Handwritten character recognition from images using CNN-ECOC. Procedia Comput. Sci. **167**(8), 2403–2409 (2020)
9. Rashid, S.F., Shafait, F., Breuel, T.M.: An evaluation of HMM-based techniques for the recognition of screen rendered text. In: 2011 International Conference on Document Analysis and Recognition, China, pp. 1260–1264 (2011)
10. Kayabas, A., Topcu, A. E., Kiliç, Ö.: OCR error correction using BiLSTM. In: 2021 International Conference on Electrical, Computer and Energy Technologies (ICECET), South Africa, pp. 1–5 (2021)
11. Zhu, Z.R.: Analysis and Research of Chinese Learning App Based on Smartphone. Jilin University, China (2019)
12. Hang, P.: Analysis and suggestions based on the Chinese mobile learning software pleco for foreign students in Sichuan foreign studies university. Northern Lit. **24**, 131–132 (2018)

Author Index

B

Bian, Pengyuan I-12

C

Cai, Nina I-338
Cen, Gang II-130
Che, Xujun II-130
Chen, Hongjun I-487
Chen, Juntao II-408
Chen, Lingyu I-127
Chen, Rui II-420
Chen, Tao I-3
Chen, Yinan II-77, II-225
Chen, Zekun II-191
Cheng, Guoliang I-404
Cheng, Hao I-338
Cheng, Xinyu II-237

D

Dai, Zuxu II-330
Dang, Jianwu II-447
Deng, Chengdong II-3
Deng, Yihe II-330
Dong, Xiaoju I-259
Du, Juan II-344
Duan, Meijiao I-372

F

Fang, Wei II-31
Fang, Yanhui II-31
Fang, Youqing I-417
Feng, Xiaoning I-12
Feng, Yihua I-338
Feng, Yong II-13
Fu, Yuesheng II-379

G

Gao, Chao I-233
Gao, Qiang I-298, I-363
Gao, Xiaofeng II-62, II-139

Gao, Yiming I-78, II-237
Gou, Husheng I-89, I-127
Gu, Jiaming II-130
Guo, Dingdong II-237
Guo, Haifeng I-145, I-207, I-221
Guo, Manyuan II-202

H

Han, Haiyun II-168
Han, Longjie I-37
Han, Qilong I-182, I-192, I-433, II-255
Hong, Qian II-40
Hou, Lin I-3
Hu, Jiajing I-60
Hu, Yalin I-37

J

Jia, Jingwen I-417
Jia, Mei I-165
Jin, Anan I-350
Jin, Zhaoning II-116

K

Kang, Zhong I-279
Kong, Linghe II-420

L

Lan, Guo I-25
Lang, Dapeng I-233
Leng, Chongyang I-182
Li, Fengming I-89
Li, Gege II-269
Li, He I-404
Li, Hefan I-165
Li, Junqiu I-60
Li, Lijie II-202, II-255, II-304, II-318
Li, Mengmeng I-145, I-207, I-221
Li, Tie I-165
Li, Xian I-49
Li, Xiang I-350

Li, Xin   II-255
Li, Xuewei   II-237
Li, Yang   I-329
Li, Yanzi   I-78
Li, Yuanhui   II-168, II-225
Li, Zhixin   I-389
Li, Zhuping   I-207
Liang, Peng   II-213
Lin, Bo   II-213
Lin, Junyu   I-89, I-127
Lin, Xiaofei   II-174
Liu, Chao   II-364
Liu, Fangwei   I-127
Liu, Lihan   II-62
Liu, Peng   I-433
Liu, Pingshan   I-455
Liu, Xia   II-77, II-433
Liu, Xinyu   I-89, I-127
Liu, Yujia   I-475
Long, Yang   I-389
Lu, Dan   I-182
Lu, Yan   II-447
Lu, Zhongyu   I-89
Luo, Enze   II-304
Luo, Wanyu   I-475
Lyu, Shanxiang   I-338
Lyu, Wanli   I-417

M
Ma, Jiquan   II-3
Ma, Baoying   I-109
Ma, Fuxiang   II-344
Ma, Tianjiao   I-165
Ma, Wenhai   I-192
Ma, Xiaowen   I-279
Ma, Xiujuan   II-344
Ma, Zhiqiang   I-192
Mei, Aohan   II-191
Meng, Qingyu   I-109
Meng, Yulong   I-127

N
Ning, Beixi   I-12
Niu, Man   II-139

O
Ou, Zhipeng   II-168
Ou, Zhi-peng   II-433
Ouyang, Xue   II-95

P
Pan, Haiwei   I-109

Q
Qi, Zhenheng   II-364
Qiao, Jiyun   I-487
Qiao, Yulong   I-3
Qin, Jiaji   I-233
Qin, Ying   II-95

S
Sang, Yu   I-165
Shen, Wuqiang   I-338
Shen, Yahui   I-3
Shi, Chenxiao   II-202
Shu, Xinya   II-139
Song, Jun   I-338
Sun, Han   I-298, I-363
Sun, Hui   II-364
Sun, Ruilu   II-379
Sun, Yemin   I-455

T
Tang, Huiyi   I-338
Tao, Wenjian   II-304
Tian, Fei   II-408
Tian, Xuefei   I-259
Tian, Yuxin   II-420

W
Wan, Xinrui   I-487
Wang, Hong   II-13
Wang, Hongzhi   I-145, I-207, I-221
Wang, Maoning   I-279, I-372
Wang, Rongbing   II-13
Wang, Tianxin   II-153
Wang, Xinran   II-153
Wang, Yangping   II-447
Wang, Yanqiang   II-213
Wang, Yanqing   I-78, I-475, II-62, II-139,
       II-153, II-237
Wang, Ye   II-202, II-255, II-304, II-318
Wang, Youwei   I-329
Wang, Yu   II-3
Wang, Yue   II-40
Wang, Yuhua   I-49
Wang, Zhanquan   II-420
Wang, Zhifang   I-298, I-363
Wang, Zhiping   I-372

Wei, Jiahan   II-116
Wei, Jianguo   II-447
Wei, Yanjie   I-221
Weiquan, Fan   I-25
Wu, Han   II-393
Wu, Lin   II-95
Wu, Mengjiao   I-487
Wu, Shuhui   II-130
Wu, Yao   I-487

X

Xi, Mingjun   I-455
Xian, Dan   I-309
Xiong, Qingzhi   I-350
Xu, Hongyan   II-13
Xu, Hui   I-37
Xu, Jin   I-329
Xu, Peng   I-89
Xu, Xiujuan   II-285
Xu, Yiqin   I-475
Xu, Yiqing   II-153
Xue, Xianbin   I-404

Y

Yang, Dequan   II-191
Yang, Donghua   I-145, I-207, I-221
Yang, Fan   I-338
Yang, Guogui   II-95
Yang, Jianhua   II-318
Yang, Peng   I-246
Yang, Qingpei   I-487
Yang, Weizhen   II-31
Yang, Xiao   I-433
Yang, Yuhai   I-417
Yang, Yutong   I-259
Yang, Zhigang   I-3
Yao, Jian   II-318
Ye, Tiansheng   I-207
Ye, Xiaojun   I-12
Ye, Zhonglin   II-174, II-269
Yin, Shushu   I-165

Yong, Jiu   II-447
Yu, Wenqian   II-344
Yu, Zhiyan   II-116
Yu, Zhongshu   I-455
Yuan, Maocai   I-109
Yuan, Peikai   II-364

Z

Zhan, Jinmei   II-408
Zhang, Haitao   I-192
Zhang, Han   II-433
Zhang, Haoran   I-145
Zhang, Kejia   I-109
Zhang, Qinghui   I-60
Zhang, Tianyu   II-213
Zhang, Xinfeng   I-127
Zhang, Yanling   I-259
Zhang, Yonggang   II-13
Zhang, Ziying   I-49
Zhao, Baokang   II-95
Zhao, Haixing   II-174, II-269
Zhao, Hang   II-3
Zhao, Jing   II-191
Zhao, Wen   II-168
Zhao, Xiaowei   II-285
Zheng, Bo   I-145, I-221
Zheng, Jingjing   II-213
Zheng, Kaichao   I-404
Zhou, Bin   II-344
Zhou, Huan   II-95
Zhou, Jie   II-285
Zhou, Lin   II-269
Zhou, Meng   I-259
Zhou, Shijie   I-246
Zhou, Xiangyang   I-246
Zhou, Yang   II-116
Zhu, Jianming   I-329
Zhu, Keying   II-130
Zhu, Yiwen   I-78
Zhu, Yuanyuan   II-62